INTRODUCTION TO ELECTRONICS

INTRODUCTION TO ELECTRONICS
Second Edition

The Late H. Alex Romanowitz
Russell E. Puckett

Second edition revised by

Russell E. Puckett, P.E.
College of Engineering, University of Kentucky

JOHN WILEY & SONS, INC.
New York · London · Sydney · Toronto

AVAILABLE FOR USE WITH THIS TEXTBOOK

PUBLISHED BY JOHN WILEY & SONS, INC.

Solutions Manual for Introduction to Electronics, Second Edition

by

H. A. Romanowitz and R. E. Puckett

PUBLISHED BY BUCK ENGINEERING CO., INC., LAB-VOLT EDUCATIONAL SYSTEMS

Laboratory Manual: Introduction to Electronics, Second Edition

by

H. A. Romanowitz and R. E. Puckett

MANUFACTURED AND DISTRIBUTED BY
BUCK ENGINEERING CO., INC.,

LAB-VOLT EDUCATIONAL SYSTEMS:

Student Experiment System SES 505

Copyright © 1968, 1976, by John Wiley & Sons, Inc

All rights reserved. Published simultaneously in Canada.

No part of this book may be reproduced by any means, nor transmitted, nor translated into a machine language without the written permission of the publisher.

Library of Congress Cataloging in Publication Data:

Romanowitz, Harry Alex, 1901–1971
   Introduction to electronics.

   Includes bibliographies and index.
   1. Electronics. 2. Electronic circuits.
I. Puckett, Russell E., joint author. II. Title.
TK7815.R595   1976   621.381   75-23060
ISBN 0-471-73264-8

Printed in the United States of America

10 9 8 7 6 5 4 3 2 1

In memory of the friendship and guidance of
Dr. H. Alex Romanowitz

PREFACE

Since the first edition of this book was published, many changes have occurred in the electronics field. Not only have new devices and circuits been developed, but their applications can be found in new and different areas. There are hand-held calculators, digital wristwatches, numerical-control production machinery, cardiovascular monitors, and a host of other commonplace applications. In all of them, though, the fundamental components of electric circuits—resistors, capacitors, and inductors—and basic electronic circuits are necessary for their operation. Rectifiers and power supplies, amplifiers and oscillators, and electronic control circuits still distinguish these new applications.

Electronic circuits and their applications involve many concepts that are fundamental in electronics. New developments have not replaced existing applications of electronic circuits, nor have they replaced fundamental circuit concepts. These will continue to be a part of any study of basic electronic circuits. For this reason, a substantial portion of the first edition has been retained although many topics have been eliminated. However, this second edition offers some new topics that broaden and expand the first edition. For example, much of the discussion of vacuum tubes is no longer included, while a new chapter is devoted to field-effect transistors. The characteristics and applications of solid-state devices are emphasized in this edition.

The reader will need some knowledge of algebra and trigonometry, and an introduction to the calculus would be helpful. A background in basic physics at high school level is also expected, as well as previous study of dc and ac electric circuits. Chapter 1 includes some basic electric circuit analysis that will be used in later chapters without further elaboration, but this material is intended as a review rather than an introduction. Skills in mathematical processes and a good working knowledge of Ohm's and Kirchhoff's laws of electric circuits as reviewed here are prerequisite to the study of later chapters. Those whose background is sufficiently strong can begin their study of electronic devices and circuits with Chapter 2.

The topical sequence chosen for the contents goes from physical principles that govern the behavior of a particular device to the analysis of its behavior in operating circuits. Examples of numerical problems and drill problems that reinforce the topic under discussion can be found throughout the book. Additional problems are provided at the end of each chapter, with challenges to the best students in some of them. Answers to the problems are included in the Appendix.

A *Solutions Manual* comprising detailed solutions to all problems in the book is available for use with it. In addition, a *Laboratory Manual* has been designed to accompany the textbook. It includes experiments that have been tested and proved practical. An instructor may select from them according to the desired coverage and the time assigned for laboratory instruction.

Adaptation of material from the writings of many authors—too numerous to mention individually—is acknowledged. Manufacturers of electronic equipment, who cooperated by furnishing photographs and other materials, have contributed greatly to the preparation of this edition. Their assistance is truly appreciated.

I gratefully acknowledge the debt owed to my family for their patience and forbearance during the preparation of this work. Special thanks are due my wife, Dorothy, whose understanding made it possible to complete the job.

Russell E. Puckett

Lexington, Kentucky

CONTENTS

Table of Useful Constants   xvii

1
Some Useful Circuit Concepts   1

1-1	Symbols and Units	1
1-2	Laws of Electric Circuits	4
1-3	Sinusoidal Voltage and Current	8
1-4	Relative Phase Angles of Sinusoidal Waveforms	10
1-5	Effective Value of Sinusoidal Waveforms	11
1-6	Phasor Algebra	11
1-7	Generation of Sinusoidal Waveforms by Rotating Phasor	15
1-8	Rules for Complex Number Calculations	17
1-9	Impedance in AC Circuits	19
1-10	Analysis of AC Circuits	28
1-11	Determinants	30
1-12	Use of Determinants in Circuit Analysis	32
1-13	Circuit Analysis by Loop Currents	33
1-14	The Superposition Principle	35
1-15	Thévenin's Theorem	36
1-16	Norton's Theorem	38
1-17	Fourier Series for Periodic Waveforms	39
1-18	Time Constant	41
1-19	Decibels	43
1-20	Volume Unit	46

ix

x   Contents

1-21	Maximum Power Transfer Theorem	46
1-22	Series Resonance	48
1-23	Sharpness of Resonance, Q	50
1-24	Half-Power Currents and Frequencies	52
1-25	Parallel Resonance (Antiresonance)	53
1-26	Parallel Resonance with Resistance in Both Branches	55
1-27	Q of a Parallel-Resonant Circuit	56

2
Diodes   65

2-1	Atomic Arrangement in Materials	65
2-2	Pattern of Energy Levels	67
2-3	Conductors, Insulators, and Semiconductors	68
2-4	The Germanium Atom	70
2-5	Covalent Bonding	71
2-6	Adding Impurities to Semiconductors	72
2-7	N-type and P-type Semiconductors	74
2-8	The Fermi Level of Energy	75
2-9	Preparation of Semiconductor Metal	75
2-10	Forming a P-N Junction	79
2-11	Impurity Contact Junctions	80
2-12	Packaging	82
2-13	P-N Junction Diode	84
2-14	Majority-Carrier Currents	85
2-15	Minority Carriers	85
2-16	Potential Hill at the Barrier Region	86
2-17	Forward Biasing of Diodes	87
2-18	Reverse Biasing of Diodes	87
2-19	Avalanche Breakdown	89
2-20	Zener Diode	90
2-21	Diode Ratings	92
2-22	Diode Characteristics	92
2-23	Comparison of Germanium and Silicon Diodes	95
2-24	Diode Symbols for Circuit Diagrams	96
2-25	Diode in a Circuit	98
2-26	High-Vacuum Diode with Thermionic Cathode	101
2-27	Characteristics of the High-Vacuum Diode	103

3
Applications of Diodes   111

3-1	Diode as a Voltage Limiter (Clipper)	111
3-2	Diode as a Voltage Clamper	113
3-3	Diode as a Logic Circuit Element	114
3-4	Voltage Regulation with Zener Diodes	117
3-5	Instrument Protection	120
3-6	Half-Wave Rectification	121
3-7	The Nature of Rectifier Output Voltage	123
3-8	Output Current and Power of a Half-Wave Rectifier Circuit	124
3-9	Full-Wave Rectification	126
3-10	Bridge Rectifier Circuit	129
3-11	Power Transformer	130
3-12	Comparison of Full-Wave and Half-Wave Rectifiers	132
3-13	Smoothing Filters	133
3-14	Shunt-Capacitor Filter	135
3-15	Bleeder Resistor	139
3-16	Ripple Factor	139
3-17	Practical Values of Filter Elements	140
3-18	An AC Voltmeter	141
3-19	The Diode Detector or Peak Rectifier	143
3-20	Voltage Doubler	145
3-21	Semiconductor Stack Rectifiers	146

4
Junction Transistors   157

4-1	Types of Transistors	158
4-2	The Junction Transistor	158
4-3	Equalization of Fermi Levels in a P-N Junction	161
4-4	Effects of Bias on Energy "Hills" at a P-N Junction	162
4-5	Symbols and Bias Polarities for Transistors	163
4-6	Interaction of Internal Currents	165
4-7	Reverse-Bias Leakage Current	168
4-8	Common-Terminal Connections	169
4-9	Characteristic Curves for Transistors	172
4-10	Letter Symbols for Transistor Circuits	174
4-11	Characteristic Data and Curves; Alpha and Beta Factors	175
4-12	Some Important Characteristics and Maximum Ratings	179
4-13	Simple Transistor Circuit; The DC Load Line	180
4-14	Steady-State Operation; The Q-Point	183

xii *Contents*

4-15	Transistor Cutoff and Saturation	185
4-16	Collector Dissipation	186
4-17	Effects of Temperature on the Q-Point	188
4-18	Q-Point Stabilization	191
4-19	Locating the Q-Point	194
4-20	Amplifying Action in a Transistor	196
4-21	Design Considerations	197

5
Field-Effect Transistors (JFET)   205

5-1	The Junction Field-Effect Transistor, JFET	205
5-2	Depletion Region in the Channel	206
5-3	Symbols and Bias Polarities for Junction FET	207
5-4	Static Characteristics of JFET	208
5-5	Electrical Ratings	210
5-6	Load Line for JFET Circuit	211
5-7	Self-Bias of JFET	212
5-8	Effect of Source Resistor on Q-Point	213
5-9	High Impedance DC Voltmeter Using FET	215
5-10	MOS Field-Effect Transistor (MOSFET)	216
5-11	Symbols and Characteristics of MOSFET	218
5-12	Precautions Regarding MOSFET	219

6
Vacuum Tubes   223

6-1	Physical Characteristics of the Triode	224
6-2	Function of the Grid	224
6-3	Types of Triode Tubes	226
6-4	Plate-Characteristic Curves of a Triode	227
6-5	Letter Symbols for Vacuum Tubes and Circuits	228
6-6	Three Dynamic Factors	229
6-7	Grid-Plate Transfer Characteristics of a Triode	233
6-8	Grid Bias	233
6-9	Grid Bias Arrangements	234
6-10	Grid-Leak Bias	235
6-11	The Triode in a Circuit	236
6-12	Simple Circuit Equations	237
6-13	The Screen-Grid Tetrode and the Pentode	239
6-14	Effect of Cathode-Biasing Resistor on Q-Point	242
6-15	Graphical Analysis of Triode-Tube Performance	244

7
Small-Signal Equivalent Circuits   255

7-1	The AC Load Line	255
7-2	Theory of Hybrid Parameters	260
7-3	Transistor Equivalent Circuit Using h-Parameters	262
7-4	Mathematical Analysis of Transistor Amplifier	263
7-5	Manufacturer Designation of h-Parameters	265
7-6	Calculation of h-Parameters from Manufacturer's Data	265
7-7	Formulas for Calculating h-Parameters	267
7-8	Actual and Equivalent Circuits of Vacuum-Tube Amplifier	268
7-9	Simple Amplifier Equations	270
7-10	AC Load Line for Vacuum Triode	271
7-11	Output Power	273
7-12	Variation of Vacuum-Tube Parameters	274
7-13	Circuit Models for JFET and MOSFET	274

8
Analysis of Amplifier Operation   279

8-1	The Basic Single-Stage Amplifier	279
8-2	Two-Stage Voltage Amplifier	281
8-3	Operating Conditions; Classes of Amplifiers	282
8-4	Distortion in Amplifiers	284
8-5	Frequency Response of R-C Coupled Amplifiers	286
8-6	Analysis of Common-Emitter Amplifier	287
8-7	Comparison of the Three Connections	290
8-8	Analysis of Small-Signal Two-Stage Transistor Amplifier	294
8-9	Analysis of R-C Coupled Tube Amplifier	297
8-10	Analysis of Transistor-Amplifier Operation Outside Midband	311
8-11	Improving Amplifier Response at Low Frequencies	312
8-12	Bandwidth Reduction in Cascaded Stages	314
8-13	Intercommunication System Using Vacuum Tubes	315
8-14	Intercommunication System Using Transistors	318

9
Power Amplifiers   327

9-1	Single-Stage Power Amplifier	328
9-2	AC Load of a Power Amplifier	329
9-3	Transistor Power Amplifiers	332
9-4	Output Power and Efficiency	334

xiv Contents

9-5	Class B Push-Pull Operation	336
9-6	Push-Pull Power Amplifier Using Tubes	341
9-7	Class B Push-Pull Amplification with Tubes	347
9-8	Complementary Symmetry	349
9-9	Transistor Heat Sinks	350

10
Special-Purpose Amplifiers   357

10-1	Differential Amplifier	357
10-2	Differential Operation	359
10-3	Common-Mode Rejection Ratio	360
10-4	Biasing for Maximum Signal Swing	361
10-5	Input and Output Impedances	362
10-6	The Darlington Circuit	365
10-7	Operational Amplifiers	369
10-8	Applications of Operational Amplifiers	370
10-9	Amplifiers with Tuned Loads	372
10-10	Double and Stagger Tuning	376

11
Feedback Amplifiers and Oscillators   381

11-1	Basic Considerations of Feedback	381
11-2	Negative (Inverse) Feedback in Amplifiers	383
11-3	Effect of Voltage Feedback on Input and Output Impedances	385
11-4	Feedback in Transistor Amplifiers	388
11-5	Criterion for Sustained Oscillations	393
11-6	Resistance-Capacitance Oscillators	397

12
Basic Modulation and Detection in Communications   403

12-1	Amplitude Modulation (A-M)	403
12-2	Frequency Modulation (F-M)	404
12-3	Pulse-Code Modulation	405
12-4	Demodulation (Detection)	405
12-5	Classification of Radio Waves	405
12-6	Separation of Radio Carriers	407
12-7	Amplification of Modulated Carriers	407

12-8	Fundamental Modulation Process	408
12-9	Analysis of Amplitude Modulation (A-M)	409
12-10	Generation of Sidebands in A-M	412
12-11	Bandwidth Requirements in A-M	413
12-12	Power Contained in A-M Sidebands	414
12-13	Methods of Generating A-M	416
12-14	Modulated Class C Amplifier	417
12-15	Balanced Modulator and Single-Sideband (SSB) Operation	420
12-16	Frequency Conversion and Mixing	421
12-17	Single-Sideband Receiver	423
12-18	The Envelope Detector for A-M Signals	424
12-19	Analysis of Frequency Modulation (F-M)	427
12-20	Bandwidth of F-M Signals	428
12-21	Basic F-M Radio Receiver	432
12-22	The F-M Discriminator	433
12-23	Pulse-Code Modulation (PCM)	436

13
Electronic Power Conversion — 445

13-1	The Semiconductor Controlled Rectifier (SCR)	446
13-2	Construction of the SCR	446
13-3	Theory of Operation; Two-Transistor Analogy	446
13-4	Effect of the Gate	450
13-5	Ratings and Characteristics of the SCR	452
13-6	Gate Trigger Characteristics	455
13-7	Full-Wave Phase Control of the SCR	459
13-8	RMS and Average Load Voltage Calculations	459
13-9	Bilateral SCR, the TRIAC	462
13-10	Photovoltaic Action; The Solar Cell	463
13-11	Light Amplification by Laser	466
13-12	Coherency of Laser Radiation	467
13-13	Types of Lasers	468

14
Integrated Circuits — 473

14-1	Fabrication of Integrated Circuits	474
14-2	Use of Photomasks	475
14-3	The Epitaxial Layer	477
14-4	Diffusion of Additional Layers	478

14-5	Metalization to Form Electrical Connections	480
14-6	Cost Factors	481
14-7	IC Packaging	483
14-8	Component Ratings and Unit Values	484
14-9	Thin-Film and Multichip Construction of ICs	485
14-10	Circuits Constructed as ICs	487

Appendix 489

A-1	Calculation of Effective (RMS) Value of a Sinusoidal Waveform	489
A-2	Determinants	490
A-3	Derivation of Charging Current in a Capacitor	492
A-4	Discharge Current in a Capacitor	492
A-5	Voltage Applied to an Inductor	493
A-6	Current Decay in an Inductor	494
A-7	Proof of Maximum Power Transfer Theorem	495
A-8	Temperature Conversion Factors	495
A-9	Bessel Function Coefficients	496
A-10	Required Gain of Phase-Shift Oscillator	496
A-11	General-Purpose Curves and Data	498
A-12	Sine Wave Template	515

Answers to Problems 517

Index 527

Table of Useful Constants

Charge of an electron, q_e	1.602×10^{-19} C
Mass of an electron, m_e	9.108×10^{-31} kg
Mass of a proton	1.672×10^{-27} kg
Velocity of light	2.9976×10^8 m/s
Planck's quantum of action ($\hbar$)	6.624×10^{-34} J-s
One electronvolt of energy	1.602×10^{-19} J
Boltzmann's gas constant	1.3803×10^{-23} J/°K
Angstrom unit, Å, AU	10^{-10} m
Permittivity of free space, ϵ_0	8.854×10^{-12} F/m
Permeability of free space, μ_0	$4\pi \times 10^{-7}$ H/m
Base, system of natural logarithms, ϵ	2.71828
$\log_{10} \epsilon$	0.4343
$\log_{10} \pi$	0.4971
$\ln \pi$	1.4473

1
SOME USEFUL CIRCUIT CONCEPTS

A good working knowledge of the principles governing electric circuits is essential to formal study of electronic circuits and devices. The physical laws and mathematical methods used in electric circuit analysis are also applicable to electronic circuits.

The study of electronics requires a familiarity with the characteristic behavior of the electric circuits containing resistance, inductance, and capacitance elements. Students also should be prepared to analyze circuits when these elements are connected in series and parallel combinations, and when they are excited by either dc or ac single-phase sinusoidal waveforms. Some of the more useful circuit concepts are reviewed in this chapter. Those whose background knowledge adequately covers these concepts may begin with Chapter 2.

1-1
Symbols and Units

Standard symbols and names have been adopted for physical quantities associated with electrical devices. The use of such standards facilitates discussion of the operation of electronic circuits and lets us illustrate them simply in schematic diagrams of the interconnected circuit elements. The names of physical quantities are based on an international system of basic unit names, their abbreviations, and symbols for each unit and multiples of it. Table 1-1 lists units for electrical quantities encountered in this book.

Table 1-1
Electrical Units and Their Abbreviations

Quantity	Unit	Symbol
Time	Second	s
Electric current	Ampere	A
Angular velocity	Radian per second	rad/s
Area	Square meter	m^2
Electric capacitance	Farad	F
Electric field strength	Volt per meter	V/m
Electric inductance	Henry	H
Electric potential difference	Volt	V
Electric resistance	Ohm	Ω
Electromotive force	Volt	V
Energy	Joule	J
Frequency	Hertz	Hz
Magnetic flux	Weber	Wb
Power	Watt	W
Quantity of electricity (charge)	Coulomb	C
Velocity	Meter per second	m/s

Unabbreviated unit names are not capitalized. However, when the unit name is derived from the name of a person, its symbol is capitalized. For example, the hertz, named for Henrich Rudolf Hertz (1857–1894), is abbreviated Hz. Numerical prefixes and their symbols usually are abbreviated as lowercase letters; exceptions are noted in Table 1-2. As an example, a 1-megohm resistor

Table 1-2
Multiple and Submultiple Units and Their Abbreviations

Multiplying Factor	Prefix	Symbol
10^{12}	tera-	T
10^9	giga-	G
10^6	mega-	M
10^3	kilo-	k
10^{-1}	deci-	d
10^{-2}	centi-	c
10^{-3}	milli-	m
10^{-6}	micro-	μ
10^{-9}	nano-	n
10^{-12}	pico-	p
10^{-15}	femto-	f

may be abbreviated 1.0 MΩ; a capacitance of 0.01 microfarad may be abbreviated 0.01 μF.

Some of the more common circuit symbols used in this book are shown in Fig. 1-1. These symbols are used in schematic diagrams to represent the devices interconnected in the circuit and to indicate polarity of voltages and direction of current within the circuit.

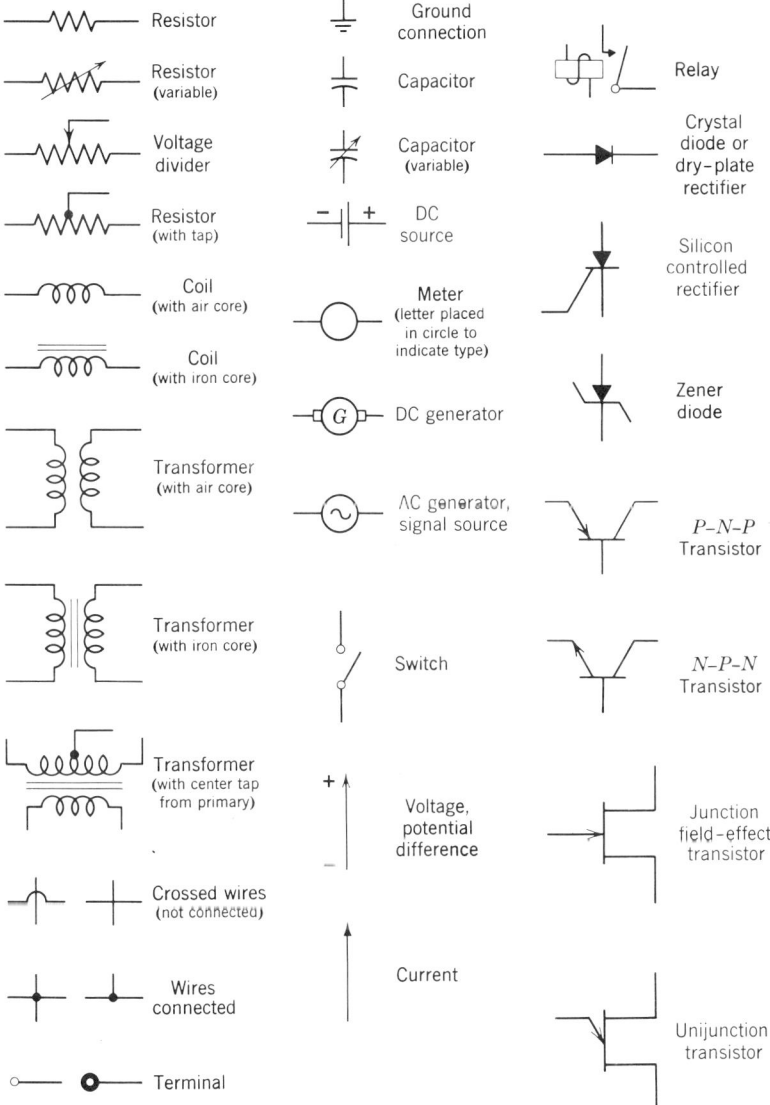

Figure 1-1
Symbols used in circuit diagrams.

4   Some Useful Circuit Concepts   Ch. 1

1-2
Laws of Electric Circuits

An analysis of electronic circuit characteristics can be based on three physical laws of electric circuits: *Ohm's law* and *Kirchhoff's laws* for voltage and current.* These laws are applicable to static and dynamic circuits, for both dc and ac operation. At every instant in time, they govern what happens to voltages and currents in an electronic circuit. Furthermore, any analysis of circuit operation that violates these laws is invalid and useless. We shall review these three laws of electric circuits briefly.

Ohm's Law

This basic law relates the interdependence of voltage, current, and resistance to current in an electric circuit. Ohm's law states that *the resistance R of an electric circuit is directly proportional to the voltage E_R across the circuit and inversely proportional to the current I_R in the circuit.* Expressed mathematically, Ohm's law may be written†

$$R = \frac{E_R}{I_R} \quad [\text{ohms} = \text{volts/amperes}] \quad (1\text{-}1)$$

Of course, these symbols may be manipulated algebraically to yield alternate forms of the equation:

$$I_R = \frac{E_R}{R} \quad \text{or} \quad E_R = I_R R \quad (1\text{-}2)$$

In order for current to exist in an electric circuit, there must be a closed path for the current. When current passes through a resistance element, a voltage will be developed across the element. The potential difference from one end of the element to the other can be determined from Ohm's law. As an example, refer to Fig. 1-2a. Here we see a dc voltmeter connected across part of a circuit. The voltmeter indicates 24.8 Vdc across the resistance. Because of Ohm's law, we know there is current in the resistor. Substituting the given numerical values for voltages and resistance, into one form of Ohm's law, Eq. 1-2, we may calculate the value of the current:

$$I_R = \frac{E_R}{R} = \frac{24.8 \text{ Vdc}}{1000 \, \Omega} = 0.0248 \text{ A, or } 24.8 \text{ mAdc}.$$

In which direction is current passing through the resistor? The current must

---

*These concepts are named for Georg Simon Ohm (1787–1854) and Gustav Robert Kirchhoff (1824–1887), in honor of their classical work in electricity.

†In this book, mathematical symbols for physical quantities will be capital letters for constant quantities and lowercase for variable quantities.

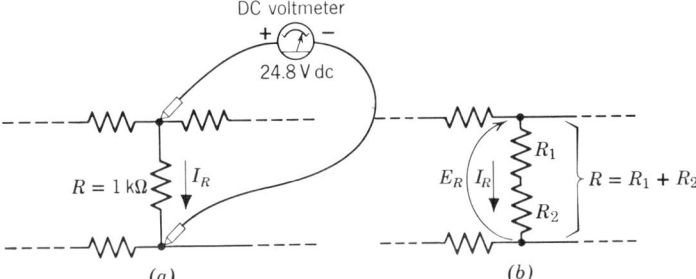

Figure 1-2
Circuit to show example of Ohm's law. (a) Voltage developed across a resistor caused by a current in the resistor. (b) Total resistance $R_1 + R_2$ develops voltage E_R caused by I_R in both resistors.

be entering the upper end of the resistor, because that end has the positive potential.

It is important to understand that Eqs. 1-1 and 1-2 apply to a single resistance element in a circuit, to a combination of elements in part of a circuit, or to a complete circuit. Where a combination of elements between two separate points can be represented as having a single resistance value (R), then Ohm's law applies to the combination as a single resistance. Factors representing voltage and current in these equations apply to the difference in potential across the resistance (E_R) and to the current in the resistance (I_R). Refer to Fig. 1-2b. In steady-state dc operation, the voltage–current relationship will be determined solely by ohmic resistors.

In circuits where the current is changing, the resistance to current is called *impedance*, and circuit elements may consist of capacitors and inductors as well as resistors. Impedance is represented by the letter Z and has units of ohms (Ω). It can be used to replace R in the equations for Ohm's law, provided voltage and current factors are interpreted to include phase relationships that exist in a dynamic ac circuit. These concepts are reviewed beginning in Section 1-9.

In writing the equation for either dc or ac circuits, the voltage is that developed across the resistance and the current is that in the resistance, whether there is a single element or a combination of elements. However, although Ohm's law applies to both dc and ac circuits, it must be applied separately for dc and ac quantities where they exist simultaneously in the same circuit.

Kirchhoff's Laws

Many electronic circuits contain unknown quantities that cannot be analyzed with Ohm's law alone. To determine the behavior of such circuits, we also

6   Some Useful Circuit Concepts   Ch. 1

employ Kirchhoff's laws for current and voltage. Together with Ohm's law, these comprise the basic set of tools needed to analyze electronic circuits.

Kirchhoff's current law says *the total current entering any junction of circuit elements is equal to the total current leaving that junction.* If we consider current leaving as the negative of current entering, we may restate the law as: the total current entering a junction of circuit elements is equal to zero. In mathematical terms, the two statements of the law are equivalent. Let us look at the circuit in Fig. 1-3 to see why.

In Fig. 1-3, the junctions of the circuit elements are labeled A, B, C, and D. We see that current I_1 is entering junction A, while currents I_2 and I_3 are leaving. According to Kirchhoff's current law,

$$I_1 = I_2 + I_3 \qquad \text{(current entering = current leaving)} \qquad (1\text{-}3)$$

Algebraic manipulation of this equation shows that the second statement of the current law is also correct:

$$I_1 - I_2 - I_3 = 0 \qquad \text{(current entering = 0)} \qquad (1\text{-}4)$$

Now consider junction C. Currents I_2 and I_3 are entering the junction and I_1 is leaving. For junction C, Kirchhoff's current law may be written

$$I_2 + I_3 = I_1 \qquad \text{(current entering = current leaving)}$$

which is equivalent to Eq. 1-3. Similar reasoning shows that this equation may be changed to the form of Eq. 1-4.

Trivial applications of the current law occur at junctions B and D. At each junction the current entering is the same current leaving, because there is only one path for current at these junctions.

The use of Kirchhoff's current law will be developed further after we have reviewed the voltage law. *Kirchhoff's voltage law* says that in *traversing a closed electric circuit, the sum of voltage rises equals the sum of voltage drops.* Here, a *voltage rise* is an increase in potential difference, from − to + between two separate points along the circuit path, and a *voltage drop* is a decrease,

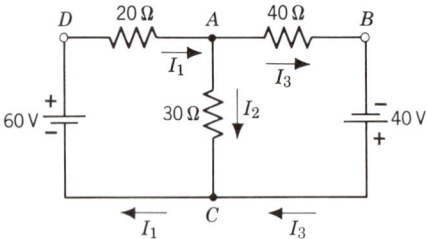

Figure 1-3
Circuit to illustrate Kirchhoff's laws.

from + to − between two separate points. Since, in mathematical terms, *voltage drops can be expressed as negative voltage rises*, the voltage law is sometimes stated in another way: in traversing a closed electric circuit, the sum of all voltage rises equals zero.

Returning to the circuit in Fig. 1-3, we can identify three separate, but related, closed circuits. One includes all elements around the outside of the network. A second closed loop, starting at junction D, includes the resistor from D to A, then the resistor to C, and the return to D through the 60 V source. The third closed network includes the resistor from A to B, the 40 V source to C, and the return to A through the 30-Ω resistor.

Any two of these closed circuits can be used to analyze the voltage behavior of the network. Applying the voltage law to the left-hand loop and starting clockwise at C, we can write the voltages in terms of known sources, using Ohm's law for voltages across the resistors:

$$60 \text{ V} = I_1(20 \text{ }\Omega) + I_2(30 \text{ }\Omega) \quad \text{(rises = drops)} \quad (1\text{-}5)$$

Starting at B and moving clockwise in the right-hand loop,

$$40 \text{ V} + I_2(30 \text{ }\Omega) = I_3(40 \text{ }\Omega) \quad \text{(rises = drops)} \quad (1\text{-}6)$$

Algebraic changes in these two equations show that taking voltage drops as negative rises in going clockwise around the loops will lead to the alternate form of Kirchhoff's voltage law:

$$60 \text{ V} - I_1(20 \text{ }\Omega) - I_2(30 \text{ }\Omega) = 0 \quad \text{(rises = 0)} \quad (1\text{-}7)$$

$$40 \text{ V} + I_2(30 \text{ }\Omega) - I_3(40 \text{ }\Omega) = 0 \quad (1\text{-}8)$$

We have only two equations for three unknown currents. To analyze this circuit, it is necessary to have a third independent equation containing the currents that is also valid for the circuit. We already have it, in Eq. 1-3, the expression for Kirchhoff's current law. Thus, Eqs. 1-7, 1-8, and 1-3 may be solved simultaneously to obtain numerical values for the currents. Algebraic solution gives the values:

$$I_1 = 2.076 \text{ A} \qquad I_2 = 0.615 \text{ A} \qquad I_3 = 1.461 \text{ A}$$

Drill Problems

D1-1 Show that the algebraic results given above for the three currents in Fig. 1-3 are correct.

D1-2 Suppose the voltage sources in Fig. 1-3 are doubled. What effect will this have on the values of the three currents?

D1-3 Suppose the resistor values are doubled in the circuit of Fig. 1-3. What effect will this have on the values of the three currents?

1-3
Sinusoidal Voltage and Current

It is advisable for you to have clearly in mind some terms and concepts for sinusoidal voltage and current waveforms before proceeding with analysis of circuits containing them. A few terms used regularly in this book are defined here for current; similar definitions apply to voltage and other physical variables.

An *oscillating current* alternately increases and decreases in magnitude with respect to time in accordance with some physical law.

A *periodic current* is an oscillating current with magnitudes that recur at equal time intervals.

The *period* of a periodic current (symbol T) is the smallest time increment that separates recurring magnitudes of the current. It is sometimes said to occur within 360° of electrical rotation, referring to machinery that can be used to generate the current.

Waveform of a periodic current refers to the shape of the curve representing its amplitude with respect to time.

Frequency (symbol f) is the number of periods occurring in unit time, usually in one second.

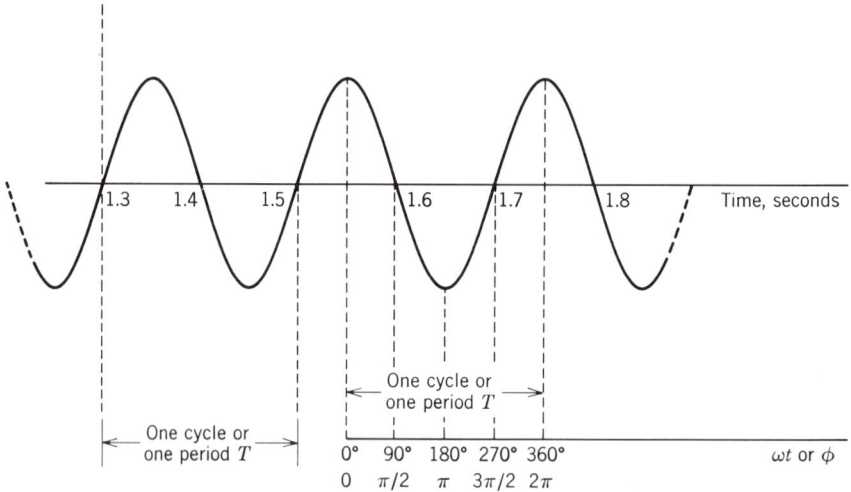

Figure 1-4
Sinusoidal waveform, showing defined terms.

Sec. 1-3   *Sinusoidal Voltage and Current*   9

A *cycle* is one complete set of magnitudes of a periodic current, or those magnitudes that occur during one period.

The *angular velocity* of a periodic current (symbol ω) equals the frequency multiplied by 2π, or $\omega = 2\pi f$. The factor 2π represents the number of radians in a complete circle, or within 360° of electrical rotation.

A *sinusoidal current* has the characteristic waveform of the sine wave with respect to time, and reverses direction regularly with successive half-waves of the same shape and area.

The term "cycles per second," which was used in earlier literature as the unit of frequency of sinusoidal voltage and current, has generally been replaced by the term *hertz*, abbreviated Hz. Electric power in this country is 60 Hz, radio broadcast frequencies range from 550 kHz to 1600 kHz (kHz = kilohertz), and television frequencies in the UHF band lie between 450 MHz and 1000 MHz (MHz = megahertz).

One cycle of a sinusoidal waveform is a complete set of positive and negative values. This alternating shape is shown in Fig. 1-4, where some of the terms defined above are marked for easy identification. It would be helpful to compare the definitions and these marked portions of the sinusoidal waveform. Note that one horizontal axis is plotted with respect to time, while the other is specified in terms of angular measure. What is the frequency of this waveform? (*Hint:* $T = 1/f$).

Each instantaneous value of a sinusoidal voltage waveform occurs in accordance with the equation of sines

$$e = E_m \sin \omega t = E_m \sin \phi = E_m \sin 2\pi f t \qquad (1\text{-}9)$$

where e = instantaneous value, volts (V)

E_m = maximum amplitude, volts (V)

$\omega = 2\pi f$ = angular velocity, radians/second (rad/s)

t = time, seconds (s)

$\phi = \omega t$ = angular change, electrical degrees (°) or radians (rad)

f = frequency, hertz (Hz)

Alternating currents and voltages that follow a cosine wave are also called sinusoidal waves. Since $\cos \omega t = \sin(\omega t + 90°)$, a cosine wave is simply a sine wave displaced along the axis by 90°. Cosine and sine waveforms are compared in Fig. 1-5.

10 Some Useful Circuit Concepts   Ch. 1

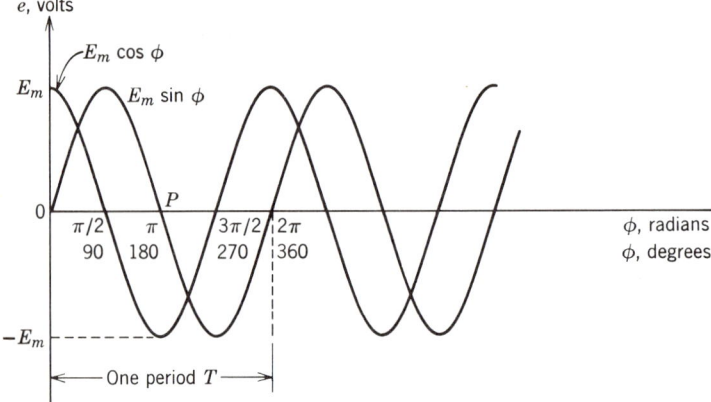

Figure 1-5
Comparison of cosine and sine voltage waveforms.

Drill Problems

D1-4 What is the frequency of a current expressed as $i = 10 \sin(377t + 90°)$? What is its period?

D1-5 Show that $\sin \phi = \cos(\phi - 90°)$.

1-4
Relative Phase Angles of Sinusoidal Waveforms

Phase is a term used to describe the position of one point on a sinusoidal waveform relative to some reference point on the waveform. The choice of reference point is arbitrary, but it is usually taken as the last previous zero point where the waveform is increasing from negative to positive amplitudes, such as the zero reference point in Fig. 1-5. The phase value is the fraction of a period between the two points and thus will range from 0 to 1.0 over one cycle. The phase at P in Fig. 1-5 is 0.5.

Phase angle is the angle obtained by multiplying the phase by 360° if it is to be expressed in electrical degrees, or by 2π to get the angle in radians. Thus, the phase angle of the point P in Fig. 1-5 is expressed as either 180° or π radians, when measured from the zero reference point.

The *relative phase angle* between two sinusoidal waveforms that have the same period (same frequency) is the angle separating similar values of the two waveforms. For example, the relative phase angle or angular phase difference between the sine and cosine waves is 90°. The cosine wave is said to *lead* the sine wave, because it passes through its values before the sine wave experiences similar values. In the same way, the sine wave is said to *lag* the cosine wave by 90°.

The terms "phase" and "phase difference" are commonly used loosely by practicing engineers as equivalent to phase angle and relative phase angle as defined above. Thus, the phase difference of a sine wave and cosine wave is said to be 90°, or their angular phase difference. As will be seen later, it is possible for voltage and current waves to be out of phase in ac circuits and to have a phase difference other than 90°.

1-5
Effective Value of Sinusoidal Waveforms

A voltage with sinusoidal waveform will cause a current of sinusoidal waveform in a circuit containing resistors, inductors and capacitors that do not change their values, i.e., elements that maintain a constant impedance. In the analysis of ac circuits, it is often useful to be able to assign an *effective value* to sinusoidal voltage and current waveforms. This is a measure of the equivalent dc voltage or current that would cause the same heating in a resistor as the ac waveform being analyzed. The effective values of sinusoidal voltage and current waveforms are

$$E_{\text{eff}} = \frac{E_m}{\sqrt{2}} = 0.707 E_m \quad \text{and} \quad I_{\text{eff}} = \frac{I_m}{\sqrt{2}} = 0.707 I_m \quad (1\text{-}10)$$

where E_m and I_m are the maximum amplitudes of the ac voltage and current, respectively. The subscript "eff" is often omitted from the symbol for effective value.

The effective value of sinusoidal waveforms is sometimes called the root-mean-square or *rms value*. Why this name is used can be explained by referring to the Appendix, where the effective value of a sinusoidal waveform is calculated.

1-6
Phasor Algebra

Calculations involving dc voltage and current or ac circuits at a given instant in time are simple algebraic operations. However, in ac circuits the relationships between effective values of ac voltage and current are usually more important than those between instantaneous or maximum values.

In 1893–94, Dr. C. P. Steinmetz introduced a method for the analysis of ac circuits that has become common engineering practice. His method is generally known as *phasor algebra* but is sometimes referred to as "vector algebra" or "complex algebra." Phasor algebra can be used to analyze complex circuits containing ac sources of voltage and constant impedances operating in a *steady-state* condition. This means that changes in voltage and current occur in

12 Some Useful Circuit Concepts   Ch. 1

Figure 1-6
Real numbers plotted along a straight line.

some repetitious manner, such as the periodic changes in sinusoidal waveforms.

Phasor algebra consists of mathematical operations performed on *complex numbers*, numbers consisting of a "real" part and an "imaginary" part. Let's review briefly how these names were chosen for the two parts.

Numbers representing integers and fractions, as well as irrational numbers such as π and $\sqrt{2}$, can be plotted along a continuous straight line, as in Fig. 1-6. Negative numbers are pictured on the line extended to the left of zero. Any positive or negative number that can be placed on this line is said to be a "real" number.

In the eighteenth century, mathematicians attempted to obtain the solution to the equation

$$x^2 + 1 = 0$$

Although the solution could be expressed as $x = \sqrt{-1}$, this number could not be placed along the line in Fig. 1-6. This number was therefore said to be "imaginary" because it did not fit the concept of "real" numbers. Hence, the terms "real" and "imaginary" are but labels to distinguish different number

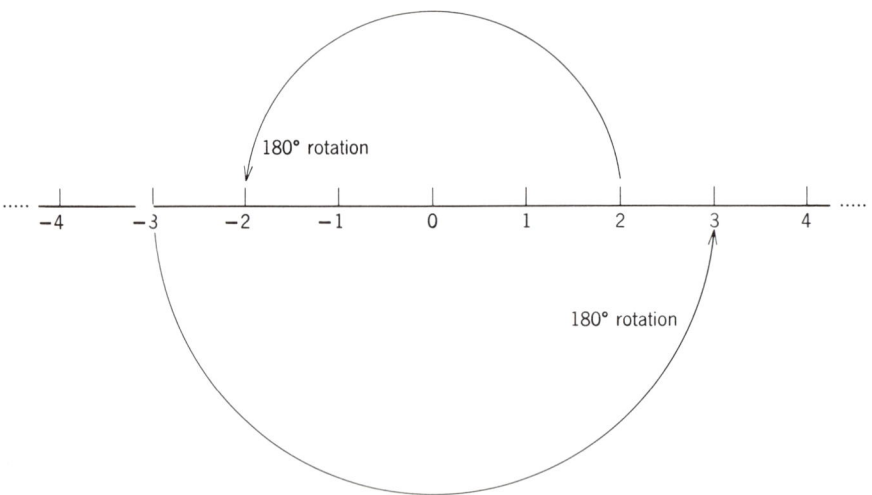

Figure 1-7
Real numbers rotated 180° about zero when multiplied by -1.

Sec. 1-6   Phasor Algebra 13

concepts. Mathematicians symbolize $\sqrt{-1}$ with the letter i, but since we use i for current, we will use $\sqrt{-1} = j$ for our studies.

Refer now to the real number line shown in Fig. 1-7. If the number $+2$ is multiplied by -1, the product is the number -2, which is a rotation of $+2$ through 180° about zero. Similarly, if the number -3 is multiplied by -1, it can be understood as $+3 = -3$ rotated by 180°. That is,

$$+2(-1) = -2 \quad \text{and} \quad -3(-1) = +3 \tag{1-11}$$

We adopt a special notation to express Eqs. 1-11 in terms of angular rotation:

$$+2\underline{/180°} = -2 \quad \text{and} \quad -3\underline{/180°} = +3 \tag{1-12}$$

Equations 1-11 can also be expressed in terms of j in the following manner. First, writing the factor -1 in terms of its square root, Eqs. 1-11 become

$$+2(\sqrt{-1})(\sqrt{-1}) = -2 \quad \text{and} \quad -3(\sqrt{-1})(\sqrt{-1}) = +3$$

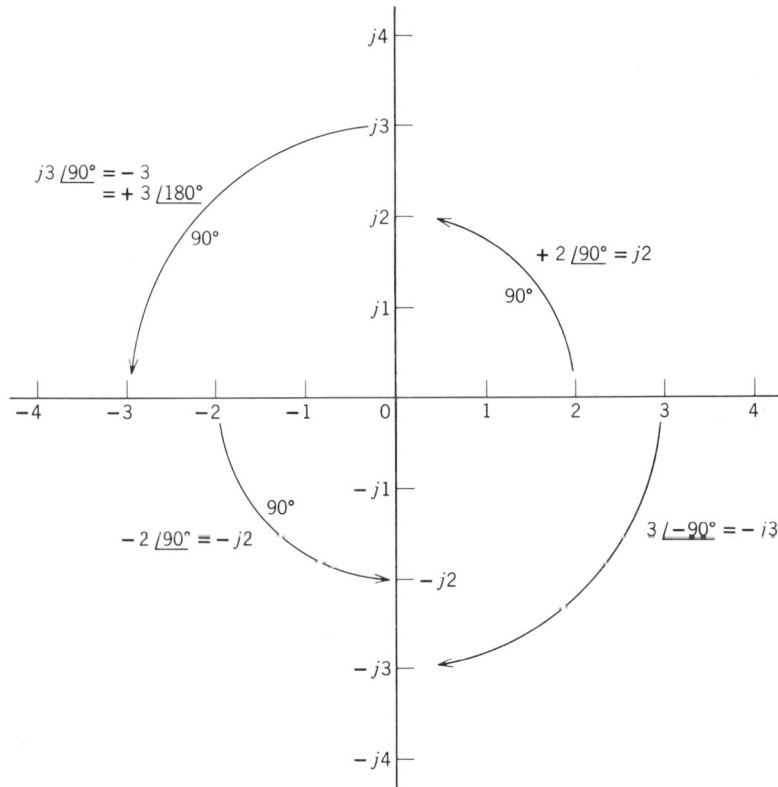

Figure 1-8
Rotation by 90° by the j-operator.

14 Some Useful Circuit Concepts   Ch. 1

Substituting j for $\sqrt{-1}$ gives

$$+2(j)(j) = -2 \quad \text{and} \quad -3(j)(j) = +3 \quad (1\text{-}13)$$

Since it takes two multiplications by j to equal rotation through 180°, we might conclude that only one multiplication by j would lead to rotation through 90°, with a second j completing the 180° rotation. In our shorthand notation,

$$+2(j)(j) = +2\underline{/90°} \times +1\underline{/90°} = +2\underline{/180°} = -2$$

where the j-factor is taken as $+1\underline{/90°}$. Any real number operated on by this j-factor is rotated about zero by 90°. In a similar way, any imaginary number operated on by the j-factor is also rotated through 90° to become a real number as the resultant. Figure 1-8 illustrates rotation by the j-factor, with a vertical j axis passing through 0 of the real number line.

Because real numbers are rotated by 90° to become imaginary numbers, it is reasonable to construct a graph that includes both real and imaginary numbers, as in Fig. 1-9. The two intersecting axes then define a plane, with real numbers

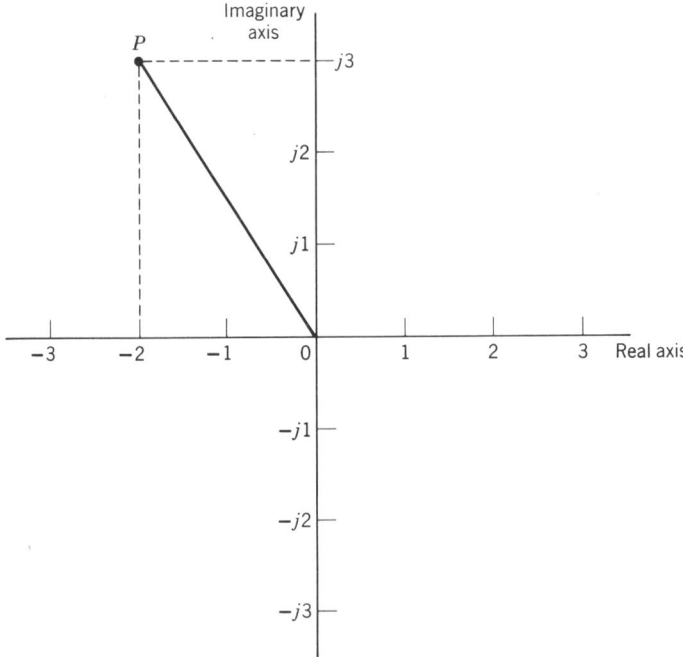

Figure 1-9
The complex plane of numbers.

along the horizontal axis and imaginary numbers along the vertical axis. This is the *complex plane* of numbers. Any point in the plane may be defined in terms of a real part and an imaginary part, for example, the point $P = -2 + j3$ in Fig. 1-9. The line joining P and the origin of the graph is called a *phasor*.

Following the idea of a rotational operator, the j-factor, we can recognize that $j^2 = -1$ and

$$j(j3) = j^2(3) = -1(3) = -3 = 3\underline{/180°}$$

Similarly, $j^3 = -j$ and $j^4 = +1$.

In Fig. 1-9, the phasor P can be expressed as a *complex number*, consisting of a real part and an imaginary part: $P = -2 + j3$. All other points (phasors) on the complex plane can be expressed in the same way. For example, a phasor A might be written as $A = 0.12 - j0.0093$. The two parts of a complex expression do not have to be integers; they may have any numerical values.

Drill Problems

D1-6 Write these expressions in terms of positive real numbers and the j-factor: (a) $2\underline{/180°}$; (b) $2\underline{/90°}$; (c) -2; (d) $-j2$.

D1-7 Sketch a graph of the complex plane of numbers and locate these points on the plane: (a) $3 + j4$; (b) $-j5$; (c) -2; (d) $-5 + j12$; (e) $3 - j4$; (f) $-6 - j8$.

1-7
Generation of Sinusoidal Waveforms by Rotating Phasor

Sinusoidal waveforms may be developed by a phasor that rotates on the complex plane. In Fig. 1-10, the phasor has a magnitude of 3 units and is rotating at a constant angular velocity ω, so that $\omega t = \phi$. In other words, it rotates through the same number of degrees during similar periods of time. Starting at the time when this phasor is crossing the positive real axis, and measuring angles counterclockwise from this axis, the projection of the phasor on the vertical axis generates a sine wave. At the instant shown in the figure, the rotating phasor is located at point A and the magnitude of the sine wave is $3 \sin \phi_A$.

The cosine wave is developed by a similar process, using a projection of the phasor on the horizontal axis. The 90-degree phase difference between the sine and cosine waves is again evident here; one projects on the real axis, the other on the imaginary axis of the complex plane.

The phasor A at the instant shown may be written as $A = 3\underline{/\phi_A}$. This shorthand notation, which consists of the maximum amplitude and an angle

16 Some Useful Circuit Concepts   Ch. 1

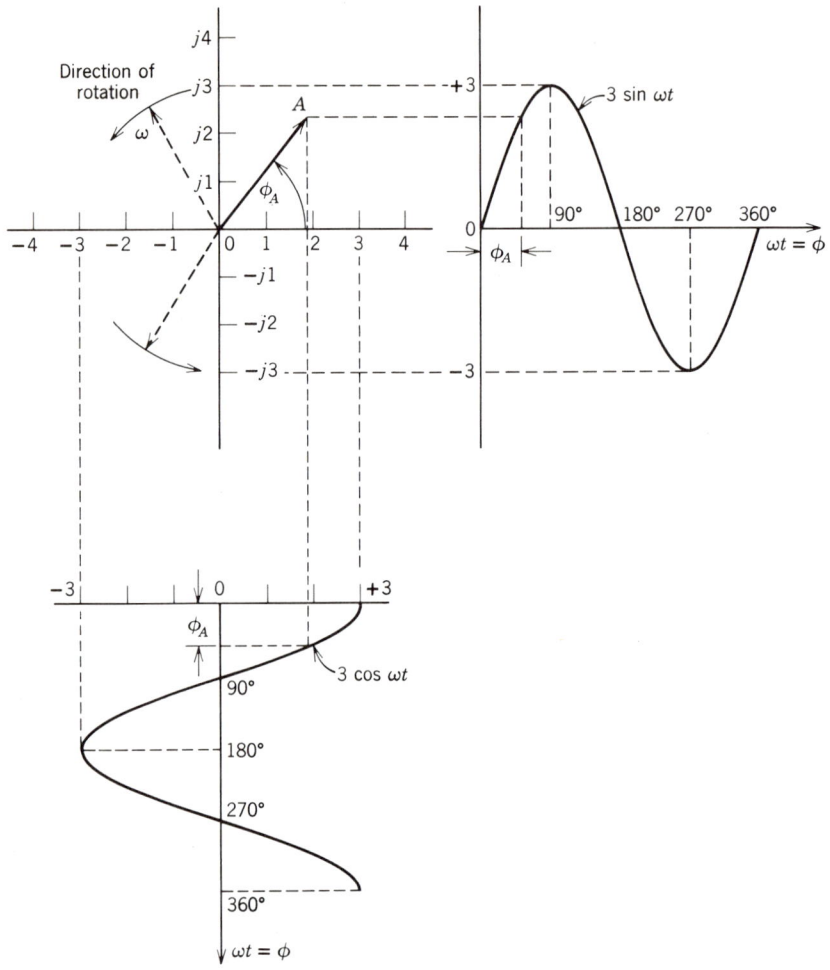

Figure 1-10
Generation of sine and cosine waves by rotating phasor.

measured from the positive real axis, will be useful in handling complex numbers. It is called the *polar form* of the phasor A. Phasor A also may be expressed in terms of its real and imaginary parts, $A = 3\cos\phi_A + j3\sin\phi_A$, which is called its *rectangular form*. We will see that these two forms of phasors are useful for manipulating complex numbers in ac circuit analysis.

1-8
Rules for Complex Number Calculations

Phasors may be added, subtracted, multiplied, and divided, but special rules apply to these operations because they involve complex rather than real numbers. The sum of two phasors A and B is shown in Fig. 1-11a as a graphical addition of their real and imaginary parts. It is seen that their addition yields the diagonal of a parallelogram, or the third side of the triangle formed by moving the tail of B to the point of A (and the tail of A to the point of B) and drawing C from the origin. The same phasor C is formed whether B is added to A or A is added to B; the diagonal is the same. Algebraically, the phasor addition is

$$A + B = (x_A + jy_A) + (x_B + jy_B)$$
$$= (x_A + x_B) + j(y_A + y_B) = C \qquad (1\text{-}14)$$

Figure 1-11b illustrates the subtraction of two phasors. Algebraically, subtraction is the same as adding the negative of one phasor,

$$D = A - B = A + (-B)$$
$$= (x_A + jy_A) + (-x_B - jy_B)$$
$$= (x_A - x_B) + j(y_A - y_B) \qquad (1\text{-}15)$$

It should be noted, however, that it does make a difference which phasor is subtracted from the other. From Fig. 1-11b, it is seen that $E = B - A$ is the negative of $D = A - B$. This can also be shown from the algebra,

$$E = B + (-A)$$
$$= (x_B - x_A) + j(y_B - y_A)$$
$$= -(x_A - x_B) - j(y_A - y_B)$$
$$= -D \qquad (1\text{-}16)$$

Multiplication and division of complex numbers are handled easily when the phasors are expressed in their polar forms. These operations may be performed on the rectangular forms of the phasors, but the polar forms lead to results more quickly. Given two phasors to be multiplied,

$$A = A_m \underline{/\phi_A} \quad \text{and} \quad B = B_m \underline{/\phi_B}$$

then

$$A \times B = A_m \times B_m \underline{/\phi_A + \phi_B} \qquad (1\text{-}17)$$

The magnitude of the phasor product is the product of the individual magnitudes, and the resultant angle is the sum of the individual angles of the two phasors.

18   Some Useful Circuit Concepts   Ch. 1

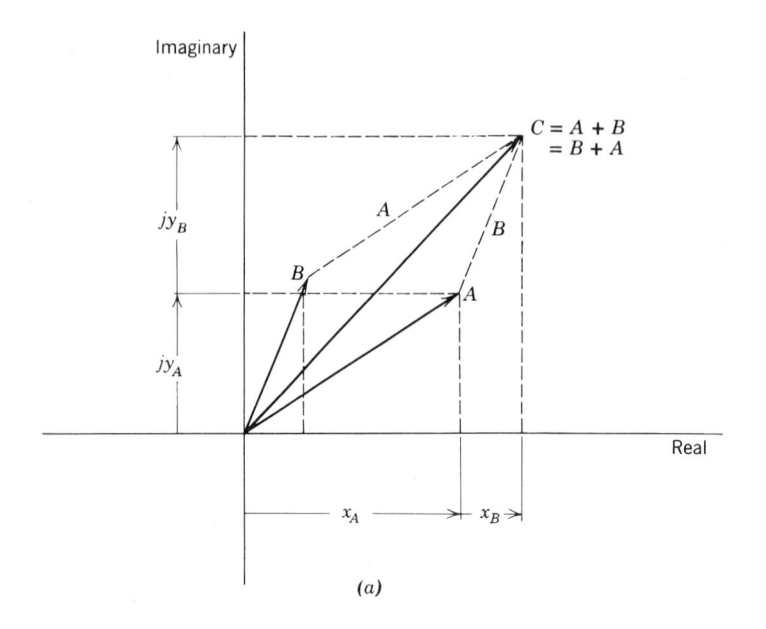

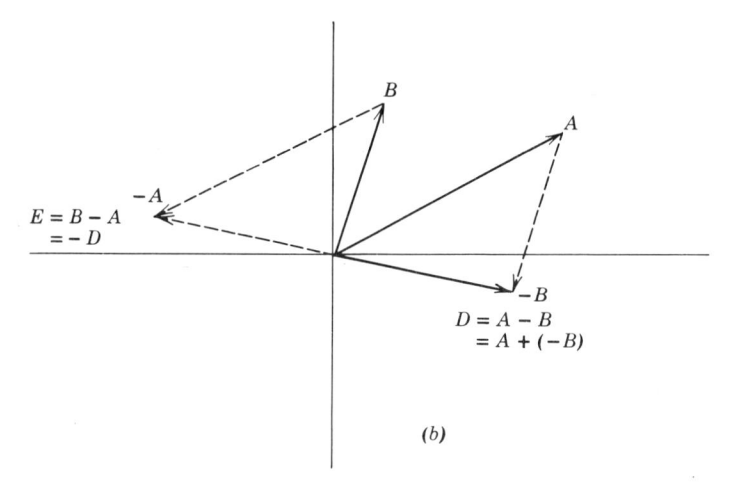

Figure 1-11
(a) Addition of phasors. (b) Subtraction of phasors.

In a similar way, two phasors may be divided:

$$\frac{A}{B} = \frac{A_m}{B_m} \underline{/\phi_A - \phi_B} \tag{1-18}$$

The phasor quotient is the quotient of the two magnitudes, and the resultant

Sec. 1-9 *Impedance in AC Circuits* 19

angle is the difference between the angles of numerator and denominator phasors.

To examine the multiplication of phasors in their rectangular form, let us consider the two phasors

$$A = 10\underline{/30°} \quad \text{and} \quad B = 5\underline{/60°}$$

Expressed in rectangular form, they are

$$A = 10 \cos 30° + j10 \sin 30° = 8.66 + j5.0$$
$$B = 5 \cos 60° + j5 \sin 60° = 2.5 + j4.33$$
(1-19)

To multiply A and B,

$$\begin{aligned} A \times B &= (8.66 + j5.0)(2.5 + j4.33) \\ &= (8.66)(2.5) + j^2(5.0)(4.33) + j(5.0)(2.5) + j(4.33)(8.66) \\ &= 21.6 - 21.6 + j(12.5 + 37.5) \\ &= 0 + j50.0 = 50.0\underline{/90°} \end{aligned}$$

Note that this is the same as the polar result: $10\underline{/30°} \times 5\underline{/60°} = 50\underline{/90°}$.

The slide rule provides an easy method for converting phasors from polar form to rectangular and the reverse procedure.*

Drill Problems

D1-8 Add these phasors algebraically, and sketch their addition on the complex plane:

(a) $3 + j4$ and $5\underline{/30°}$

(b) $-4 - j5$, $-2 + j6$, and $6 - j3$

D1-9 Show that division of phasors A and B of Eq. 1-19 yields the phasor $1.73 - j1 = 2\underline{/-30°}$.

1-9
Impedance in AC Circuits

The behavior of the basic circuit components of resistance, capacitance, and inductance must be understood before the study of ac circuit analysis is undertaken. Mathematical relationships between voltages and currents in circuits containing these components will establish a background for the

---

*See, for example, Alfred L. Slater, *The Slide Rule—A Complete Manual*, Holt, Rinehart and Winston, New York, 1967.

20 Some Useful Circuit Concepts Ch. 1

analysis of ac circuits. We will consider at first only those circuits in which pure sinusoidal waveforms of only one frequency are present. Later, we will extend the analysis to cases where more complicated waveforms exist in the circuits. The relations between sinusoidal voltages and currents in these circuit components will lead us to simplified means of analysis using phasor algebra.

The Resistance Component

Assume that a sinusoidal current $i = I_m \sin \omega t$ is in the resistor R of Fig. 1-12a. Since by Ohm's law, $v_R = iR$,

$$v_R = I_m R \sin \omega t = V_m \sin \omega t \tag{1-20}$$

The current and voltage waveforms have no phase difference; that is, they are *in phase*. The ratio of their maximum amplitudes is

$$\frac{V_m}{I_m} = \frac{V}{I} = R \tag{1-21}$$

The time occurrence of voltage and current is shown in Fig. 1-12b. Their phase relationship is given in Fig. 1-12c, assuming that $\omega t = 0$ represents the instant in time when the sine wave is at 0° and increasing through positive values. (Refer to Section 1-4.) At later instants, the phasors representing current and voltage would be rotated counterclockwise, but they still would be in phase regardless of the time of observation. For example, when current is at −23°, the voltage will be at −23°. During each cycle of sinusoidal variations in amplitude, these phasors will be rotated through a full 360° circle on the complex plane.

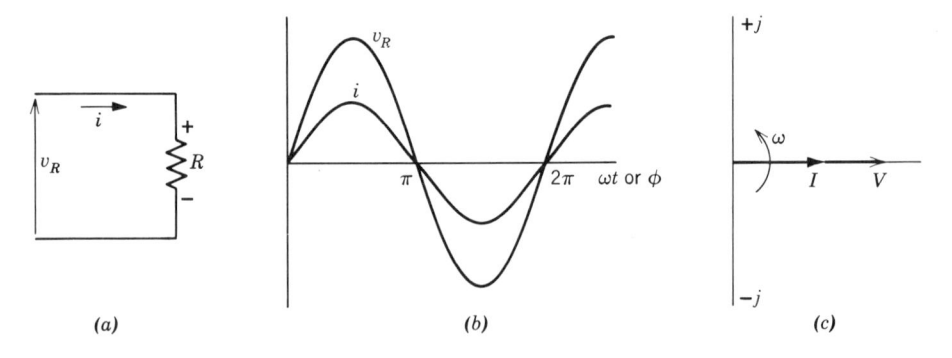

Figure 1-12
(a) Sinusoidal voltage and current in resistor R. (b) Waveforms of i and v_R. (c) Phase relationship of i and v_R on complex plane.

Sec. 1-9 *Impedance in AC Circuits* 21

The Inductance Component

Assume a sinusoidal current $i = I_m \sin \omega t$ in the inductor of Fig. 1-13a. From basic considerations, the voltage developed across the inductance is

$$v_L = L \frac{di}{dt} = I_m \omega L \cos \omega t$$
$$= I_m \omega L \sin (\omega t + 90°)$$
$$= V_m \sin (\omega t + 90°) \qquad (1\text{-}22)$$

Then $V_m = I_m \omega L$ and

$$\frac{V_m}{I_m} = \frac{V}{I} = \omega L = 2\pi f L \qquad (1\text{-}23)$$

The factor ωL is used so often in ac circuit analysis that it has been given the name *inductive reactance* and symbolized X_L. The *reactance* label implies that it is not an ohmic resistance, although its numerical value can be obtained from the ratio of a voltage maximum and a current maximum (also the ratio of rms values of these sinusoids). It should also be noted that the maximum amplitudes of voltage and current are 90° apart in time (see Fig. 1-13b). This relationship is important: *the voltage across an inductive reactance and the current in the reactance are always 90° apart in time.*

In Fig. 1-13c, phasors represent the current and voltage associated with the inductance. As they rotate at a constant ω, the waveform in (b) will be generated with respect to time.

Since the voltage phasor can be plotted along the $+j$ axis and the current along the positive real axis, it is reasonable to write the phasors as a ratio that defines X_L:

$$\frac{jV}{I} = \frac{V\underline{/90°}}{I\underline{/0°}} = X_L\underline{/90°} = jX_L = j\omega L \qquad (1\text{-}24)$$

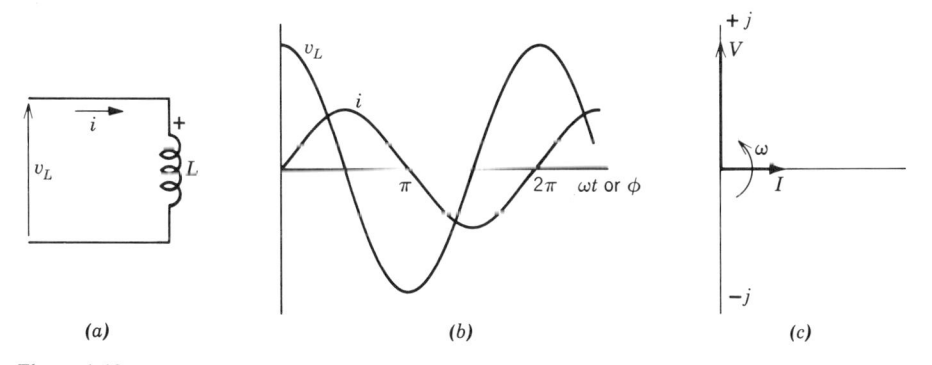

Figure 1-13
(a) Sinusoidal current and voltage associated with inductance component. (b) Waveforms of v_L and i. (c) Phasors representing v_L and i.

22 Some Useful Circuit Concepts   Ch. 1

It should be evident that the voltage is *leading* the current—that is, the voltage reaches a maximum first. Conversely, since the current passes through its positive peak magnitude after the voltage does, it may be said that the current *lags* the voltage by 90°. This relationship is usually expressed as: *current lags voltage in an inductive reactance.*

Inductive reactance, X_L, in the ac circuit is analogous to ohmic resistance in the dc circuit. Inductive reactance is a measure of impedance to ac current in the inductance component. It is the ratio of the voltage maximum and current maximum amplitudes, without regard to their relative phase difference. Inductance reactance (impedance) has units of ohms, just as resistance has units of ohms.

The Capacitance Component

A similar analysis can be made for the capacitance component in Fig. 1-14a, which has a current $i = I_m \sin \omega t$. From basic behavior of this circuit element, the voltage across its terminals may be expressed

$$v_C = \frac{1}{C} \int I_m \sin \omega t \, d(\omega t) = \frac{-I_m}{\omega C} \cos \omega t$$

$$= \frac{+I_m}{\omega C} \sin(\omega t - 90°)$$

$$= V_m \sin(\omega t - 90°) \tag{1-25}$$

Then $V_m = I_m/\omega C$ and

$$\frac{V_m}{I_m} = \frac{V}{I} = \frac{1}{\omega C} = \frac{1}{2\pi f C} \tag{1-26}$$

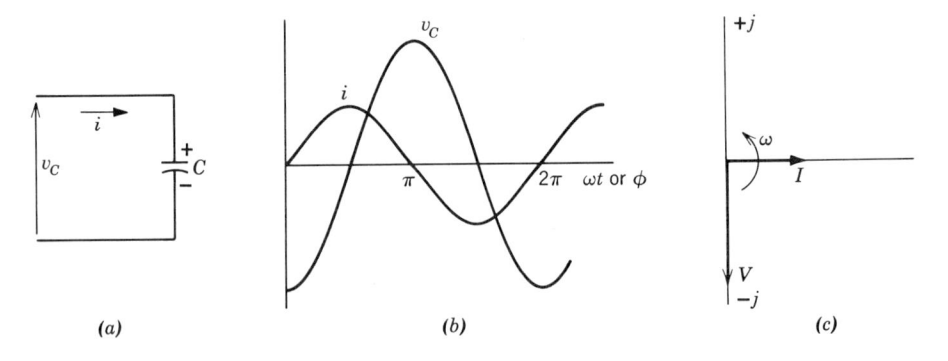

Figure 1-14
(*a*) Sinusoidal current and voltage associated with capacitance component. (*b*) Waveforms of v_C and i. (*c*) Phasors representing v_C and i.

Sec. 1-9　　　　　　　　　　　　　　　　　　　　　　　*Impedance in AC Circuits*　23

The factor $1/\omega C$ is called the *capacitive reactance* and symbolized by X_C. The value of capacitive reactance is the numerical ratio of voltage and current maximum amplitudes (or the ratio of rms values). Here, as in the inductive circuit component, the voltage and current are 90° apart. (Refer to Fig. 1-14b.)

Phasors are drawn in Fig. 1-14c to represent the voltage and current associated with the capacitance component. As these phasors rotate counterclockwise on the complex plane, they will generate the waveforms in (b).

The voltage and current phasors may be written as a ratio that defines X_C:

$$\frac{-jV}{I} = \frac{V\underline{/-90°}}{I\underline{/0°}} = X_C\underline{/-90°} = -jX_C = -j\left(\frac{1}{\omega C}\right) \qquad (1\text{-}27)$$

Dividing both sides of the last equation by $-j$ gives us

$$X_C = \frac{1}{\omega C}$$

In a capacitance component, the current *leads* the voltage by 90°. It may be said that the *current leads voltage in a capacitive reactance*. Again we see that capacitive reactance (impedance) is simply the ratio of voltage and current maximum amplitudes, without regard to their relative phase difference.

Drill Problems

D1-10　A voltage $v = 100 \sin(377t + 90°)$ is applied across a 100-Ω resistor. (a) What current will exist in the resistor? (b) What is the frequency of the current? (c) Sketch phasors on the complex plane to represent voltage and current at the instant when $t = 0$.

D1-11　A pure inductance of 10H contains a current $I = 10$ mA at a frequency of 159 Hz. (a) What impedance value is associated with the inductance at this frequency? (b) What happens to the impedance value as the frequency of the current is increased? (c) What voltage is required across the inductance to sustain this current? (d) Express both current and voltage in terms of maximum amplitude and frequency, as in Eq. 1-22.

D1-12　(a) What capacitance will have the same reactance at 159 Hz as the inductance in D1-11? (b) What happens to the reactance value as the frequency is increased? as capacitance is increased?

Series Combination of Resistance and Inductance

We now look at a series combination of resistance and inductance, which may be represented by R and L in Fig. 1-15a. Assume a current $i = I_m \sin \omega t$

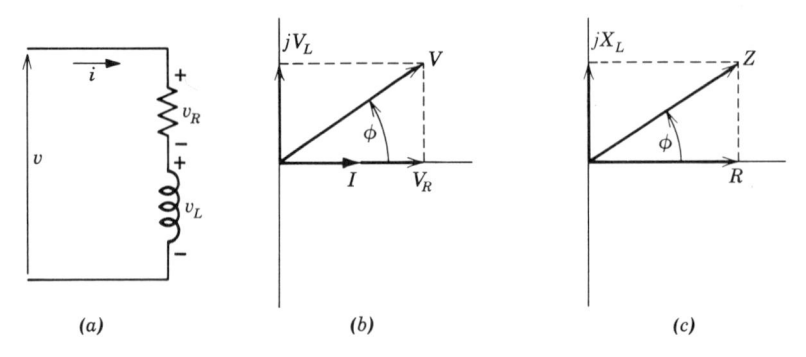

Figure 1-15
(a) Circuit consisting of series R and L. (b) Phasor relationships between voltages and current. (c) Phasor relationships between resistance, inductive reactance, and circuit impedance.

exists in the circuit. From Kirchhoff's voltage law,

$$v = v_R + v_L$$
$$= iR + L\frac{di}{dt} \tag{1-28}$$

Substituting for i as we did in Eqs. 1-20 and 1-22 gives

$$v = I_m R \sin \omega t + I_m \omega L \sin(\omega t + 90°) \tag{1-29}$$

We know that these voltages and the current can be represented on the complex plane as phasors that generate sinusoidal waveforms. We also know that v_R is in phase with i and that v_L leads i by 90°. Therefore, we may sketch phasors as shown in Fig. 1-15b.

Now, since $V_R + jV_L = V$ and the component voltages define a right triangle, V is the hypotenuse of this triangle, and

$$V^2 = V_R^2 + V_L^2$$

or

$$V = \sqrt{V_R^2 + V_L^2} \tag{1-30}$$

Equation 1-29 may be written as a complex number in this way:

$$V = IR + jI\omega L$$
$$= I(R + j\omega L)$$

and

$$\frac{V}{I} = Z \tag{1-31}$$

from which the impedance $Z = R + j\omega L$. These components are plotted on the complex plane in Fig. 1-15c. Note that Z is now the hypotenuse of the right triangle formed by R and X_L. The total impedance to current in this circuit is then the phasor sum of the resistance and inductive reactance. The magnitude of Z may be calculated from the relation

$$Z = \sqrt{R^2 + X_L^2}$$

The angle ϕ between V and the real axis and also between Z and the R component is

$$\phi = \tan^{-1}\frac{V_L}{V_R} = \tan^{-1}\frac{X_L}{R} \qquad (1\text{-}32)$$

It is then evident that the applied voltage v leads the current i by the angle ϕ, which is less than 90°. Depending upon the relative values of X_L and R, the angle ϕ may have any value between 0° and 90°.

From Eq. 1-31 and Fig. 1-15c, it is seen that the impedance Z can be expressed as a phasor in terms of ϕ:

$$\frac{V\underline{/\phi}}{I\underline{/0°}} = Z\underline{/\phi} = \sqrt{R^2 + X_L^2}\underline{/\tan^{-1}(X_L/R)} \qquad (1\text{-}33)$$

Thus, in accordance with the convention we have adopted for ac circuit analysis, the angle associated with the circuit impedance indicates the phase difference between applied voltage and circuit current.

Drill Problems

D1-13 What is the impedance of 100-Ω resistor in series with 0.1-H inductor at 159 Hz? Express your answer (a) as $Z = R + jX_L$ and (b) as $Z\underline{/\phi}$.

D1-14 Sketch a phasor diagram of voltage and current when 100 V is applied to the impedance in D1-13.

Series Combination of Resistance and Capacitance

The circuit of Fig. 1-16a will be used to analyze circuit behavior when resistance (R) and capacitance (C) are connected in series. From a combination of Eqs. 1-20 for R and 1-25 for C, and assuming $i = I_m \sin \omega t$, Kirchhoff's voltage law gives

$$V = V_R + (-jV_C)$$
$$V = IR - jIX_C \qquad (1\text{-}34)$$

from which

$$\frac{V}{I} = R - jX_C = Z$$

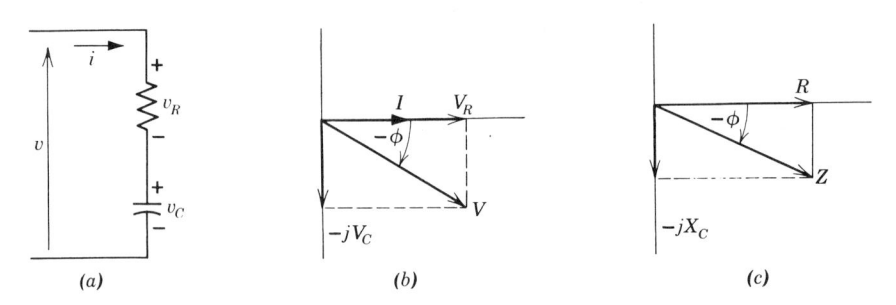

Figure 1-16
(a) Circuit containing R and C. (b) Phasor diagram relating voltages and circuit current. (c) Phasors representing R, X_C, and Z.

Phasors representing the voltages and current are plotted in Fig. 1-16b. Again I and V_R are in phase, as they must be, and the current leads the capacitance voltage by 90°. The applied voltage is the hypotenuse of a right triangle formed by V_R and $-jV_C$, and the angle $-\phi$ represents the phase difference between the current and applied voltage.

Phasors drawn in Fig. 1-16c represent R, X_C, and Z for this circuit. Note that the impedance Z defines a right triangle formed by R and $-jX_C$ and is located at the angle $-\phi$ from R. The angle may be defined

$$-\phi = \tan^{-1}\frac{X_C}{R} \quad \text{or} \quad \phi = \tan^{-1}\frac{-X_C}{R} \tag{1-35}$$

Drill Problems

D1-15 What is the impedance of a 0.1-μF capacitance at 100 kHz?

D1-16 What value of resistance in series with the impedance of D1-15 would yield a circuit impedance having $\phi = -45°$?

R, L, and C in Series

Having reviewed phasor relationships for series R-L and R-C circuits, we will now extend the technique to R-L-C series circuits. The circuit of Fig. 1-17a contains this combination of components, in which there is a current $i = I_m \sin \omega t$.

Phasor diagrams are drawn in Fig. 1-17b and c for this circuit. Here we have assumed that the voltage across L is greater than the voltage across C. This assumption is valid if X_L is greater than X_C at the frequency represented by ω.

The angle ϕ for this circuit is

$$\phi = \tan^{-1}\frac{X_L - X_C}{R} \tag{1-36}$$

Sec. 1-9   Impedance in AC Circuits   27

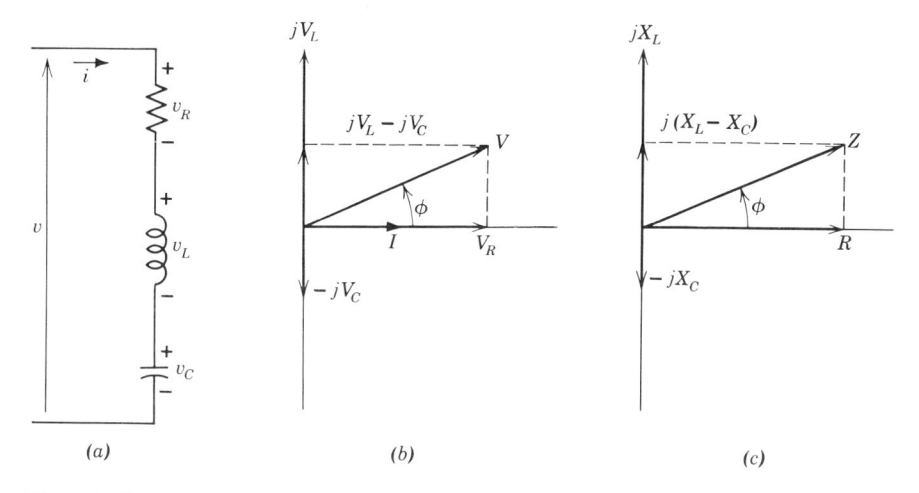

Figure 1-17
(a) Circuit with series R, L, and C components. (b) Phasor diagram of voltages and current. (c) Phasors representing impedance components.

It will be a negative angle if X_C is greater than X_L. The current will lead the applied voltage if ϕ is a negative angle and lag if ϕ is positive. It should be noted, however, that ϕ may have values only between $-90°$ and $+90°$, and will be $0°$ only if the reactances are equal in magnitude at the applied frequency.

In general, the impedance expressions may be written as phasors for the three circuit components:

$$\left.\begin{aligned} Z_R &= R + j0 = R\underline{/0°} \\ Z_L &= 0 + jX_L = X_L\underline{/90°} \\ Z_C &= 0 - jX_C = X_C\underline{/-90°} \\ Z_{\text{series}} &= R + j(X_L - X_C) \\ &= \sqrt{R^2 + (X_L - X_C)^2}\underline{/\tan^{-1}(X_L - X_C)/R} \end{aligned}\right\} \quad (1\text{-}37)$$

In summary, it is important to recognize that *impedance in ac circuits acts the same as resistance in dc circuits*. Impedance restricts the amount of current that can exist in combinations of resistance, inductance, and capacitance. It can be calculated as the ratio of maximum sinusoidal voltage to maximum current, or as the ratio of their rms values. The magnitude of impedance is independent of the phase difference between applied voltage and circuit current, although this phase difference is determined by the components making up the circuit impedance as well as the frequency of the sinusoidal variations.

The major difference between dc circuits and ac circuits is the relative phase difference between applied voltage and circuit current in ac circuits. In dc

28   Some Useful Circuit Concepts   Ch. 1

circuits, the voltage and current are constant quantities and there is no relative phase difference. In ac circuits, voltage and current have sinusoidal waveforms and their peak values may occur at different times depending upon the reactive impedance components present in the circuit. However, we have seen that Ohm's law and Kirchhoff's laws are useful tools for analyzing the behavior of ac circuits.

1-10
Analysis of AC Circuits

In electric circuits in which single-frequency sinusoidal waveforms are present, we may analyze the operation of the circuit by means of the phasor notation developed in the preceding section. By applying Ohm's and Kirchhoff's laws, we may determine voltage drops across circuit elements, how much current exists in different branches of the circuit, what phase differences there are for voltages and currents, how the frequency of the waveforms affects circuit operation, and so on. All these factors may be important in a given circuit, or only one of them may be of interest. We will now develop a procedure for analysis of ac circuits that will be useful in later chapters where we deal with electronic circuits.

The circuit in Fig. 1-18 will be used to show how to analyze an ac circuit using Ohm's and Kirchhoff's laws and the phasor notation for impedances. Ohm's law relates the voltage across a single impedance to the current in it, and Kirchhoff's laws relate voltages in a closed loop and currents entering a junction of circuit elements, just as we discussed for dc circuits.

In Fig. 1-18, junctions of circuit elements are labeled A, B, C, and D. Current I_1 is shown entering junction A, while currents I_2 and I_3 are leaving. From Kirchhoff's current law,

$$I_1 = I_2 + I_3 \quad \text{(current entering = current leaving)}$$

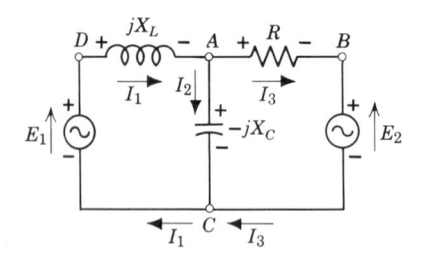

Figure 1-18
Circuit to determine a procedure for ac circuit analysis. Voltage sources E_1 and E_2 have the same frequency.

Sec. 1-10   Analysis of AC Circuits

or

$$I_1 - I_2 - I_3 = 0 \quad \text{(current entering} = 0) \quad (1\text{-}38)$$

Applying the voltage law to the left-hand loop and starting clockwise at C, we can write

$$E_1 = jX_L I_1 + (-jX_C)I_2 \quad \text{(rises = drops)} \quad (1\text{-}39)$$

Starting at B and moving clockwise in the right-hand loop,

$$-E_2 + (-jX_C)I_2 - RI_3 = 0 \quad \text{(rises = 0)} \quad (1\text{-}40)$$

Equations 1-38 through 1-40 describe the relationships between voltage, current, and impedance in this circuit. Since I_2 occurs in both voltage equations, we may substitute for I_2 from Eq. 1-38 and rearrange the terms:

$$jX_L I_1 + (-jX_C)(I_1 - I_3) = E_1$$

and

$$(-jX_C)(I_1 - I_3) - RI_3 = E_2$$

or, multiplying through by -1,

$$(jX_C)(I_1 - I_3) + RI_3 = -E_2$$

Collecting terms for I_1 and I_3 leads to

$$\begin{aligned} j(X_L - X_C)I_1 + jX_C I_3 &= E_1 \\ jX_C I_1 + (R - jX_C)I_3 &= -E_2 \end{aligned} \quad (1\text{-}41)$$

In order to simplify the solution to these equations, let us assume that the impedances are $R = 10\,\Omega$, $X_L = 20\,\Omega$, and $X_C = 10\,\Omega$, and that each voltage source is $10\underline{/0°}$ V. Then Eq. 1-41 becomes

$$\begin{aligned} j10I_1 + j10I_3 &= 10\underline{/0°} \\ j10I_1 + (10 - j10)I_3 &= -10\underline{/0°} \end{aligned} \quad (1\text{-}42)$$

Subtracting the second equation from the first gives

$$-(10 - j10)I_3 + j10I_3 = 10\underline{/0°} - (-10\underline{/0°})$$

or

$$(-10 + j20)I_3 = 20\underline{/0°}$$

from which the current is

$$\begin{aligned} I_3 &= \frac{20\underline{/0°}}{-10 + j20} = \frac{20\underline{/0°}}{22.35\underline{/106.6°}}\ \text{A} \\ &= 0.895\underline{/-106.6°}\ \text{A} \\ &= -0.4 - j0.8\ \text{A} \end{aligned}$$

30 Some Useful Circuit Concepts   Ch. 1

Substituting this value for I_3 in the first equation permits a solution for I_1:

$$j10I_1 + j10(0.895\underline{/-106.6°}) = 10\underline{/0°}$$

$$j10I_1 + 8.95\underline{/-26.6°} = 10\underline{/0°}$$

$$I_1 = \frac{10\underline{/0°} - 8.95\underline{/-26.6°}}{10\underline{/90°}}$$

$$= \frac{10.0 - 8.0 + j4.0}{10\underline{/90°}} = \frac{2.0 + j4.0}{10\underline{/90°}}$$

$$= \frac{4.475\underline{/63.4°}}{10\underline{/90°}}$$

$$= 0.4475\underline{/-26.6°} \text{ A}$$

$$= 0.4 - j0.2 \text{ A}$$

1-11
Determinants

In the analysis of electronic circuits, many calculations become necessary to determine the relationships between voltage, current, and impedance components of the circuits. However, a "shorthand" procedure can be used to reduce the effort required to obtain solutions for complicated networks. This procedure is based on our previous use of Ohm's and Kirchhoff's laws for writing network equations and on determinants for solving algebraic equations. First, let us review how to solve simultaneous equations with determinants. (A detailed development of the method is included in the Appendix.)

As an example, we can represent an electric circuit by the equations

$$2I_1 - 3I_2 = -4$$
$$-I_1 + 6I_2 = 5$$

These equations may also be written as

$$(2)I_1 + (-3)I_2 = -4$$
$$(-1)I_1 + (6)I_2 = 5$$

The determinant, Δ, for these equations is made up from the coefficients of I_1 and I_2:

$$\Delta = \begin{vmatrix} 2 & -3 \\ -1 & 6 \end{vmatrix} = 2(6) - (-1)(-3) = 12 - 3 = 9$$

We may use this value of Δ to solve for I_1 and I_2:

Sec. 1-11

Determinants

$$I_1 = \frac{\begin{vmatrix} -4 & -3 \\ 5 & 6 \end{vmatrix}}{\Delta} = \frac{(-4)(6) - (5)(-3)}{9} = \frac{-24 + 15}{9} = \frac{9}{9} = 1.0$$

$$I_2 = \frac{\begin{vmatrix} 2 & -4 \\ -1 & 5 \end{vmatrix}}{\Delta} = \frac{2(5) - (-1)(4)}{9} = \frac{10 + 4}{9} = \frac{14}{9} = 1.55$$

Solutions for the two variables I_1 and I_2 are obtained as the ratio of two determinants, where Δ is the denominator. The numerator is formed from Δ by replacing a coefficient column with the constant column. When I_1 is being calculated, its coefficients are replaced by the constants. When three or more equations are involved, the procedure becomes more complicated, but it follows the same concept of solving a ratio of determinants.

As an example for three variables and three equations, we will use this set:

$$(-2)I_1 + (1)I_2 + (-1)I_3 = -2$$
$$(3)I_1 + (-2)I_2 + (2)I_3 = 3$$
$$(0)I_1 + (1)I_2 + (-3)I_3 = 1$$

The determinant Δ is made up of the coefficients of the variables I_1, I_2, and I_3:

$$\Delta = \begin{vmatrix} -2 & 1 & -1 \\ 3 & -2 & 2 \\ 0 & 1 & -3 \end{vmatrix}$$

$$= (-2)(-2)(-3) + (1)(2)(0) + (-1)(1)(3)$$
$$- (-1)(2)(0) - (1)(3)(-3) - (-2)(1)(2)$$
$$= -12 + 0 - 3 + 0 + 3 + 4 = -8$$

Current I_1 is solved

$$I_1 = \frac{\begin{vmatrix} -2 & 1 & -1 \\ 3 & -2 & 2 \\ 1 & 1 & -3 \end{vmatrix}}{\Delta}$$

$$= \frac{(-2)(-2)(-3) + (1)(2)(1) + (-1)(1)(3) - (-1)(-2)(1) - (1)(3)(-3) - (-2)(1)(2)}{-8}$$

$$= \frac{-12 + 2 - 3 - 2 + 9 + 4}{-8} = \frac{-2}{-8} = 0.25$$

Drill Problem

D1-17 Solve for I_2 and I_3 in the example above.

1-12
Use of Determinants in Circuit Analysis

The properties of determinants in solving algebraic equations lead to a simple method for analyzing electric circuits. What we are looking for is a procedure that can be used for any circuit without having to write out equations. This section shows how determinants can help reduce the labor of circuit analysis.

We will use the circuit in Fig. 1-19. Ohm's and Kirchhoff's laws for the circuit lead to the equations:

$$I_1 - I_2 - I_3 = 0$$

$$E_1 = 5I_1 - j5I_1 + 10I_3 + j5I_3 = 10\underline{/0°} = 10$$

$$-E_2 = -j5I_3 - 10I_3 + j10I_2 + 5I_2 = -10\underline{/90°} = -j10$$

Now substituting for I_3, since $I_3 = I_1 - I_2$,

$$5I_1 - j5I_1 + 10(I_1 - I_2) + j5(I_1 - I_2) = 10$$
$$-j5(I_1 - I_2) - 10(I_1 - I_2) + j10I_2 + 5I_2 = -j10 \qquad (1\text{-}43)$$

Collecting terms in I_1 and I_2:

$$(5 - j5 + 10 + j5)I_1 + (-10 - j5)I_2 = 10$$
$$(-j5 - 10)I_1 + (j5 + 10 + j10 + 5)I_2 = -j10$$

A simpler form of these equations combines real and imaginary parts of the impedances,

$$(15 + j0)I_1 + (-10 - j5)I_2 = 10$$
$$(-10 - j5)I_1 + (15 + j15)I_2 = -j10 \qquad (1\text{-}44)$$

Solving for each of the unknown currents as ratios of determinants gives

$$I_1 = \frac{\begin{vmatrix} 10 & -(10 + j5) \\ -j10 & 15 + j15 \end{vmatrix}}{\begin{vmatrix} 15 & -(10 + j5) \\ -(10 + j5) & 15 + j15 \end{vmatrix}} \qquad I_2 = \frac{\begin{vmatrix} 15 & 10 \\ -(10 + j5) & -j10 \end{vmatrix}}{\begin{vmatrix} 15 & -(10 + j5) \\ -(10 + j5) & 15 + j15 \end{vmatrix}}$$

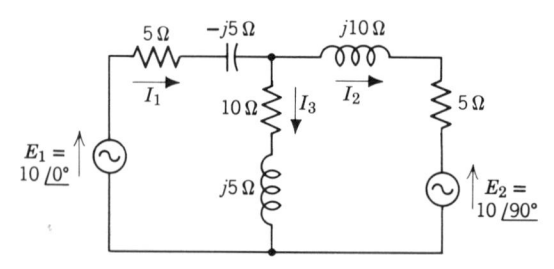

Figure 1-19
Circuit for development of determinant method in circuit analysis.

Sec. 1-13 *Circuit Analysis by Loop Currents* 33

The value of Δ is

$$\Delta = 15(15 + j15) - (10 + j5)(10 + j5)$$
$$= 225 + j225 - 100 + 25 - j50 - j50$$
$$= 150 + j125$$
$$\Delta = 195\underline{/39.8°}$$

Using this value to calculate I_1 gives

$$I_1 = \frac{10(15 + j15) - j10(10 + j5)}{\Delta}$$
$$= \frac{150 + j150 - j100 + 50}{\Delta} = \frac{200 + j50}{\Delta}$$
$$= \frac{206.5\underline{/14.0°}}{195\underline{/39.8°}}$$
$$= 1.06\underline{/-25.8°} \text{ A}$$

Drill Problems

D1-18 Calculate the value of I_2 in Fig. 1-19 and express as a phasor in polar form.

D1-19 Make a phasor sketch on the complex plane showing the relative phase differences between the voltage sources and the three currents I_1, I_2, and I_3 for the circuit in Fig. 1-19.

D1-20 Determine the voltage drop across the 5-Ω resistor in the left-hand loop of the circuit.

1-13
Circuit Analysis by Loop Currents

In the analysis of the circuit of Fig. 1-19, the current I_3 could be written in terms of the other two branch currents as $I_3 = I_1 - I_2$. This circuit is repeated in Fig. 1-20a for easy reference.

Now we will assume that two currents, I_1 and I_2, are circulating in the left-hand and right-hand loops, respectively, as shown in Fig. 1-20b. Note that the net current in the center branch is indeed the difference of I_1 and I_2, because one of these currents is passing downward and the other upward through the center branch. In our previous analysis, the current in this branch was labeled I_3, but in Eq. 1-43 we eliminated I_3, substituting $I_1 - I_2$ as given by Kirchhoff's law. If we had defined the two separate loop currents, no substitution would have been necessary. We could have written Eq. 1-43 in terms of these two loop currents.

34 Some Useful Circuit Concepts

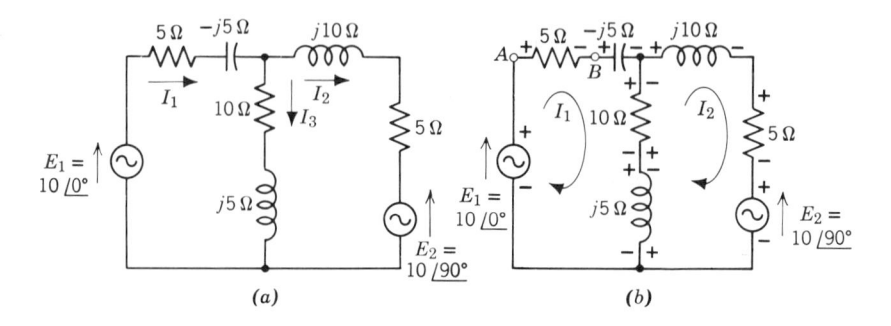

Figure 1-20
(a) Circuit of Fig. 1-19 repeated for convenience, and (b) same circuit with loop currents indicated.

From Kirchhoff's voltage law applied to each of the loops in turn, we get

Loop 1: $5I_1 - j5I_1 + 10(I_1 - I_2) + j5(I_1 - I_2) = 10\underline{/0°} = 10$

Loop 2: $j10I_2 + 5I_2 + j5(I_2 - I_1) + 10(I_2 - I_1) = -10\underline{/90°} = -j10$

Combining terms in I_1 and I_2 gives

Loop 1: $\qquad (15 + j0)I_1 + (-10 - j5)I_2 = 10$

Loop 2: $\qquad (-10 - j5)I_1 + (15 + j15)I_2 = -j10$

(1-45)

These are the same equations we developed as Eqs. 1-44. We see that the individual loop currents may be used to analyze circuits.

If we examine the results in Eqs. 1-45, we see that there is a pattern to the coefficients of I_1 and I_2. This pattern is diagrammed below in the form of the determinant of the two equations for loop currents:

Loop 1: Sum of *voltage drops in the direction of* I_1 = Sum of *source voltage rises* in loop 1

Loop 2: Sum of *voltage drops in the direction of* I_2 = Sum of *source voltage rises* in loop 2

The corresponding determinant is

	I_1	I_2	Voltage sources
Loop 1:	Sum of impedances in loop 1 carrying I_1	—Sum of impedances in loop 1 also carrying I_2	Sum of source rises in loop 1
Loop 2:	—Sum of impedances in loop 2 also carrying I_1	Sum of impedances in loop 2 carrying I_2	Sum of source rises in loop 2

The important feature of this "shorthand" method of writing equations in determinant form is the negative sign on two of the impedance sums. This minus sign occurs because the equations contain voltage drops in the direction of the current in the loop. But there are also voltage rises (negative drops) because of the other current in these impedances, which is passing in the opposite direction. For confirmation refer to the equations that led to Eq. 1-45.

By inspection of Fig. 1-20b, it should be possible to write the determinant form without first writing the equations as given in Eq. 1-45. It takes this form:

$$\begin{array}{cc} I_1 & I_2 \\ \begin{vmatrix} 15 + j0 & -(10 + j5) \\ -(10 + j5) & 15 + j15 \end{vmatrix} & \begin{array}{l} 10\underline{/0°} = 10 \\ -10\underline{/90°} = -j10 \end{array} \end{array}$$

Currents and voltages then may be calculated as in Section 1-12, without first writing the equations according to Ohm's and Kirchhoff's laws. However, it should be understood that the determinant "shorthand" method expresses the same relationships as the circuit equations.

Drill Problem

D1-21 Write the determinant form of the loop equations for this circuit:

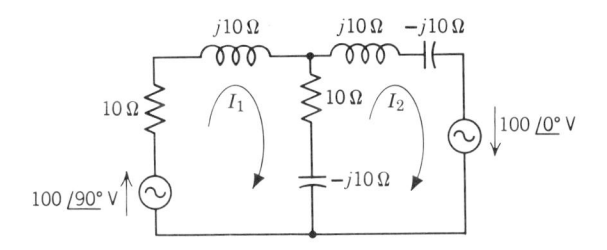

1-14
The Superposition Principle

It is often necessary to analyze circuit behavior when both dc and ac voltage sources are acting simultaneously in the circuit, especially when the circuit contains electronic devices and components. The *superposition principle* is useful in these cases. Briefly stated, it says that as long as circuit components maintain constant impedance values, whether dc or ac voltage sources are active singly or in combination, the circuit may be analyzed for each voltage source separately and the results superposed to yield the behavior when all sources are acting at the same time. For example, a circuit may contain a dc source and an ac source, as in Fig. 1-21, or it may contain two or more ac sources having different frequencies.

For this circuit we will determine the current from the dc source first, and assume that both voltage sources are ideal, that is, that they contain no internal

36   Some Useful Circuit Concepts   Ch. 1

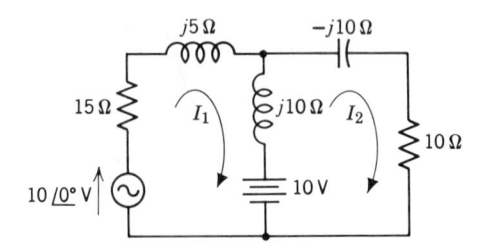

Figure 1-21
Circuit for demonstrating the superposition principle.

impedance. The reactive impedances are given for the frequency of the ac voltage source. For dc voltages, the inductive and capacitive reactances are zero and infinite, respectively. This means that an inductive reactance is zero for dc (frequency = zero) and a capacitive reactance is infinitely large for dc (no dc current will pass). Then the dc current will be limited only by the 15-Ω resistor on the left:

$$I_{dc} = \frac{10 \text{ V}}{15 \text{ }\Omega} = 0.67 \text{ A, counterclockwise in loop 1}$$

The ac currents may be found from the determinant of the circuit:

$$I_1 = \frac{\begin{vmatrix} 10/\underline{0°} & -(j10) \\ 0 & 10+j0 \end{vmatrix}}{\begin{vmatrix} 15+j15 & -(j10) \\ -(j10) & 10+j0 \end{vmatrix}}, \quad I_2 = \frac{\begin{vmatrix} 15+j15 & 10/\underline{0°} \\ -(j10) & 0 \end{vmatrix}}{\begin{vmatrix} 15+j15 & -(j10) \\ -(j10) & 10+j0 \end{vmatrix}}$$

From the superposition principle, current I_1 consists of two parts, with the ac value calculated above superposed on 0.67 Adc counterclockwise. Current I_2 has no dc component; it consists of the ac value calculated above.

1-15
Thévenin's Theorem

Circuit analysis may be used to determine the effect of changing some component in the circuit, perhaps through some range of values that may be of interest. If the calculations necessary for each of these values had to be made individually, and including all sources and impedances again and again, the effort would be substantial even though determinant solutions could be used. Thévenin's theorem permits such calculations to be made with a minimum of calculations for each value of interest.

In an electric circuit, any two terminals may be considered *output terminals*. Any impedance or any combination of impedances and voltage sources that

Sec. 1-15   Thévenin's Theorem   37

may be connected to these two terminals may be considered the *load* on the circuit between these two terminals. In applying the theorem for circuit analysis, the component that is to take on different values is referred to as the load.

Thévenin's theorem is stated in this way: the current in the load will be the same whether it is connected in the original circuit or the remainder of the circuit is replaced by the open-circuit voltage between the output terminals, in series with the impedance measured between the open-circuit output terminals. This impedance is measured (or calculated) after all voltage and current sources in the circuit have been replaced by their internal impedances. (An ideal voltage source has zero internal impedance; an ideal current source has infinite internal impedance.) The series combination of the open-circuit voltage and this impedance is often referred to as the "Thevenin equivalent" circuit. Once these equivalent values have been determined, they will not change because of changes in the load; only the load current and voltage drop will change as the load changes.

Now we will look at the circuit in Fig. 1-22a to see how this theorem may be used. For purposes of illustration, the 5-Ω resistor between A and B is considered the load.

The equivalent circuit (Fig. 1-22b) may be considered to consist of two parts, the open-circuit voltage in series with the open-circuit impedance. Let us determine the impedance first. First we replace all circuit voltage sources with their internal impedances, zero for each one in this circuit. Then we determine the impedance between A and B with the 5-Ω load removed.

$$Z_{eq} = -j5 + \frac{(10 + j5)(5 + j10)}{10 + j5 + 5 + j10} = -j5 + \frac{50 + j^250 + j25 + j100}{15 + j15}$$

$$= -j5 + \frac{125/90°}{21.2/45°} = -j5 + 5.9/45° = -j5 + 4.16 + j4.16 = 4.16 - j0.84$$

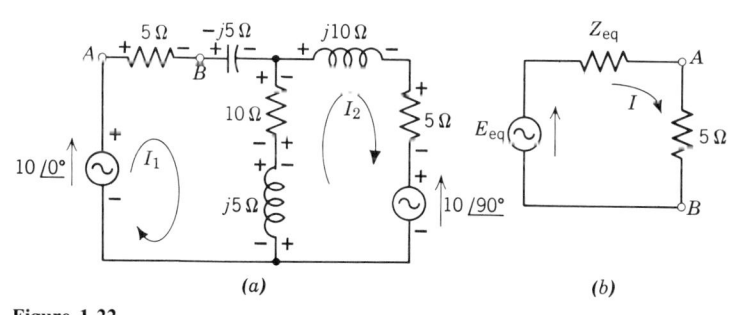

$$= 4.25/-11.4° \text{ ohms}$$

Figure 1-22
(a) Two-loop circuit with load 5 Ω connected between output terminals A and B. (b) Thévenin equivalent circuit for the 5-Ω load.

38 Some Useful Circuit Concepts   Ch. 1

The open-circuit voltage is positive at A with respect to B by the combination of the $10/\underline{0°}$ source and the voltage across $10 + j5$ caused by I_2:

$$E_{eq} = 10/\underline{0°} + \frac{(10 + j5)(-10/\underline{90°})}{10 + j5 + 5 + j10} = 10 + \frac{(11.2/\underline{26.5°})(-10/\underline{90°})}{21.2/\underline{45°}}$$

$$= 10 - 5.28/\underline{71.5°} = 10 - 1.675 - j5 = 8.325 - j5 = 9.71/\underline{-31°} \text{ V}$$

The current I in the equivalent circuit of Fig. 1-22b may be calculated:

$$I = \frac{E_{eq}}{Z_{eq} + 5} = \frac{9.71/\underline{-31°}}{4.16 - j0.84 + 5} = \frac{9.71/\underline{-31°}}{9.17/\underline{-5.2°}} = 1.06/\underline{-25.8°} \text{ A} \quad (1\text{-}46)$$

Compare this value with the result for I_1 in Section 1-12, which was calculated by determinants.

For other values of load impedance, the simple relationship of Eq. 1-46 may be used by substituting the other values for the 5-Ω load. It should also be evident that the load may be reactive as well as resistive, because the equivalent circuit was obtained without regard to the nature of the load. The equivalent circuit is valid for any load connected between A and B, whether it is a single resistor, a complex impedance, or a combination of sources and impedances. The equivalent circuit is independent of the load.

1-16
Norton's Theorem

Another circuit analysis tool that is often useful in electronic circuits is Norton's theorem. It is based on the ability of a generator to supply constant current to a load. Certain kinds of electronic circuits can supply a current that is practically constant over a wide range of load, and Norton's theorem is useful in analyzing their performance.

Norton's theorem states that an electric network between two points may be replaced by an equivalent short-circuit current source shunted by the open-circuit impedance between those two points, as far as the load between the points is concerned. The load current will be the same whether the original circuit or its "Norton equivalent" is used to analyze circuit behavior.

We will again use the circuit of Fig. 1-22a and determine the Norton equivalent for the 5-Ω resistor considered as the load between A and B. We already have the value for the open-circuit impedance: $Z_{eq} = 4.16 - j0.84 = 4.25/\underline{-11.4°}$ (Section 1-15). The short-circuit current will be determined when the load is removed, by calculating the current that will pass from A to B in a short-circuit connection between them.

In order to simplify the calculation, we will use the Thévenin equivalent circuit in Fig. 1-22b. This simple circuit has already been shown to be equivalent to the network without the 5-Ω load.

Sec. 1-17 · · · Fourier Series for Periodic Waveforms · 39

When a short-circuit is connected between A and B, current will flow from A to B:

$$I_{sc} = \frac{E_{eq}}{Z_{eq}} = \frac{9.71/-31°}{4.25/-11.4°} = 2.29/-19.6° \text{ A}$$

The Norton equivalent between A and B is shown in Fig. 1-23, with the 5-Ω load connected. The load current, I, may be calculated using the concept of current division by parallel impedances:

$$I = I_{sc}\left(\frac{Z_{eq}}{Z_{eq} + 5}\right) = \frac{(2.29/-19.6°)(4.25/-11.4°)}{9.17/-5.2°} = 1.06/-25.8° \text{ A}$$

This result should not be too surprising. We have determined it already using determinants and Thévenin's equivalent for this circuit (Sections 1-12 and 1-15). We also have seen how Norton's equivalent may be determined from a Thévenin equivalent circuit. This interchange of equivalent circuits may be useful in later studies of electronic circuits.

1-17
Fourier Series for Periodic Waveforms

Many electronic circuits contain voltages and currents that repeat their waveforms at regular intervals and therefore are said to have periodic waveforms. The simplest type of periodic voltage or current is the sinusoid, for example, $e = E_m \sin \omega t$. Only the most elementary circuit, however, would have such simple waveforms. Electronic circuits often contain waveforms that are periodic but by no means as simple as pure sine and cosine waves. Some examples of periodic waveforms that may occur in electronic circuits are the human voice and musical tones (as in radio transmission), similar signals occurring in television receivers, pulses of information flowing between parts of a digital computer, and telemetry signals representing the internal temperature of a space vehicle.

Waveforms that are periodic may be expressed as the sum of mathematical terms involving the sine and cosine of harmonic frequencies. This sum of sine

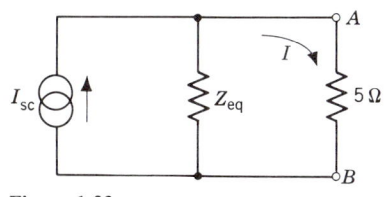

Figure 1-23
Norton's equivalent circuit for the network of Fig. 1-22a.

40 Some Useful Circuit Concepts   Ch. 1

and cosine terms is called the *Fourier series** of the waveform and may be expressed as follows:

$$e = a_0 + a_1 \cos \phi + a_2 \cos 2\phi + a_3 \cos 3\phi + \cdots + a_n \cos n\phi + \cdots$$
$$+ b_1 \sin \phi + b_2 \sin 2\phi + \cdots + b_n \sin n\phi + \cdots \quad (1\text{-}47)$$

where n can be as large as you wish. It is clear that such a waveform in an electronic circuit could be handled by the superposition principle.

In Eq. 1-47, a_0 is called the dc term of the series and its value is the average of the voltage e. The other a terms are maximum values of the cosine harmonic frequencies, and the b terms are maximum values of the sine harmonic frequencies.

The half-sine wave of voltage obtained from a rectifier, such as is used in many radio and TV circuits, is sketched in Fig. 1-24a. This waveform is periodic and may be expressed in a Fourier series. The mathematical procedures for determining the values of a terms and b terms are explained in reference books on the calculus.† It can be shown that the Fourier series for a half-sine waveform is

$$e = E_m \left[\frac{1}{\pi} + \frac{1}{2} \sin \omega t - \frac{1}{\pi} \sum_{n=\text{even}}^{\infty} \left(\frac{\cos n\omega t + 1}{n^2 - 1} \cos n\omega t \right) \right]$$
$$= 0.318 E_m + 0.5 E_m \sin \omega t - 0.212 E_m \cos 2\omega t - 0.0424 E_m \cos 4\omega t - \cdots$$

Similarly, the output voltage from a full-wave rectifier, sketched in Fig. 1-24b, can be given as a Fourier series,

$$e = E_m \left[\frac{2}{\pi} - \frac{2}{\pi} \sum_{n=\text{even}}^{\infty} \left(\frac{\cos n\omega t + 1}{n^2 - 1} \cos n\omega t \right) \right]$$
$$= 0.636 E_m - 0.424 E_m \cos 2\omega t - 0.085 E_m \cos 4\omega t + \cdots$$

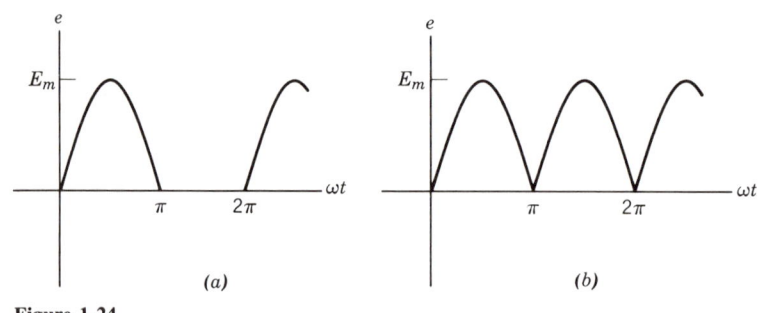

Figure 1-24
Periodic voltage waveforms obtained at the output of (a) a half-wave rectifier, (b) a full-wave rectifier.

* Named for Jean Baptiste Joseph Fourier (1768–1830), the French mathematician.
† See, for example, J. D. Strange and B. J. Rice, *Analytic Geometry and Calculus with Technical Applications*, Wiley, New York, 1970, pp. 407–417.

Equation 1-47 is a basic expression, not only for waveforms containing only curved lines but also for square, triangular, and other periodic waveforms. A single musical note played on a violin is very complex, consisting of many harmonic frequencies sounding at the same time. But such a sound is also periodic and may be expressed in the form of the Fourier series. Square pulses, such as may be generated by a telephone dialing mechanism, contain several harmonic frequency components. If this square pulse is to be handled by an electronic circuit, it is important that the circuit be capable of handling all of the frequencies. The Fourier series, together with the principle of superposition, may be used to analyze circuit behavior for such a signal.

The effective value of a complex wave may be obtained by a simple square root process that involves all the components in its Fourier series. For the voltage wave in Eq. 1-47,

$$E_{\text{eff}} = \left[a_0^2 + \frac{a_1^2 + a_2^2 + \cdots + a_n^2 + \cdots + b_1^2 + \cdots + b_n^2 + \cdots}{2} \right]^{1/2}$$

(1-48)

The effective value of the output voltage of a full-wave rectifier is approximately

$$E_{\text{eff}} = \left[(0.636 E_m)^2 + \frac{(0.424 E_m)^2 + (0.085 E_m)^2}{2} \right]^{1/2} = 0.706 E_m$$

This calculation is only approximately correct because only the first three terms of the Fourier series were used. However, terms involving the higher harmonic frequencies have such small magnitudes that their effect on the result is negligible.

Drill Problems

D1-22 Why are the components in the Fourier series for the half-wave and full-wave rectifier voltages only even-numbered harmonics?

D1-23 Calculate the next three terms in the Fourier series for the full-wave rectifier voltage given in the text above.

1-18
Time Constant

It is common in electronic circuits for a capacitor to be charged and discharged through a resistance, and for a coil and a resistance in series to have increasing or decreasing voltage applied to them.

When a constant voltage E is applied to a series combination of resistance R and capacitance C, current will start flowing immediately (at $t = 0$) and its

42 Some Useful Circuit Concepts

instantaneous value will be given by*

$$i = \frac{E}{R}\epsilon^{-t/RC} \qquad (1\text{-}49)$$

in which ϵ is the base of the system of natural logarithms. If a charged capacitor with capacitance C is connected in series with a resistance R, the current at any instant will also be given by Eq. 1-49. Figure 1-25 shows current and voltage plotted against time for these conditions. Obviously, the larger the capacitance, the longer it will take to charge the capacitor to full voltage E and the longer it will take to discharge it. In both cases the current will eventually become zero; that is, it will stop flowing. At $t = 0$, $i = E/R$, the initial value in both cases.

If $t = RC$ in Eq. 1-49, the exponent becomes -1 and

$$i = \frac{E}{R\epsilon} = \frac{E}{2.71828R} = 0.367\frac{E}{R} \qquad (1\text{-}50)$$

At $t = \infty$, $i = 0$ and the current has gone through 100 per cent of its change.

Note—and this is *most important*—that at $t = RC$ the current has gone through 0.633, or 63.3 per cent, of its *total change*. RC is called the *time constant* of the circuit. It is the time *in seconds* that is required for charging or discharging current to go through 63.3 per cent of the change from initial value to final value.

When a constant voltage E is applied to a coil and a resistance R in series, the current is given by

$$i = \frac{E}{R}(1 - \epsilon^{-(R/L)t}) \qquad (1\text{-}51)$$

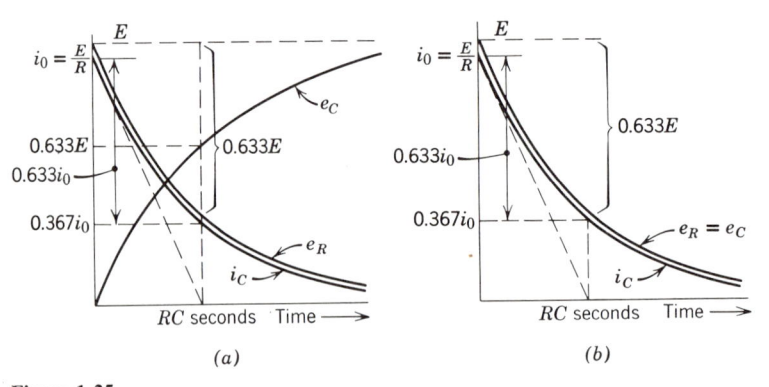

Figure 1-25
(a) Voltage and current curves for charging of capacitor through a resistor. (b) Voltage and current curves for discharge of capacitor through a resistor.

* Derivations of Eqs. (1-49), (1-51) and (1-52) are given in the Appendix.

Sec. 1-19   *Decibels*   43

In L/R seconds the current goes through 63.3 per cent of its total change from zero initial value to E/R final value. L/R is the *time constant* expressed in seconds. When $t = L/R$, the exponent becomes -1.

If an inductance L carrying a current is connected across a resistance R so that all the energy in the inductance is dissipated in R, the instantaneous current is given by

$$i = \frac{E}{R} \epsilon^{-(R/L)t} \tag{1-52}$$

in which E is the product of the current and the resistance. As in the previous case, the *time constant* is L/R, the time required for the current to go through 63.3 per cent of the total change it will experience in falling from its initial value of E/R at $t = 0$ to zero at $t = \infty$.

In all these cases the current reaches its final value in a matter of seconds or a fraction of a second, depending on the values of R, L, and C. Infinite time is only *theoretically* indicated.

1-19
Decibels

The *decibel* came into use as the result of a desire for some accurate means of measuring and comparing the power required to produce sounds. The human ear responds to sound intensity in a logarithmic fashion. It readily recognizes the doubling of sound power, but it can hardly distinguish two sounds of the same pitch when their powers differ by as much as 20 per cent.

The common logarithm of the ratio of two powers (they do not have to be sound powers) *is called the bel*, after Alexander Graham Bell, who invented the telephone. Because the bel represents a relatively large power ratio, the *decibel*, which is *one-tenth of a bel*, is more often used. The equation used to calculate decibels is

$$\mathrm{dB} = 10 \log_{10} \frac{P_2}{P_1} \tag{1-53}$$

Suppose we supply a power P_1 to a system, which may be an amplifier or any other transmission network, and get out of it a power P_2. If $P_2 > P_1$, the *dB value obtained from Eq. 1-53 is a positive value* and we have that much *dB gain*. If $P_2 < P_1$, the dB value will be *negative* and we have a *dB loss* (a negative gain). When $P_2 < P_1$ we may use the expression

$$\mathrm{dB} = 10 \log_{10} \frac{P_1}{P_2} \tag{1-54}$$

which will give us the same numerical answer as will Eq. 1-53 but without a minus sign. We must, of course, remember that we have a loss when the output power (P_2) is less than the input power. All transmission networks not employing some type of power amplifier have some losses and therefore deliver less power than they take in.

Example. Compare 10 W with 5 W in decibel units.

Solution:

$$10 \log_{10} \frac{10}{5} = 10 \log_{10} 2 = 3 \text{ dB}$$

That is, 10 W power is 3 dB above 5 W power. Considering a loss from 10 W input to 5 W output, we have

$$\begin{aligned} \text{dB} &= 10 \log_{10} \tfrac{5}{10} = 10 \log_{10} \tfrac{1}{2} \\ &= 10(\log_{10} 1 - \log_{10} 2) \\ &= 10(0 - 0.3) = -3 \end{aligned}$$

That is, 5 W is 3 dB lower than 10 W. *Every time a power value is halved, its decibel value is decreased by 3. Every time a power value is doubled, its decibel value is increased by 3.* (We are using 0.3 for the value of $\log_{10} 2$ instead of its five-place value, 0.30103. In practical calculations 0.3 is commonly used.)

There is another advantage in using decibels to denote amplification. If one stage of an amplifier multiplies power by 8 and feeds a second stage that multiplies it again by 4, the overall power amplification is 32. That is, we multiply the power amplifications together to get the overall amplification. But when decibels are used they are simply added. The amplification of 8 corresponds to 9 dB, and the amplification of 4 corresponds to 6 dB, so that there is an overall amplification of 9 + 6 = 15 dB. Checking this by using the overall power amplification of 32, we have

$$\text{dB} = 10 \log_{10} 32 = 10 \times 1.5 = 15$$

Zero power does not have a place on a decibel scale, because the power ratio would be either infinite or zero. The logarithm would be either plus or minus infinity.

It is convenient to choose a small value of power and to say that it corresponds to zero decibel or to zero decibel level. Then all other power values may conveniently be expressed in decibels with reference to this level. One *milliwatt* is usually used as a *zero decibel level*. Sometimes 6 mW is used. The power output of a microphone may be as low as 50 dB below 1 mW. Its output would then be rated at -50 dB referred to the 1-mW level. To determine the actual power output of such a microphone, we substitute the known values into Eq. 1-53 and solve for P_2:

Sec. 1-19   Decibels   45

$$-50 = 10 \log_{10} \frac{P_2}{0.001}$$

$$\log_{10} \frac{P_2}{0.001} = -5$$

$$\frac{P_2}{0.001} = 10^{-5}$$

$$P_2 = 1 \times 10^{-8} \text{ W} = 0.01 \text{ }\mu\text{W}$$

Example. One of the huge hydroelectric generators in the Hoover Dam power station delivers 82,500 kW. How many decibels is this above the 1-mW (0.001-W) level?

Solution:

$$dB = 10 \log_{10} \frac{82.5 \times 10^6}{0.001} = 10 \log_{10}(82.5 \times 10^9)$$
$$= 10(1.916 + 9) = 109.16$$

The amplification of voltage may also be expressed in decibels. One important restriction in doing so, however, is that *the voltages must exist across equal resistances.* Current amplifications may also be expressed in decibels *if the currents flow through equal resistances.*

If P_1 and P_2 exist in two equal resistances $R_1 = R_2 = R$, we may write

$$dB = 10 \log_{10} \frac{P_2}{P_1} = 10 \log_{10} \frac{E_2^2/R}{E_1^2/R}$$
$$= 10 \log_{10} \left(\frac{E_2}{E_1}\right)^2$$

which may be written

$$dB = 20 \log_{10} \frac{E_2}{E_1} \qquad (1\text{-}55)$$

For current amplification, we have

$$dB = 10 \log_{10} \frac{P_2}{P_1} = 10 \log_{10} \frac{I_2^2 R}{I_1^2 R}$$
$$= 20 \log_{10} \frac{I_2}{I_1} \qquad (1\text{-}56)$$

It is again emphasized that before using Eqs. 1-55 and 1-56 the student should be sure that the voltages are across, or the currents are in, equal resistances. These relations also hold for impedances, but with the corresponding restric-

1-20
Volume Unit

In radio and television broadcasting, power level is expressed in terms of *volume units* (vu) and measured by a *volume level meter*. The zero reference is 1 mW. Accordingly,

$$vu = 10 \log_{10} \frac{P}{0.001} \qquad (1\text{-}57)$$

in which P is the amount of power measured in watts.

The similarity of this equation to Eq. 1-53 indicates that a power level at any *positive number of volume units above 1 mW* is a power level of the same number of *decibels above 1 mW, if 1 mW is used as a 0 dB reference*. For example, if P in Eq. 1-57 is 0.01 W,

$$vu = 10 \log_{10} \frac{0.01}{0.001} = 10$$

A power level of 10 vu is evidently equal to 10 dB on the decibel scale *where $0\,dB = 0.001\,W$*. A voltmeter calibrated to read vu will thus have its zero point at the position of the pointer when the meter is across a 600-Ω load carrying 1 mW of power. The *voltage* indication is evidently calculated as follows:

$$\frac{E^2}{600} = 0.001$$

$$E = \sqrt{0.6} = 0.774 \text{ V}$$

As mentioned earlier, 3 dB and 6 dB are commonly used values. When power is reduced one-half, the new level is spoken of as 3 dB down or 3 vu down. When it is reduced to one-fourth, its value is 6 dB or 6 vu down. A doubling of power by an amplifier is a 3-dB (or 3-vu) gain, while a power gain of 6 dB (or 6 vu) means a quadrupling of power.

1-21
Maximum Power Transfer Theorem

When electrical energy is to be delivered to a load, there are times when the load should receive *maximum power* rather than maximum voltage or current. One is the driving of a loudspeaker by an electronic amplifier.

The load resistance R_L in the circuit of Fig. 1-26 will receive maximum power if R_L has *only one particular ohms value* and not any other. Since any

Maximum Power Transfer Theorem

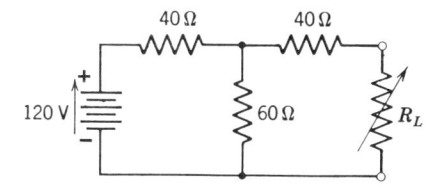

Figure 1-26
Circuit considered for condition for maximum power into the load R_L.

four-terminal network may be represented by an equivalent circuit by using Thévenin's theorem, the circuit of Fig. 1-26 will be used in discussing maximum power transfer. The Thévenin equivalent circuit has a supply E' of 72 V and a series impedance Z' of 64 Ω.

If one used a slide-wire resistance as a load so that its ohms value could be varied, thus varying R_L, a wattmeter showing the power in R_L *would have a maximum reading* only when the slide wire was set at 64 Ω. This can be proved by using simple calculus.* We shall be content with calculating the power transferred to the load for a few values of R_L above and below 64 Ω, and show that if a curve were plotted with power in R_L on the vertical axis and with ohms resistance of R_L on the horizontal axis, the peak of the curve would be at $R_L = 64$ Ω.

The voltage across R_L will always be given by

$$E_L = \frac{R_L}{64 + R_L} \times 72 \text{ V}$$

and the power in R_L by

$$P_L = \frac{E_L^2}{R_L} = \frac{72^2 R_L}{(64 + R_L)^2} \tag{1-58}$$

It has been seen that in a dc circuit, maximum power is delivered to a load when the load resistance is made equal to the generator resistance. In ac circuits the load and generator *impedances* must be equal in magnitude and they must be *conjugates of each other. The conjugate of an impedance is another impedance of the same magnitude but opposite phase angle.*

$$R_1 + jX_1 \quad \text{and} \quad R_1 - jX_1$$

are conjugate impedances, R_1 having a single real value and X_1 having a single real value.

To prove the maximum power transfer theorem for the ac case, we recall that power is given by I^2R, no matter what reactance values are present. When a

---

* Calculus proof is given in the Appendix.

48 Some Useful Circuit Concepts Ch. 1

source, or generator, that *develops or generates* a voltage E and has an *internal impedance* $Z_G = R_G + jX_G$ is connected to a *receiver circuit* of $Z_R = R_R + jX_R$, the current that flows is

$$I_R = \frac{E_G}{\sqrt{(R_G + R_R)^2 + (X_G + X_R)^2}} \underline{/\theta}$$

$$\tan \theta = \frac{X_G + X_R}{R_G + R_R} \tag{1-59}$$

The power, P_R, delivered to the receiver is

$$P_R = I_R^2 R_R = \frac{E_G^2 R_R}{(R_G + R_R)^2 + (X_G + X_R)^2} \tag{1-60}$$

With R_G and R_R already *fixed and equal*, the *power delivered to the receiver will be maximum when X_G and X_R are equal and of opposite sign.* This makes the denominator a minimum, and P_R will be a maximum. The conclusion is that, for a source of ac power to deliver maximum power to a load, *the complex load impedance must be the conjugate of the generator impedance.* This means that if $Z_G = R_G + jX_G$, then Z_R must be $R_R - jX_R$, and $R_R = R_G$, $X_R = X_G$. Also, if $Z_G = R_G - jX_G$, then Z_R must be $R_R + jX_R$, where $R_R = R_G$ and $X_R = X_G$.

1-22
Series Resonance

In a series circuit consisting of inductance, capacitance, and resistance, it is found that when an alternating voltage of *constant magnitude and frequency* is applied, the *current increases* as the *difference* between the inductive reactance and the capacitive reactance is *decreased*. If the capacitive reactance is the larger, it can be reduced by increasing the capacitance until the values of the two reactances in ohms are made equal. By reducing the capacitance, its reactance can be increased to equal a larger inductive reactance.

For fixed values of L and C, a frequency value exists which will produce series *resonance*, in which case the reactances are equal:

$$2\pi f L = \frac{1}{2\pi f C}$$

$$f = \frac{1}{2\pi \sqrt{LC}} \tag{1-61}$$

Obviously, for a fixed frequency, a value of L (henrys) or of C (farads) is given by Eq. 1-61 if one of them is known.

Example. An example of series resonance will be explained with reference to Fig. 1-27. $L = 200$ mH, $C = 0.05 \,\mu$F, $R = 100 \,\Omega$, $E = 10$ V rms. The fre-

Sec. 1-22   Series Resonance   49

quency of E required for series resonance will be determined, as will values of I, E_R, E_L, and E_C.

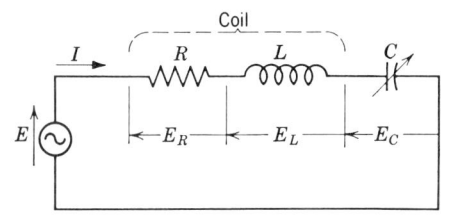

Figure 1-27
Series circuit in which X_C can be made equal to X_L at a fixed frequency.

Solution:

$$f = 1/(2\pi\sqrt{LC}) = 1/(2\pi\sqrt{0.200 \times 0.05 \times 10^{-6}}) = 1591 \text{ Hz}$$
$$I = E/Z = E/R \quad \text{since } X_L = X_C$$
$$I = 10/100 = 0.1 \text{ A}$$
$$X_C = 1/(2\pi \times 1591 \times 0.05 \times 10^{-6}) = 2000 \text{ }\Omega$$
$$X_L = 2\pi \times 1591 \times 0.2 = 2000 \text{ }\Omega$$
$$E_R = 0.1 \times 100 = 10 \text{ V}$$
$$E_L = IX_L = 0.1 \times 2000 = 200 \text{ V}$$
$$E_C = IX_C = 0.1 \times 2000 = 200 \text{ V}$$

Note that although only 10 V is applied to the circuit, 200 V exists across L and also across C. This is a "resonance rise in voltage" of a 20-to-1 ratio.

The impedance diagram and the phasor diagram are shown in Fig. 1-28. If L and R were values for a coil, the voltage across its terminals would be as shown in the phasor diagram.

When the frequency of the voltage applied to a series circuit composed of L, R, and C is below resonant frequency, the current that flows leads the applied voltage. This means that the total impedance of the circuit is capacitive. When the frequency of the applied voltage is above resonant frequency, the total impedance is inductive, and so for all frequencies greater than resonant frequency the current lags the applied voltage. Since the net reactance of a coil and a capacitor connected in series is given by their algebraic sum,

$$X = X_L + X_C$$

in which X_C is *negative*, it is readily seen in Fig. 1-29 that X will be negative (capacitive) below resonance, and positive (inductive) above resonance. At

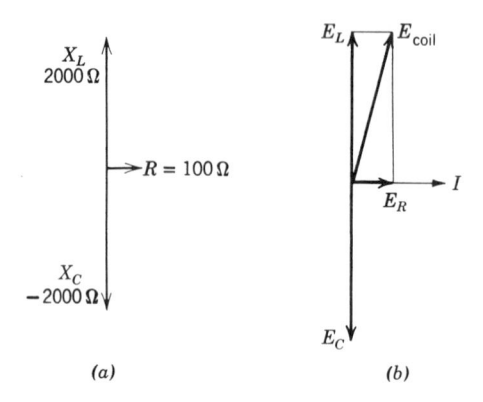

Figure 1-28
Diagrams for circuit of Fig. 1-27. (a) Impedance diagram. (b) Phasor diagram.

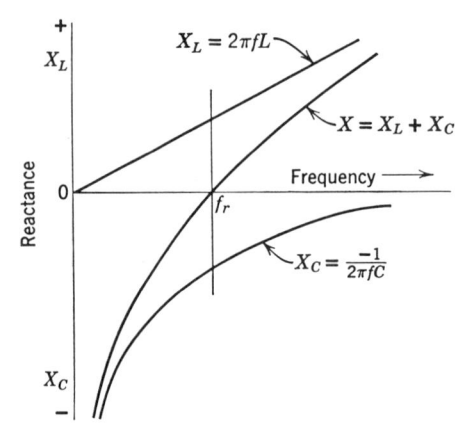

Figure 1-29
Reactances of an inductance L and a capacitance C plotted against frequency. They are equal in magnitude at f_r, the resonant frequency.

resonant frequency, X_L and X_C cancel each other and the circuit is resistive, as has been shown. The current is, in that case, in phase with the applied voltage.

1-23
Sharpness of Resonance, Q

The curves of Fig. 1-30 show the effects of changing the resistance in a series-resonant circuit. When the resistance is increased, the current at resonance decreases, as expected, but the curve also loses its sharpness in the

Sec. 1-23 **Sharpness of Resonance, Q** 51

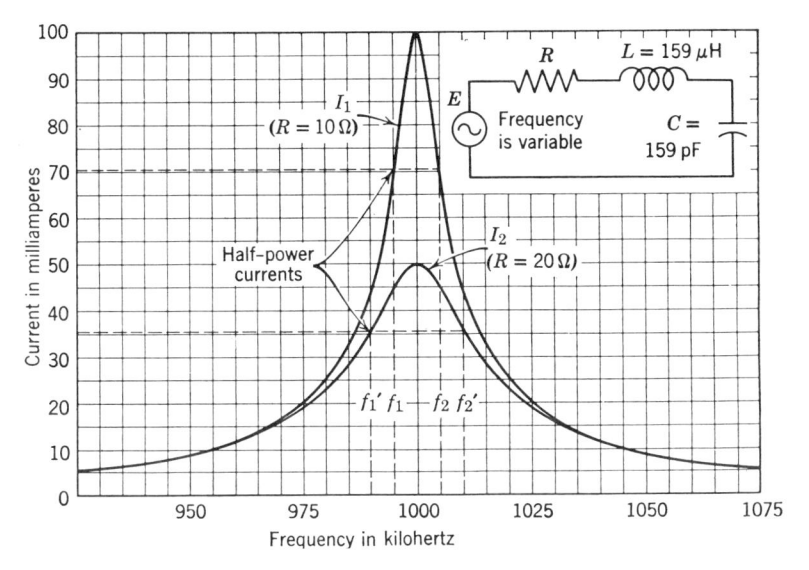

Figure 1-30
Resonance curves for series R-L-C circuit shown; currents and half-power points when $R = 10\,\Omega$ and when $R = 20\,\Omega$.

region of the peak. The smaller the resistance can be made, the *sharper the peak*, and the *larger* will be the *changes in current* for small changes in frequency. That is, the circuit will *discriminate* in favor of currents whose frequencies are in the neighborhood of the resonant frequency.

The degree of sharpness, or of discrimination, is expressed in terms of *half-power points* on the current vs. frequency curve. The sharpness of resonance is also denoted by Q, *the figure of merit of the inductance coil* used in a resonant circuit. Q is defined as the *ratio of the coil's inductive reactance to its resistance*:

$$Q = \frac{X_L}{R} \qquad (1\text{-}62)$$

The current at resonance is given by

$$I_r = \frac{E}{R} \qquad (1\text{-}63)$$

but

$$R = \frac{X_L}{Q}$$

so that

$$I_r = \frac{E}{X_L} Q \qquad (1\text{-}64)$$

52 Some Useful Circuit Concepts   Ch. 1

1-24
Half-Power Currents and Frequencies

Since power is given by I^2R, one-half the maximum power in the circuit is obtained when I is decreased to $I_{p/2}$ so that

$$(I_{p/2})^2 R = \tfrac{1}{2} I_r^2 R \tag{1-65}$$

$$I_{p/2} = \frac{I_r}{\sqrt{2}} = 0.707 I_r \tag{1-66}$$

Let us now calculate the two values of frequency at which the current becomes 0.707 of the resonant value. Let $Z_{p/2}$ be the impedance of the circuit at the half-power point:

$$I_{p/2} = \frac{E}{Z_{p/2}} = 0.707 \frac{E}{R} \tag{1-67}$$

From this,

$$R = 0.707 Z_{p/2} \tag{1-68}$$

This is true when R equals the net reactance $(X_L - X_C)$, as is obvious in a phasor diagram for Fig. 1-30. Since $X = R$ at these points,

$$Z_{p/2} = \sqrt{R^2 + X^2} = \sqrt{2R^2} = \sqrt{2}R$$

$$R = \frac{Z_{p/2}}{\sqrt{2}} = 0.707 Z_{p/2} \tag{1-69}$$

It can be shown that *when the voltage applied to a series-resonant circuit is kept constant while its frequency is increased, X_L will be increasing at the same rate as X_C* as the frequency passes through its resonance value (Fig. 1-29). This means that during a small increase in frequency, Δf, near the f_r value, ΔX_L is accompanied by ΔX_C of the same value in ohms. Since $\Delta X_L = 2\pi \Delta f L$, we can say that the total *change in reactance* due to Δf is $2\Delta X_L = 4\pi \Delta f L$.

Consider now the change in frequency required to go from f_r to the half-power frequencies:

$\Delta f = f_r - f_1,$ where f_1 is the *lower* half-power frequency

$\Delta f = f_2 - f_r,$ where f_2 is the *upper* half-power frequency

Also, since X changes from *zero at f_r* to $X = R$ at these points,

$$4\pi(f_r - f_1)L = R \tag{1-70}$$
$$4\pi(f_2 - f_r)L = R \tag{1-71}$$

$$f_r - f_1 = \frac{R}{4\pi L} \tag{1-72}$$

$$f_2 - f_r = \frac{R}{4\pi L} \tag{1-73}$$

Adding the last two equations gives

$$f_2 - f_1 = \frac{R}{2\pi L} = \Delta f \qquad (1\text{-}74)$$

This shows that *the smaller the ratio of R to L, the narrower will be the frequency band between the half-power points.* The *ratio* of this *bandwidth* to the *resonant frequency* is a measure of the sharpness of resonance of the circuit.

$$\frac{\Delta f}{f_r} = \frac{f_2 - f_1}{f_r} = \frac{R}{2\pi f_r L} = \frac{1}{Q} \qquad (1\text{-}75)$$

It must be pointed out that R is the total resistance of the series-resonant circuit, including the resistance of a generator if one is present.

Obviously, the larger Q is made, the narrower will be the frequency band between the half-power points and the sharper will be the circuit. The circuit effectively rejects currents whose frequencies are outside the "pass band," as $f_2 - f_1$ may be called.

1-25
Parallel Resonance (Antiresonance)

Consider the parallel circuit of Fig. 1-31. The phasor diagram represents the situation in which the frequency of the applied voltage E has been adjusted so that the input current I to the parallel circuit is *in phase with the applied voltage.* This is the *unity power factor condition.* It will be shown that, to bring

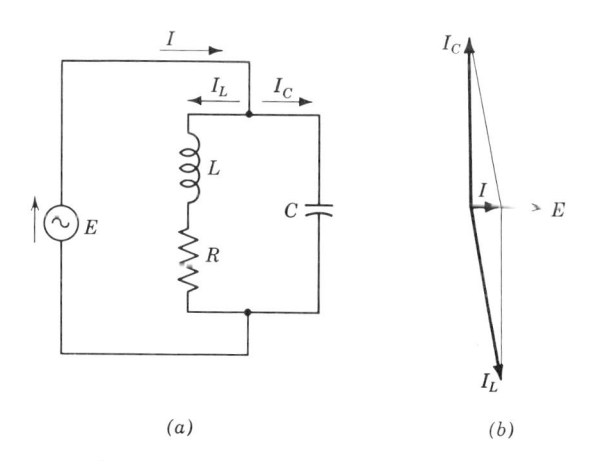

Figure 1-31
(*a*) Parallel-resonant circuit. (*b*) Phasor diagram of circuit in (*a*).

54 Some Useful Circuit Concepts   Ch. 1

this condition about, X_L must be made *practically equal to* X_C in ohms value, and X_L must be *very much larger than R*. It will also be found that under these conditions (i.e., at parallel resonance) the whole circuit will present to the applied voltage source an impedance that is purely resistive and that has a value *much larger* than the resistance R of the inductive branch. Assuming the series resistance of the capacitor to be zero helps to simplify the analysis a great deal; it is negligible in practical circuits.

In the analysis of parallel circuits, the admittances of the parallel branches may be added. The total admittance of the circuit of Fig. 1-31 is

$$Y_T = Y_L + Y_C = \frac{1}{Z_L} + \frac{1}{Z_C} \tag{1-76}$$

$$Y_T = \frac{1}{R + jX_L} + \frac{1}{-jX_C} = \frac{R + jX_L - jX_C}{(R + jX_L)(-jX_C)}$$

$$Z_T = \frac{1}{Y_T} = \frac{X_L X_C - jRX_C}{R + j(X_L - X_C)} \tag{1-77}$$

Rationalizing the denominator,

$$Z_T = \frac{X_L X_C - jRX_C}{R + j(X_L - X_C)} \times \frac{R - j(X_L - X_C)}{R - j(X_L - X_C)}$$

$$= \frac{RX_L X_C - jR^2 X_C - jX_L^2 X_C - RX_L X_C + jX_L X_C^2 + RX_C^2}{R^2 + (X_L - X_C)^2}$$

$$Z_T = \frac{RX_C^2}{R^2 + (X_L - X_C)^2} + j\frac{X_L X_C^2 - X_L^2 X_C - R^2 X_C}{R^2 + (X_L - X_C)^2} \tag{1-78}$$

Proper choice of X_L or X_C, when one of them is known, can be made so that the j term is zero. This will result in the total impedance being *resistive only*; this means *unity power factor* operation. When the reactive component of the total impedance is zero,

$$X_L X_C^2 - X_L^2 X_C - R^2 X_C = 0$$
$$R^2 = (X_C - X_L)X_L \tag{1-79}$$

Although this is precisely the requirement for unity power factor, in high-frequency circuits such as those in radio and television X_L is very much larger than R. This is expressed as

$$X_L \gg R$$

With this in mind, it will now be shown that Eq. 1-79 can be reduced for practical applications to $X_C = X_L$. From Eq. 1-79,

$$R^2 = X_C X_L - X_L^2$$
$$X_L^2 + R^2 = X_C X_L \tag{1-80}$$

If $X_L \gg R$, then R^2 may be dropped as negligible in comparison with X_L^2:

Sec. 1-26   *Parallel Resonance with Resistance in Both Branches*   55

$$X_L^2 = X_C X_L$$
$$X_L = X_C \qquad (1\text{-}81)$$

To show the common sense of this procedure, suppose that

$$X_L = 2000 \ \Omega \quad \text{and} \quad R = 25 \ \Omega$$

Using the exact equation (1-79),

$$X_C = \frac{R^2}{X_L} + X_L$$

$$X_C = \frac{625}{2000} + 2000 = 2000.3$$

Therefore, when $X_L \gg R$, X_C may be taken equal to X_L to produce unity power factor in a parallel-resonant circuit.

The impedance of the parallel circuit at *resonant frequency* is a pure resistance given by

$$Z_r = \frac{R X_C^2}{R^2 + (X_L - X_C)^2} = \frac{X_C^2}{R} \qquad (1\text{-}82)$$

since $X_L = X_C$. It is possible to express this impedance in terms of R, L, and C by expressing X_C^2 as $X_L X_C$:

$$Z_r = \frac{X_L X_C}{R} = \frac{2\pi f_r L}{R} \times \frac{1}{2\pi f_r C} = \frac{L}{RC} \ \Omega \qquad (1\text{-}83)$$

The resonant frequency of a parallel circuit with small resistance is given to a high degree of accuracy by the same equation as that used for series resonance, namely,

$$f_r = \frac{1}{2\pi \sqrt{LC}} \qquad (1\text{-}84)$$

1-26
Parallel Resonance with Resistance in Both Branches

It is easily proved in more advanced books that if there is resistance in *both* branches of a circuit like that in Fig. 1-31 the impedance at parallel-resonant frequency is

$$Z_T = \frac{L}{C(R_L + R_C)} \qquad (1\text{-}85)$$

where R_L is the resistance in the inductive branch and R_C is the resistance in the capacitive branch. Note that this equation can be converted to one that contains only X_C or to one that contains only X_L, by first multiplying both

56   *Some Useful Circuit Concepts*   Ch. 1

numerator and denominator by $2\pi f_r$, thus getting X_L and X_C in the equation, and then writing either X_L^2 or X_C^2 in the numerator since they are equal ($1/2\pi f_r C = 2\pi f_r L$) at resonance. That is, at unity power factor resonance,

$$Z_T = \frac{2\pi f_r L}{2\pi f_r C(R_L + R_C)} = \frac{X_L^2}{R_L + R_C} = \frac{X_C^2}{R_L + R_C} \tag{1-86}$$

1-27
Q of a Parallel-Resonant Circuit

The Q of a parallel-resonant circuit is the *ratio of the reactance of either branch to the total resistance of the two branches.*

$$Q_p = \frac{\omega L}{R} = \frac{1}{\omega CR} \quad \text{at resonance} \tag{1-87}$$

Since the total impedance Z_T is given by either reactance squared divided by the resistance, it is evident that

$$Z_T = \frac{\omega^2 L^2}{R} = \omega L Q_p = \frac{Q_p}{\omega C} \quad \text{at resonance} \tag{1-88}$$

In high-frequency circuits (radio frequencies and higher), the resistance of a coil increases with frequency, so that Q is more nearly constant than R. The fundamental equation of *any two-branch parallel circuit* gives the product of the branch impedances divided by their sum as the *equivalent impedance*, which we are now calling Z_T.

$$Z_T = \frac{Z_L Z_C}{Z_L + Z_C} = \frac{Z_L Z_C}{Z_S} \tag{1-89}$$

in which Z_L is the impedance of the L branch and Z_C is the impedance of the C branch, regardless of the resistance values in the branches. Z_S is the series impedance of the two branches.

In practical circuits, Q_p is often relatively high (50 or higher), and in such cases it is permissible to neglect the resistance values in the numerator of Eq. 1-89 but not those in the denominator. When this is done, the numerator becomes L/C and, since $\omega_r L = 1/\omega_r C$, we can write

$$Z_L Z_C = (\omega_r L)^2 = \frac{1}{(\omega_r C)^2}$$

and the total impedance becomes

$$Z_T = \frac{(\omega_r L)^2}{Z_S} = \frac{1}{Z_S(\omega_r C)^2} \tag{1-90}$$

Problems

1-1 Express these electrical quantities in the units called for:

1200 ohms	_____ kilohms	_____ megohms
33 kilohms	_____ ohms	_____ megohms
2.2 megohms	_____ ohms	_____ kilohms
150 picofarads	_____ microfarads	_____ farads
0.02 microfarad	_____ picofarads	_____ farads
0.000120 farad	_____ microfarads	_____ picofarads
0.015 millihenry	_____ henries	
63 millihenries	_____ henries	
148 millivolts	_____ millivolts	_____ volts
0.38 millivolt	_____ microvolts	_____ volts
120 volts	_____ millivolts	_____ microvolts

1-2 Three resistors are connected in series across a source of 120 Vdc. $R_1 = 48\,\Omega$, $R_2 = 12\,\Omega$, and $R_3 = 60\,\Omega$. (a) How much current does each resistor carry? (b) How much current is in the complete circuit? (c) What is the voltage drop across each resistor?

1-3 The three resistors of Problem 1-2 are connected in parallel across the 120 V source. (a) Calculate the amount of current in each resistor. (b) How much current is supplied from the voltage source?

1-4 A voltage-dropping resistor is connected in series with the heaters of five radio tubes to permit their operation directly from a 120-V power line. Assume that each heater requires 0.3 A to operate properly. Four of the heaters operate at 6.3 V and the fifth at 12.6 V. Sketch the circuit and calculate the resistance value needed to permit the circuit to operate correctly.

1-5 Part of an electrical circuit is shown in Fig. P1-5. Measurements show that the current I exists as marked, and the milliammeter indicates 10 mA. The internal resistance of this milliammeter is 50 Ω. The gal-

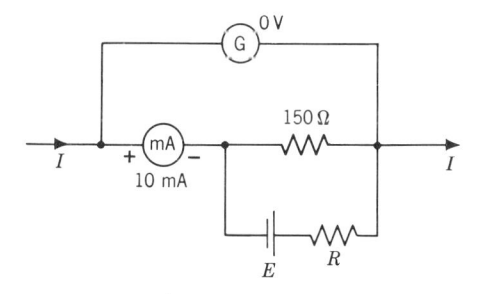

Figure P1-5

vanometer (G) indicates that there is no voltage drop across it, although it has internal resistance of 1000 Ω to develop a voltage drop when current is in it. (a) What is the magnitude of I? (b) If $E = 2$ V, what must be the value of R? (c) What is the magnitude and direction of current in the 150-Ω resistor?

1-6 Calculate the potential difference between points A and B in Fig. P1-6. Then assume that a wire having no resistance is used to connect A and B. How much current will flow in the wire? What will be its direction between A and B?

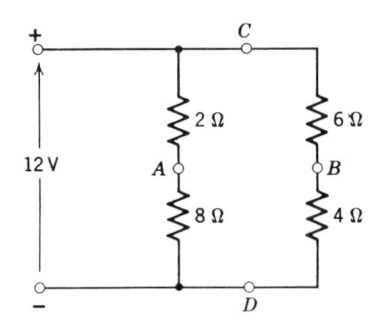

Figure P1-6

1-7 Solve for the currents in the circuit of Fig. P1-7, using Kirchhoff's laws and Ohm's law.

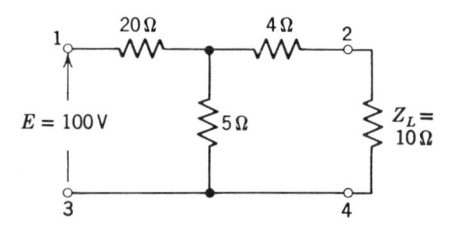

Figure P1-7

1-8 What is the frequency of a current expressed as $i = 10 \sin(377t - 37°)$ A? What is its period? What is its rms value? Sketch at least one cycle of this current and mark significant points in time.

1-9 Two currents differ in phase by 90°. Their maximum instantaneous values are 12 A and 5 A. (a) What is the maximum instantaneous value of their sum? (*b*) What is the angle between the phasor representing their sum and that representing the 12-A current?

1-10 Two voltages are expressed as complex numbers: $E_1 = 86.6 + j50$ V, $E_2 = 50 - j86.6$ V. (*a*) Express their sum and difference in complex form, $a + jb$. (*b*) Express the sum and difference in polar form, A/ϕ.

1-11 The sum of E_1 and E_2 given in Problem 1-10 is applied to an impedance $Z = 12 + j5 \, \Omega$. How much current will flow, and what will be its phase with respect to the applied voltage?

1-12 Three currents at a circuit junction are expressed as $I_1 = 15$, $I_2 = 10/60°$, and $I_3 = 20/90°$, all in amperes. (*a*) Calculate their sum. (*b*) Sketch the three currents and their sum on the complex plane. (*c*) What is the phase of the sum with respect to each of the currents?

1-13 A coil has an inductance of 50 mH. What is its reactance at (*a*) 1 kHz, (*b*) 1 MHz, (*c*) 60 Hz, (*d*) 159 Hz?

1-14 Calculate the magnitude and angle of the voltage across the inductor in Fig. P1-14, if the applied voltage and capacitor voltage are as shown. (*Hint:* There are two possible answers; find both.)

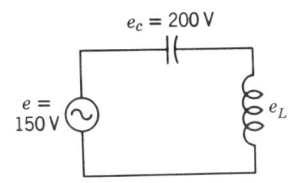

Figure P1-14

1-15 What two circuit components in series will have an impedance of $10 \, \text{k}\Omega/60°$ at a frequency of 1 kHz?

1-16 Determine the impedance of the circuit in Fig. P1-16 at 159 Hz.

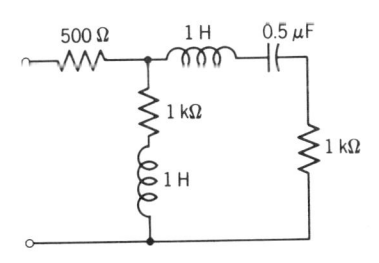

Figure P1-16

1-17 Evaluate and express in rectangular form, $a + jb$:

$$\frac{(3 + j4)(5/\pi/3 \text{ rad})(12/30°)}{20(\cos 60° + j \sin 60°)(15/\pi/6 \text{ rad})}$$

1-18 A $0.2\,\mu\text{F}$ capacitor is in series with a 10-Ω resistor. What is the series impedance at 8 kHz?

1-19 Solve for the current in the 1000-Ω resistor of Fig. P1-19, using loop currents and determinants.

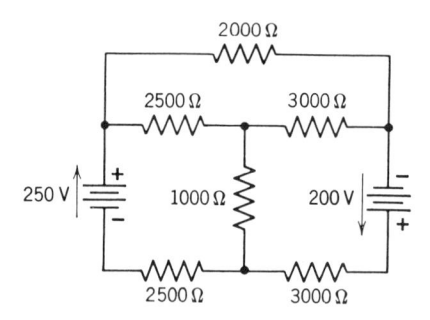

Figure P1-19

1-20 An impedance has a voltage drop of $100 \sin(\omega t - 20°)$ V when it passes a current of $5 \sin(\omega t + 30°)$ A. (a) What is the magnitude of this impedance? (b) What is its resistance? (c) What is its reactance? (d) What two circuit elements make up this impedance?

1-21 In the circuit of Fig. P1-21, ohmic values of resistors and reactances have been determined at 159 Hz. (a) What is the impedance between points A and B? (b) How much current would be supplied from a source of $100/0°$ V connected between A and B? (c) What is the phase difference between this applied voltage and the current?

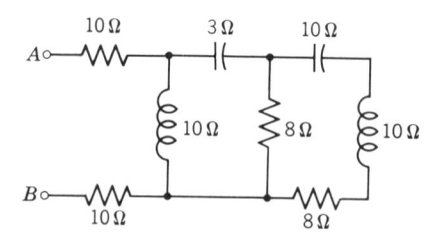

Figure P1-21

Problems 61

1-22 What is the magnitude, phase, and polarity of the voltage drop across the 30-Ω resistor in Fig. P1-22?

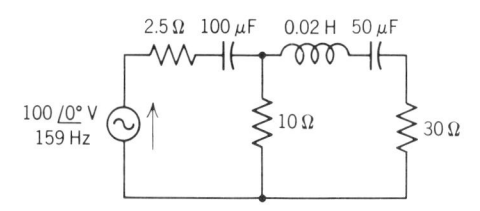

Figure P1-22

1-23 How much current is in the 10-Ω resistor of Fig. P1-23?

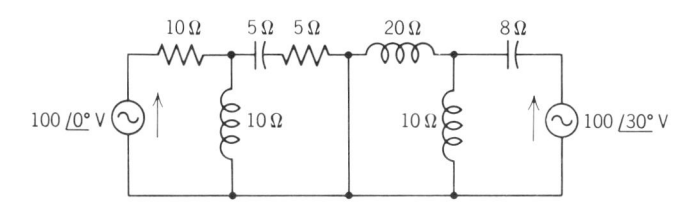

Figure P1-23

1-24 Solve for the current in the resistor connected between E and F in Fig. P1-24, using the superposition principle.

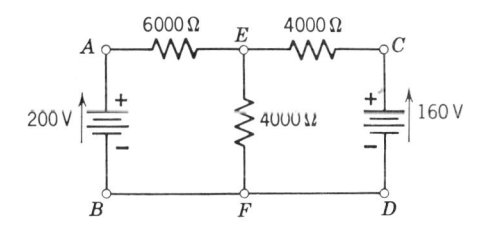

Figure P1-24

1-25 For the circuit in Fig. P1-25, determine the Thévenin equivalent circuit, taking the 20-Ω resistor as the load.

62 Some Useful Circuit Concepts Ch. 1

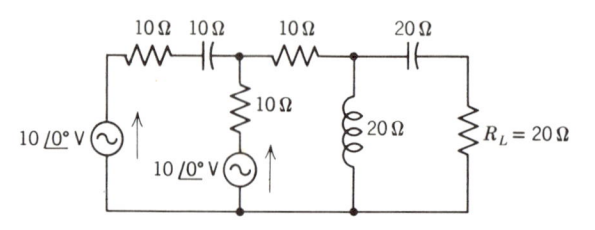

Figure P1-25

1-26 Determine and sketch the Norton equivalent circuit from the result in Problem 1-25. Calculate the load current, magnitude, polarity, and phase with respect to E_1.

1-27 In the circuit of Fig. P1-27, $E = 100$ V, $L = 0.04$ H, and $R = 1$ kΩ. (a) Determine the initial and final values of current, e_L', E_R at $t = 0$. (b) What is the initial rate of rise of current? (c) What is the time constant and what are the values of i, e_R, and e_L' at $t = $ the first time constant?

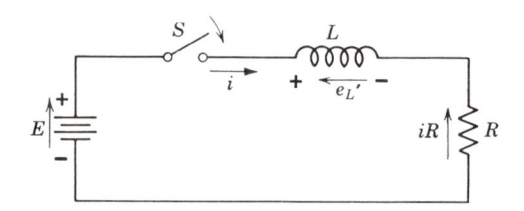

Figure P1-27

1-28 Repeat Problem 1-27 for L values of 0.01, 0.08, and 0.125 H, keeping the same value of R. Sketch current vs. time curves for each inductor value.

1-29 A generator has an internal impedance $Z = 120 + j50$ Ω. It is expected to deliver maximum power to a load at a frequency of 400 Hz. What should be the load impedance? What circuit elements make up this impedance?

1-30 Sketch the circuit for which this determinant system is valid:

$$\begin{array}{ccc} I_1 & I_2 & I_3 \end{array}$$

$$\begin{vmatrix} 25 & -5 & 0 \\ -5 & 20 & -5 \\ 0 & -5 & 15 \end{vmatrix} \begin{matrix} 10\text{ V} \\ -10\text{ V} \\ 0\text{ V} \end{matrix}$$

1-31 Make a graphical sketch of this equation:

$$e = 10(\sin \omega t + \tfrac{1}{3}\sin 3\omega t + \tfrac{1}{5}\sin 5\omega t + \tfrac{1}{7}\sin 7\omega t).$$

1-32 A series circuit consists of $R = 100\,\Omega$, $L = 15.9\,\text{mH}$, and $C = 159\,\text{pF}$. (a) What is the resonant frequency? (b) What is the Q of the circuit?

1-33 A signal of 1 V at the resonant frequency of the circuit in Problem 1-32 is applied to the series elements. (a) Compute the voltages across R, L, and C, and compare them with the applied voltage. (b) Is the voltage across C equal to Q times the applied voltage?

1-34 A signal at 8 kHz is applied to the series combination of a $0.02\,\mu\text{F}$ capacitor and 10-Ω resistor. (a) How much inductance is required to produce series resonance in this circuit? (b) What are the half-power frequencies? (c) What is the circuit Q?

1-35 A coil has inductance of 0.1 mH and 5Ω resistance. (a) What capacitance will produce parallel resonance at 1.5 MHz? (b) What is the impedance at resonance? (c) Calculate the circuit Q. (d) How much current will flow a 1-V source at the resonant frequency when it is connected to the parallel circuit?

2
DIODES

The first use of a semiconductor in electronics applications dates back to the beginning of the present century (about 1905), when a crystal of galena (lead sulfide) was used to detect signals in radio receiving sets. A fine, pointed wire called a *cat whisker* was pressed against the surface of the crystal to form a *detector*. The detector worked because it conducted current much more readily in one direction than in the opposite direction. This is the requirement of a detector of high-frequency currents used in communications and control circuits. By allowing currents to flow in only one direction through the detector (for example, in the positive direction), not only is the *average* current per cycle always in the positive direction, but the current also contains low-frequency variations that convey intelligence.

Semiconductors are used extensively in the form of crystal diodes and as transistors. The general behavior of semiconductors is treated here to provide a good background for an understanding of their operation as crystal diodes and, later, as transistors. Because the junction-type diode is superior in many ways to the point-contact diode, it will be explained in some detail.

Vacuum-tube diodes are briefly described for their historic interest and because some older electronic equipment still in use contains vacuum-tube diodes.

2-1
Atomic Arrangement in Materials

Matter, in general, is composed of atoms that are made up of electrons rotating about a positively charged nucleus which contains enough charge to make the complete atom electrically neutral. The Bohr theory of atomic structure states that the electrons rotate in orbits about the nucleus in the same manner as the planets circle the sun. Rotating electrons can exist only in certain allowed, discrete shells concentric with the nucleus. This is true of atoms whether they are bound together as in solids and liquids or are in the form of a gas. The distinguishing chemical properties of different materials arise because they contain different kinds of atoms. Each kind of atom has a nucleus carrying the exact amount of positive charge required to neutralize the total negative charge of the orbiting electrons of that atom. Its electrons are arranged in *energy levels* that are unique for the element.

In the Bohr model of an atom, the energy shells are drawn concentric with the nucleus and named K shell (nearest to the nucleus) to Q shell (farthest from the nucleus). The potential energy (energy of position) of an electron increases with distance out from the nucleus. The electrons in the outermost shell of an atom are called *valence electrons*, and it is these that are important in determining the chemical properties of a material. They can pass easily from one atom to another and this movement constitutes current in the material. The passage of electrons from one atom to another in a conductor, each moving into a vacant place just left by another electron, is regarded as the mechanism of current. Valence electrons exist in the *valence band* of energy. Those in the highest energy level are at the greatest distance from the nucleus.

Hydrogen, which has the simplest atom of all elements, has only one electron rotating around its nucleus, and this is in the first (K) shell. Copper, the most widely used of all conductors, has 29 electrons arranged in shells about the nucleus: 2 in the K shell, 8 in the L shell, 18 in the M shell, and 1 in the N shell. Silicon and germanium each have four valence electrons, which accounts for their behavior as semiconductors, as will be seen.

When a material forms into a solid, the atoms most often are arranged in a regular geometric array, thus defining a crystal structure. Such a structure has symmetry in three dimensions, with the atoms arranged in definite *unit cells*, each of which looks the same throughout space in the *crystal lattice* of the solid. Two examples of crystal arrangements are shown in Fig. 2-1.

Since we shall be primarily concerned with crystalline solids, some of their special characteristics are of interest.

1. The valence electrons of an isolated atom describe a radial dimension for the atom. Consequently, we should expect the atoms in a crystal to be separated by about twice this distance.
2. The outer orbits of neighboring atoms overlap at the atomic separation expected in (1). This produces a distortion of the outer orbits, so that it becomes difficult to say which electron goes with which atom. In this way, some valence electrons are shared by several neighboring atoms.
3. The valence electrons will take on a systematic motion and thus join each atom to its neighbors while maintaining each atom electrically neutral.

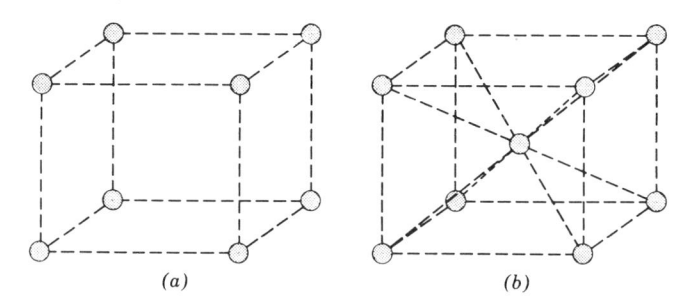

Figure 2-1
Two examples of atoms arranged in a solid crystal. (a) Simplest cubic lattice. (b) Body-centered cubic lattice.

2-2
Pattern of Energy Levels

The energy levels of atoms begin with energy values in the K shell for hydrogen and extend through discrete levels called L, M, N, O, P, and Q, for the 104 known elements. The maximum number of electrons that each level can accommodate is determined according to a unique principle in physics called the *Pauli principle*.

It has been determined that the K shell can contain only 2 electrons; the L shell, 8 electrons; M, 18; N, 32; O, 50; P, 72; and Q, 98. However, the largest known number in the O shell is 31; in the P shell, 10; and in the Q shell, 2. Hydrogen, which is the first one in the table of elements, has already been mentioned. Helium, the second element, has only two electrons rotating about its nucleus and they are both in the K shell. Lithium, number three, has three electrons, but only two of them are in the K shell and the third is in the L shell. There are only two electrons in the K shell of each of the remaining 101 elements. Neon has eight in the L shell and two in the K shell as expected. Silicon, the fourteenth element, has 2 K, 8 L, 4 M. Germanium, element 32, has 2 K, 8 L, 18 M, 4 N.

Immediately following the first element in the periodic table that accommodates eight electrons in one shell (the L shell), the next element in the table starts the next shell with the one added electron that distinguishes it from its predecessor. The L shell of every element in the table beyond and including neon has only eight electrons. As soon as the M shell of an element has acquired eight electrons (the first one to do this is argon), the very next element, potassium, starts one electron in the N shell. This process continues on to element 86, radon, which has eight electrons in the P shell.

It has been explained that as soon as eight electrons are contained in the outermost shell, further electron orbits must be established in the next outer shell whether or not empty places (energy levels) remain in that shell. The stable condition represented by this so-called "octet" theory is important in the explanation of the behavior of semiconductor materials used in electronic devices.

2-3
Conductors, Insulators, and Semiconductors

Certain elements and their solid-state crystals are classed as conductors, semiconductors, or insulators, based on their relative ability to carry electric current. The crystal structure of highly conductive materials, like copper, gold, silver, and aluminum, is such that the outer electrons are shared by *all* of the atoms in the material. These electrons are actually free to wander from atom to atom, over a very wide temperature range. In most metals, each atom supplies one such free electron, making the number of free electrons per cubic centimeter of the order of 10^{23}. These highly conductive materials have one valence electron, which can readily accept enough energy to free it from its parent atom and enable it to move about in the material. Thus copper is classed as a good conductor of electricity. Silver is even better than copper as a conductor, because its valence electron is in the fifth (O) energy shell, far removed from the binding influence of its nucleus. Copper is more generally used because it is less expensive than other metals that are better conductors.

Whether a material behaves as a conductor, semiconductor, or insulator can be explained by the permissible energy levels of the atoms making up the material. There is a *forbidden region*, or *forbidden energy gap*, between each of the allowed energy bands from K to Q. Further, an energy gap (forbidden region) may exist between the energy of the outermost valence electrons and the energy needed for conduction from one atom to another. When some outside influence such as an electric field or heat is applied, a valence electron may acquire sufficient energy to jump through the forbidden region and on into the *conduction band* of energy. Here it will have enough energy to free itself from any influence of its positive nucleus and become a carrier of electricity, ready to take the place of another electron that just left its own atom in the conductor in the same manner.

Energy can be given to electrons only in discrete amounts (*quanta*), according to the Bohr theory. Hence, there are certain energy values that an electron simply cannot acquire. As a rough example, if a man were climbing a vertical ladder that had rungs 1 ft apart, he could stand with both feet in a stable position only at 1-ft intervals above the base. He could not stand $1\frac{1}{2}$ ft or $2\frac{1}{4}$ ft above the base nor at any height involving a fraction of a foot. The discrete quantum of energy that can be acquired by an electron is given by hf, where h is Planck's constant (6.624×10^{-34} J-s) and f is the frequency of the energy in cycles/second or hertz (Hz).

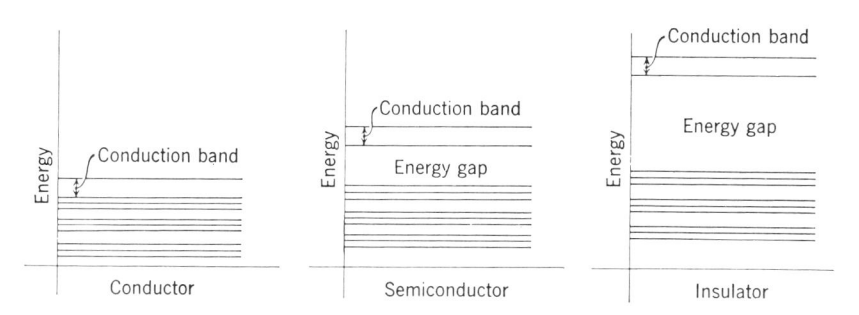

Figure 2-2
Relative magnitudes of energies in shells and gaps.

Figure 2-2 shows relative energy bands in conductors, semiconductors, and insulators. In conductors, there is no gap between the valence band and the conduction band. Some researchers are convinced that they overlap.

In contrast to the action of conductors, insulators are poor carriers of electricity. The structure of solid insulators is such that almost all the electrons remain bound to their parent atoms. The forbidden energy gap of these materials is very wide, so the conduction band remains almost empty. Few electrons can acquire enough energy to cross the forbidden region. Sulfur is a good insulator. Although it has six valence electrons in the M band, it has a very wide energy gap. It is very difficult to impart to these electrons sufficient energy for them to bridge the forbidden gap and enter the conduction band.

Semiconductors, on the other hand, are neither good conductors nor good insulators at normal room temperatures. More precisely, they are insulators at very low temperatures and reasonably good conductors at high temperatures. The elements germanium and silicon are the most important semiconductors used in electronics. Their use in crystal diodes almost excludes other semiconductors, although other materials are finding increasing use. However, these two elements appear to be the only ones suitable for the fabrication of transistors.

The element germanium has four electrons in the valence band of its atom. Its conductivity lies in the range somewhere between that of insulators and conductors. The conductivity of a good insulator, such as quartz, may be as low as 10^{-16} mho/m, while that of a conductor may be as high as 10^7 to 10^8 for a typical metal. The conductivity of copper is 5.8×10^7 mho/m. Semiconductors have conductivities in the range from 10^{-3} to 10^5 at 300°K (27°C). In terms of energy gaps, conductors may require as little as 0.05 electronvolts (eV) to jump the forbidden region, while insulators require many electronvolts. Germanium and silicon require 0.7 eV and 1.1 eV, respectively. An electronvolt is the energy acquired by 1 electron falling through 1 V of potential difference.

70 *Diodes*

Drill Problems

D2-1 Calculate the energy, in joules, an electron will acquire in falling through 1 V of potential difference; 10 V of potential difference.

D2-2 How many joules of energy are represented by 0.7 eV? by 1.1 eV?

D2-3 The frequency range of visible light extends from about 4×10^{14} Hz to 7×10^{14} Hz. Calculate the minimum frequency of photons that can excite electrons from the valence to conduction bands in germanium. Can energy at visible-light frequencies excite the electrons?

D2-4 Make a similar calculation as in D2-3, for silicon.

2-4
The Germanium Atom

Germanium and silicon are used in the manufacture of crystal diodes. These two elements are also the dominant material in transistors. The atomic structure of germanium is studied here in some detail, largely because of its universal use as a basic semiconductor material.

The germanium nucleus of 32 positive charges is surrounded by 32 electrons revolving in the *K*, *L*, *M*, and *N* shells, which contain 2, 8, 18, and 4 electrons, respectively. Each shell (except the *N* shell) is filled with the maximum number of electrons that it can accommodate. Because only the valence electrons are involved in electrical interactions with other atoms, they are the only ones that will be considered in the behavior of semiconductors.

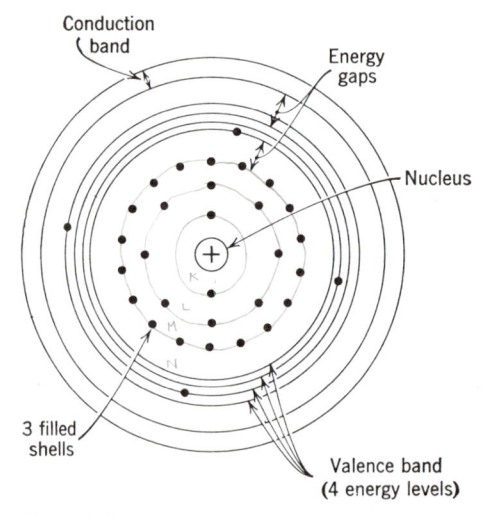

Figure 2-3
Bohr model of germanium atom, two-dimensional view.

Sec. 2-5   Covalent Bonding   71

Figure 2-3 represents the Bohr model of a germanium atom in two dimensions. The black dots around the nucleus and inside the first large circle represent electrons in the filled K, L, and M shells. Each valence electron is shown having a unique energy level in the valence band. The forbidden energy gaps occur above and below the valence band.

2-5
Covalent Bonding

The four valence electrons of a germanium atom are required to make the atom electrically neutral, but they are in excess of those needed for a preferred energy state. Completely filled bands represent the preferred state, which is also the minimum energy state of an atom. As in all physical systems, atoms prefer a state of minimum energy. When an aggregation of atoms are brought together under favorable conditions, they will combine in a state of minimum energy. This state is achieved, for germanium atoms, when each atom shares its valence electrons with four of its neighbors. The chemical bonds that attach neighboring atoms are called *covalent bonds* or *electron-pair* bonds. Under this condition each atom *effectively* has eight valence electrons and it is closer to a preferred state than when there is no sharing of electrons. Covalent bonding holds the atoms together to form the crystal lattice of germanium.

Covalent bonding in a germanium crystal is shown schematically in Fig. 2-4. Each bar connecting two adjacent atoms represents an electron-pair bond.

Covalent bonding action is happening simultaneously to each of the valence electrons of an atom as it encounters a companion valence electron from the atoms above, below, and on the other sides of itself in the crystal. It has been determined that about fifteen times as much energy is needed to free a covalent bonded electron as is required to free an unbonded valence electron.

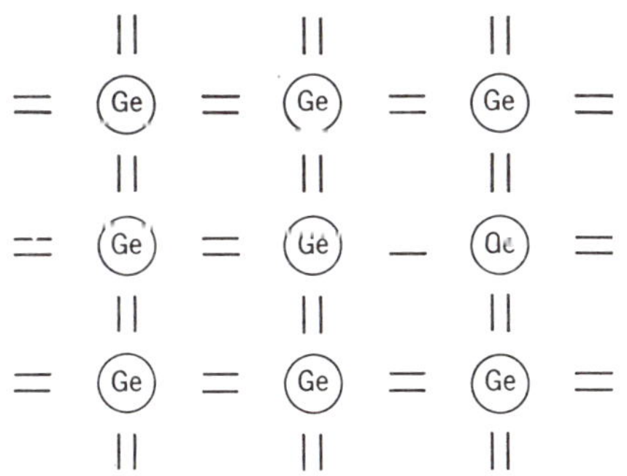

```
      ||         ||         ||
 = (Ge)  =  (Ge)  =  (Ge) =
      ||         ||         ||
 = (Ge)  =  (Ge)  —  (Ge) =
      ||         ||         ||
 = (Ge)  =  (Ge)  =  (Ge) =
      ||         ||         ||
```

Figure 2-4
Valence bonds between atoms in a germanium crystal.

In addition to nuclear attraction, two closely located electrons experience a bonding force that effectively lowers their potential energy, since it makes a still larger energy gap for them to cross before they can enter the conduction band and become free electrons. This bonding force is due to magnetic fields produced by the spinning of electrons on their own axes. When two of these electrons come close to each other, their magnetic fields join in additive fashion. This creates a force that tries to keep the electrons from moving farther apart. These bonding effects tend to stabilize the germanium atoms, owing to the fact that each has effectively eight electrons in its outer orbit.

The germanium atoms arranged in covalent bonds in their crystal structure have important physical properties that make germanium useful in electronic devices. The discussion presented here for germanium applies equally well, in a qualitative sense, to silicon and other crystalline materials.

2-6
Adding Impurities to Semiconductors

Germanium atoms arranged in a solid-state crystal will receive enough energy from heat at room temperature to break some of the valence bonds and permit some electrons to act as free charge carriers in the crystal. Thermal agitation of the electrons results in interactions among them and endows the crystal with electrical conductivity. The conditions in the crystal when an electron escapes from its parent atom are shown in Fig. 2-5. A free electron (minus sign) has produced an imperfection (plus sign) in the crystal. This process is called *thermal ionization* of the crystal.

One electron is now missing from one of the valence bonds, and a *hole*, or vacancy for an electron, exists in the crystal. The missing electron can be easily replaced by another neighbor electron, restoring electrical neutrality to this atom, but the hole has now moved to the neighbor atom that lost its electron.

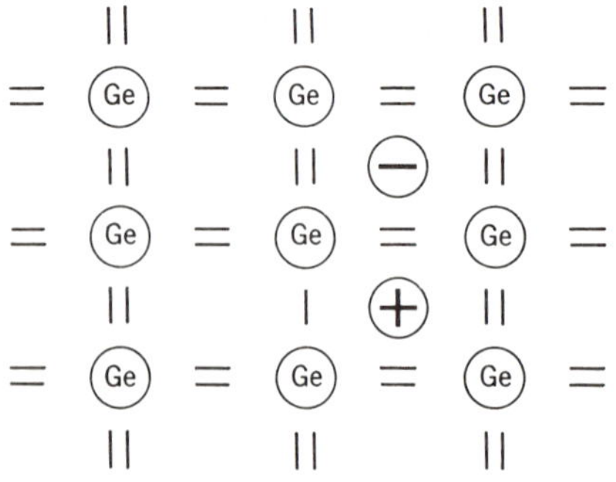

Figure 2-5
Thermal ionization of a germanium crystal.

An electron jumping into the hole has effectively moved the hole to a new position in the crystal. Thus, there are now two mobile carriers of current that can move about in the crystal, thermally ejected electrons and the holes they have left behind. The hole carries with it a positive charge equal in magnitude to the electronic charge. If there are enough of these charge carriers available in the crystal, the germanium can act as a fairly good conductor.

Thermal ionization of solid materials is strongly dependent upon the temperature to which they are subjected. Minor variations in the ambient temperature cause radical variations in the number of charge carriers present in the material, thereby varying the electrical conductivity over a wide range. To be useful in electronic devices, the conductivity of the materials should be controllable under any required operating conditions. For this reason, germanium and other semiconductor materials are "doped" by the addition of minute quantities of other elements referred to as "impurities." The presence of impurity atoms in the crystal allows its conductivity to be increased and controlled by the amount of doping and reduces its dependence on temperature.

Intrinsic (pure) germanium crystals contain, as a result of thermal ionization, electrons and holes as current carriers. These crystals are useful in the manufacture of resistors that vary their resistance with temperature changes, but this is about the limit of their usefulness. Semiconductor diodes and transistors require a combination of materials in which the simultaneous existence of an excess of electrons is possible in part of the combination and an excess of holes in the other part. Ideally, semiconductors used in crystal diodes and transistors should contain only one kind of charge carrier. However, manufacturing limitations do not permit this perfection in the fabrication of semiconductor crystals. In practice, the fraction of impurity atoms can be controlled in the range from about 1 part in 10^5 to 1 part in 10^7.

Impurity atoms are added to intrinsic germanium to make the crystal contain either extra electrons as carriers or extra holes as carriers. An impurity that provides extra electrons is called a *donor* element; one that provides extra holes is called an *acceptor* element.

Donor and acceptor elements are chosen from materials whose atoms have five and three valence electrons, respectively. When small amounts of these impurity materials are added to germanium under the proper conditions, they greatly change the electrical properties of the germanium crystal, but its metallurgical properties are not changed perceptibly. When an impurity atom enters the lattice structure of the host crystal, its valence electrons form covalent bonds with adjacent germanium atoms. Donor impurities with five valence electrons will have one electron that is very loosely bound to its parent atom, because it will not be required to complete the covalent bonding with neighboring germanium atoms. Acceptor impurities, on the other hand, do not have enough valence electrons to complete the covalent bonding. An effective

74   Diodes   Ch. 2

hole exists in the lattice because the acceptor atom has only three valence electrons. In each case, however, the complete crystal is electrically neutral.

2-7
N-Type and P-Type Semiconductors

A semiconductor in which donor impurities predominate is called an N-type semiconductor, the N signifying the negative charge of loosely bound electrons in the material. Acceptor impurities in the material provide holes for electrons, and a semiconductor containing predominantly acceptor atoms is called a P-type semiconductor, P signifying the positive charge of the holes.

N-type semiconductors use the elements arsenic, antimony, and phosphorus as impurities. Other pentavalent atoms can be used as donor elements, but some are less suitable because of other characteristics that become important in the manufacture of semiconductors. Trivalent atoms, such as those of aluminum, gallium, boron, and indium are used as acceptor elements in the fabrication of P-type semiconductors.

If donor and acceptor elements are added to a crystal in equal amounts, the free electrons of donor atoms can fill the holes of acceptor atoms. No current carriers are present except those occurring in thermal ionization. This process of cancelling the effects of donor and acceptor impurities is called *compensation* and is important in the fabrication of semiconductors. Adding further amounts of either N-type or P-type impurities to a compensated crystal produces an N-type or P-type crystal. Thus, a crystal can be changed during manufacture from an N-type to a P-type and back again, by adding successive

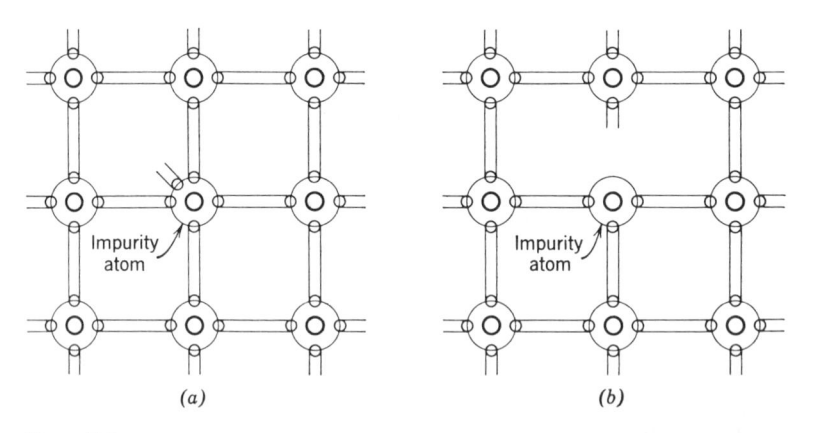

Figure 2-6
Germanium lattice containing impurity atoms. (a) Pentavalent donor impurity with 5 valence electrons. (b) Trivalent acceptor impurity with 3 valence electrons.

amounts of the necessary impurities. It should be remembered that impurity elements make up a very small fraction of the total material, in the range from 1 part in 10^5 to 1 part in 10^7. Their important characteristics are electrical, and their presence hardly affects the metallurgical properties of the crystal.

Figure 2-6 illustrates the lattice pattern of germanium containing impurity elements. A donor impurity supplies an extra electron in the covalent bonding, and an acceptor impurity supplies a hole.

Some important properties of impurity materials that are commonly used in fabricating germanium and silicon semiconductors are listed in Table 2-1.

2-8
The Fermi Level of Energy

It is desirable to have a particular energy level as a reference to which all electron energies in a solid may be compared. A kind of average value, perhaps midway between the valence and conduction bands, might be a good choice. In practice, it is chosen at an energy value, called the *Fermi level*, for which there is a 50 *per cent probability of occupancy*. For a pure semiconductor crystal, the Fermi level is midway between the valence and conduction bands. However, in doped crystals, donor and acceptor impurities provide energy levels within the forbidden region of a pure, undoped crystal. The Fermi level is shifted slightly by the presence of impurities. In any case, it is the *most probable energy value* that will be found among the electron energies.

2-9
Preparation of Semiconductor Metal

Depending on the type of semiconductor being formed, the preparation and structure of the metal may produce a nearly perfect single crystal of the material or a polycrystalline state. Crystal diodes and transistors depend for their operation on a single-crystal structure whose preparation will be discussed here. Since these devices perform as the result of added impurities in the crystal, and any defects in the crystal behave much the same as impurities, the preparation of germanium and silicon metal for transistor and diode manufacture has become a major industry.

In order to have the desired control over the concentration of impurities in P-type and N-type semiconductors (from 1 part in 10^5 to 10^7), the raw metals are purified so that the impurity concentration is as low as 1 part in 10^{11}. Afterwards, control of the dominant impurity can be obtained by doping the material during the manufacture of single crystals.

Table 2-1
Summary of Properties of Common Impurities Added to Germanium and Silicon Semiconductors

Element	Atomic Number	Number of Valence Electrons	Function	Majority Carrier	Ionization Energy to Generate Carriers (eV) in Ge[a]	in Si[a,b]	Type of Semiconductor
Boron (B)	5	3	Acceptor	Hole	0.0104	0.045	P-type
Aluminum (Al)	13	3	Acceptor	Hole	0.0102	0.057	P-type
Gallium (Ga)	31	3	Acceptor	Hole	0.0108	0.065	P-type
Indium (In)	49	3	Acceptor	Hole	0.0112	0.16	P-type
Tin (Sn)	50	4	Solder (neutral)	—	—	—	—
Phosphorus (P)	15	5	Donor	Electron	0.0120	0.044	N-type
Arsenic (As)	33	5	Donor	Electron	0.0127	0.049	N-type
Antimony (Sb)	51	5	Donor	Electron	0.0096	0.039	N-type
Silicon (Si)	14	4	Host	(Equal)	$1.205 - 2.8 \times 10^{-4} T$		Intrinsic
Germanium (Ge)	32	4	Host	(Equal)	$0.782 - 3.9 \times 10^{-4} T$		Intrinsic

(T in °K)

[a] R. A. Smith, *Semiconductors*, Cambridge University Press, London, 1959.
[b] E. M. Conwell, *Proc. I.R.E.*, **46**, 1281–1300, 1958.

Germanium is usually obtained as a byproduct of zinc refining in the United States and from certain coal dusts in Europe. It is available commercially in compound form as germanium dioxide and germanium tetrachloride. The initial purification of silicon typically involves chemical reactions that produce silicon tetrachloride or dioxide. These compounds can be processed to obtain metallic germanium or silicon of relatively high purity. The purity of the metal can be developed further by a process called *zone refining.*

The zone-refining technique for metal purification makes use of the fact that many impurities are more soluble when the metal is in a liquid state. If a short section of a germanium bar is melted and the melted region caused to move slowly from one end of the bar to the other, most of the impurities present in the metal tend to remain dissolved in the liquid state. Thus, as the melted region is moved from one end of the bar to the other, the impurities tend to concentrate at one end of the bar. Such heating of small bars of metal is most easily done by the induction heating process. The heat necessary to produce a narrow molten zone in the bar is accomplished by coils encircling the bar and carrying radio-frequency energy to induce eddy currents in the bar.

A simplified zone-refining apparatus is shown in Fig. 2-7. A bar of low purity germanium or silicon is placed in the apparatus so that the induction heating coil surrounds it at the right end of the figure. As radio-frequency energy is applied, the right end of the bar will melt. Once melting has started, the coil is moved slowly to the left at such a rate that the molten zone moves along the bar

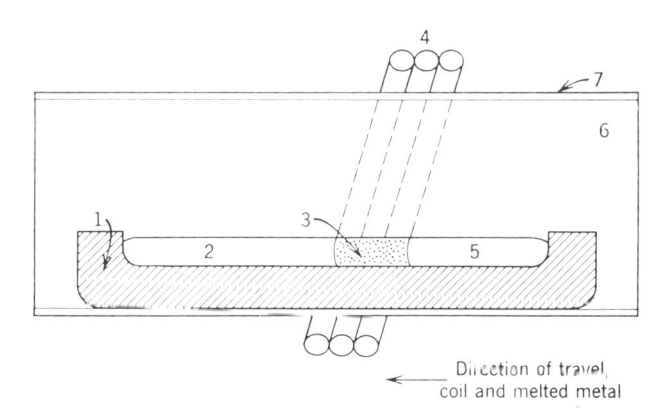

1. Graphite boat
2. Low purity metal
3. Molten zone
4. Induction coil
5. High purity solid metal
6. Inert atmosphere
7. Quartz container

Figure 2-7
Zone-refining apparatus for purifying germanium metal.

at the same rate. In this way, the impurities in the liquid state move to the left end of the bar. As the coil progresses from one end to the other, the purity of the material to the right of the molten zone is increased as it returns to a solid state. Several passes of the bar through the apparatus may be necessary to yield the desired purity of the germanium.

It is important that the metal be protected from other contaminants during the refining process. Graphite and quartz are used in the apparatus to hold the metal, and the container is either filled with an inert gas or held as a vacuum. This prevents the introduction of additional impurities into the melt.

After purification to the degree necessary, the metal is ready for doping and forming into a single crystal. One common method for growing a single crystal is shown in Fig. 2-8. The crucible contains metal at a temperature a few degrees above its melting point. A small piece of single crystal of the metal, previously prepared, is lowered into the liquid metal. This small piece is referred to as a

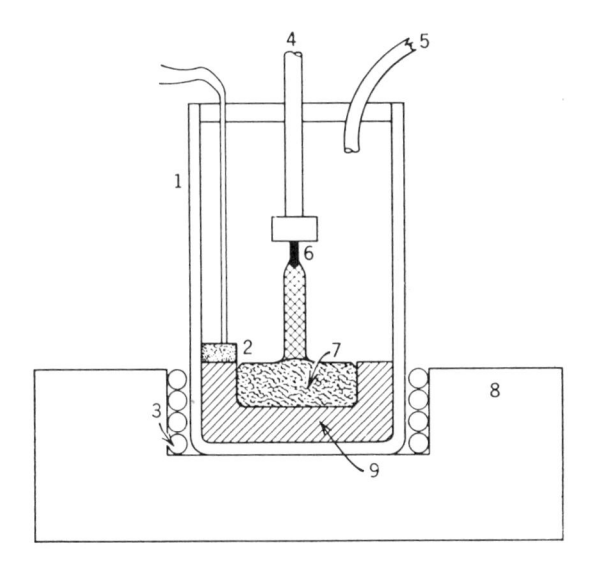

1, Quartz container
2, Thermocouple, to measure temperature
3, Heating coils
4, Pulling rod
5, Gas inlet
6, Seed crystal
7, Molten metal
8, Insulation
9, Graphite crucible

Figure 2-8
Furnace for growing single-crystal germanium.

seed crystal, because as it is slowly withdrawn from the melt, it permits the formation on it of another crystal having the same uniform lattice structure. If the temperature conditions are properly set and controlled, a single crystal of the metal can be *grown* onto the seed crystal until all of the metal is contained.

A more recent technique for both refining and growing single crystals has been introduced. Called the *floating zone* method, it is very similar to zone refining except that the graphite boat is eliminated. Instead, clamps at both ends of the bar hold it in a vertical position. Absence of the graphite container helps to reduce possible contamination during the process. The metal in the molten zone is held in place by surface tension of the liquid metal. Doping materials can be added at one end of the bar, and they will be distributed uniformly through the crystal by a single cycle of zone refining. Thus, the metal can be refined and grown into a doped, single crystal in one operation. This technique has been used successfully in producing high-quality silicon semiconductors.

2-10
Forming a *P-N* Junction

During the growth of single crystals of semiconductor metals, appropriate impurities may be added to the melt to form either a *P*-type or an *N*-type semiconductor. Their concentrations can be controlled and the growth of *P-N* junctions in the crystal permitted with temperature and rate control. Alternate doping with *P*-type and *N*-type impurities produces several junctions, each of which has the required change of conductivity across the junction. Each junction will behave as a *P-N* junction diode. When diodes are formed during crystal growth in this way, they are called *grown-junction diodes.*

The grown-junction technique for forming a *P-N* junction is used to manufacture crystal diodes and transistors. The fundamental difference in these two devices is the fact that the diode has only one *P-N* junction, whereas the transistor has *two*, separated by a very narrow region. A separate chapter is devoted to transistors, so only the diode will be discussed here.

There are two major examples of the grown process: rate-grown and grown-diffused. These processes are illustrated in Fig. 2-9 and discussed in the following sections.

In the rate-grown process, molten germanium metal in the crucible contains both donor and acceptor impurities. The donor element is sensitive to the rate of growth, so that the amount of this impurity in the crystal varies as the growing conditions change. While the temperature of the melt is dropping, the crystal grows very rapidly as the metal solidifies. Then heating power is rapidly applied, causing the growth to stop, and the crystal starts to remelt. As the metal is allowed to cool again, melting stops and the crystal begins to grow. At the boundary where the growth rate is zero, the acceptor impurity will

80 *Diodes* Ch. 2

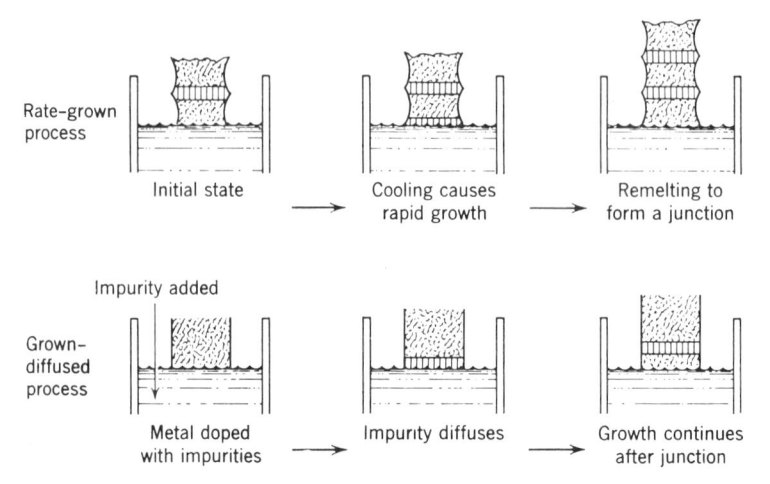

Figure 2-9
Processes for forming grown-junction diodes.

predominate, forming a *P* region across the germanium crystal. A junction is thus formed between the *N* region and the *P* region, forming a grown-junction diode. Repeating this process several times as the crystal grows will yield *N-P-N* structures which can be used to manufacture transistors.

The grown-diffused process is started with a crystal that is doped to some desired conductivity. *N*-type and *P*-type impurity elements are added to the molten material. The addition of other impurities to the melt while the crystal is growing increases the total impurity concentration. By taking advantage of the different diffusion rates of donor and acceptor elements and by controlling the added amounts of each, a junction area can be formed in the crystal. This method is used to manufacture transistors in silicon, where the higher diffusion of acceptor elements produces a narrow *P* region between two *N* regions. Thus, the product is actually two junctions separated by the narrow *P* region, leading to an *N-P-N* transistor structure. Of course, machine processing may be used to slice the junction area in order to form two *P-N* junction diodes.

These processes can be used for making crystal diodes, but other methods are preferable. They are discussed in the following section.

2-11
Impurity Contact Junctions

While the grown processes are suitable for forming a *P-N* junction in a semiconductor crystal, they are better suited for manufacturing transistor structures, which require two junctions spaced so closely together that they

Sec. 2-11 *Impurity Contact Junctions* 81

interact with each other. A definite single junction is desirable for use in a crystal diode.

Three basic methods for producing single junctions in a semiconductor crystal are: alloy, diffusion, and point-contact. These techniques involve contact between solid forms of semiconductor and impurity, rather than growth from a molten batch of doped metal. These methods are illustrated in Fig. 2-10.

In the alloy process, a wafer of semiconductor doped with the desired impurity has a small pellet of impurity material pressed against it. Heat is applied to melt the pellet and some of the wafer and yields an alloy where the materials meet. After the heat is removed the solution becomes solid again. At the alloy–semiconductor boundary, a heavy concentration of donors or acceptors is formed. The boundary then behaves as a *P-N* junction diode. This method is also employed in the manufacture of transistors, using two pellets each, in multiple graphite molds.

In diffusion processes, a wafer of semiconductor that has been appropriately doped is heated in a gaseous atmosphere containing the opposite impurity. Control of the temperature and pressure causes some of the gaseous impurity to diffuse into the wafer. When germanium is used as the host material, donor elements diffuse more rapidly than acceptor elements, but either type of impurity may be used to form a diffusion layer on the wafer. After the diffusion cycle is complete, cutting and etching of the wafer yields a *P-N* junction diode.

Early forms of diodes were of the point-contact type. In their manufacture, a thin wafer of *N*-type semiconductor, a few millimeters square and a fraction of

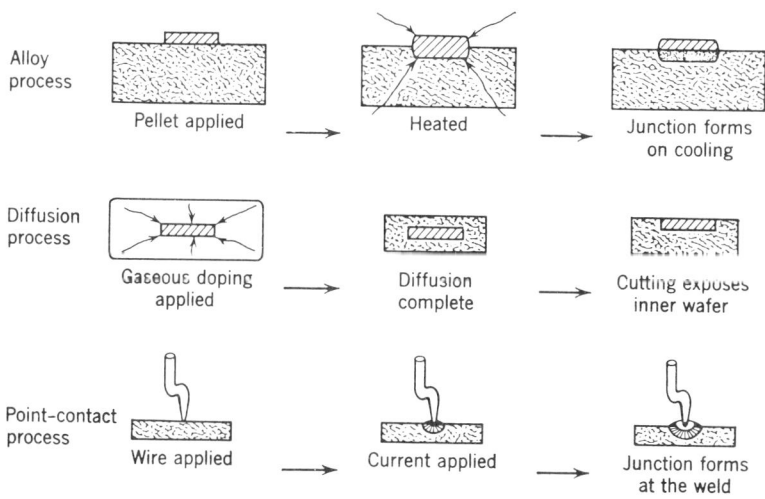

Figure 2-10
Junction formed by impurity contact methods.

a millimeter thick, has a fine pointed wire called a *cat whisker* pressed against one surface. The wire is usually phosphor-bronze, which is quite stiff. Its tip is welded to the crystal surface by passing a heavy current through the wire during a short time interval. Atoms of the phosphorus, an acceptor impurity, diffuse from the wire tip into the N-type crystal and form a P-N junction in the region close to the point of contact.

Uniformity in characteristics of crystal diodes is desired, whether they are manufactured by one process or another. The ability to manufacture diodes of the same type and characteristics is determined in part by the methods used to form the junction. The rate-grown process, for example, will yield several junctions from the same crystal and they will be similar if proper control of the variables is maintained. A point-contact type of diode, on the other hand, will be affected by the placement of the contact wire as well as by variables associated with the welding process. Uniformity in lead attachment and encapsulation of the finished product can contribute effectively to promote uniformity of the junctions in crystal diodes.

2-12
Packaging

Ohmic contacts are needed to attach leads to crystal diodes. Materials that will not produce extraneous P-N junctions are used to connect the N and P regions to external leads. Unless care is taken in forming the ohmic contacts, the diode may not function well at high currents and high temperatures. Some of the diffused junctions have only a small area for attaching contacts. This is a common cause of the formation of spurious junctions.

Connecting leads may be made of aluminum, gold, indium, nickel, or other suitable metals. Gold is especially useful because it can be readily doped either P or N type.

Alloying, soldering, and welding are used for attaching leads to the crystal structure. Gold and aluminum can be alloyed with germanium and silicon. They are extensively used in the manufacture of crystal diodes and transistors. Fluxless soldering is the preferred method in some cases, particularly for attaching leads to the pellet of alloy-process diodes. Close control of the manufacturing processes is necessary to maintain a balance between good electrical contact and mechanical ruggedness.

Encapsulation is a process that seals the finished diode and its connecting leads into a suitable structure that permit it to be handled and inserted into a practical circuit. Its primary purpose is to assure reliability of the device when it is used as a circuit component. It protects the junction and its ohmic contacts from mechanical damage and provides a seal against damaging environmental effects. Diodes that are designed to carry high currents and dissipate large

Sec. 2-12 Packaging 83

amounts of power during operation are mounted on metal-stud conductors and require external leads made of large wire. Low-power diodes are encapsulated in small metal cans, in glass cylinders, or in plastic. Examples of encapsulated germanium and silicon diodes are shown in Fig. 2-11.

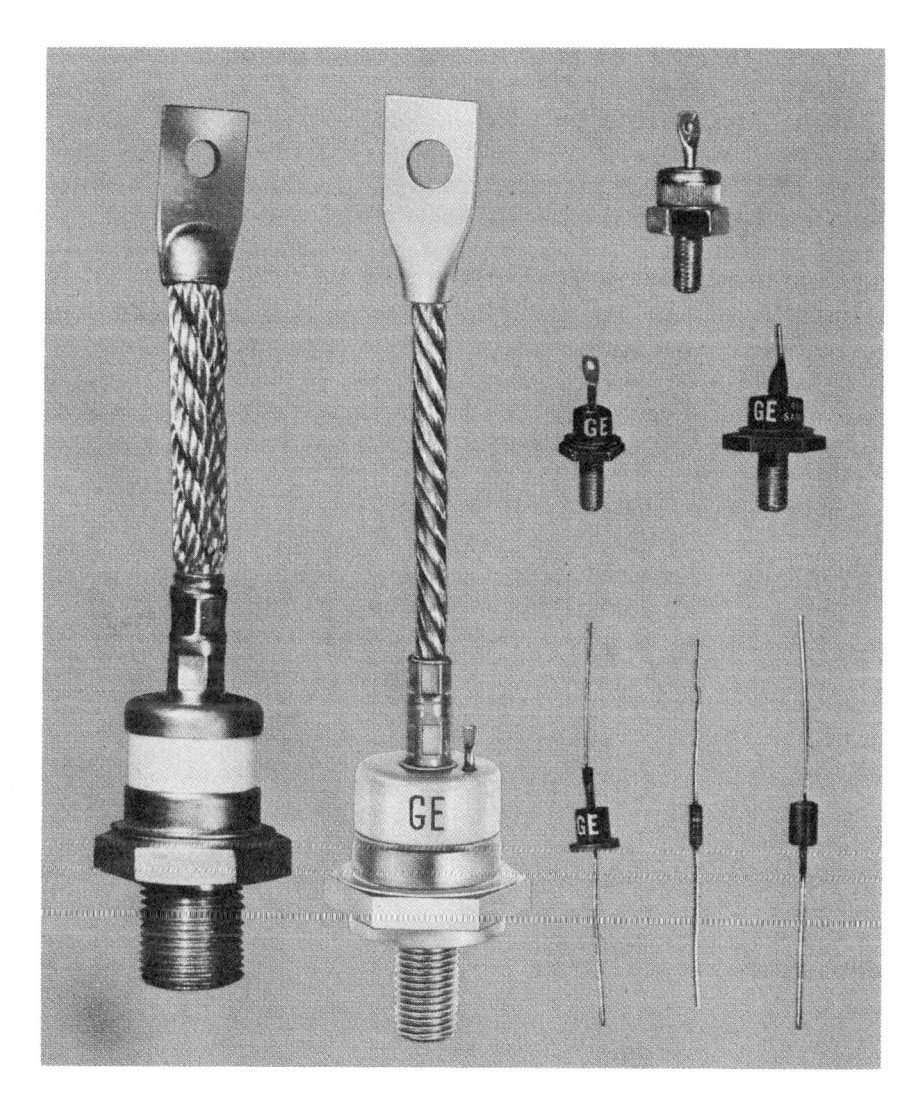

Figure 2-11
Typical silicon and germanium diodes. (Courtesy of General Electric Co., Syracuse, N.Y.)

2-13
P-N Junction Diode

When a germanium crystal is doped during manufacture in such a way that a definite boundary exists between the *P*-type and *N*-type semiconductors, the result is a *diode* with very useful properties. (A diode is an electronic device having two external electrodes to which electrical connections may be made.) Such an arrangement is called a *P-N junction diode*. It includes the boundary, or *P-N* junction, and some of the *N*-type and *P*-type semiconductor material on each side of the boundary.

Some of the electrons in the valence band receive thermal energy, and as a result, electrons and holes move about in the material. On the *P* side of the junction there is a high concentration of free holes, and on the *N* side, a high concentration of free electrons. These mobile charge carriers, owing to their motions, spread out and diffuse into opposite sides of the junction. Each kind of charge then finds itself surrounded by charges of opposite sign, and several recombinations take place. Figure 2-12 shows the junction area after diffusion and recombination have occurred. Note that the boundary region soon becomes devoid of holes on the *P* side and of electrons on the *N* side. The result is an accumulation of positive charges at the border on the *N* side and of negative charges on the *P* side. These are referred to as *bound charges* by some authors. In reality, they are positive and negative impurity *ions* bound in the crystal lattice, and they cannot move about.

Two equal and opposite charges separated a small distance are called a *dipole*, and in the situation shown in Fig. 2-12 a *dipole layer* exists at the *P-N* junction. Such a layer is present in all *P-N* junctions. It is often called a *barrier region* or *depletion region* because of the absence of free carriers near the junction. The separation of bound ions in the crystal results in a potential difference across the junction, shown as V_B in Fig. 2-12. This potential

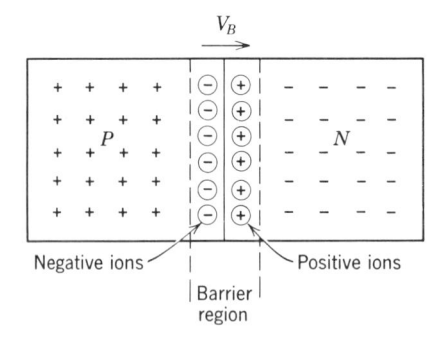

Figure 2-12
Formation of a dipole layer and barrier potential V_B at a *P-N* junction.

difference is about 0.3 V in germanium and about 0.6 V in silicon. The barrier voltage will limit current in the diode, as we shall soon see.

The formation of a barrier region at the junction results in an equilibrium condition such that free electrons on the N side cannot go over to the P side because of the opposing forces of its *negative ions*. Free holes on the P side cannot go over to the N side because of the opposing forces of its *positive* ions.

2-14
Majority-Carrier Currents

The "missing electron" in covalent bonding of P-type germanium constitutes a *hole* that may be neutralized electrically by an electron jumping into it from a nearby bond and thus effectively moving the hole to a new position in the crystal. Energy received from an electric field or from heat could cause the electron to make the jump from the nearby bond to its new position. The motion of the hole, *which acts the same as a positive charge*, constitutes current in the crystal. Similarly, excess electrons in N-type germanium are relatively free to move about in the crystal when they receive energy from an external source such as an electric field or heat. Their motion from atom to atom constitutes current, a net motion of electric charge in the material.

Conduction in P-type germanium at ordinary temperatures is by means of holes that are created when an acceptor impurity is added. The hole acts as a positive charge, and, since the doped germanium has an excess of these "missing negative charges" in its atoms, the impurity has in effect contributed net positive charge to the conduction process. Holes are in the majority, and in this case they are called *majority carriers* of current. Similarly, electrons supplied by donor impurities in N-type germanium are spoken of as the *majority carriers*.

2-15
Minority Carriers

Thermal ionization takes place in doped germanium just as it does in intrinsic germanium. This process yields the same number of holes as electrons, whether it happens in N-type or P-type germanium. In N-type semiconductors, a *minority-carrier* current flows owing to the motion of thermally generated holes. The free electrons arising from thermal ionization receive thermal energy which elevates them into the conduction band, and this increases the *majority* carriers. Holes that are left behind in the valence band may be neutralized by other free electrons falling into the lower potential-energy level occupied by the hole. Others, not eliminated by "recombination"

86 Diodes   Ch. 2

in this way, are driven by an applied electric field toward the negative-potential terminal and constitute a minority-carrier current.

Thermally generated electrons in P-type semiconductors may contribute a minority-carrier current. Because of the added thermal energy, electron-hole pairs are generated and the additional free electrons are elevated into the conduction band. Some of them neutralize holes in the crystal, and others are driven by an applied electric field toward the positive terminal of the crystal and constitute a minority-carrier current.

Minority carriers in N-type material are holes, while those in P-type material are electrons. Minority carriers occur as a result of thermal ionization in the crystal. At very low temperatures few minority carriers are generated. However, as the ambient temperature of the crystal increases, more and more electron–hole pairs are generated. This property of semiconductors becomes important in their application in crystal diodes and transistors operating at high temperatures. Desirable characteristics of these devices are impaired by minority-carrier currents, so it is important to note that crystal diodes and transistors sometimes require extra care in their application to assure that their operating temperature is kept low. More will be said of this in further discussions of their applications.

2-16
Potential Hill at the Barrier Region

Bound charges in the barrier region establish an electric field across the junction, and a potential difference exists because of unlike charges making up the barrier region. The potential distribution inside the diode, with the P region taken as a reference potential, is shown as a *potential hill* in Fig. 2-13.

The height of the potential hill at the P-N junction is determined by an equilibrium between two factors: (1) thermal ionization generating electron–hole pairs on both sides of the junction, and (2) the diffusion of

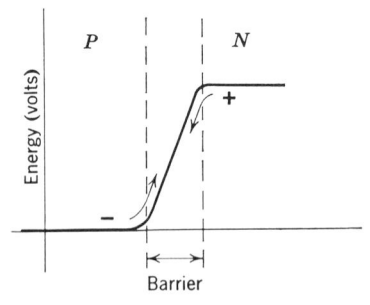

Figure 2-13
Potential hill through barrier region of P-N junction.

carriers across the junction against the potential barrier due to their having acquired sufficient kinetic energy from thermal generation. Under equilibrium conditions with no external circuit for current, these two factors cancel their effects as far as internal current is concerned.

2-17
Forward Biasing of Diodes

If a voltage is applied to the diode as shown in Fig. 2-14, holes in the P region move across the junction into the N region, and electrons in the N region move across the junction into the P region. These carriers recombine with opposite charges in the barrier region, lowering the potential hill at the P-N junction. External application to the diode of a voltage with this polarity, positive to P side, negative to N side, is called *forward biasing* of the diode. Lowering the potential hill in this way makes it easy to conduct current in the diode. Forward biasing reduces the strength of the barrier potential. The result is a current (flow of both electrons and holes) that is much greater than that which the resistance of the semiconductor alone would allow.

When conducting current under forward biasing conditions, semiconductor diodes have only a small voltage drop (a few tenths of a volt) and thus offer very low impedance to current. The internal resistance is almost entirely due to the presence of the potential hill, depending very little on the crystal material on either side of the junction area. A current-limiting resistor such as R in Fig. 2-14 is needed to limit the current in the diode.

2-18
Reverse Biasing of Diodes

When the positive terminal of the applied voltage is connected to the N side and the negative terminal to the P side, the diode is said to be *reverse-biased*.

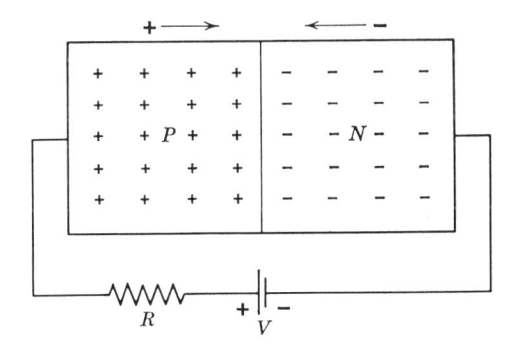

Figure 2-14
Forward-biased P-N junction.

88   Diodes   Ch. 2

This condition is illustrated in Fig. 2-15. The barrier region contains no free carriers and it has much higher resistance than the remainder of the crystal on either side of the barrier. In fact, the *P* and *N* regions have such comparatively low resistance that they act much like metal contacts between the external connecting leads and the barrier at the common boundary.

The external voltage source attracts electrons and holes away from the junction, producing a wider ion layer at the junction. This results in an increase in the potential hill that charge carriers must overcome in traversing the *P*-*N* junction. The ion layers in the barrier region produce an electric field such as an applied battery potential (shown dotted in Fig. 2-15) would produce. This field is in such a direction as to make it still more difficult for *majority carriers* (electrons in the *N* region) to cross the border from *N* to *P* and for holes in the *P* region to cross from *P* to *N*. The current that predominates is due to minority carriers, as will soon be explained.

Remember that the applied bias voltage is trying to produce conduction in the diode. The *N* side of the crystal, having been made *positive*, is trying to "attract" electrons from the *P* side through the junction, but *N*-type semiconductors already have a large excess of electrons. Any migration of electrons toward the positive terminal merely widens the positive-ion layer in the *N*-side.

The *P* side of the junction has a scarcity of electrons, and although they would willingly go up the potential hill, there are relatively few available to contribute to current. Of those that receive enough thermal energy to break loose from the valence band in the *P* material, some may reach the barrier before recombining with a hole that has been created by a similar thermal ionization in the *P* region. Upon reaching the foot of the potential hill at the barrier, an electron goes up into the *N* region and becomes a *minority carrier* there. Because its arrival adds one more electron than is needed for equilibrium

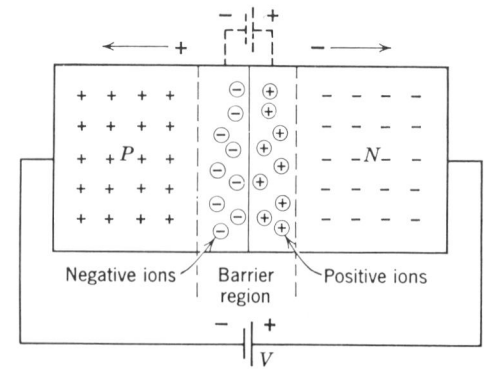

Figure 2-15
Reverse-biased *P-N* junction.

in the N region, an electron will be delivered to the positive terminal of the biasing battery.

The P side now has an excess hole left by this electron. Another electron is accepted from the negative terminal of the biasing battery, and the excess hole is neutralized to restore equilibrium to the P side. Thus, current in the reverse-biased diode is by *minority carriers*. Its magnitude is much smaller than current in a forward-biased diode.

An electron–hole pair may be generated in the N region in the same way described for the P region. Then a minority carrier (hole) will "slide down" the potential hill at the barrier and "take on" an electron in the P region. The negative terminal of the battery will supply another electron to keep the P region in equilibrium, and because there is now an excess of electrons in the N region, one will go to the positive terminal of the battery.

It is important to note that while current inside a diode may be composed of both electrons and holes, all *external* current consists of electrons only. This is not too surprising because *conductors* (which have only free electrons for conduction) are used in the external circuitry. The internal material is doped *semiconductor*, which contains both holes and electrons.

2-19
Avalanche Breakdown

When reverse bias is increased beyond values for normal operation of the diode, minority-carrier electrons acquire sufficient energy in crossing the boundary region to release valence-band electrons out of the semiconductor atoms by collisions in the N region. These add greatly to the normal electron flow, rapidly increasing current with increasing reverse-bias voltage, as shown in Fig. 2-16. This phenomenon is called *avalanche breakdown*. Of course, there is simultaneous hole flow across the barrier into the P region. Electrons are taken readily from the negative-bias terminal, and they neutralize the holes.

The shape of the characteristic curve may be explained as follows. With forward biasing (N side negative), the effect is to lower the potential hill at the barrier. Of course, the battery is now trying to move electrons from the N to the P side, and they find it easy to surmount the much smaller hill and thus get through the barrier. The current increases in a nonlinear fashion with increasing voltage. The higher the bias voltage, the weaker the barrier, until a voltage is reached at which the barrier has been counterbalanced by the bias battery and disappears. The current–voltage curve from that point on is essentially a straight line, and the current is limited only by the series resistance in the circuit.

With reverse biasing, the current is almost independent of the amount of biasing voltage up to the value necessary to start the breakdown process. The

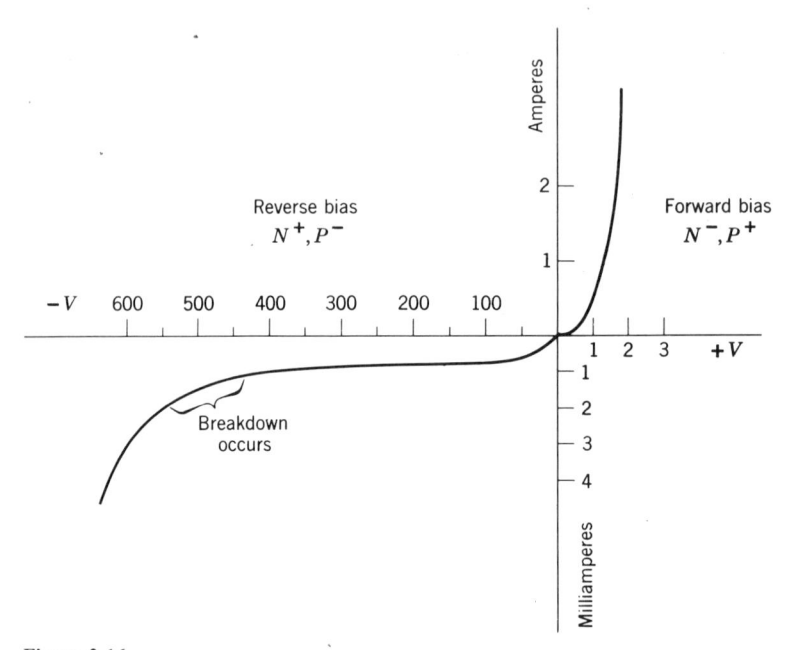

Figure 2-16
Average current–voltage characteristic curve of a P-N junction germanium diode. Note the change in scales for forward and reverse biasing. (Curve exaggerated to show effects of bias.)

reason is that, at a given temperature, the rate of production of carriers available for migration depends on the concentration of impurity, which is fixed when the crystal is doped. It will be noted that reverse current is much smaller than forward current, typically in the ratio 1:1000 at the same bias voltage.

2-20
Zener Diode

Another type of breakdown process that can supply carriers to support conduction when the voltage across the junction is large is called the *Zener effect*. It arises when the barrier potential becomes so large that electrons may be ripped from covalent bonds in the barrier region. Thus, electron–hole pairs are created in the ion layers at the junction. The hole moves quickly into the P region under the influence of the electric field in the barrier region, and the electron is swept into the N region. These moving charges constitute an increase in reverse current in the junction. When the electric field is strong enough to break a single covalent bond, it is sufficiently strong to break many. Once the Zener effect has started, large increases in reverse current may continue with the same applied bias potential. In a diode in which Zener breakdown can occur, its current–voltage curve has a sharp corner at the breakdown region, rather than the smooth transition of Fig. 2-16.

Research has shown that the Zener effect occurs when the diode is heavily doped and when the breakdown occurs at a low voltage—about 10 V or so. When the breakdown occurs in lightly doped crystals or at high voltages, the mechanism is something other than the Zener effect.

Operation in the breakdown region does no damage to the diode, provided the reverse current is limited to a maximum value set by the manufacturer. This value is determined by the power-dissipating capabilities of the crystal structure and greatly depends on the physical size of the diode. By controlling the doping concentration and other factors, diodes can be designed to have any desired breakdown voltage in the range from about 2 V to over 200 V. Diodes which are designed to operate in the breakdown mode are called *Zener diodes* or *breakdown diodes*.

Figure 2-17 shows the typical voltage–current curve for a Zener diode. This one breaks down at 6.8 V reverse bias. Note that this potential is constant even though the reverse current covers a significant range of values.

A crystal diode operating in its Zener region performs many useful and important tasks. The current passing through it may be required to change over a relatively wide range. While this is happening the voltage across the diode will

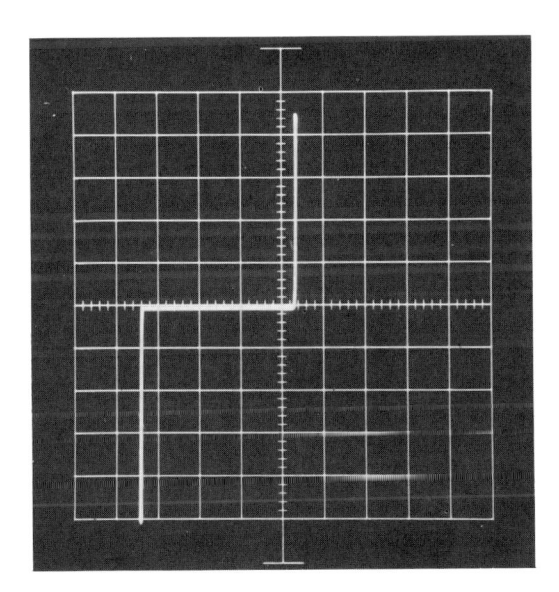

Figure 2-17
Typical Zener diode voltage–current curve; this one is a silicon unit that breaks down at 6.8 V reverse bias. Calibration of the display is: horizontal, 2 V per division; vertical, 2 mA per division. Current and voltage axes have their origin at the center of the display.

change very little, in some cases by an insignificant amount. We can readily see that this would be a means of holding the voltage between two points practically constant if the Zener diode were connected to the two points. Much more will be said about the operation of Zener diodes when diode applications are discussed.

2-21
Diode Ratings

Ratings are the limiting values assigned by a manufacturer to various parameters of a diode and to environmental factors in its operation. If these assigned values are exceeded, the result may be permanent damage to the diode or impairment of its performance or life.

The load ratings of semiconductor diodes are based on their ability to dissipate heat losses and thus not exceed the junction temperature limits specified by the manufacturer. Losses are usually given in watts, determined by electrical characteristics such as forward voltage drop during conduction and leakage current in the reverse direction. Thermal characteristics are specified as a certain number of degrees temperature rise per watt at the junction, measured from some reference temperature such as that of the metal stud or of ambient conditions.

Manufacturers usually specify semiconductor ratings according to the symbols and definitions shown in Table 2-2.

2-22
Diode Characteristics

Characteristics of a semiconductor diode are those measurable properties or attributes that are inherent in its design. We have looked previously at the graphical plot of current and voltage variations for a typical crystal diode. Such a display of characteristics is useful in the analysis of circuits containing the diode as a component. However, manufacturers usually present the important characteristics of semiconductor diodes by tabulating typical values at specified operating points, rather than a continuous plot point by point.

The characteristics that are usually specified by a manufacturer are given in Table 2-3.

As one might expect, the characteristics of the point-contact diode are similar to those of the junction diode. However, the point-contact diode, because of its design, is inferior to the junction diode in many of its characteristics. For example, it cannot carry nearly as much current because of heat generated in its comparatively small active area. Further limitations are: (1) reverse current increases significantly with reverse voltage instead of remaining practically

Table 2-2
Ratings

Term	Symbol	Definition of Rating
Peak reverse voltage or Peak inverse voltage	PRV or PIV	Maximum allowable *instantaneous* reverse voltage that may be applied across the diode under other specified conditions. Voltage may be a sine wave, but a continuous dc rating, defined below, is not specified. While this rating does not refer to a "breakdown" voltage value, it should not be exceeded on an instantaneous, repetitive basis.
Transient peak reverse voltage	PRV_{trans}	Maximum allowable instantaneous value of reverse voltage that may be applied on a "one-shot," nonrecurring basis. Its duration and other conditions are specified with the rating.
Reverse dc voltage, for continuous operation	V_{RDC}	Maximum reverse voltage that the diode may block over long periods of time.
Maximum dc output current	I_F	Maximum dc forward current that may be allowed under stated conditions of temperature and reverse voltage. This value will be the *average* current in the forward direction, when the load is resistive or inductive. Derating factors are usually applied for capacitive loads and other conditions.

Table 2-3
Characteristics

Term	Symbol	Definition of Characteristic
Forward voltage drop	V_F	Value of instantaneous forward voltage drop during conduction of load current, and under any other specified conditions.
Reverse (or leakage) current	i_R	Instantaneous value of reverse current at the stated conditions of temperature and voltage.

constant, and (2) its peak inverse voltage (PIV) rating is much lower than that of a junction diode. Figure 2-18 illustrates the basic differences between a point-contact and a junction diode, by comparing their ratings and characteristics.

Figure 2-18
Comparison of point-contact and junction diode characteristics.

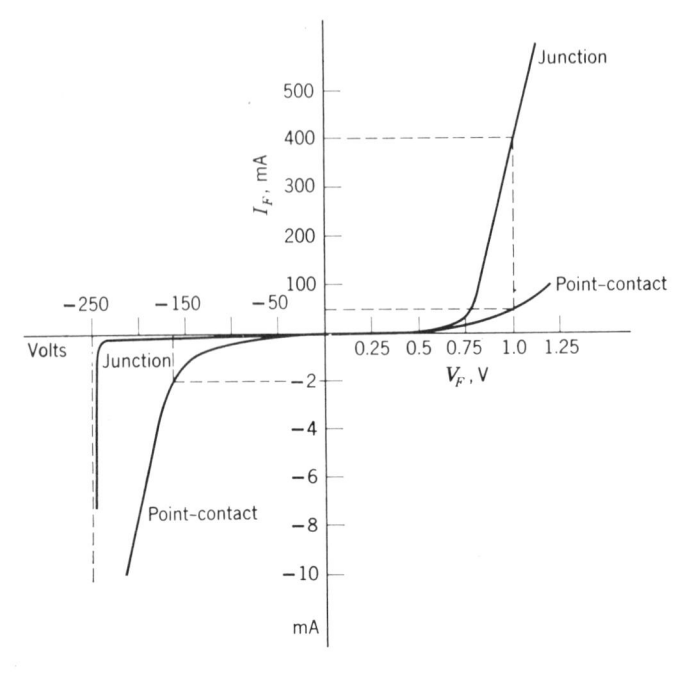

	Silicon, 25°C		
Junction			Point-contact
240 V	PIV		160 V
200 V	V_{RDC}		150 V
600 mA (25°C)	I_F		60 mA
50 mA (200°C)	I_F		10 mA
1 V (400 mA)	V_F		1 V (50 mA)
0.1 mA (200°C, 160 V)	i_R		2 mA (200°C, 160 V)

The average current–voltage characteristic of a typical point-contact diode closely resembles the portion of the curve for a junction diode in forward conduction, except for the smaller maximum current. On the reverse-bias side of the curve, leakage current starts to increase rapidly with increasing reverse voltage. This illustrates the lack of a small reverse-saturation current, which is so important in many applications of diodes in practical circuits. The unilateral conductivity of semiconductor diodes, the ability to pass current readily in one direction and to effectively block it in the reverse direction, is their most important property. Point-contact diodes are inferior to junction diodes in this

2-23
Comparison of Germanium and Silicon Diodes

Either germanium or silicon may be used as the host material in a semiconductor crystal diode, but silicon has many favorable properties that make it superior to germanium. In general, silicon units cost much more than comparable germanium units, but they can be designed to perform in many applications where germanium units simply cannot meet the specifications.

Silicon can operate at temperatures up to 200°C (392°F), while germanium is limited to about 105°C (221°F). Silicon diodes can be designed to carry currents up to 250 A and to operate under reverse voltages to 1000 V/cell. Germanium is limited to a maximum of about 400 V PIV per unit, although diodes can be connected in series to obtain higher voltage ratings when circuit conditions require them. When so connected, each diode should be shunted with a resistance of 100 kΩ or more to equalize the diode reverse voltages that would otherwise divide in proportion to the individual diode reverse resistances.

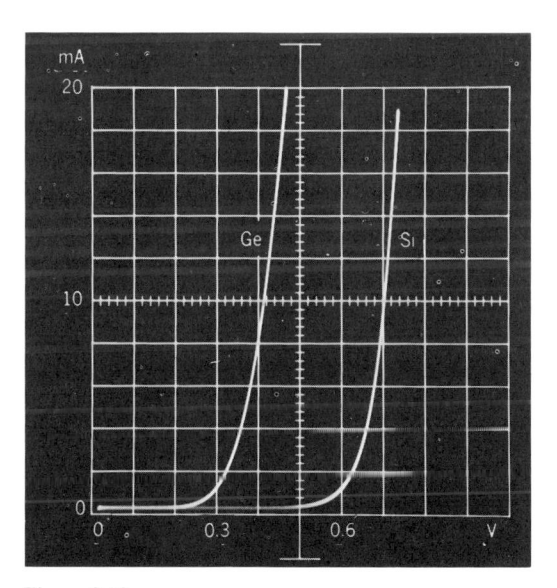

Figure 2-19
Comparison of germanium and silicon diodes operating with forward bias. Forward voltage V_F is displayed along the horizontal axis, calibrated as 0.1 V per division; forward current I_F along the vertical axis, 2 mA per division; zero at the lower left corner.

96 Diodes   Ch. 2

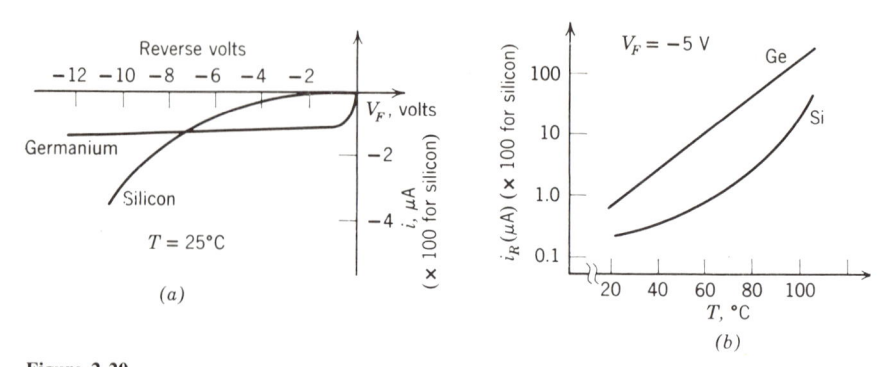

Figure 2-20
(a) Reverse-bias characteristics for silicon and germanium diodes; and (b) effects of temperature on saturation current. Note the change in current scale for silicon diode.

Germanium is superior to silicon in one important property: it has lower forward voltage drop. The lower cost of germanium units for use in low-power installations may make this property an advantage. Typically, silicon diodes exhibit about 0.6 V forward voltage drop, while germanium units have a forward voltage drop of about 0.3 V when operated under similar conditions. Figure 2-19 illustrates and compares forward characteristics of similar silicon and germanium diodes.

The reverse-bias characteristics of crystal diodes deviate from a nominal reverse saturation current and depend on both the reverse voltage and junction temperature. Figure 2-20 shows the relative shapes of reverse characteristics and changes that occur because of their temperature dependence.

Drill Problems

D2-5 A certain diode has these specifications supplied by the manufacturer: PIV 200 V, V_{RDC} 200 V, I_F (100°C case temperature) 1.5 A, V_F (1.5 A) 1.2 V, i_R (150°C, 200 V) 0.1 mA. Would you expect it to be silicon or germanium?

D2-6 A derating factor is supplied for the diode: "Derate I_F for case temperature above 100°C by the factor -2.5 mA/°C." What is the maximum permissible forward current at 200°C case temperature?

D2-7 What temperature in degrees Fahrenheit is: 25°C, 100°C, 150°C, 200°C? (Temperature conversions are given in the Appendix.)

2-24
Diode Symbols for Circuit Diagrams

The curves representing characteristics of crystal diodes, such as those in Fig. 2-18, are usually not available from the manufacturer, although ratings and

Sec. 2-24   Diode Symbols for Circuit Diagrams   97

characteristics may be furnished in tabular form for typical operating values. It is important in analyzing circuits containing diodes to be able to investigate the effect of the diodes in circuit behavior. Linear circuit analysis techniques are easier to work with than are nonlinear techniques. For this reason a crystal diode is often represented in circuit analysis by a linear model that describes its characteristics over the operating region of currents and voltages. A reasonably accurate model for a crystal diode is given in Fig. 2-21, together with various symbols used in circuit diagrams. Note that the arrow of the symbol points in the direction of forward current.

The ideal diode, whose characteristic is presented in Fig. 2-21b, has no leakage current and no forward voltage drop. The forward voltage drop of an actual diode is taken into account by the internal battery shown in Fig. 2-21a. Its potential difference is approximately the barrier voltage V_B, 0.3 V for germanium and 0.6 V for silicon. This potential is the level of voltage that must

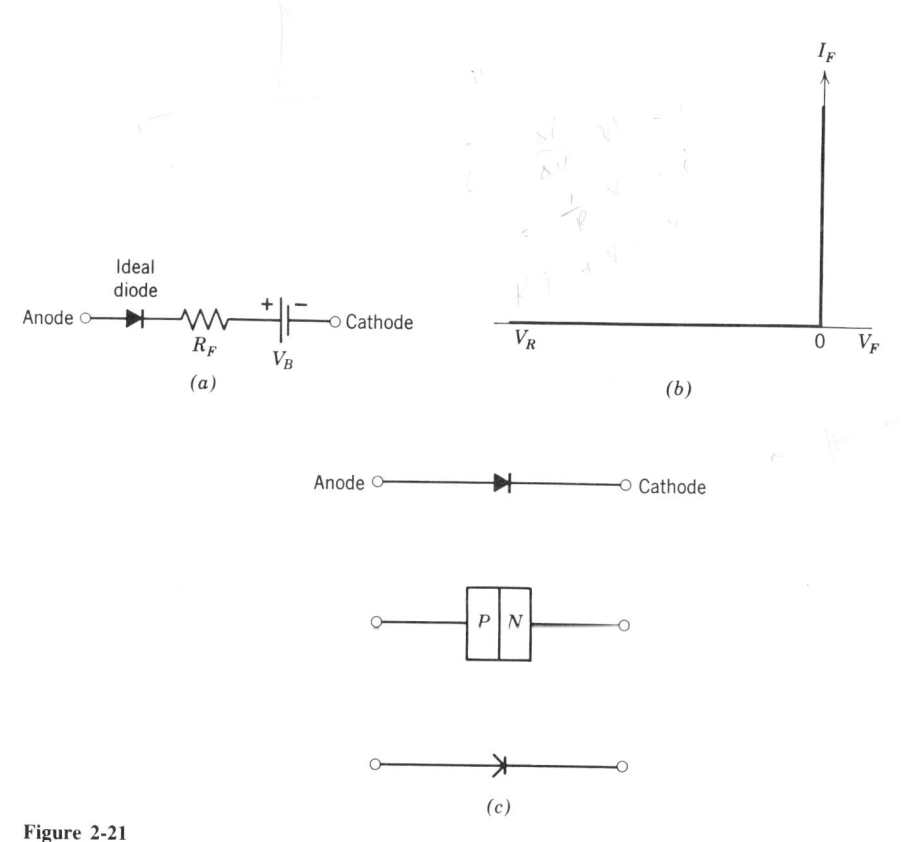

Figure 2-21
Circuit model and symbols for crystal diode. (a) Circuit model. (b) Characteristics for ideal diode. (c) Symbols.

be exceeded by forward bias before appreciable forward current will pass. The internal linear resistance element (R_F) represents the approximately linear variation of forward current with forward voltage drop. When a crystal diode is operated at reverse voltages below its breakdown potential and within other maximum ratings, this model is sufficiently accurate for analysis of circuits containing diodes. The model more nearly approximates the behavior of junction diodes than point-contact diodes, and the applications discussed in subsequent sections are assumed to use junction diodes.

In many of the circuit diagrams used to illustrate typical applications of crystal diodes, the symbol may be that of an ideal diode, but it should be analyzed in circuit behavior as in Fig. 2-21. In some applications, it may not make much difference if an ideal diode is assumed. In such cases the forward voltage drop and reverse current are so small that they can be neglected and the diode can be represented by an ideal element.

Drill Problems

D2-8 Determine values for the circuit model of a crystal diode whose characteristic is given by the curve for a junction diode in Fig. 2-18.

D2-9 Under what conditions can the diode of Fig. 2-18 be assumed an ideal element?

2-25
Diode in a Circuit

The voltage–current characteristic curve of an *ideal diode* shows that when it carries forward current, its voltage drop is zero regardless of how large the current may become. Thus, an ideal diode acts as a switch. It behaves as a short circuit to current in the forward direction and as an open circuit when the voltage drop across the diode is negative, that is, when reverse bias is applied to the diode. No reverse current passes through the ideal diode regardless of the magnitude of the reverse voltage. Whenever the voltage polarity changes, the diode changes from open-circuit to short-circuit behavior or vice versa.

This ideal model of a diode is, of course, not representative of any physical device but has utility in the analysis of some circuits containing diodes, as we shall see. A more accurate model is given in Fig. 2-22a and its characteristic is shown in Fig. 2-22b. This model will be used to describe the operation of a diode as a rectifier, that is, to change an alternating voltage source to a direct voltage across a load.

This circuit model of a practical diode will permit forward current only when the forward voltage across the diode terminals is greater than the internal voltage drop, V_B. When forward voltage exceeds V_B, forward current is

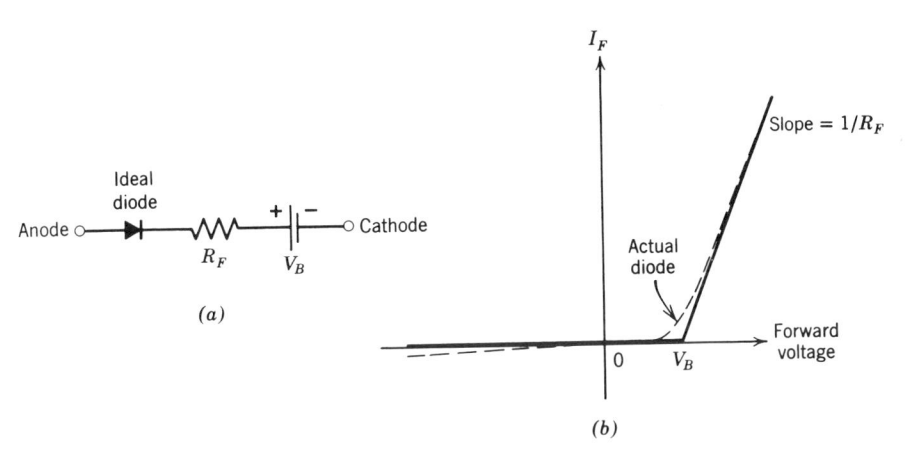

Figure 2-22
Diode circuit model and its characteristic curve. (a) Circuit model. (b) Characteristic curve.

limited by the linear resistance, R_F. For voltages less than V_B, no current is permitted in the diode. Thus, in a series circuit containing this diode, an open-circuit condition exists until forward voltage is greater than V_B and for greater voltages the diode behaves like a linear resistance.

In order to illustrate the operation of a crystal diode as a rectifier, we consider the operation of the circuit in Fig. 2-23. This simple combination of circuit elements is called a *half-wave rectifier*. It will be studied in some detail in a later chapter. Here, our interest is in the behavior of the diode rather than the circuit as a whole.

A sinusoidal voltage source, $e_s = E_s \sin \omega t$, is applied to the circuit containing the diode and a resistance load, R_L. When e_s is positive and greater than V_B, it produces a forward current in the diode and R_L. Since this is the forward direction for current in the ideal diode, it acts as a short circuit and the voltage across R_L will be determined by the current. For values of e_s less than V_B, no current will be supplied to the load. The load voltage waveform (Fig. 2-23) is given by

$$e_L = \frac{R_L}{R_F + R_L}(e_s - V_B) \quad \text{for } e_s > V_B$$

and

$$e_L = 0 \quad \text{for } e_s < V_B$$

(2-1)

Current in the circuit may be expressed as

$$I_F = \frac{e_s - V_B}{R_F + R_L} \quad \text{for } e_s > V_B$$

$$I_F = 0 \quad \text{otherwise}$$

(2-2)

100 Diodes Ch. 2

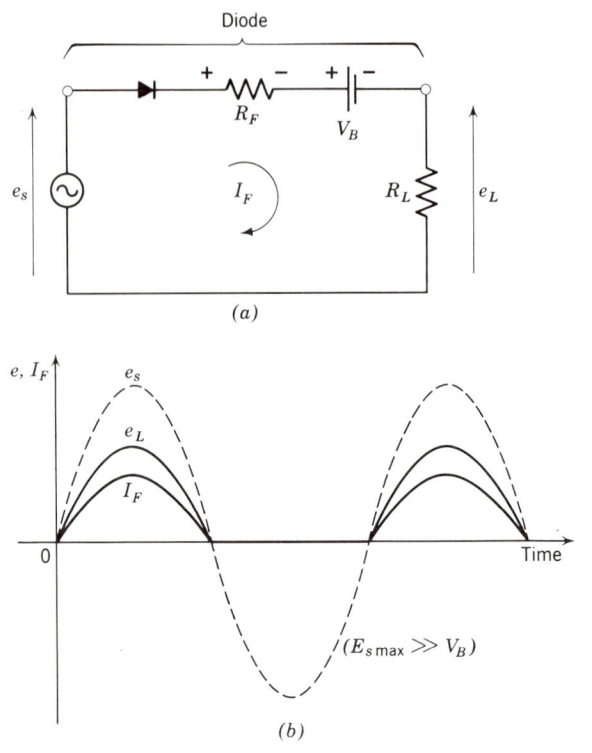

Figure 2-23
Diode rectifier circuit and waveforms. (a) Circuit. (b) Waveforms.

If $E_{s,\,max} \gg V_B$, then the current is approximately

$$I_F \approx \frac{E_{s,max}}{R_F + R_L} \sin \omega t$$

when it exists during each positive half-cycle. During negative half-cycles of e_s, reverse current is assumed to be zero and no load voltage will be developed during these time periods.

Now let us consider operation of the diode when a complex voltage waveform, one that consists of both ac and dc components, is applied to it. The circuit in Fig. 2-24 contains a diode, a resistor load, a battery, and a sine wave voltage source. Forward current will be conducted only when $V_{AB} > V_B$, that is, when V_{AB} exceeds 0.6 V. Current I_F will then have the waveform of V_{AB}, and so will the load voltage, e_L. Current cutoff occurs during the period when V_{AB} is less than V_B.

Further reasoning leads to the conclusion that the magnitude and polarity of E_1 can be used to control the period of current cutoff, from always to never. If

Sec. 2-26   *High-Vacuum Diode with Thermionic Cathode*   101

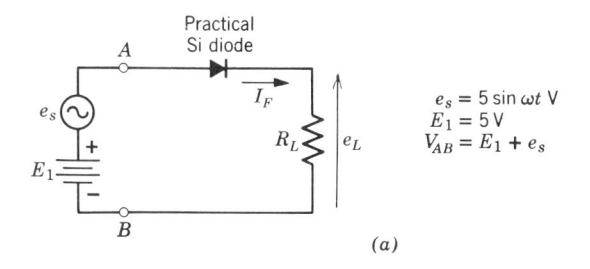

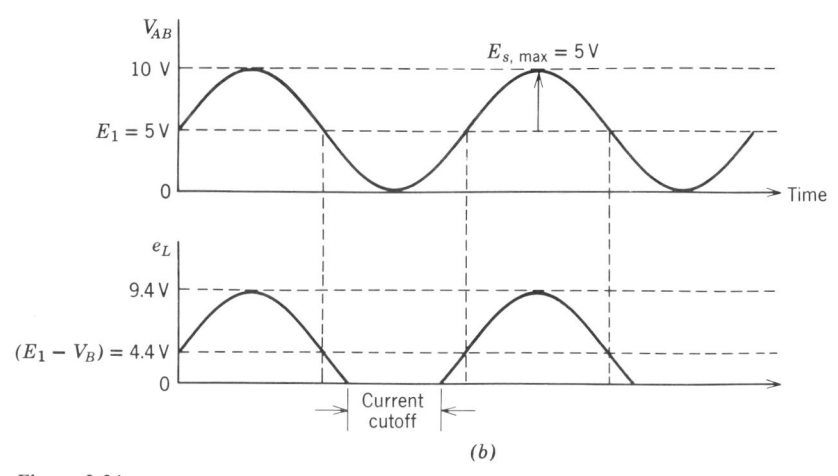

Figure 2-24
Diode circuit with complex input voltage waveform. (*a*) Schematic diagram. (*b*) Waveforms.

$E_1 > V_B + E_{s,\text{max}}$, current I_F will exist at all times. If E_1 is made more negative than $-E_{s,\text{max}}$, then I_F will always be zero. Applications are discussed in Chapter 3.

2-26
High-Vacuum Diode with Thermionic Cathode

An electronic tube that has only an anode (plate) and a cathode as active electrodes is called a diode. A vacuum diode also includes a cathode heater, which does not enter directly into the operation of the tube but does influence its behavior. These elements are enclosed in a glass or metal envelope that has been evacuated of air and other gases during manufacture to leave a very high vacuum in the interelectrode space.

Cathodes designed for use in high-vacuum diodes are in two general classes: (1) filamentary, or directly heated, and (2) equipotential, or indirectly heated. In

102 *Diodes* Ch. 2

the filamentary construction a wire is heated by passing an electric current through it, and electron emission takes place directly from the wire or a special coating on it. The *filament* is the *cathode element* of the diode when the cathode is directly heated; the heating current passes through the filament wires, from which electrons are emitted. The directly heated cathode structure may appear as in Fig. 2-25a. It is built to have good emission properties and mechanical ruggedness.

Directly heated cathodes operate with the best economy of heating power when they are coated with certain metallic oxides. Barium oxide is commonly used with a small amount of strontium oxide added to give the coating mechanical strength and longer life. Filaments are made of materials that may be fabricated to operate over a wide temperature range without losing their electrical and mechanical properties. Nickel and tungsten are widely used for the current-carrying filament, because their metallurgical properties permit a good metallic bond for the oxide coatings.

Some directly heated filaments are made of thoriated-tungsten wire that has been given a special heat treatment. They are not quite so efficient as oxide-coated cathodes, but they are much more efficient than cathodes made of pure tungsten. Pure tungsten must be used when operation is at extremely high temperatures.

Figure 2-25b shows the construction of indirectly heated cathodes. A loop of tungsten wire insulated with a refractory material is inserted into a thin, hollow cylinder, usually made of nickel. The outside of the cylinder is coated with the emitting material. Electric current in the wire heats the cylinder to emitting temperature. Since no heating current is in the cylinder, the cathode is an equipotential surface. This means all points on the cathode are at the same electrical potential. Cathodes of this type exhibit heat storage that reduces the disturbing effect of temperature variations that may occur in the heater.

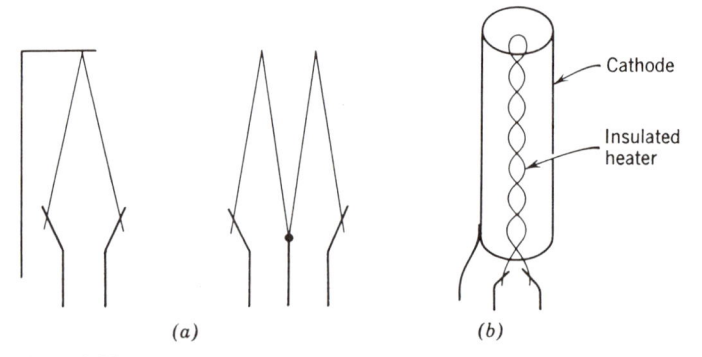

Figure 2-25
Thermionic cathode construction. (a) Directly heated filament. (b) Indirectly heated cathode.

Indirectly heated cathodes have some important advantages, but because of current limitations their use is restricted to the relatively small tubes. Some of their advantages are: (1) The entire emitting surface is at one potential, so that no potential difference exists along the surface as is the case when direct heating is employed; (2) operation of two or more cathodes at different potentials is possible while the same heating current is being used; (3) alternating current may be used for heating without introducing disturbing effects that may result from stray electric fields due to the continuously changing current and from nonuniform plate-to-cathode potentials; and (4) these cathodes have higher emission efficiency.

2-27
Characteristics of the High-Vacuum Diode

It is desirable to understand how the plate current in a diode is affected by cathode temperature. To determine this experimentally, consider the tube connected in a circuit (as shown in Fig. 2-26) with a particular value of dc voltage, E_b, between the plate and the cathode.

A change in the heater-cathode current, I_f, will change the temperature of the cathode, and because of this there will be a change in the plate current, I_b, of the diode.

Curve 1 of Fig. 2-27 shows how the plate current of a high-vacuum diode depends on cathode heating current, and therefore on cathode temperature, at a constant value of plate-to-cathode voltage, E_{b1}. When a higher value of constant plate voltage is used, a curve like curve 2 is obtained. The curves coincide for more than half their length.

Note that no plate current is indicated at low values of heater current. The

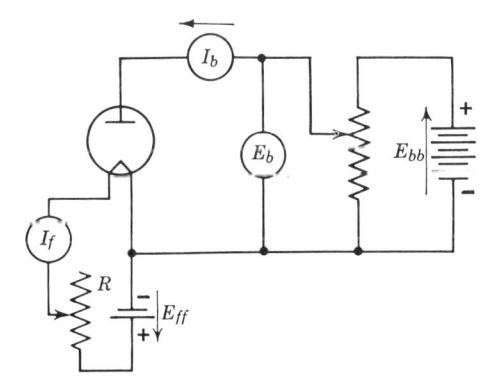

Figure 2-26
Measurement of diode characteristics.

104 Diodes   Ch. 2

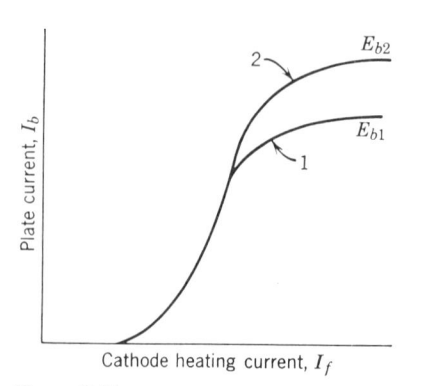

Figure 2-27
Diode characteristics. Cathode temperature varied, plate voltage having constant values.

reason is that the cathode is not hot enough to give off electrons in sufficient quantities to constitute a significant current.

The plate current rises rapidly after the emission temperature has been reached and soon levels off if the increase in heater current is continued. The leveling off is due to the fact that electrons repel one another, and so they gang up in the space in front of the cathode and form a *space-charge cloud*. This negative space charge limits the rate of flow of electrons toward the plate to a value determined partly by the electric field between the plate and cathode, and it drives the excess electrons back into the cathode. The plate current is said to be *space-charge-limited*. Naturally, the higher the plate voltage, the stronger will be the electric field in the space between the plate and the space charge; hence more electrons per second will reach the plate, manifesting a larger current in the tube. The plate current is said to be *temperature-limited* when existing conditions make the curves coincide, as in Fig. 2-27. Evidently, in that range of cathode temperature, the higher voltage, E_{b2}, does not cause any more current to flow than does the lower voltage, E_{b1}, because, even at the lower value of plate voltage, the electric field in the space is able to force the electrons toward and into the plate as fast as they are emitted from the cathode. *Temperature saturation* is said to exist where the curves have leveled off, because further increase in cathode temperature does not increase the current.

The other condition, that of constant heater current, is used normally in electron-tube operation. The dependence of plate current on plate voltage is shown in Fig. 2-28 for two constant values of heater current. Again, the lower portions of the curves coincide. In this region the current is limited by space charge, and at voltages corresponding to the lower values of current (up to point *A* on the graph) the cathode temperatures are high enough to provide electrons as fast as the electric field forces, just off the cathode, can take them away. At plate voltages higher than that corresponding to the current at *A*,

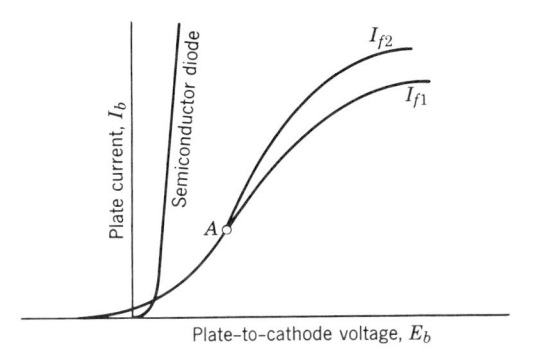

Figure 2-28
Diode characteristics. Plate voltage varied, cathode temperature having constant values.

however, the current begins to be temperature-limited and an increase in filament or heater current from I_{f1} to I_{f2} will cause the plate current to increase even though the plate-to-cathode voltage is held constant.

Where the curves begin to level off, the condition known as *voltage saturation* begins. Evidently, when the curves become practically horizontal, the only way to get more plate current is to increase the cathode temperature, that is, increase the filament or heater current.

The voltage–current curve for a semiconductor diode is also shown for comparison. The vacuum diode requires a much larger applied voltage for significant current.

Questions

2-1 Discuss the physical concept of an isolated atom, using the Bohr model, in terms of a nucleus, electrons, energy levels, and shells.

2-2 Tell all you can about valence electrons.

2-3 What is a quantum of energy? How is it related to Planck's constant? In what units is Planck's constant expressed?

2-4 Why are copper and silver good conductors of electricity?

2-5 Aluminum is a light-weight electrical conductor. Why is it not more universally used to take advantage of its light weight?

2-6 Why is sulfur a good electrical insulator?

2-7 Why are elements like germanium and silicon called semiconductors? How do the Bohr models of these atoms differ from the models of atoms of conductors and insulators?

2-8 What is meant by covalent bonding?

106 *Diodes* Ch. 2

2-9 What is a donor element in a semiconductor crystal, and what purpose does it serve? Name two donor elements.

2-10 What is an acceptor element in a semiconductor crystal, and what purpose does it serve? Name two acceptor elements.

2-11 Describe the concept of a "hole" in semiconductor theory.

2-12 Compare *N*-type and *P*-type germanium.

2-13 What is a *P-N* junction? How is it formed? Why can't a satisfactory *P-N* junction be formed by forcing together pieces of *N*-type and *P*-type crystals?

2-14 Both *N*-type and *P*-type semiconductors have *majority carriers*. Explain what they are.

2-15 Discuss the terms conduction band, valence band, and forbidden band.

2-16 What is a *minority carrier*? Describe minority-carrier current in *N*-type and *P*-type semiconductors.

2-17 Discuss the internal behavior of a *P-N* junction under (*a*) forward bias, and (*b*) reverse bias. Relate this behavior to a plot of current–voltage characteristics of a junction diode.

2-18 Describe the potential hill across the barrier region of a *P-N* junction and its effect on current through the junction.

2-19 What is meant by avalanche breakdown?

2-20 Why is it necessary to "grow" *P-N* junctions, rather than form them as welded joints between *P*-type and *N*-type semiconductors?

2-21 What is the basic principle of "zone refining"?

2-22 Why is zone refining accomplished in an inert atmosphere or in vacuum?

2-23 What characterizes a "single crystal" of semiconductor material?

2-24 Explain the difference between a grown-junction diode and an alloy-junction diode.

2-25 What are the reasons for strict control over manufacturing processes to produce uniform characteristics in any particular type of diode?

2-26 Give some reasons why encapsulation of crystal diodes is desirable.

2-27 Why do manufacturers of semiconductor diodes provide ratings as well as characteristics of their products?

2-28 What are the limitations of a point-contact diode compared with a junction diode?

2-29 Why is a large value of reverse leakage current undesirable?

2-30 Compare the properties of silicon and germanium for use as the host materials in semiconductor crystal diodes.

Problems

2-1 An electron in intrinsic germanium at room temperature must be excited with about 0.7 eV of energy to go from the valence band to the conduction band of energy. How many joules is this?

2-2 How many joules of energy are represented by 1.1 eV?

2-3 Calculate the energy an electron will acquire in falling through 1 V of potential difference. Express the energy in joules and in electronvolts.

2-4 Repeat Problem 2-3 for a potential difference of 100 V.

2-5 Suppose the electron of Problem 2-1 receives the 0.7 eV from a photon that strikes it and gives up its energy. Calculate the frequency of the photon.

2-6 Visible-light frequencies are in the range from about 4×10^{14} to 7×10^{14} Hz. Determine the minimum frequency of photons that may excite electrons from the valence band to conduction bands in germanium at room temperature. Is this frequency in the visible range?

2-7 Repeat Problem 2-6 for silicon at room temperature.

2-8 Using the data of Table 2-1, calculate the change in ionization energy of intrinsic germanium when its temperature is increased from 0°K to room temperature (27°C).

2-9 Repeat Problem 2-8 for intrinsic silicon.

2-10 What is the ionization energy of intrinsic germanium at room temperature (27°C)?

2-11 Which material decreases its ionization energy faster with increases in temperature—intrinsic germanium or intrinsic silicon? Show clearly how you know.

2-12 A germanium P-N junction diode has its bias suddenly changed from 1.5 V in the forward direction to 200 V in the reverse direction. Sketch circuit diagrams for these two cases, showing the correct polarity for external bias batteries. Mark the direction of current in the diode in each case.

2-13 If the diode of Problem 2-12 has the characteristic of Fig. 2-16, what are the magnitudes of current before and after the change?

2-14 A manufacturer specified a certain diode with these values:

        PIV   200 V
        V_{RDC} 200 V
        I_F 1.5 A (at 100°C case temperature)
        V_F 1.2 V (at $I_F = 1.5$ A)
        i_R 0.1 mA (at 200 V, 150°C)

108 Diodes   Ch. 2

Would you expect the diode to be constructed of germanium or of silicon? Why?

2-15 The diode of Problem 2-14 has a derating factor for operation above 100°C: "Derate I_F linearly for case temperature above 100°C by the factor -2.5 mA/°C, to a maximum of 200°C." What is the maximum permissible forward current at 200°C case temperature?

2-16 Determine values for the circuit model of the junction diode whose characteristic curve is given in Fig. 2-18.

2-17 Justify or deny that the diode of Problem 2-16 may be represented by a linear resistance in series with an ideal diode. Sketch the equivalent circuit for the diode, for both forward and reverse conduction. What are the values of resistance in the equivalent circuit? Are they constant values for all possible ranges of bias voltages?

2-18 Assuming ideal diodes in the circuit of Fig. P2-18, make a graphical plot of I vs. E.

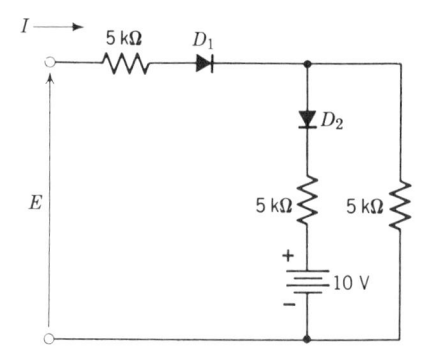

Figure P2-18

2-19 Repeat Problem 2-18 for D_2 reversed in the circuit.

2-20 A linear model of a semiconductor diode consists of a 1-Vdc voltage source in series with a 100-Ω resistor and an ideal diode. (*a*) Sketch the schematic diagram of the model and mark the anode and cathode terminals. (*b*) Sketch the voltage–current characteristic curve that is valid for this model of the diode. (*c*) A signal $e = 100 \sin 377t$ is applied to the diode in series with a 1-kΩ load. What peak current passes in the diode? What must be the PIV rating of the diode?

2-21 In the circuit shown in Fig. P2-21, diode D is passing current. Label the voltage source, diode D, and R_L with correct polarity marks indicating the voltages across these components, at the time D passes current.

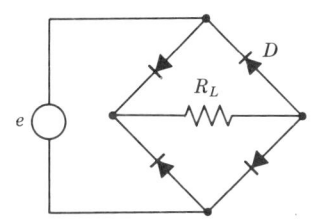

Figure P2-21

2-22 Assume that the voltage source in Fig. P2-21 is sinusoidal and that the diodes may be considered ideal in the operation of this circuit. Sketch the waveform of the voltage across R_L.

2-23 For the circuit shown in Fig. P2-23, sketch the waveform of voltages across R_L for the following voltage sources and assuming the diode is ideal: (a) $E_1 = 0$ V, $e_2 = 5 \sin 377t$. (b) $E_1 = 10$ V, $e_2 = 5 \sin 377t$. (c) $E_1 = -10$ V, $e_2 = 20 \sin 377t$.

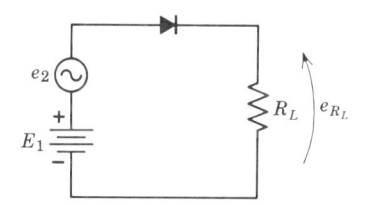

Figure P2-23

2-24 For the conditions in part (a) of Problem 2-23, sketch voltage waveforms for the voltage across R_L, across the diode, and across the sources. In your sketch show that the sum of voltage rises at every instant in time equals zero, that is, that Kirchhoff's voltage law holds for this circuit at all times.

3
APPLICATIONS OF DIODES

It is natural that we should want to learn how P-N junction diodes and vacuum diodes perform in some typical circuit applications. In this chapter we shall see how they are used to control voltages, to protect other circuit components, to act as switches, and to rectify alternating current—that is, to obtain a direct current from an alternating current source. Each of these applications is important in a multitude of electronic circuits, from a simple laboratory voltmeter to complex systems of digital computers and industrial control. The electrical valve-like action of a diode results in current flow in one direction only, which is an important feature of many applications in electronic equipment.

3-1
Diode as a Voltage Limiter (Clipper)

The diode is a useful circuit element in applications requiring voltage levels to remain within prescribed amplitude levels. A circuit that uses a diode as a voltage limiter is illustrated in Fig. 3-1a. This circuit is designed to prevent the output voltage from exceeding the value of the dc source voltage marked E_1. The circuit operation is easy to understand. When diode D_1 is reverse biased by the external source, no current can pass through the diode and the output voltage will follow any variations of the external source. However, when this value reaches the amplitude of $E_1 + V_B$, the diode will be forward biased. Current can then pass through the diode, and the output voltage will be equal to

112   Applications of Diodes   Ch. 3

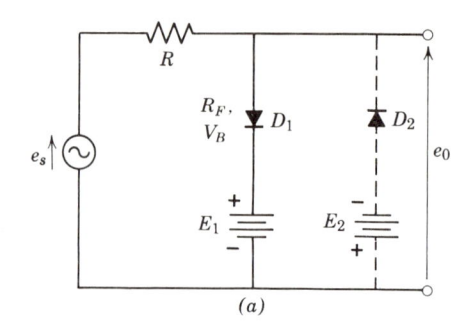

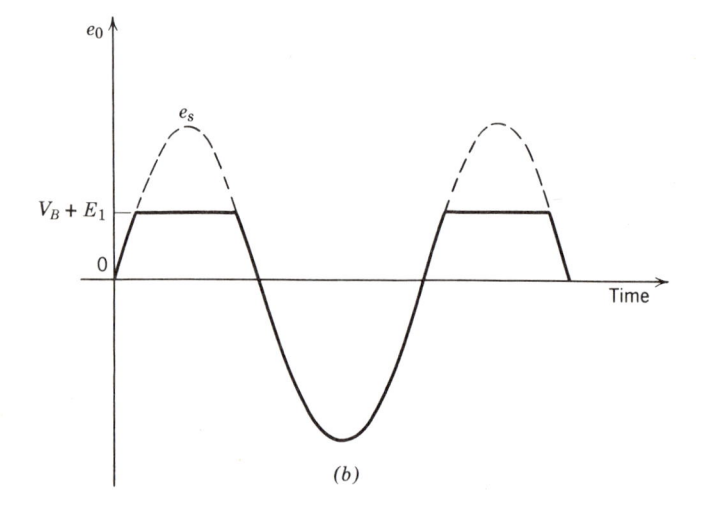

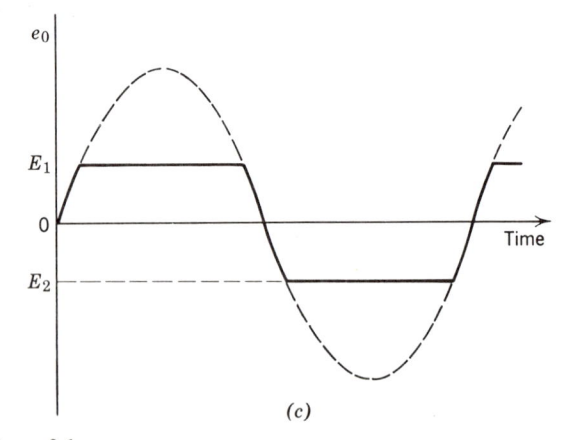

Figure 3-1
Diode voltage limiter (clipper) circuit and waveforms. (*a*) Circuit. (*b*) D_1 operative. (*c*) D_1 and D_2 operative, and $V_B \ll E_1$ and E_2.

$E_1 + V_B$ during periods of forward biasing. Refer to Fig. 3-1b. If the internal voltage drop of the diode is very small compared to the dc source, then the output voltage will be practically limited to the value of E_1.

It is sometimes necessary to limit voltage variations in both positive and negative ranges about some zero reference voltage. The additional circuitry (D_2 and E_2) and waveform of the output voltage are shown in Figs. 3-1a and 3-1c. The diode limiter might be used to alter the waveform of a sinusoidal input voltage. If E_1 and E_2 are chosen equal in magnitude and much smaller than the input amplitude, then the output will be very nearly a square wave of voltage.

Drill Problem

D3-1   Low-power silicon diodes are used in the circuit of Fig. 3-1 to generate a nearly square wave of voltage from an input sinusoidal source of 100 v rms. (a) Assuming the diodes identical and ideal elements, what must be the magnitude of dc limiting sources E_1 and E_2, if the square wave is to be 20 V, peak to peak? (b) What must be the PIV rating of the diodes? (c) What value of R is required to limit diode current to 100 mA?

3-2
Diode as a Voltage Clamper

In many electronics applications, it is necessary or desirable to eliminate either all positive or all negative amplitudes of input voltage waveforms. The circuit that can perform this alteration of an input waveform is called a *clamper circuit*. The diode voltage limiter of Fig. 3-1 can be extended to this application. Simply by making E_1 equal to zero, the positive output voltage will be limited to the internal voltage drop of the diode. The negative voltages will be "clamped" by D_2 and E_2 at the output terminal. If the other voltages in the circuit are very large compared to the internal diode drop, which is usually less than 1 V, then it can be said that the output voltage is *clamped at zero volts*. Only negative voltages will appear at the output. The diode limiter might also be said to be a clamper, where its output is clamped to positive E_1 and negative E_2. References in the literature to the use of diodes in these applications sometimes employ the terms *voltage limiter* and *voltage clamper* synonymously. [The term "limiter" is also employed for a circuit used for a different purpose in frequency-modulated (FM) communications receivers.] Of course, the action of this particular voltage clamper circuit is similar to that of the diode limiter described, the only difference being the amplitude to which the output voltage is restricted.

3-3
Diode as a Logic Circuit Element

Electronic computers have been in use since about 1940, and those of modern design employ thousands of semiconductor diodes that perform logical functions at tremendous speeds. In operation, the diodes change from low-current to high-current conduction states (and the reverse) in a few millionths of a second. Operating as switches in this way, they can be used as circuit elements in the design of functional circuits for logical operations, such as addition and subtraction, in computers. We can learn how such control functions are provided by studying circuits called *switching gates*. They are called gates because a signal current *may* or *may not* pass through them, depending on conditions existing at the time the signal is applied. Another control signal will determine whether all input signal will appear as an output current or voltage from the gate circuit.

In computer circuits it is possible to use diodes to perform certain operations in logic, *involving the thought processes* of AND or OR. Circuits designed to combine signals in these relationships are gates in a general sense and are called AND circuits and OR circuits.

The circuit of Fig. 3-2a is an AND circuit, designed so that its output will be a negative value equal to $-E$ when all inputs are more negative than $-E$. If any one input signal is more positive than $-E$, its associated diode will be forward biased, raising the output voltage at O to equal the input signal. In a practical circuit, the input signals are generated by voltage pulses varying from ground potential to a value more negative than E. The output voltage at O will be equal to $-E$ only when input A AND input B (AND input C, etc.) are negative. Thus, this circuit is referred to as a *negative* AND *circuit*.

The circuit of Fig. 3-2b is called a *negative* OR *circuit*. With the diodes as shown, a negative input *on any terminal* will cause the output to be negative.

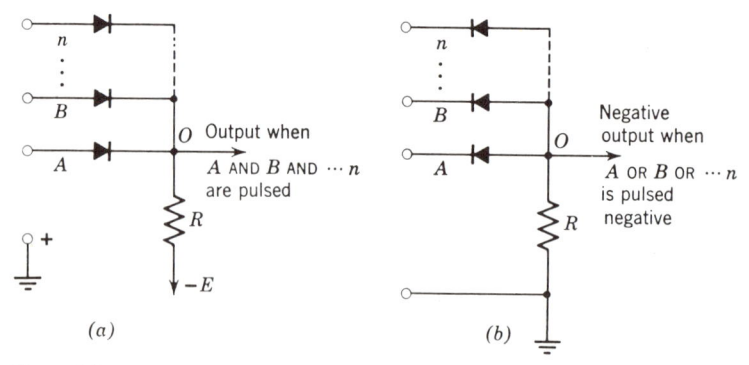

Figure 3-2
Simple diode gating circuits. (*a*) AND circuit. (*b*) OR circuit.

Sec. 3-3   *Diode as a Logic Circuit Element*   115

Thus, input A OR input B (OR input C, etc.) will yield the desired output. All diodes are in the forward conducting state. Note that the output cannot have positive amplitudes, because this would require that the diodes conduct heavily in the reverse direction. As an example of the circuit operation, suppose that a negative pulse (with respect to ground) is applied to one input terminal, say input A. This will make diode A carry more current, which must pass through the load resistance, lowering the potential of the output terminal O from ground potential to some value below ground potential. This reverses the bias on diode B (and other input diodes), putting it in the reverse-conduction state. If more than one input has a signal impressed on it, the circuit behaves as if only one is active. *Any one input signal* will activate the circuit, whereas in the AND circuit *all inputs* were required to yield an output.

Reversing the diodes and reference polarities in these circuits will yield *positive* AND and OR circuits. Whether an output signal should be positive or negative with *respect to a reference ground* will be determined in an application by other factors in their operation. The theory of logic circuit design is well beyond the scope of this book, but an understanding of diode performance in logic circuits should not be too difficult for the student who masters the foregoing material.

A positive AND logic circuit with two inputs is shown in Fig. 3-3. It is a positive AND circuit because the output will be at its most positive value only when positive signals are applied simultaneously to both inputs (A and B). An output signal in this case will be a voltage of about 12 V above ground potential. In general, the variation of the voltage at the output terminal must be large enough to give the desired effect in a circuit or device connected to the output terminal. Any smaller variation that will not accomplish the desired effect

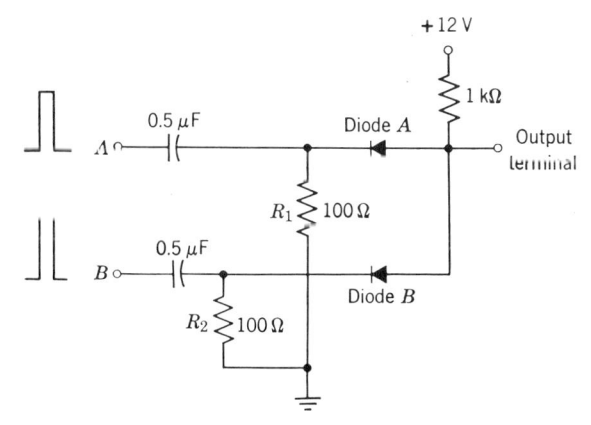

Figure 3-3
Forward conduction (positive) AND circuit with two inputs.

would not be classed as an output signal, practically speaking. The effect desired may be the operation of an electromechanical relay or the magnetization of a bit of metal in the form of a magnetic film or computer memory core.

The circuit is designed so that the diodes operate in the forward-conduction region. Assuming the diodes to have a forward drop of about 1 V, the current through the 1-kΩ resistor causes a drop of about 10 V, making the potential of the output terminal about 2 V above ground. The top terminals of R_1 and R_2 are then at about 1 V above ground. These voltage values are approximate because of the variable forward drop of the diodes and the voltage-dividing action of the resistors in the circuit.

Now suppose that two voltage pulses of short duration (a few microseconds) are applied *simultaneously* to terminals A and B, raising the top terminals of R_1 and R_2 to 12 V above ground. This would *reverse the polarity* across both diodes and thus change their operation to reverse-conduction conditions. The current in the 1-kΩ resistor reduces to practically zero (except for reverse leakage current), causing the output potential to rise to about 12 V, thus providing a positive output voltage or signal.

When only one input signal is applied, the diode in the other input circuit will continue to pass current, but it is not disturbed enough to alter the voltage drop across it or its associated 100-Ω resistor. These two voltages are small compared with the drop across the 1-kΩ load, and their sum will remain at about 2 V as before. If the input signal is applied to input A, for example, only diode A will be changed to the reverse-conduction state. Diode B and its 100-Ω resistor will continue to operate at practically the previous voltage drops, with the result that the output voltage will remain essentially constant at 2 V.

This is more readily seen when we consider that, since the current through diode B and R_2 is essentially unaffected by the signal on input A, their voltage drops do not change appreciably and therefore the output voltage changes very little. Diode B and R_2 are said to *clamp* the output potential to the presignal value, about 2 V in this case. This means there will be no appreciable output signal when only one input is applied. But, as explained above, if both input signals are large enough and applied *in coincidence*, both diodes will be changed from forward to reverse conduction. This will almost completely shut off current in the 1-kΩ resistor, resulting in a substantial rise in the output voltage, from 2 V to 12 V in this case.

The positive OR logic circuit, given in Fig. 3-4, produces an output *when either input circuit* receives a positive signal large enough to reverse bias the diode of the other input circuit. In the circuit as shown, both diodes are forward biased when no input signals are present. Again, assume that the forward drop of the diodes is about 1 V. Notice that, with the 1-kΩ load resistor connected to the negative side of the bias supply, the ground potential is positive, making each diode conduct heavily because of its forward biasing. In this condition, the output terminal will be at about -2 V with respect to ground.

Sec. 3-4   *Voltage Regulation with Zener Diodes*   117

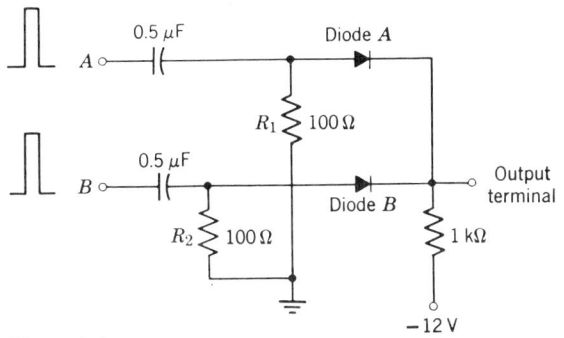

Figure 3-4
Forward conduction (positive) OR circuit with two inputs.

Now assume that a positive pulse (with respect to ground) is applied to one of the input terminals, say terminal A. Current will increase in diode A and in the 1-kΩ resistor, raising the potential of the output terminal *from* −2 V to a value *above ground potential*, i.e., a positive value. This reverses the polarity of diode B, putting it in reverse conduction. The *output signal is the potential increase* at the output terminal.

Drill Problems

D3-2 Assuming ideal diodes in Fig. 3-2a, $R = 1$ kΩ, $-E = -10$ V, how much current flows in R when the output is negative? What is the value of the output voltage under this condition? How much current flows in R when the output voltage is at ground potential?

D3-3 Assume ideal diodes in Fig. 3-2b, $R = 1$ kΩ, and output voltage to be either ground potential or −10 V. How much current flows in R when the output is at ground potential? If two inputs are simultaneously −10 V, what will be the value of output voltage?

3-4
Voltage Regulation with Zener Diodes

The phenomenon of *avalanche breakdown* in diodes was discussed in Section 2-20 in relation to the operation of Zener diodes. A device in which a sudden change from low-current conduction to high-current conduction can be produced is obviously a useful tool for switching operations. This is easily accomplished in a semiconductor diode whose characteristic permits avalanche breakdown, called Zener operation. It should be particularly noted that *for Zener operation, the diode must be biased for reverse conduction.*

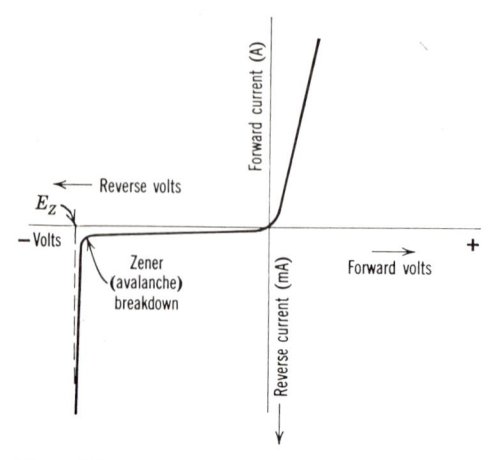

Figure 3-5
Silicon diode curve showing critical Zener voltage E_z, and almost constant voltage for large change in current in Zener region.

Figure 3-5 is presented to show the behavior of a silicon Zener diode in the reverse-bias region of its characteristic. The value of voltage at which Zener breakdown occurs is called the *Zener point*, designated E_z in the diagram. (Refer also to Fig. 2-17.)

Zener diodes are available with their avalanche–conduction region at various values of Zener points. For example, one manufacturer makes diodes with Zener points at the following approximate voltage values: -4.7, -5.6, -6.8, -8.2, -10, -12, -15. Referring to Fig. 3-6, these diodes are represented by the curve designated by the diode type number 1Z followed by the E_z value. For example, 1Z15 has $E_z = -15$ V. It is shown that not only does the current increase abruptly when the Zener voltage is approached but also the voltage across the diode remains almost constant even though the current through it is allowed to increase substantially. This feature can be exploited in circuits requiring voltage regulation.

A very simple voltage regulator circuit using a Zener diode is shown in Fig. 3-7. Assume nothing is connected to the output terminals. The current-limiting resistor, R, should be chosen so that the reverse (Zener) current in the diode will not exceed the safe operating value when the fluctuating dc voltage, applied at the left, is at its maximum value.

Now, when a load is connected to the output terminals it will take current, but the voltage drop across the diode will not change. Keeping this fact in mind, let us consider two conditions: first, the applied dc voltage does not change when the output current is flowing, and second, the applied dc voltage decreases with increase in load current.

Sec. 3-4 Voltage Regulation with Zener Diodes 119

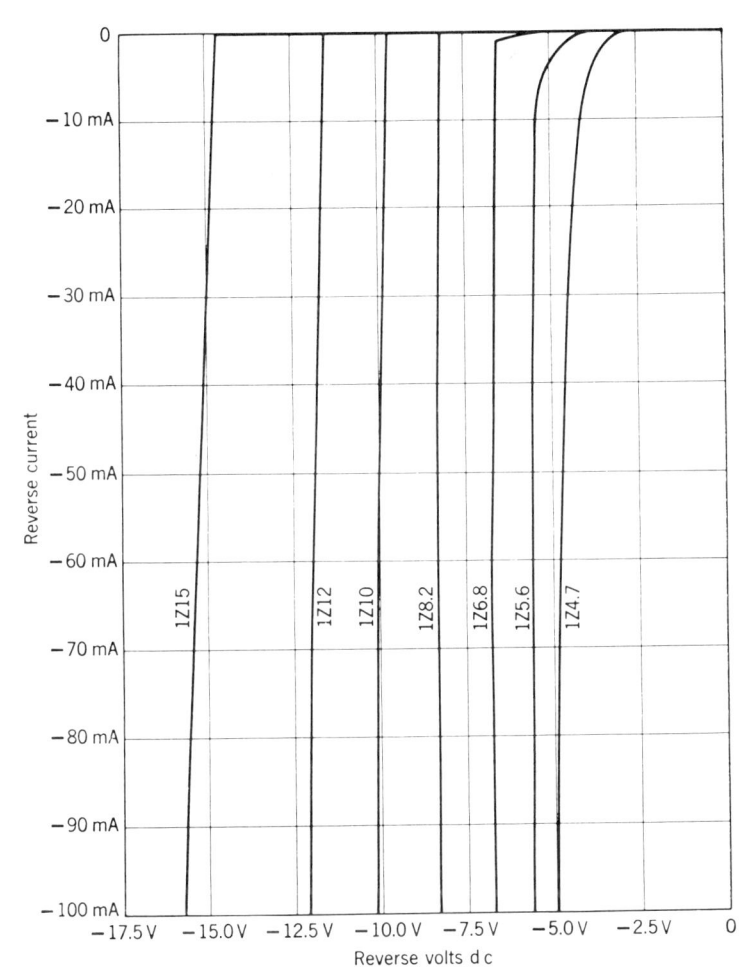

Figure 3-6
Reverse-current curves of typical Zener diodes. (Courtesy of International Rectifier Corp., El Segundo, Calif.)

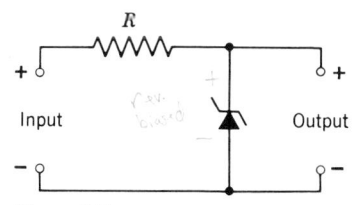

Figure 3-7
Simple voltage regulator using a Zener diode.

120   Applications of Diodes   Ch. 3

Table 3-1
Zener Voltage and Power Dissipation Ratings for Commercial Zener Diodes

Zener Voltages (V)				Power Dissipation
2.7	10.5	23.5	68	150 mW
3.9	11	24	75	200
4.7	11.7	27	82	250
5.6	12	30	91	400
6.2	12.7	33	100	500
6.35	13	36	110	600
6.8	14.5	39	120	750
8.2	15	40.8	130	1 W
8.3	15.7	43	150	1.5
8.4	16	47	160	1.6
8.7	18	49.6	180	3.5
9	19	51	200	5
9.1	20	56		10
10	22	62		50

Adapted from Rufus P. Turner, *Diode Circuits*, Howard W. Sams & Co., Inc., Indianapolis, Ind. 1967.

With unchanging input voltage, a constant voltage across the diode means a constant voltage across the series resistor R, and therefore a constant input current. A demand for output current or for an increase in output current will be accommodated by the diode without a change in the voltage. With a decreasing input voltage, the current through R will decrease, and the diode will still maintain a constant voltage across the load, but it will be carrying less current than in the first condition because it will "give up" some of its current as before.

Two or more diodes in series will provide larger values of regulated output voltage. Each must be capable of carrying the current required. Refer to Table 3-1 for voltage and power ratings of commercial Zener diodes. The power rating is another statement of the maximum current capability of the Zener diode: $P_{max} = E_Z I_{max}$.

3-5
Instrument Protection

The delicate moving element of a meter may be protected by a silicon diode arranged as shown in Fig. 3-8. Silicon diodes have a characteristic like the

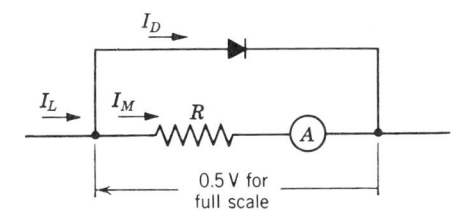

Figure 3-8
Meter protection by a silicon diode.

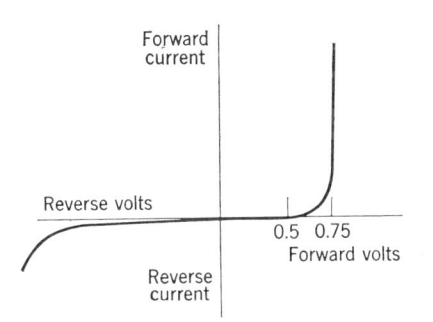

Figure 3-9
Silicon diode characteristic.

curve in Fig. 3-9. As long as the line current I_L does not exceed the amount that gives full-scale deflection of the meter, the current through the diode is zero. At higher values of line current, the voltage drop across the diode increases and it begins to conduct. The diode's characteristic is such that its forward resistance will be small enough to prevent more than 50 per cent overload current going through the meter.

3-6
Half-Wave Rectification

An electronic tube conducts current only when its anode (plate) is electrically positive with respect to its cathode. When an alternating voltage is used to supply the plate current of a tube, the current will flow only during the positive half-cycles because the plate cannot furnish electrons to flow to the cathode during negative half-cycles of the alternating voltage. A transistor behaves in much the same way, requiring a dc power supply for its operation in useful circuits.

In order that a transistor may conduct current continuously, without being shut off periodically, the power supply must furnish a direct voltage that does

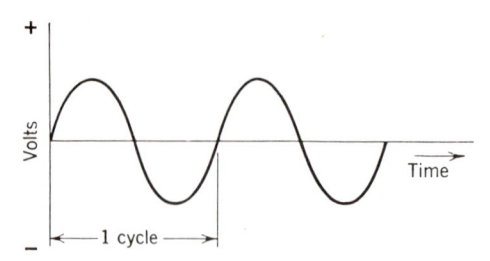

Figure 3-10
Sine curve representing an alternating voltage.

not become zero. Because all commercial electricity is available at alternating voltage only, it must be *rectified* in order to supply transistors with continuous dc power. The following sections consider the operation of crystal diodes and diode tubes used as rectifiers in *power supplies* needed in electronic circuits. Various rectifier circuits will be studied to see how they convert ac voltages and currents into dc voltages and currents. The part that *smoothing filters* play in their operation will be discussed.

We have observed that during each cycle an alternating voltage, as shown by the sine curve of Fig. 3-10, rises to a maximum value, falls to zero and reverses, reaches a maximum in the reverse direction, and then returns to zero. When a voltage with this waveform is applied to a resistor, the current flowing in the resistor will be alternating and of the same waveform as the voltage.

When an alternating voltage as represented in Fig. 3-10 is applied to a diode rectifying element (crystal diode or diode tube) connected in series with a resistor, the variations in the voltage across the resistor (and the current through it and the diode) are as shown in Fig. 3-11. The waves are drawn on different vertical scales. The diode acts as a *half-wave rectifier* by allowing current to flow only during the positive half of each alternating-voltage cycle.

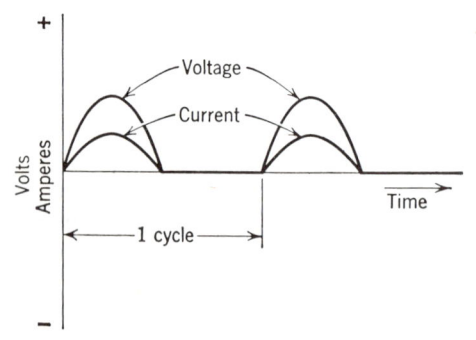

Figure 3-11
Rectified voltage and current through load resistor.

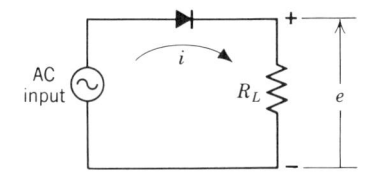

Figure 3-12
Half-wave rectifier circuit.

Current can flow and thus establish voltage across the resistor (Fig. 3-12) only during the positive half-cycles when the anode is more positive than the cathode. During the negative half of each cycle the anode is negative with respect to the cathode and the diode will not conduct because of its reverse bias.

3-7
The Nature of Rectifier Output Voltage

If a dc milliammeter is connected in the half-wave rectifier circuit of Fig. 3-12, it will show a steady reading even though pulses of current (see Fig. 3-11) are flowing through it. One current pulse occurs in each cycle. It is interesting that the reading of the meter will be only 0.318 of the maximum value to which the current rises in each pulse; for example, if the maximum value of the current supplied by the half-wave rectifier is 10 mA, the dc milliammeter will read 3.18 mA. This can be verified by Fourier analysis of the current pulse, a half-sine wave, and assuming no voltage drop across the diode.

The mathematical expression representing the complete half-sine wave of voltage during one cycle of the input signal can be determined as

$$e = 0.318 E_m + 0.5 E_m \sin 2\pi f t - 0.212 E_m \cos 2\pi(2f)t \\ - 0.0424 E_m \cos 2\pi(4f)t - \cdots \qquad (3\text{-}1)$$

It is important to understand that each term of Eq. 3-1 represents a component of voltage and that *all these components exist simultaneously.* They add together from instant to instant, producing the half-sine wave as their total effect.

A half-sine of voltage is found to be made up of several components. (1) an average, dc component with a constant value 0.318 of the maximum value, (2) an ac component with a maximum value equal to 0.5 of the original half-sine-wave maximum and the same frequency as the unrectified sine wave, (3) an ac component of twice the fundamental frequency with a maximum value equal to 0.212 of the original sine-wave maximum, and (4) many high-frequency components of decreasing amplitudes. The dots following the fourth term of Eq. 3-1 indicate additional terms of lesser importance.

124 Applications of Diodes   Ch. 3

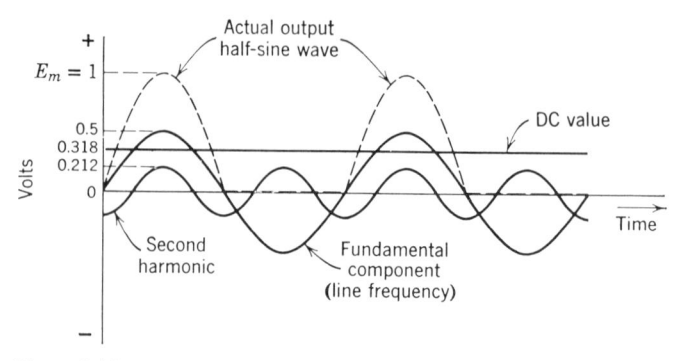

Figure 3-13
Components of half-sine wave.

Figure 3-13 shows the dc fundamental frequency, and second harmonic components in the half-sine wave of a half-wave rectifier. Harmonics higher than the second are not shown because of their small amplitudes. The vertical distance from any point on the horizontal axis to the dashed half-sine curve can be approximated by adding together the vertical distances from that point to the component graphs. Due regard must be paid to the negative sign when distances are measured downward from the horizontal axis. The contribution made by the fourth harmonic (only one-fifth as large as the second harmonic) is negligible. Contributions from higher harmonics will be correspondingly smaller.

Drill Problems

D3-4 Show by graphical addition that the Fourier components of a half-sine wave add together to yield the curve as their total effect.

D3-5 A germanium diode is rated for $I_F = 100$ mA and a maximum forward current of 350 mA. It is to be used in a half-wave rectifier circuit supplied by a sinusoidal source of 100 V rms. (*a*) What value of load resistance can be used to receive maximum power without exceeding the diode ratings? (*b*) Calculate dc load voltage and current when the load resistance of (*a*) is used.

3-8
Output Current and Power of a Half-Wave Rectifier Circuit

It can be shown mathematically that the peak instantaneous value of the current wave through the load resistance is given by

$$I_m = \frac{E_m}{R_F + R_L} \text{ A} \qquad (3\text{-}2)$$

where E_m is the peak instantaneous value of the ac input voltage and R_F is the resistance of a crystal diode.

The voltage impressed upon R_L is represented by Eq. 3-1, the *first term* of which is the *dc component*. Since $0.318 = 1/\pi$, it may be written E_m/π. The direct current in the load is then

$$I_{dc} = \frac{1}{\pi} \frac{E_m}{R_L} = \frac{I_m}{\pi} \text{ A} \qquad (3\text{-}3)$$

and the dc power in the load is

$$P_{dc} = I_{dc}^2 R_L = \left(\frac{I_m}{\pi}\right)^2 R_L = \frac{E_m^2 R_L}{\pi^2 (R_F + R_L)^2} \text{ W} \qquad (3\text{-}4)$$

Power is lost in the diode because the current must flow through R_F, the diode resistance. The *effective value* of the *half-sine-wave current* can be shown to be (over the full period)

$$I_{rms} = I_m/2$$

so that the ac power in the circuit is given by

$$P_{ac} = \left(\frac{E_m}{2(R_F + R_L)}\right)^2 [R_F + R_L] = \frac{E_m^2}{4(R_F + R_L)} \text{ W} \qquad (3\text{-}5)$$

The *efficiency of power conversion* is then obtained by dividing the dc power by the ac power. To get the expression for efficiency, we divide Eq. 3-4 by Eq. 3-5 and get

$$\text{Power conversion efficiency} = \frac{R_L}{R_F + R_L} \times \frac{4}{\pi^2} (100)\% \qquad (3\text{-}6)$$

If the diode resistance is zero,

$$\text{Maximum theoretical power conversion efficiency} = \frac{4}{\pi^2} \times 100 = 40.6\%$$

The peak value of current that the diode must be able to pass safely is, from Eq. 3-3,

$$I_{peak} = I_m = \pi I_{dc} \text{ A} \qquad (3\text{-}7)$$

The *peak-inverse voltage* to which the diode is subjected *is equal to the peak of the ac voltage applied to the circuit*. This means that when the diode is not conducting (during negative half-cycles), the full voltage is impressed across the diode in reverse polarity. Should the diode conduct when the polarity is reversed, its rectifying properties would fail and the diode might be destroyed.

126 Applications of Diodes   Ch. 3

Data given by manufacturers include the safe value of peak-inverse voltage and also ratings of average and peak current. *Peak-inverse voltage* is defined as the maximum instantaneous voltage applied in the direction opposite that of normal current conduction.

After the full-wave rectifier has been described, a comparison of the advantages of the two types will be made.

Drill Problem

D3-6 A half-wave rectifier is supplied from a 60Hz, 120-V rms source. A diode having internal resistance of 500 Ω is used to supply dc power to a load resistance of 2 kΩ. (*a*) Calculate the maximum value of diode current; of load current. (*b*) What is the dc power in the load? (*c*) What is the rms value of the output current? (*d*) Calculate the power conversion efficiency. (*e*) What must be the PIV rating of the diode?

3-9
Full-Wave Rectification

A full-wave rectifier circuit and the waveforms of rectified voltage and current in a resistance load are shown in Figs. 3-14 and 3-15. This circuit is called full-wave because it utilizes the full wave of input voltage to supply power to the load. Each diode is *forward biased* on alternate half-cycles of the input voltage, and each conducts current through the load in the same direction. Thus, the full wave of input voltage is rectified and produces two half-sine waves of load voltage and current of each cycle of input sine wave.

Fourier analysis of the load voltage waveform shows that it may be expressed in this form:

$$e = 0.636 E_m - 0.424 E_m \cos 2\pi(2f)t - 0.085 E_m \cos 2\pi(4f)t - \cdots \quad (3\text{-}8)$$

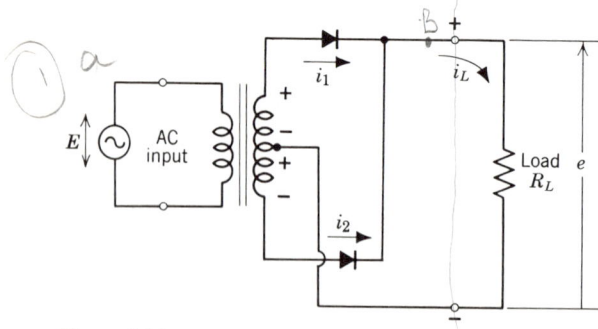

Figure 3-14
Full-wave rectifier circuit.

Sec. 3-9   Full-Wave Rectification   127

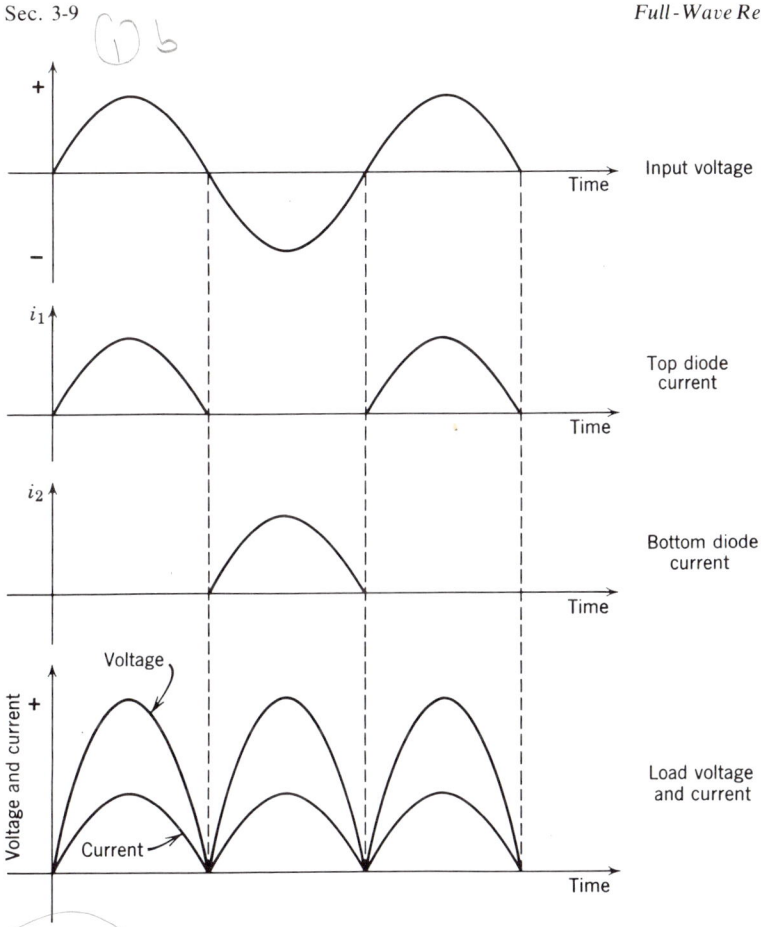

Figure 3-15
Voltage and current waveforms in full-wave rectifier with resistance load.

Note that the dc component is twice as large as that produced by the half-wave rectifier. This should be expected because there are two half-sine-wave portions of the voltage during each input cycle and their average value is twice that of one half-sine wave during the same time. A result that might not be expected is the *absence of a component with input frequency*. Only its second and higher harmonic components appear in the load resistance. The second harmonic and high-frequency terms are twice their values in the half-wave case. This rapid reduction in the ac components (the magnitude of the third term is only $0.085 E_m$) helps to improve the power-conversion efficiency and reduces the amount of filtering necessary to get a constant dc power for the load. It therefore cuts down the size and cost of filters.

128   Applications of Diodes   Ch. 3

If an *unfiltered* rectified sine wave is impressed on a pure resistance load, the load will *simultaneously* carry dc current and ac components at the frequencies given by Eq. 3-1 or 3-8. The peak value of each component of current will be given by the coefficient of the corresponding sine or cosine term divided by the total circuit resistance, $R_F + R_L$.

The components of the full-wave rectifier output are shown in Fig. 3-16. Waves for terms of the expression that represent higher harmonics have been neglected.

Equations for current, power, and efficiency will now be obtained. As explained above, and also from Eq. 3-8

$$I_{dc} = 0.636 \frac{E_m}{R_L} = \frac{2}{\pi} I_m \text{ A} \tag{3-9}$$

so that the dc power in the load is given by

$$P_{dc} = 4\left(\frac{I_m}{\pi}\right)^2 R_L \text{ W} \tag{3-10}$$

The effective value of the fully rectified sine wave of current is equal to the instantaneous maximum divided by $\sqrt{2}$:

$$I_{rms} = \frac{I_m}{\sqrt{2}} \text{ A}$$

The ac power is

$$P_{ac} = \frac{I_m^2}{2}(R_F + R_L) \text{ W} \tag{3-11}$$

Dividing Eq. 3-10 by Eq. 3-11 gives the power conversion efficiency:

$$\text{Efficiency} = \frac{8}{\pi^2}\left(\frac{R_L}{R_F + R_L}\right) \times 100\%$$

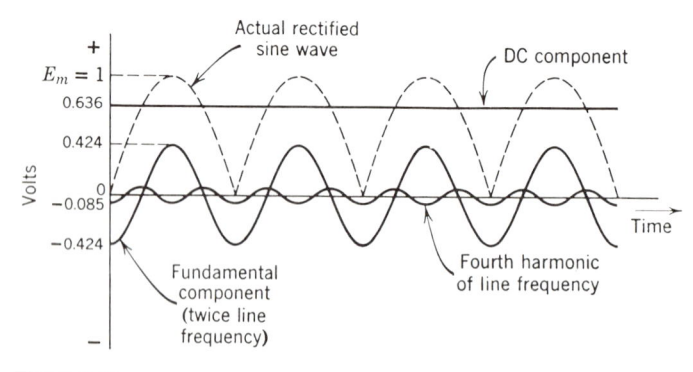

Figure 3-16
Components of full-wave rectifier output.

Sec. 3-10 **Bridge Rectifier Circuit** 129

The maximum theoretical power conversion efficiency is again found by assuming $R_F = 0$:

$$\text{Maximum theoretical efficiency} = \frac{8}{\pi^2} \times 100 = 81.2\%$$

The peak value of current that the diode must be able to pass safely with resistance load is, from Eq. 3-9,

$$I_{peak} = I_m = \frac{\pi}{2} I_{dc}$$

The peak-inverse voltage on the diode that is not conducting is twice the peak value of the input ac voltage, *provided the voltage drop through the conducting diode is assumed to be zero.* Thus

$$\text{Peak-inverse voltage} = 2E_m = 2\sqrt{2} E_{rms} = 2.82 E_{rms}$$

Although R_F is certainly not zero, the possible peak-inverse voltage applied to each diode on alternate half cycles when the load current is small (and it can be zero if the load circuit is open) should be considered to be *twice* the instantaneous peak value of the applied ac voltage.

Drill Problem

D3-7 Repeat D3-6 for two diodes in a full-wave rectifier circuit, assuming the transformer supplies each diode with 120-V rms input. Compare the results with those for the half-wave rectifier.

3-10
Bridge Rectifier Circuit

Four diodes may be connected in a bridge circuit as shown in Fig. 3-17 to achieve full-wave rectification. This type of circuit is widely used because it

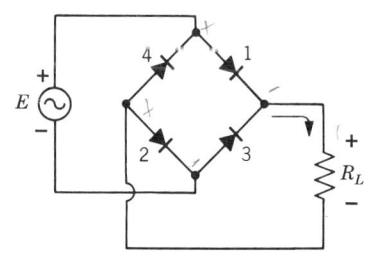

Figure 3-17
Bridge-type rectifier circuit.

does not need a center-tapped transformer to direct diode currents to the load. It is well adapted to semiconductor rectifiers but is difficult to construct using tube-type rectifiers, because their cathodes would be at different potentials and they could not be connected in parallel to a single filament supply. The full-wave bridge rectifier is widely used in industrial electronic applications.

An interesting application of the full-wave bridge rectifier circuit is found in the ac voltage section of many laboratory multimeters. In order to use the D'Arsonval meter, which responds to dc currents, to measure ac voltages, it is necessary to rectify the ac signal. The resultant pulsating current can then be used to drive the meter movement. If the frequency of the ac signal is high enough, the mechanical inertia of the meter will cause it to respond to the average current through it, indicating a constant value on its scale where the pointer rests. This constant value can be marked on the scale to represent the rms value of a sine wave signal, because the average current and the rms voltage are both related to the amplitude of the input sine wave:

$$I_{avg} = \frac{0.636 E_m}{R_{series}} \quad \text{and} \quad E_{rms} = 0.707 E_m$$

The operation of the circuit in Fig. 3-17 is easy to understand. When the source E has the polarity shown, diode 1 will pass current through the load and back to the source through diode 2. When the source changes polarity, diode 3 and diode 4 pass current to the load, which goes through the load in the same direction as before. Thus, the circuit behaves as a full-wave rectifier.

All of the relationships developed in Section 3-9 apply to the bridge rectifier circuit. Diodes used in the circuit must be chosen to withstand the applied voltage in the reverse-bias mode.

3-11
Power Transformer

In many electronic circuits it is necessary to have dc voltages that are greater than may be obtained by rectifying ac line voltage. In other cases it may be desirable to have voltages that are lower in amplitude. For these cases a power transformer is often used to increase or decrease the line voltage before rectification takes place. It is also convenient to reduce the ac line voltage for supplying the filaments of vacuum tubes used in a circuit. A properly designed power transformer can easily step up or step down ac voltage to any desired practical value.

The transformer is composed of at least two windings of conducting wire wound on an iron core and insulated from the core and from each other. When a voltage is applied to one of the windings, voltages are induced in the other windings. The windings usually have different numbers of turns of wire. The

Sec. 3-11   Power Transformer   131

number of turns of wire in a given winding determines the magnitude of voltage that is induced in it.

In order to step up a voltage of 115 to 575 (5-to-1 ratio), it is necessary to use a transformer that has five times as many turns in one winding as in the other. The 115-V ac supply is connected to the winding with the smaller number of turns, and 575 V alternating current is obtained at the ends of the other winding. The winding with the smaller number of turns is called the low-voltage winding, and the winding with the larger number of turns is called the high-voltage winding.

To use a power transformer in a full-wave rectifier circuit, it is necessary that the high-voltage winding be center-tapped, as shown in Fig. 3-14. Most power transformers have at least one winding, of relatively few turns of wire, that supplies filament (or heater) current for the cathode. The voltage of the filament winding may be 12.6, 6.3, 5, 2.5 V, or some other low value. The most commonly used values with small rectifier tubes are 6.3 and 5 V. These low-voltage windings are made of comparatively heavy wire because filament currents are usually much greater than the plate currents supplied by the high-voltage winding. Transformers are built to operate at a definite voltage on each winding. In the case described above, the transformer could be used to "step-down" 575 V to 115 V.

Consider a transformer that has a rated voltage applied to one of its windings, the primary, while the other winding, the secondary, has nothing connected to it; that is, the secondary is "open." An *exciting current*, being alternating in nature, flows in the primary and sets up an alternating magnetic field in the iron core. That is, the magnetic field increases to maximum strength in phase with the current, falls to zero, and then builds up strength in the reverse direction in step with the current.

This changing flux induces a voltage in the secondary winding of the transformer. The induced voltage will be proportional to the number of turns. Letting the subscript 1 denote the primary winding and 2 the secondary winding, and using the letter N to denote the number of turns, we may write

$$\frac{E_2}{E_1} = \frac{N_2}{N_1} \qquad (3\text{-}12)$$

When the second winding of a transformer, usually called the secondary, is closed through an appropriate load, current will be supplied to the load, and the current will flow in the secondary winding. This current produces a magnetic field that opposes the magnetic field produced by the current put into the first winding, which is called the primary winding. This opposing field builds up in strength as the current in the secondary winding builds up. The reduction in the net magnetic flux results in an increase in current flowing into the primary from the source of ac supply.

We readily see why the input current to the primary of the transformer must increase when the secondary is connected to a load. The reason is that the power taken by the load must come from somewhere, and the only place is the source of supply serving the primary. In order to get more power into the primary, the current must increase since the voltage is constant. The flux produced by the secondary current opposes the flux produced by the primary current, thus reducing the counter EMF of self-induction of the primary winding. This results in an increase of primary current. The CEMF of self-induction determines the value of the current in the primary winding. The smaller the CEMF, the larger the primary current.

The currents in the windings are related to the turns in the following manner:

$$\frac{I_2}{I_1} = \frac{N_1}{N_2} \quad (3\text{-}13)$$

$$N_1 I_1 = N_2 I_2 \quad (3\text{-}14)$$

Equation 3-14 shows that the ampere turns (NI) of the primary are equal to the ampere turns of the secondary.

From Eq. 3-12 we obtain

$$\frac{N_1}{N_2} = \frac{E_1}{E_2} \quad (3\text{-}15)$$

Substituting Eq. 3-15 into Eq. 3-13, we get

$$\frac{I_2}{I_1} = \frac{E_1}{E_2}$$

$$E_1 I_1 = E_2 I_2 \quad (3\text{-}16)$$

Equation 3-16 states that the primary voltamperes equal the secondary voltamperes. Equations 3-15 and 3-16 are not exact because they neglect such things as leakage flux, magnetizing component of current, and power losses in the core and in the windings. They are sufficiently exact for practical purposes, however.

Power transformers for use in dc power supplies for transistor circuits likely will have a stepdown ratio for voltages. Transistors often operate at voltage levels well below ac power-line voltage at 115 Vac, some as low as 5 Vdc.

3-12
Comparison of Full-Wave and Half-Wave Rectifiers

When compared with the half-wave rectifier, the full-wave rectifier has the following advantages:

1. It is capable of considerably higher efficiency (theoretically double).

2. Its dc output power is four times as great, for a given load resistance, because the dc component of load current is twice as large.
3. Output requires much less filtering because the ac components of output current are, on the whole, much smaller in amplitude.
4. If a transformer must be used, a smaller one may be used for a given power output. This is deduced from Eqs. 3-4 and 3-10. By equating them for equal power conditions, it can be seen that the required transformer current for full-wave rectification is only one-half for half-wave rectification.
5. The dc component of load current produces an alternating magnetic flux in the transformer because it flows in opposite directions in the two halves of the secondary windings. This does not polarize the core, whereas it is polarized by the unidirectional flux in the half-wave case. This is another reason for using a smaller transformer in the full-wave circuit.

The peak-inverse voltage applied to the diode in a half-wave rectifier is equal to E_m, the maximum instantaneous value of the input voltage. In the full-wave rectifier, the diodes are subjected to twice this amount of peak-inverse voltage, i.e., $2E_m$.

The half-wave rectifier has an advantage, if the ac voltage need not be stepped up or down, in that its circuit does not require a center-tapped transformer. Also, in applications that must operate on either alternating or direct current, as in some small radio receivers, the half-wave rectifier is useful whereas the full-wave rectifier is not. Half-wave rectifiers have application also in television receiver circuits.

3-13
Smoothing Filters

It is possible to reduce greatly the amplitudes of the ac components in the output voltage of a rectifier by means of one or two sections of a *smoothing filter*. Its action provides a path from the rectifier output to the load for dc components and blocks the ac components or short-circuits them around the load. Since the ac components are eliminated from the load current, only the dc component produces a voltage across the load.

Smoothing filters used with rectifiers that supply power to electronic tubes and transistors are usually made of iron core inductances and capacitors. In circuits where the load currents are very small, or if the load current need not be extremely smooth, only capacitors and resistors may be used in the filter. When the load voltage has only a small amount of ac voltage, its variation about the average (dc) value is called *ripple voltage* or simply *ripple*.

A filter usually takes one of three forms: (1) shunt-capacitor, (2) *L*-section, and (3) π-section. A shunt-capacitor filter is just what its name implies, a capacitor connected in parallel with the load resistor. The *L*-section and

134 Applications of Diodes   Ch. 3

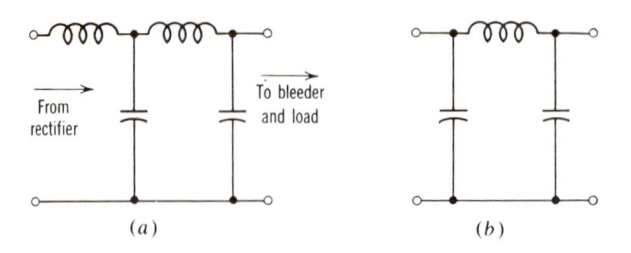

Figure 3-18
Choke-input and capacitor-input smoothing filters. (*a*) Tandem *L*-section. (*b*) π-section.

π-section filters derive their names from their appearance in a schematic diagram of the circuit. Typical filter circuits are shown in Fig. 3-18.

The tandem *L*-section filter is called a choke-input filter because the current from the diodes in the rectifier first encounters a *choke coil*, which is an inductance coil wound on an iron core. A choke-input filter often has only one *L*-section rather than two in tandem as shown. Of course, two sections will do a better filtering job, but the extra cost of circuit elements may be prohibitive in some applications and may be considered an overdesign of the circuit.

The π-section filter is called a capacitor-input filter for obvious reasons. Its operation results in a higher voltage at the load than would be obtained from a choke-input filter. The first (input) capacitor charges up to the peak of the rectifier output voltage and does not discharge very much during the remainder of each cycle while the rectifier input voltage is falling and going through its negative half-cycle. The nature of the voltage at the input and output of the capacitor-input smoothing filter is shown in Fig. 3-19.

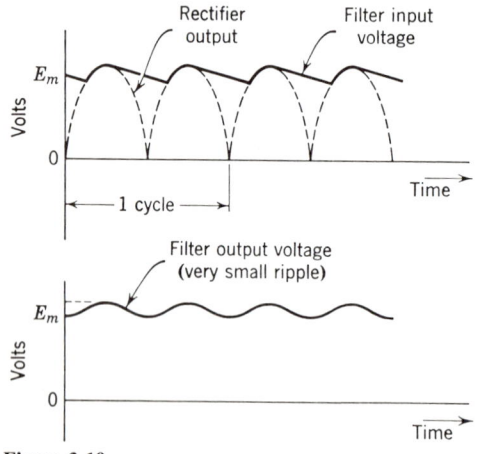

Figure 3-19
Curves for capacitor-input smoothing filter.

3-14
Shunt-Capacitor Filter

A simple filter to remove the ac ripple components in the output of a rectifier is obtained by shunting a capacitor across the load resistance, as shown in Fig. 3-20 for a half-wave rectifier. If the value of capacitance is chosen so that its reactance at the fundamental frequency is much less than the value of load resistance, the ac components will have a low-reactance path around the load resistor. Only a small ac current flows through the load and produces a small ripple voltage.

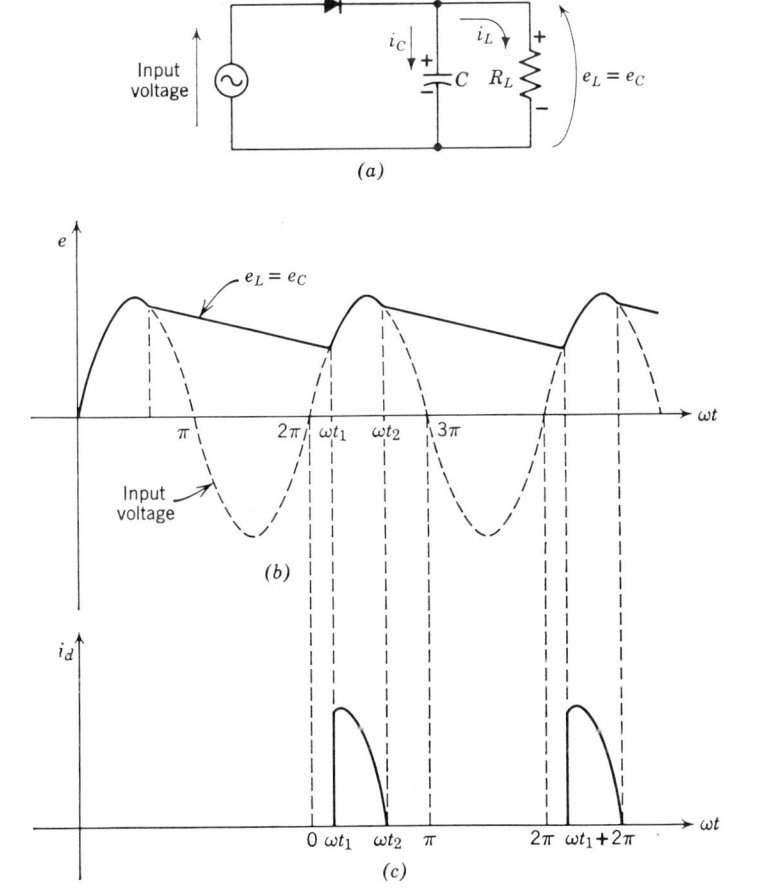

Figure 3-20
Half-wave rectifier with shunt-capacitor filter and load resistance; circuit waveforms. (a) Circuit. (b) Voltage waveforms. (c) Diode current pulses.

The action of the capacitor can be explained as follows. During the time when the rectifier output voltage is rising, the capacitor will charge to a voltage equal to the rectifier output. As the rectifier output decreases, the capacitor discharges through the load resistor. The load voltage waveform meanwhile will be an exponential curve determined by the time constant $R_L C$. When the rectifier output again rises to equal the voltage remaining on the capacitor, the diode will again be forward biased and will immediately start conducting a current equal to the existing current in the load resistor. This can easily be understood, because when the rectifier output voltage equals the voltage on the capacitor, the capacitor will be neither charging nor discharging. But the load resistor has some value of current in it. Kirchhoff's current law must hold, so the diode will pick up at this time whatever current exists in the load resistor.

The diode delivers a current pulse during each cycle to charge the capacitor, and acts as a switch to disconnect the load from the rectifier. If the discharge time constant is long compared with one period of the input voltage, the load voltage will be nearly constant.

The current pulses supplied by the rectifier will have a waveform determined by the time constant of the filter and load resistance. We shall now develop a simplified expression for the diode current pulses in terms of the charging interval, ωt_1 to ωt_2, and the time constant of the filter.

Referring to Fig. 3-20, when the capacitor is charging, the voltage across it equals the input voltage (assuming no forward drop in the diode), or

$$e_C = E_m \sin \omega t \qquad \text{when } \omega t_1 < \omega t < \omega t_2 \qquad (3\text{-}17)$$

During this same time interval, the diode current equals the sum of the capacitor and resistor currents,

$$i_d = i_C + i_L$$

Since the load current can be written as

$$i_L = (E_m/R_L) \sin \omega t \qquad (\omega t_1 < \omega t < \omega t_2)$$

and the capacitor current is

$$i_C = C \frac{de_C}{dt} = \omega C E_m \cos \omega t$$

then the diode current is a pulse having a form

$$i_d = E_m \left[\left(\frac{1}{R_L} \right) \sin \omega t + \omega C \cos \omega t \right] \qquad (\omega t_1 < \omega t < \omega t_2) \qquad (3\text{-}18)$$

When the input voltage starts to decrease, a magnitude will be reached at which the capacitor discharge cannot follow the decreasing voltage. At this point, ωt_2, the diode current will cease and the capacitor current will equal the load current in magnitude, and

Sec. 3-14 Shunt-Capacitor Filter 137

$$\left(\frac{1}{R_L}\right) \sin \omega t_2 = -\omega C \cos \omega t_2$$

From this,

$$\omega t_2 = \tan^{-1}(-\omega R_L C) \tag{3-19}$$

Equation 3-18 may be written as

$$i_d = \frac{E_m}{R_L} \sqrt{1 + \omega^2 R_L^2 C^2} \sin(\omega t + \phi)$$

where $\phi = \tan^{-1}(\omega R_L C) = \pi - \tan^{-1}(-\omega R_L C) = \pi - \omega t_2$. Making this substitution for ϕ and employing trigonometric identities, yields this simplified expression for the diode current pulses:*

$$i_d = \frac{E_m}{R_L} \sqrt{1 + \omega^2 R_L^2 C^2} \sin(\omega t_2 - \omega t) \tag{3-20}$$

It should be noted that Eq. 3-20 is valid only during the charging interval, ωt_1 to ωt_2. The diode current is zero at other times.

Since the waveform of diode current depends on the factor $\omega R_L C$, which includes the input frequency and time constant of the filter circuit, it should be expected that the current waveform will be variable and highly dependent on this factor. Figure 3-21 shows some typical magnitudes and waveforms of diode current.

It can be seen from Fig. 3-21 that the conduction angles for the diode are also

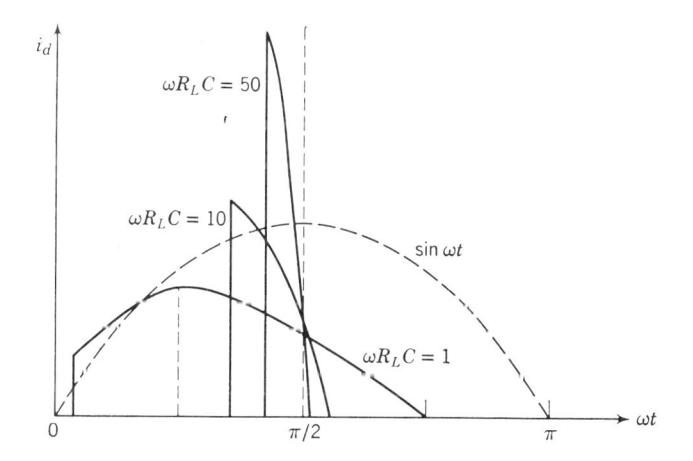

Figure 3-21
Waveform of diode current pulses for different values of $\omega R_L C$. (Curve for $\omega R_L C = 1$ is exaggerated to show its sine-wave shape.)

* After substituting, let $\omega t - \omega t_2 = \alpha$. Show that $\sin(\alpha + \pi) = \sin(-\alpha)$.

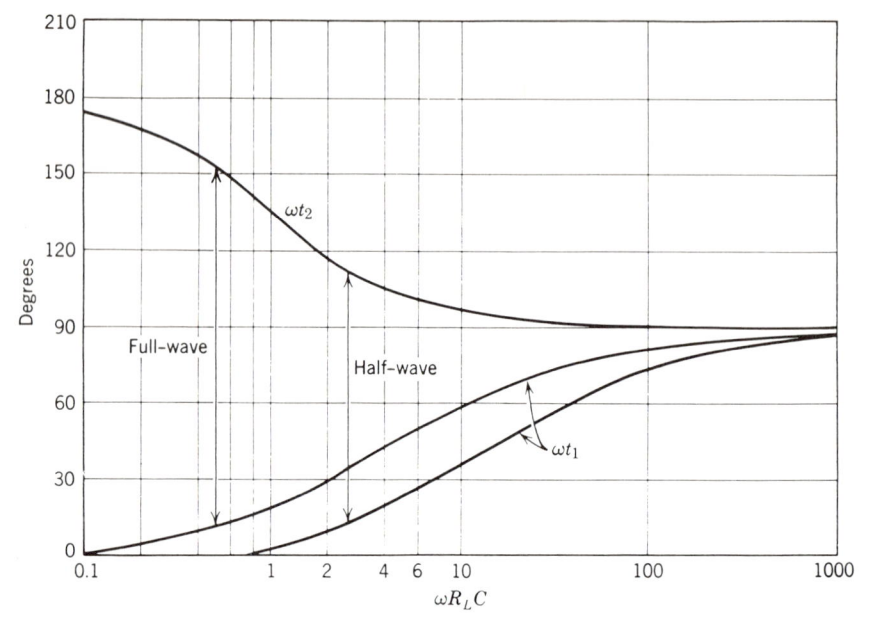

Figure 3-22
Conduction angles for shunt-capacitor filter on half-wave and full-wave rectifiers.

functions of the factor $\omega R_L C$. The relationship of $\omega R_L C$ to ωt_1 and to ωt_2 can be calculated, but the procedure is beyond the scope of this book. It can be shown graphically, as in Fig. 3-22, for both the half-wave and full wave rectifiers.

It should also be noted that the maximum amplitude of the diode current pulse increases rapidly as the factor $\omega R_L C$ increases. This means that when the load voltage has a small ripple the diode current may take on very large amplitudes. Then it is important to design the filter so that the maximum diode current rating will not be exceeded in the operation of the rectifier.

When the conduction angle is small, the inverse voltage on the diode will be approximately twice the peak input voltage. Therefore, PIV rating of the diode must be equal or greater than twice the peak input voltage.

Drill Problems

D3-8 A diode having negligible forward resistance is used in a full-wave rectifier whose load resistance is shunted by a large capacitor. Sketch and label the waveform of diode current that has a total conduction angle of 90°.

D3-9 Sketch the waveform of the voltage across the diode during one cycle of input voltage to the rectifier of D3-8.

3-15
Bleeder Resistor

Power supplies with smoothing filters are often terminated in a fixed resistance of high ohmic value, called a *bleeder resistor*. That is, a resistor of perhaps 10 to 50 kΩ or more is connected permanently to the output terminals of the smoothing filter. Its resistance is usually of such value that it carries about 5 to 10 per cent of the normal full-load current to be furnished by the power supply.

One reason for having a bleeder across the output terminals of a choke-input filter is to prevent the current output of the rectifier from becoming too small. When the load current becomes quite small, the waveform will have a high ripple content. Another reason is to provide a means of discharging the filter capacitors when the power supply is turned off. The capacitors operate at high voltages, and if there is no way for them to lose their charge, it will remain on their plates for a long time after the power supply is turned off. Thus a dangerous voltage would be maintained at the output terminals of the filter. The capacitors cannot discharge through the diodes of the rectifier because of their reverse bias.

3-16
Ripple Factor

It has been noted that the job of the power supply smoothing filter is to prevent the flow of alternating components of current through the load circuit connected to the filter. If a filter is capable of eliminating the alternating component of current that has the lowest frequency, the higher-frequency components will be eliminated at the same time. This is readily seen when we recall that the inductive reactance of the choke ($2\pi fL$) increases with frequency, and therefore the higher-frequency components will be diminished even more than the fundamental-frequency component. Likewise, the bypassing capacitor will more readily allow the higher-frequency currents to bypass the load because of its decreasing reactance with increases in frequency. The result is a load current that has a small varying component superimposed on its average value.

The action of the filter network reduces the ac components arising from rectification. Small variations in the load current may remain, however, depending upon the reactances of the filter elements. The *ripple factor* may be defined as the ratio of the rms value of these remaining ac components to the average value of the load current or voltage. Of course, the smaller the ripple factor, the better the job of filtering. Obviously, it is desirable that the ripple factor be zero so that the load current would be entirely dc.

As an example of the amount of filtering needed for half-wave and full-wave rectifiers, we will calculate the ripple factor for each circuit without any filter network. Equation 3-1 gives the output voltage of the half-wave rectifier, including its dc value and the first three ac components that a filter would be expected to remove. Higher harmonics decrease in amplitude dramatically and will be neglected in the calculation of the rms value of these ac components. Using Eq. 1-48, the rms value is approximately

$$\text{rms value} = \left[\frac{(0.5E_m)^2 + (0.212E_m)^2 + (0.0424E_m)^2}{2}\right]^{1/2} = 0.385 E_m$$

The ripple factor is $0.385/0.318 = 1.21$, which means that the ac rms components exceed the dc value of the output voltage.

Now consider the full-wave output voltage as expressed in Eq. 3-8. The rms value of its ac components is approximately

$$\text{rms value} = \left[\frac{(0.424E_m)^2 + (0.085E_m)^2}{2}\right]^{1/2} = 0.306 E_m$$

The ripple factor is $0.306/0.636 = 0.481$, about one-third that of the half-wave voltage. This result indicates that simply going from half-wave to full-wave rectification will reduce the need for filtering to obtain a good dc load voltage with little ripple.

3-17
Practical Values of Filter Elements

A power supply that rectifies ac line voltage to generate a dc voltage is designed to filter and smooth variations in its output. Values of inductors, capacitors, and resistors making up the filter network are chosen to provide the degree of filtering required. We have seen that the product $\omega R_L C$ should be large in the shunt-capacitor filter in order to make the conduction angle small and thereby have a smaller ripple. The value of ω will be determined from the lowest ac harmonic present in the rectifier, for this component makes up the greatest ripple amplitude. But what should be the values of R_L and C?

A value for R_L may be deduced from the dc power requirements. For example, a 100-V power supply may be furnishing up to 1-A load current. Then R_L is the ratio of the voltage and current, $R_L = 100\text{ V}/1\text{ A} = 100\ \Omega$. If the load takes less current at the same voltage, then R_L will be larger and tend to improve filtering at smaller currents.

Practical values of the shunt capacitor may vary from 10 to 500 μF. They are usually electrolyte types because of the need to provide large capacitance in as small a volume as possible. Similar values are also used in choke-input and π-section filters. In inexpensive ac/dc radio receivers the simple C-R-C filter

Sec. 3-18   An AC Voltmeter   141

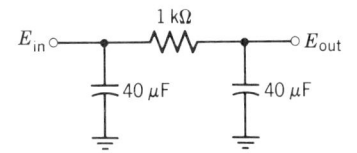

Figure 3-23
Simple C-R-C filter.

shown in Fig. 3-23 is often used. The resistor is not as effective in filtering as a choke would be, and it also reduces the dc voltage level through the filter toward the load, but it is less expensive and provides some measure of filtering. The value of R may vary from 1 kΩ to 3 kΩ in typical applications and depends to some extent on the dc power required by the load. A large load current must pass through R, tending to reduce the voltage available at the load. Typical voltages in a 117 Vac rectifier are $E_{in} = 130$ Vdc, $E_{out} = 110$ Vdc.

Inductors may be selected in the range of 5 to 10 H. A rule of thumb for the inductor value is expressed in terms of the load resistance, $L \approx R_L/1000$.

Filter design techniques are considerably more complex than is indicated here. The subject cannot be developed here, being more suitable for advanced studies in electronics.

3-18
An AC Voltmeter

A common laboratory multimeter, such as the Simpson Model 270, can be used to measure ac voltages as well as dc quantities, even though the indicating meter is a D'Arsonval moving coil type. This type of meter responds only to the *average* current passing through it. If a source of ac voltage is applied to it, the average current will be zero and the meter will read zero. Some arrangement that provides an average current must be built into the multimeter to permit the measurement of ac voltage. The meter may be calibrated to read effective values. Refer to Section 3-10.

The diode rectifier circuit of Fig. 2-23 generates a load current, called I_F, whose average value is not zero but is proportional to the ac voltage at the source. From an analysis of the pulsating load current, by Fourier series or other methods, it can be shown that the current will have an average value given by

$$I_F \approx \frac{E_{s,max}}{R_F + R_L}\left(\frac{1}{\pi}\right) \qquad (E_{s,max} \gg V_B) \qquad (3\text{-}21)$$

where R_L includes the D'Arsonval meter resistance and other series current-limiting resistances. If the voltage to be measured is much larger than the voltage drop of the diode, then the average meter current will be

$$I_F \approx \frac{0.318}{R_F + R_L} \sqrt{2} E_{s,\text{rms}} \tag{3-22}$$

in which $E_{s,\text{rms}}$ is the effective value of the sinusoidal source voltage e_s. Since all quantities are constants except the applied voltage, the meter may be calibrated to read rms values of sine waves. The use of a diode rectifier in this way allows a dc meter element to be used as an ac voltmeter.

This technique of using a diode to generate a *current* whose average value can be utilized to represent an ac voltage is employed in many electronics applications. The simple ac voltmeter is a very useful application and is representative of many similar ones. They can be found where it is important to monitor or measure an alternating parameter. Another common example is the output meter on a laboratory oscillator.

It is essential that we recognize that the meter scale will give correct readings only for the alternating waveform for which the scale is calibrated, that is, either for *half sine waves* or *full sine waves*, depending on the rectifier circuit in the multimeter. Input voltages with other kinds of waveform will generate average currents that correspond to the shapes of the waveforms but will not, in general, have the same values as those of the rectified sine waves. The ac voltmeter can be calibrated for only one waveform, and it will indicate incorrect values if an ac voltage with a different waveform is applied. When using a measuring instrument it is important to know not only the use for which it is designed but also limitations on its performance that may be imposed by parameter waveforms.

In certain electronic circuits, both ac and dc voltages are present. If an ac voltmeter is used to measure ac voltage in these circuits, any dc components in the measured voltage will change the average current in the meter, and the indicated ac voltage will be in error by an amount related to the magnitude of the dc current in the meter.

The basic circuit of a laboratory multimeter set on its 10-V ac range is sketched in Fig. 3-24. This circuit uses a full-wave bridge rectifier to establish current in the meter movement. The two 4-kΩ resistors in the full-wave bridge network develop a voltage that provides current in the meter. Variable resistors shown have been selected to permit calibration of the voltmeter so that full-scale deflection of the D'Arsonval movement (50 µA) will represent the 10 V rms value of a sine wave. Note also the use of diodes shunting the meter as overload protection.

Even though the current in the meter is pulsating, no electrical filter has been included in the circuit to achieve a steady current. If the frequency of the rectified sine wave is high enough, the mechanical inertia of the massive moving-coil mechanism provides filtering so that the meter pointer stays at one

Sec. 3-19　　　　　　　　　　　　　　　　　　　　　　*The Diode Detector or Peak Rectifier* 143

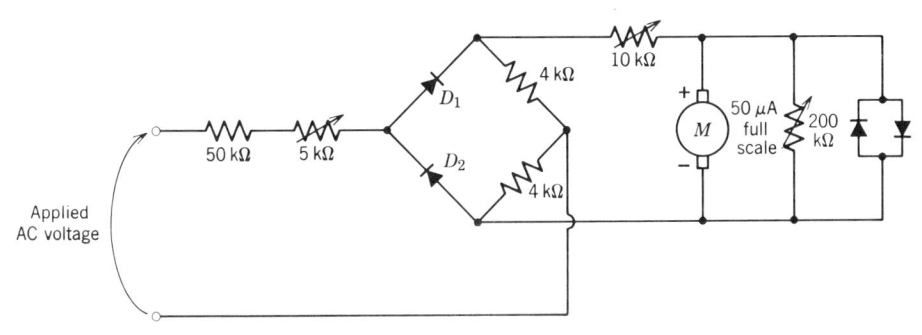

Figure 3-24
Basic circuit arrangement for 10 Vac voltmeter.

point on the voltmeter scale. This inertia causes the pointer to rest at a position that is related to the *average* current in the meter. However, the scale will be marked to indicate the rms value of the sine wave applied to the voltmeter.

Drill Problems

D3-10 An ac voltmeter is constructed using a D'Arsonval moving-coil meter and a half-wave rectifier circuit. Its scale is calibrated to indicate the rms value of a sine wave. This meter is used to measure the output voltage of an unfiltered full-wave rectifier. What value will the meter indicate on its scale? What is the rms value of the measured voltage?

D3-11 The voltmeter of D3-10 is used to measure the voltage of a waveform consisting of a 10-V rms sine wave superimposed on a 16-Vdc source. What value will the meter indicate? What is the rms value of the waveform?

3-19
The Diode Detector or Peak Rectifier

The half-wave rectifier of Fig. 3-12 delivers a pulsating current to the load. It is satisfactory for many applications. There are many others, however, where it is necessary to obtain a waveform other than the half-sine wave. We have already discussed how the circuit is used to generate a direct voltage from the pulsating current.

Half-wave rectifier operation will be discussed here to introduce another circuit that behaves in much the same way but with higher-frequency input voltages. This circuit is referred to as a *diode detector*. The name is derived from its ability to detect (identify) information signals that are carried by

144 *Applications of Diodes* Ch. 3

high-frequency voltage waveforms, such as those transmitted by commercial radio stations.

The basic circuit is shown in Fig. 3-25. It is the same as the rectifier circuit, except for the additional capacitance, C, connected across the load resistance. The capacitor acts as a reservoir for electric charge, tending to hold the output voltage at a constant value between current pulses. From the waveform of output voltage, though, it can be seen that the time constant R_LC determines how well it does its job. For small load currents the output voltage will be practically equal to the peak value of the input sinusoid. A small load current flows when the resistance is large, which means there is a long time constant. When the circuit generates an output voltage nearly equal to the peak of the input voltage, it is called a *peak rectifier.*

Assume e_s is applied at $t = 0$ and the capacitance has no charge initially. Then, as e_s increases to its positive maximum, forward current will deliver charge to the capacitance. If the series resistance R_s is small enough, the

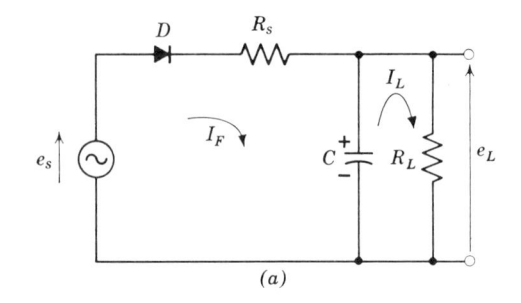

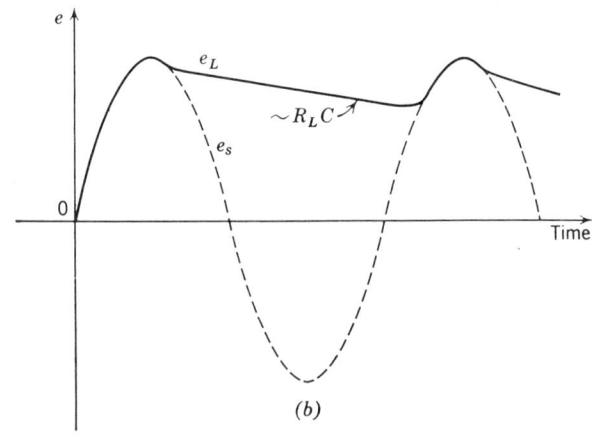

Figure 3-25
Peak rectifier or diode detector circuit.

Sec. 3-20   Voltage Doubler

voltage appearing across C will be essentially equal to e_s until it reaches its peak. When e_s starts to decrease, the diode will be reverse biased by the charge accumulated on C, and no current will pass back to the input. However, C now has a path for discharge through the load R_L. Its voltage will start to decrease at a rate determined by the time constant $R_L C$. After some time has elapsed, the input voltage will change polarity and start its positive increase again. When the value of e_s reaches the potential remaining on C, the diode will again be forward biased and current will recharge the capacitor. The cycle repeats as the input voltage goes through successive periods of its alterations.

In a communications receiver, where it is necessary to detect and separate low-frequency variations caused by changes in the *amplitude* of a high-frequency signal, the circuit behaves as a *diode detector*. In this application it is desired that the time constant of the circuit permit the output voltage to follow the low-frequency variations but not those of the high frequency. These features of the diode-detector circuit will be studied in detail in Section 12-18 that discusses its use in amplitude-modulation (AM) communications receivers.

Drill Problems

D3-12 In the circuit of Fig. 3-25, what effect on circuit operation would result from (*a*) reversing the diode in the circuit? (*b*) doubling the value of C?

D3-13 For an input signal of a 50-Hz sine wave applied to a diode detector, calculate and graphically show the effect of $R_L C$ time constants equal to: (*a*) 0.01 s, (*b*) 0.1 s, (*c*) 1 s. Assume that e_L decreases from its peak amplitude as a starting point.

3-20
Voltage Doubler

In some applications it is necessary to furnish a voltage larger than can be obtained from rectification of the power-line voltage. A step-up power transformer can be used; however, where the load current requirement is small a voltage doubler circuit (Fig. 3-26) may be used instead. Its input voltage may be either the 115-V power line or a step-up transformer.

The operation of a voltage doubler circuit is easy to understand, because it essentially consists of a voltage clamper connected to a peak rectifier. The capacitors may have values as large as 500 μF each, as in other applications of filtering. On the negative half-cycles when point 2 is positive, C_2 charges through D_2. On the positive half-cycles when point 1 is positive, the input current flows through C_2 and D_1. C_2 can discharge only through D_1 because D_2 will not conduct downward. The capacitances are usually equal: $C_1 = C_2$.

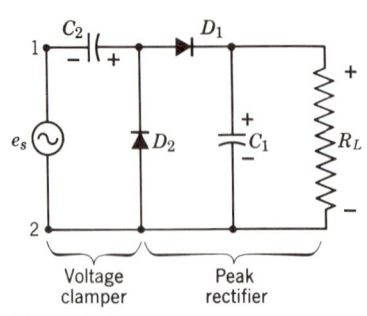

Figure 3-26
Half-wave doubler.

Experimental data taken on this circuit reveal that for values of $2\pi f C_1 R_L$ greater than 200, the dc voltage across R_L is nearly double the peak value of the input voltage for large values of R_L. R_L should be at least ten times the sum of the source resistance and the internal resistance of one diode. The peak-inverse voltage of the series diode (D_1) is the output voltage plus the drop across D_2. This is nearly double the peak supply voltage. The instantaneous peak current is of the order of 100 times the dc output current.

The ac input voltage is shown in Fig. 3-27a, and the voltage across D_1 in series with C_1 and R_L is shown in Fig. 3-27b. The current in D_1 cannot start flowing until the voltage across D_2 is greater than the voltage across C_1. Consequently, very short spurts of relatively large instantaneous values of current flow through D_1 and R_L. This explains why the instantaneous maximum peak current can be one hundred or more times the average (dc output) current. It also indicates that the capacitances must be quite large in order to store enough energy to be able to supply continuous current to the load without severe drops in voltage between successive peaks of the charging half-cycles.

Drill Problems

D3-14 Sketch the voltage waveform existing across capacitor C_2 in Fig. 3-26 for an input signal $e_s = 100 \sin \omega t$. Now suppose C_2 develops an internal short-circuit condition. Show the effect on load voltage.

D3-15 Sketch the waveform of current in D_1 for $\omega R_L C_1 = 10$; for $\omega R_L C_1 = 100$. Compare their maximum amplitudes.

3-21
Semiconductor Stack Rectifiers

Silicon and germanium rectifiers are commercially available in pre-assembled stacks complete with heat sink and electrical interconnections for a

Sec. 3-21   *Semiconductor Stack Rectifiers*   147

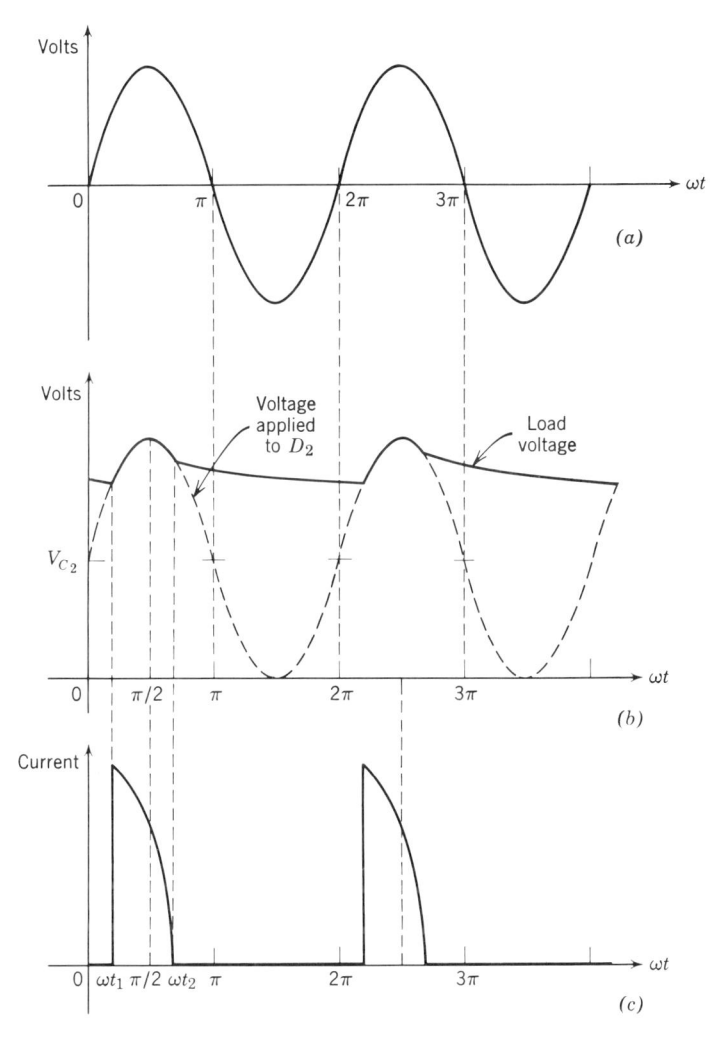

Figure 3-27
Voltage and current waveforms in voltage doubler. (*a*) Input voltage waveform; (*b*) voltages applied to D_2 and across the load; (*c*) current pulses in D_1, waveform depends on $\omega R_L C$ as in half-wave rectifier with shunt-capacitor filter.

great variety of typical applications of rectifier circuits. Standard units include half-wave, full-wave (center-tap), voltage-doubler, single-phase bridge, and arrangements for three-phase applications. Some of the stacks have semiconductor diodes connected in series and parallel for the greater voltage and current ratings needed in certain applications.

148   Applications of Diodes   Ch. 3

Figure 3-28
Typical silicon and germanium stacks. (Courtesy General Electric Co., N.Y.)

   Some typical silicon and germanium stack rectifiers are shown in Fig. 3-28. The larger units are designed with cooling fins as part of the assembly. Although much more heat is generated by the increased power in the larger units, they are practical owing to the greatly increased heat-radiating capacity provided by the fins.
   Semiconductor stack rectifiers are capable of operating at high voltages and currents. The General Electric type 4JA9013, with six fins each 7 in. × 7 in., has the following specifications:

Maximum current (free convection cooling at 55°C)	80 A/fin
Maximum current (cooling at 2000 cfpm and 55°C)	207 A/fin
Maximum PIV	800 V/fin recurrent, and 1050 V/fin transient

Questions

3-1 What purpose does a power supply serve in an electronic circuit?

Questions 3-2 through 3-7 pertain to the half-wave rectifier.

3-2 How do the dc component, the fundamental component, and the double-frequency component of the rectifier output voltage compare in magnitude with the input voltage?

3-3 What is meant by peak-inverse voltage on the diode? How does it compare with the peak input voltage?

3-4 At a given voltage input, what determines the peak value of diode current?

3-5 How does the average value of diode current compare with the peak value?

3-6 How is the rms value of the load current (resistive load) related to the dc value?

3-7 Sketch the output-voltage waveform.

Questions 3-8 through 3-12 pertain to the full-wave rectifier.

3-8 What is the frequency of the largest ac component of the output voltage of a full-wave rectifier?

3-9 How do the dc component and the two lowest-frequency ac components of the rectifier output voltage compare in magnitude with the input voltage?

3-10 How does the peak-inverse voltage compare with the instantaneous maximum input voltage?

3-11 How do the dc and rms values of load current compare with the maximum instantaneous value?

3-12 Sketch the output voltage waveform.

3-13 Compare the performance of a full-wave rectifier with that of a half-wave rectifier in five important respects.

150 Applications of Diodes   Ch. 3

3-14 What is the purpose of a smoothing filter, and how does it accomplish its purpose? How is it that, if the desired effect is produced on the ac component of lowest frequency, the other components will not cause trouble in the output?

3-15 Draw from memory a full-wave rectifier circuit with a choke-input filter and a bleeder.

3-16 Define ripple factor.

3-17 Why should a bleeder be used on a rectifier power supply?

3-18 Explain how *LC* and *RC* filters like those studied in this chapter are able to reduce the magnitudes of ac components of voltage so effectively. It is important to note that ac components of *current* are simultaneously reduced, since current = voltage ÷ impedance.

3-19 Explain the action of a diode in a voltage limiter circuit.

3-20 Discuss the need for two diodes in a voltage limiter used to alter a waveform in both positive and negative ranges of its amplitude.

3-21 The diode voltage limiter circuit is sometimes referred to as a "clipper." Why is this term descriptive of its operation?

3-22 How can a diode be used to calibrate a dc current-operated meter for measuring ac voltages?

3-23 What limitations apply to the use of a voltmeter that is calibrated for sine waves?

3-24 Explain the operation of a diode detector circuit, showing the function of each component of the circuit.

3-25 What is the basic function of a "switching gate"?

3-26 How does the AND circuit differ in operation from the action of the OR circuit?

3-27 Draw and explain the operation of a bridge-type rectifier circuit. What are its advantages? Does it have any disadvantages?

3-28 Draw a half-wave voltage-doubler circuit.

3-29 Under what conditions are power transformers used instead of doubler circuits?

3-30 Write the important equations relating the following transformer quantities: (*a*) terminal voltages and turns; (*b*) terminal voltages and currents; (*c*) currents and turns.

Problems

3-1 The circuit of Fig. 3-1 is used to generate a nearly square wave of voltage from an input sinusoidal source of 117 V rms. (*a*) Assuming the diodes to be ideal elements, what must be the magnitudes of dc limiting sources E_1 and E_2, in order to generate a square wave that is 20-V peak to peak and has similar positive and negative portions? (*b*) What changes would be made in the circuit to generate only negative portions of the square wave? (*c*) What should be the PIV rating of the diodes? (*d*) What value of R will limit diode current to 100 mA?

3-2 An ac signal of 0.5 V amplitude is superimposed on a 12-V dc signal. What is the rms value of the combination?

3-3 A diode detector circuit, as in Fig. 3-25, has an input signal of 50 Hz. Calculate the effect of $R_L C$ time constant equal to 0.01 s. Show this by a graphical sketch, assuming e_L decreases from its peak amplitude as a starting point.

3-4 In the circuit of Problem 3-3, what effect on circuit operation would result from reversing the diode connections? What would happen if the value of C were doubled? What would happen if C were to become an open circuit?

3-5 In the circuit of Fig. 3-3, two simultaneous positive voltage pulses are applied to inputs A and B. What should be their values in order to raise the output terminal voltage to 12 V?

3-6 In the circuit of Fig. 3-3, how much current flows in the 1-kΩ resistor before an input pulse is applied? How large an input pulse is needed to raise the output voltage to 4 V above ground?

3-7 A simple voltage regulator circuit is required to hold a voltage within the range from 6.5 to 7.0 V, when the input voltage varies between 10 and 20 V. Design a circuit similar to that of Fig. 3-7, using a Zener diode and a series resistor. (*a*) Choose a diode from the types given in Fig. 3-6, and calculate the required value of series resistor. (*b*) Determine the minimum power rating of the resistor. (*c*) Determine the diode current limits. (*d*) Determine the output voltage.

3-8 A junction diode has a reverse saturation current of 12 μA at reverse-bias voltage up to its Zener breakdown potential. It has a breakdown voltage of 15 V and has negligible ohmic resistance in its breakdown region. A 24-V dc source is applied to the diode as a reverse bias, in series with a 2-kΩ resistor. Calculate the diode current.

3-9 A 1000-Ω resistor has applied to it a voltage $e = 50 + 50 \sin \omega t$ V. Determine the dc current in the resistor and the required power rating for it to operate with this voltage.

152 Applications of Diodes   Ch. 3

3-10 A diode is rated for maximum current of 500 mA and average current of 200 mA. It is being considered for use in a full-wave rectifier circuit with a transformer supplying 220 V rms on each side of the center tap. (Neglect diode resistance.) (a) What value of load resistance may be used to obtain the greatest dc output without exceeding any diode rating? (b) Calculate the dc load voltage and current for the resistance chosen in (a).

3-11 Determine the value of load resistance required for maximum dc power output from a full-wave rectifier using two diodes rated at PIV = 200 V, I_{max} = 2 A (each diode), and I_{av} = 1.2 A (each diode).

3-12 A half-wave rectifier circuit is supplied from 120-V rms, 60-Hz ac lines. The circuit uses a diode that has an internal resistance of 200 Ω and supplies power to a load resistance of 5 kΩ. (a) What is the peak instantaneous value of diode current? of load current? (b) What is the dc power in the load? (c) Calculate the rms value of load current. (d) Calculate the power conversion efficiency. (e) What is the peak-inverse voltage on the diode?

3-13 Calculate the quantities asked for in Problem 3-12 for a full-wave rectifier circuit using the given diode and a center-tapped transformer that supplies a 120-V rms input to each diode. Compare the results with those for the half-wave circuit.

3-14 A half-wave rectifier is supplied with a 60-Hz sine-wave voltage of 120 V (rms) value. Calculate the dc voltage and the rms values of the fundamental component and the next two largest ac components in the output voltage.

3-15 A smoothing filter receives from a full-wave rectifier a pulsating voltage that has a fundamental component of 143 V peak. What is the value of the dc component? What is the rms value of the voltage input to the rectifier?

3-16 The rectifier-filter of Problem 3-15 delivers current to a resistance load after a dc voltage drop of 10 V in the series inductor of the filter. The rms ripple voltage across the 2125-Ω load is 2.5 V. Calculate (a) the ripple factor of the load voltage; (b) the dc voltage at the load; (c) the power delivered to the load.

3-17 A one-section RC filter has $R = 10^4$ Ω, $C = 8\ \mu$F. What is its voltage reduction ratio at 60 Hz?

3-18 The RC filter of Problem 3-17 is used with a half-wave rectifier operating on 60 Hz. The maximum instantaneous value of the filter input voltage (E_m) is 170 V. (a) Calculate the ripple factor at the load, neglecting dc voltage drop through the filter. (b) Calculate the ripple factor in the load again, allowing for a 25 per cent loss in voltage at the filter.

Problems   153

3-19 A power transformer suitable for a full-wave rectifier circuit will deliver 200 mA at 375 V to each half of a double-diode rectifier tube. At the same time it will deliver 5 A at 6.3 V rms to a filament circuit. The primary winding operates on 115 V. (*a*) What is the value of the total secondary voltamperes? of the primary voltamperes? (*b*) What is the full-load primary current? (*c*) Assuming 250 turns on the primary winding, calculate the number of turns on each of the secondary windings, assuming an ideal transformer.

3-20 A voltage divider is connected to the output terminals of a power supply filter that delivers direct current at 265 V. The various current demands and voltages are shown in Fig. P3-20. The current in R_3 is chosen to load the transformer properly. The potentials that are negative with respect to ground are needed for grid biases on tubes. No current is drawn in these cases. (*a*) Calculate the ohms values required for all the resistors and also the power dissipated in each. (*b*) If the voltage divider were made of a continuous wire-wound spool with taps, the wire would have to be large enough to carry the maximum current safely. How much power would it then be capable of dissipating? (*c*) Compare the power *rating* found in (*b*) with the sum of the separate powers in (*a*). The arrangement in (*b*) would be lower in manufacturing cost. The power rating of the actual resistors used should be at least double the calculated value if the bleeder is to be confined in a poorly ventilated space.

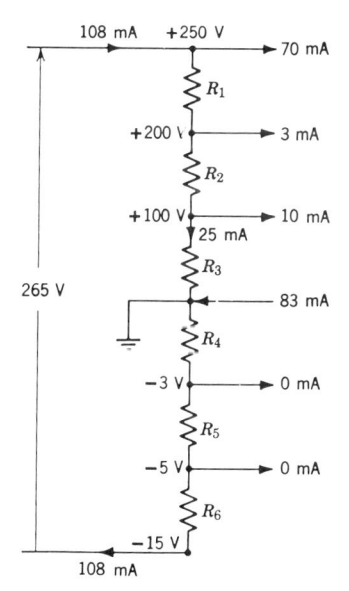

Figure P3-20
Voltage divider.

154 Applications of Diodes   Ch. 3

3-21 In Fig. 3-14, ac input = 120 V rms, and the transformer has a step-up ratio of 1:2. Assume no voltage drop across the diodes during conduction. (a) Calculate the value of resistance load to allow 200 mA average load current. (b) Suppose one of the diodes becomes inoperative. What average load current would then exist?

3-22 Explain the importance of the maximum current rating of a diode when it is used as a power rectifier with a π-section filter.

3-23 Why is "peak-inverse-voltage" rating of a diode of more significance in full-wave circuits than in half-wave circuits?

3-24 Sketch and explain the output voltage waveform of a half-wave rectifier that has a capacitor shunting the load resistance. Of what significance is the time constant of the R-C combination compared with one period of the input to the rectifier?

3-25 What must be the voltage rating of capacitors used in rectifier filter networks, compared with the rms value of the input voltage? Will this rating be the same for both full- and half-wave circuits? Explain.

3-26 Develop the general form of Eq. 3-1 by Fourier analysis.*

3-27 Verify that Eq. 3-8 represents the waveform of the output of a full-wave rectifier.*

3-28 Explain the effect on circuit operation if the diode of Fig. 3-20a were connected backwards in the circuit. Would the circuit be useful?

3-29 The supply voltage for the circuit of Fig. 3-20a is $e_s = 100 \sin 377t$. The load resistor and shunt capacitor are large enough so that there is no appreciable ripple in the load voltage. The load voltage is required to be 90 V, and the diode is rated at 10 mA maximum recurrent current. Show the changes necessary in the circuit to meet these restrictions.

3-30 A type of diode that has negligible forward resistance is used in a half-wave rectifier whose load resistance is shunted by a large capacitance. Sketch and label the waveform of diode current for a total conduction angle of 90°. Sketch the waveform of the voltage across the diode during one cycle of input voltage.

3-31 Repeat Problem 3-30 for a full-wave rectifier circuit using the same type of diode. For a 90° conduction angle, what changes would have to be made in the filter and load? If the same R and C were used in the full-wave rectifier, what would be the total conduction angle of each diode?

3-32 Sketch the waveform of diode current for a half-wave rectifier using the shunt-capacitor type of filter with $\omega RC = 10$.

* Solution requires the evaluation of trigonometric integrals.

3-33 It is desired to construct a half-wave rectifier and filter for a load resistance of 200 Ω. The supply voltage is 100 sin 377t. (a) Determine the ratings of a diode that could be used without filtering. (b) What value of capacitance is required to make $\omega RC = 10$? (c) If this value is used as a shunt-capacitor filter, will the diode chosen in (a) be adequate?

3-34 A silicon power rectifier is used in a full-wave circuit that employs a center-tapped transformer. The load resistance is shunted with a large electrolytic capacitor for filtering. Circuit conditions permit the diodes to be considered ideal without appreciable error. (a) Sketch the schematic diagram of the complete circuit, and show significant voltage polarities and current directions. (b) For 60 Hz input voltage, determine values of RC that will cause diode conduction from about 25° to 125°, referred to the input voltage as a time reference. (c) Sketch the waveform of diode current pulses for the conditions in (b). (d) Suppose it is desired to increase the dc level of load voltage. What changes could be made in the circuit to accomplish this result?

3-35 For the circuit in Fig. P3-35 and during the time when diode D is passing current, label the input voltage, diode D, and R_L with correct polarity marks. Sketch the waveform of e_{R_L} for a sinusoidal input voltage.

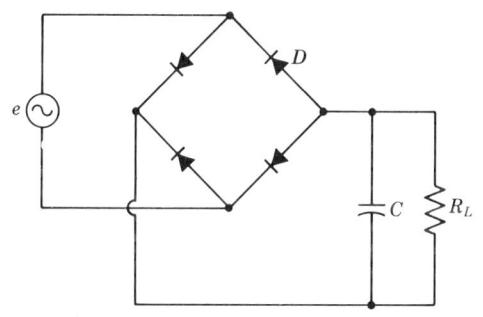

Figure P3-35

3-36 In the circuit of Fig. P3-36, which point has the best filtered dc voltage? the highest dc voltage?

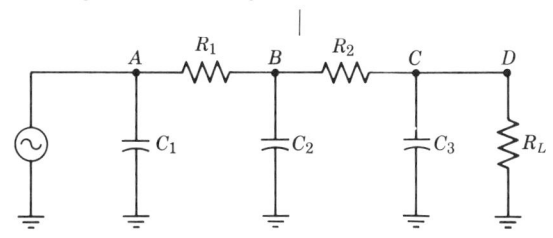

Figure P3-36

4
JUNCTION TRANSISTORS

The startling discovery in 1948 of current amplification in semiconductors, which resulted from studies by Dr. John Bardeen and Dr. Walter H. Brattain at the Bell Telephone Laboratories, led to the first *crystal triode*. Alluding to its unique properties and characteristics in controlling the internal resistance of the triode, they named their discovery *transistor*, a contraction of *trans*fer res*istor*. The name refers to the control of current in one circuit by the action of current in another circuit. We shall see that such control of current is the basic characteristic of a junction transistor.

Since that momentous discovery, whole industries have sprung up to develop and manufacture semiconductor devices. Other new industries apply these devices in minicomputers, electronic calculators, wrist watches, home appliances, automobiles, and transportation systems. Applications of electronics in communications, earth satellites, space vehicles, and high-speed digital computers, in which space and weight are at a premium, have made critical the need for small components. Many forms of transistors and other semiconductor devices are performing jobs once possible only with bulky components. Some circuit functions are performed better by transistors than by any previous electronic devices. Predictions of future developments in semiconductors include further miniaturization and the growth of other new industries to manufacture and apply these devices to our everyday needs.

There are several advantages offered by transistors. Among them are:

1. They are instantly available for use, because they require no warm-up before they will operate; in fact, they operate better at room temperature than at the elevated temperatures necessary for vacuum tubes.

2. They may be operated at very low voltages.
3. They consume low power, resulting in high efficiency.
4. They have a long life with essentially no aging effects when operated within their ratings.
5. They resist damage from shock and vibration.
6. They offer flexibility in circuit design and mechanical layout, usually mountable in any physical orientation.

The material in this chapter describes the physical structure, electrical characteristics, and theory of operation of junction transistors. The reader should be familiar with the discussion of P-N junction operation given in Chapter 2.

4-1
Types of Transistors

Space will not be devoted in this book to detailed explanations of manufacturing techniques by which transistor materials are produced and transistor fabrication is accomplished. Many references on these subjects are available in technical libraries. Instead, we will give our attention to basic properties of two types of transistors.

One type of transistor operates because of the action of both majority and minority carriers within its structure. Because of this action, it is said to be a *bipolar* device. Another type of transistor operates with majority carriers only, and is referred to as a *unipolar* device.

Bipolar devices are typified by the *junction transistor*, in which two P-N junctions control internal currents. It is the "work horse" of transistor applications and will be discussed in some detail in this chapter. Typical of the unipolar transistor is a different construction of the P-N junctions, referred to as a *field-effect transistor*. Its operation is decidedly different from that of a junction transistor, although the two may be used in similar electronic circuits. Our discussions will include both types, to show their similarities and their differences. This chapter presents the junction transistor; the field-effect transistor will be discussed in Chapter 5.

4-2
The Junction Transistor

The physical configuration of a junction transistor is two P-N junctions back-to-back. A thin layer of either P-type or N-type semiconductor separates the two junctions. The structures of two typical junction transistors are shown in Fig. 4-1. The junctions of P- and N-type semiconductors are grown or

Sec. 4-2   The Junction Transistor   159

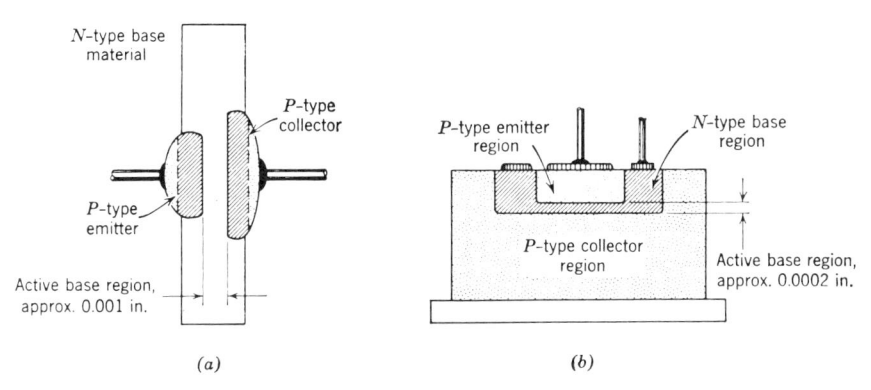

Figure 4-1
Structures of typical junction transistors. (a) P-N-P alloy-junction type. (b) P-N-P diffused-junction type.

diffused, rather than being mechanical joints of these materials, as explained in Chapter 2.

These transistors are called *P-N-P* types because of the sequence of the layers of semiconductor materials. In each of them a thin region of *N*-type material separates two regions of *P*-type material. The sequence of the layers of semiconductor materials classifies a transistor as either *P-N-P* or *N-P-N*, depending on which is the center region. Both types are used extensively in electronic circuits.

Transistor operation depends on carrier behavior in the center region, which is called the *base* region of a transistor. In Fig. 4-1, the base of these transistors consists of *N*-type semiconductor. The two *P*-regions are called *emitter* and *collector* regions of the transistors. These names will be discussed in detail in the sections that follow, as we look at the characteristic behavior of both *P-N-P* and *N-P-N* transistors.

Several junction transistors, useful in small-signal amplifiers, are shown in Fig. 4-2. Two vacuum tubes and different types of transistors are pictured together with a ball-point pen to enable comparison of physical sizes. The outside dimensions of a typical junction transistor are of the order of $\frac{3}{8}$ in. long, $\frac{1}{4}$ in. wide, and $\frac{1}{8}$ in. thick, excluding its connecting leads. Some are cylindrical in shape, about $\frac{1}{8}$ in. diameter and $\frac{1}{4}$ in. long, while others may be much smaller when they are intended for low-voltage operation. Cases may be metal or plastic encapsulation, and connecting leads may extend from the case in various configurations, although usually they extend from one end of the case. The lead arrangement often permits insertion of the transistor into a socket. Nonstandard versions of case shape and lead arrangement abound, however, and it is usually advisable to ascertain differences in any unfamiliar type of transistor. Thousands of different types have been standardized in shape and

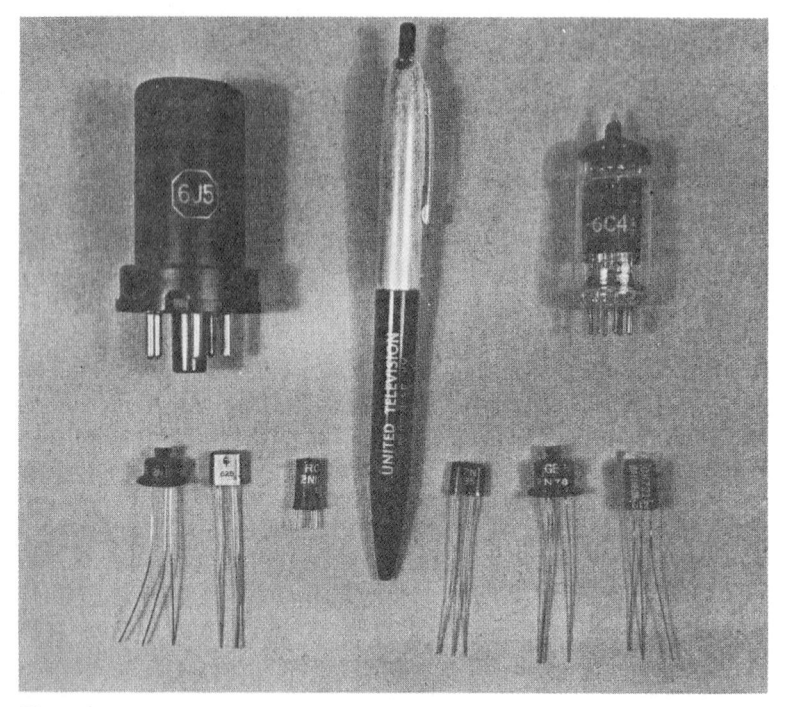

Figure 4-2
Junction transistors. (Courtesy United Electronics Laboratories, Louisville, Ky.)

electrical properties and identified by a series of numbers and letters, but there are thousands more that are identified by individual manufacturers without any standard notation.

Drill Problems

D4-1 It will be instructive for the student to obtain a transistor manual or manufacturers' data sheets, and determine for these transistors the following information: (1) N-P-N or P-N-P; (2) silicon or germanium; (3) junction, or other method of fabrication; (4) voltage and current ratings; (5) type of case and lead configuration; (6) typical forward current-transfer ratio; (7) maximum power dissipation.

2N334	2N3136
2N1308	2N3305
2N1742	2N3391
2N2043	2N3678
2N2095	2N4281

D4-2 These numbers refer to different sizes and shapes of transistor cases:

TO-5, TO-12, TO-18, TO-33, TO-39, TO-49, TO-98

Determine from data sheets or a transistor manual how they differ and which connecting lead is collector, emitter, and base.

4-3 Equalization of Fermi Levels in a P-N Junction

In the discussion of N-type and P-type semiconductors in Chapter 2, the concept of a *Fermi level* of energies was introduced. In any kind of semiconductor at room temperature or above, whether pure or doped, N-type or P-type, the electrons have a great variety of energy values and thus occupy many energy levels. In a chosen short time interval, some energy levels are occupied more frequently than others, as electrons gain or lose energy levels in maintaining an equilibrium condition. There is an energy value, called the Fermi level, for which there is a 50 *per cent probability of occupancy*. The Fermi level is the *most probable energy value* that will be found among the electron energies.

Donor impurities in a semiconductor increase the number of electron energy levels in or very near the conduction band. This raises the level in which there is a 50 per cent probability of occupancy; that is, the Fermi level is raised. *Acceptor* impurities, on the other hand, reduce the Fermi level in the semiconductor.

When N-type and P-type semiconductors are brought together as a P-N junction, the two different energy values of their separate Fermi levels will equalize at one value. This situation is illustrated in Fig. 4-3. The common Fermi level of the two types is in equilibrium across the P-N junction; in other words, no bias is applied to the junction. The net effect on the separate Fermi levels is to raise the energy levels of the P-type relative to those of the N-type.

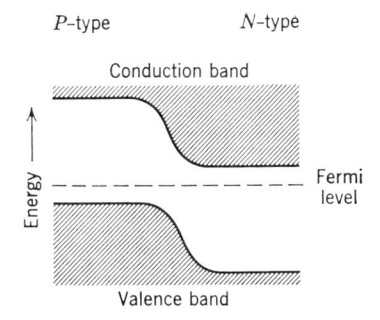

Figure 4-3
Energy levels in a P-N junction in equilibrium (unbiased).

4-4
Effects of Bias on Energy "Hills" at a *P-N* Junction

Consider first the equilibrium (no bias) state of the *P-N* junction. The *equalization action* resulting in a common Fermi level involves the addition of permissible higher-energy levels in the conduction band of the *P* region and the removal of some permissible lower-energy levels in that band. (Refer to the discussion of the *P-N* junction barrier region in Chapter 2.) In the *N* region, some high-energy levels have been lost and some low-energy levels added in its conduction band. Since the *N* region has an excess of electrons, *some of these electrons pass over to the P region*; likewise, *holes pass from the P to the N region*. Equilibrium is established when the Fermi level has reached a common value for both regions, but the *boundary face* of the *P* region has a layer of negative charge (acceptor holes have been neutralized) and the boundary face of the *N* region has a layer of positive charge (donated electrons have left impurity ions there). Thus the *potential barrier* is formed and a *potential hill exists at the junction.*

Now consider the application of *forward bias* (Fig. 4-4*a*). Since the *P* side of the barrier has a *net negative charge* in the no-bias case, the application of an external positive potential to the *P* region will lower the *Fermi level* on that side of the junction. The Fermi levels try to become aligned, and this action *lowers the potential hill* across the junction, with the result that electrons move more readily across the barrier from the *N* to the *P* region. It is important to note here that although the potential hill at the junction is favorable to electron passage from the *N* to the *P* region, there are *many fewer* electrons available for passage than in a conductor. A semiconductor current with forward-conduction bias is

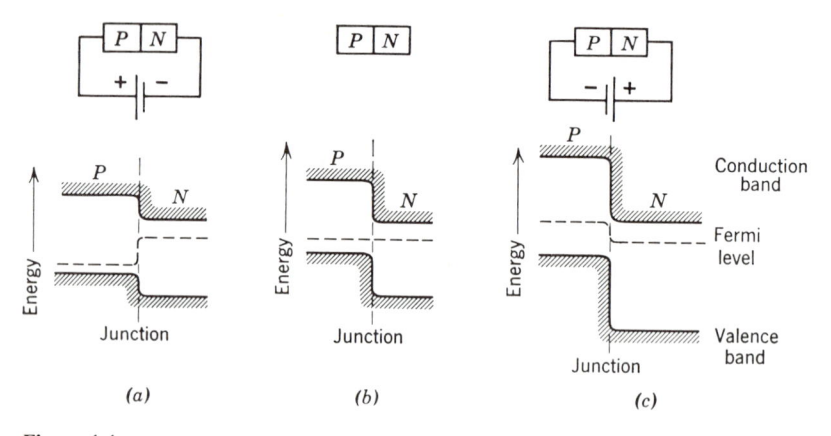

Figure 4-4
Energy diagrams of a *P-N* junction. (*a*) Forward bias, majority-carrier conduction. (*b*) No bias, equilibrium. (*c*) Reverse bias, minority-carrier conduction.

much greater than with reverse bias, for in the latter case it will be found that in order for electrons to pass through the junction they must go down a large potential hill. In doing so they lose energy.

Figure 4-4c shows the reverse-potential hill of the reverse-bias case. The effect of reverse bias is to increase the energy difference between the valence and conduction bands. Note that in Fig. 4-4c current flow requires electrons to go down the high potential hill in crossing the junction barrier from left to right. This means going against the electric field produced by the potential difference across the barrier, so comparatively few electrons have enough energy to make it and the current is extremely small. It should be remembered that the *electric field force* on an *electron* urges it to go *up a potential hill*. If we could release an electron at rest (no kinetic energy) in an electric field, it would start up the hill and gain velocity as a result of the force exerted on it by the field.

When reverse bias is applied, the majority carriers of both regions move away from the junction. Electrons in the N material move toward the positive terminal—to the right in Fig. 4-4c—and holes move toward the negative terminal. This causes a *depletion of majority carriers* on both sides of the junction, with the result that current is due to passage of *minority carriers* across the junction. Remembering that *forward bias* results in conduction by *majority carriers* (electrons from N type to P type and holes from P type to N type), and *reverse bias* results in *minority-carrier* conduction (electrons from P to N and holes from N to P) should help in understanding the mechanism of conduction across a P-N junction.

4-5
Symbols and Bias Polarities for Transistors

The standard symbols used for transistors in circuit diagrams are illustrated in Fig. 4-5. Note that the only difference between the two is in the arrow on the emitter lead, which in each case points in the direction of conventional forward current in the P-N junction between base and emitter. A simple mnemonic

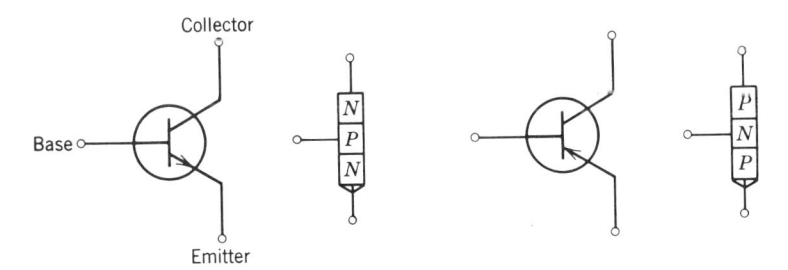

Figure 4-5
Conventional schematic symbols for transistors.

164   Junction Transistors   *also P-(N)-P*   Ch. 4

device for remembering the direction of the emitter arrow is: the arrow is *Not Pointing iN* for the *N-P-N* transistor.

A transistor acts as an electrical valve, working on the principle that an electric current in one part of a circuit can control the magnitude of another current in some other part of the circuit. In a transistor, the controlling current occurs in the base-emitter diode, and the controlled current passes from the collector to the emitter. In normal operation of a transistor as an amplifier, the base-emitter *P-N* diode is forward biased and the collector-base *P-N* diode is reverse biased. Figure 4-6 shows the external connections of biasing voltages.

In Fig. 4-6 it is seen that external biasing voltages provide forward bias for the emitter-base junction and reverse bias for the collector-base junction. Both bias conditions are necessary to operate the transistor, whether it is *N-P-N* or *P-N-P* type. The polarities of external bias sources are reversed for *N-P-N* and *P-N-P* transistors, in order to provide forward and reverse biasing for the separate junctions as required. In the upper diagrams the external sources are connected directly to the junctions, but in the lower diagrams external bias is connected directly only to the emitter-base junction. Reverse bias of the collector-base junction is provided only when the source V_{CC} is greater than source V_{BB}. Kirchhoff's voltage law holds in this circuit as in other closed circuits: the sum of voltage rises around any closed circuit will equal zero. This

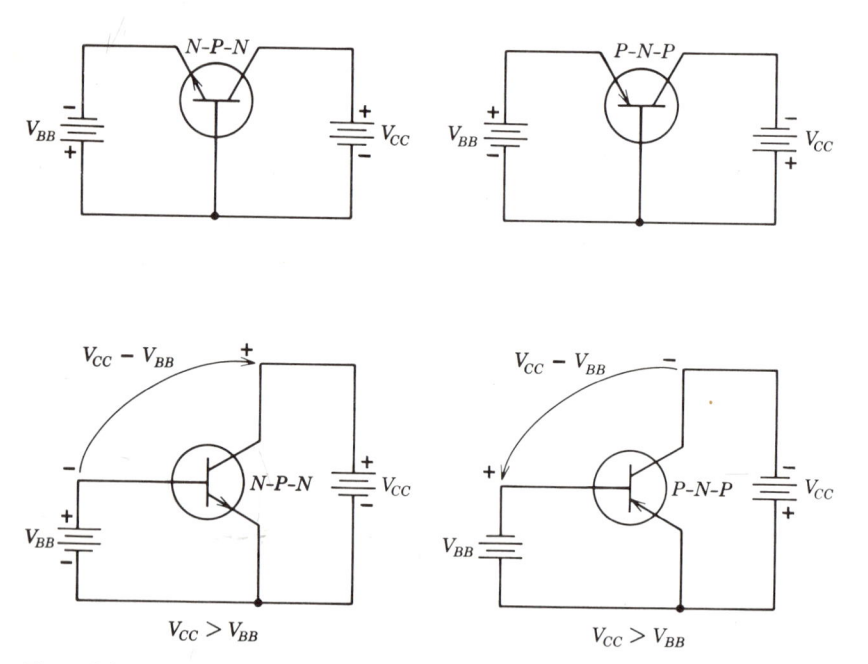

Figure 4-6
External connections of biasing voltages for transistor operation.

Sec. 4-6 Interaction of Internal Currents 165

requires that V_{CC} be greater than V_{BB} in order to provide reverse bias for the collector-base junction.

Correct bias connections for normal operation of a transistor are easily remembered. *The emitter-base diode is forward biased and the collector-base diode is reverse biased*, whether the transistor is N-P-N or P-N-P. This means that for forward bias the Positive bias source is connected to P-type semiconductor or the Negative source is connected to N-type semiconductor. Opposite connections provide reverse bias. The emitter is always biased in the forward mode and the collector is always biased in the reverse mode when the transistor is operating. If these bias conditions do not exist, then the transistor will not operate in a normal manner.

4-6
Interaction of Internal Currents

When the correct bias conditions are provided, the current in the emitter-base diode can control the current passing through the transistor from collector to emitter. The interaction of internal currents may be explained as follows, referring to the circuit in Fig. 4-7. An electron entering the emitter region from

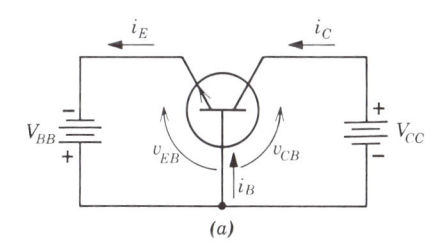

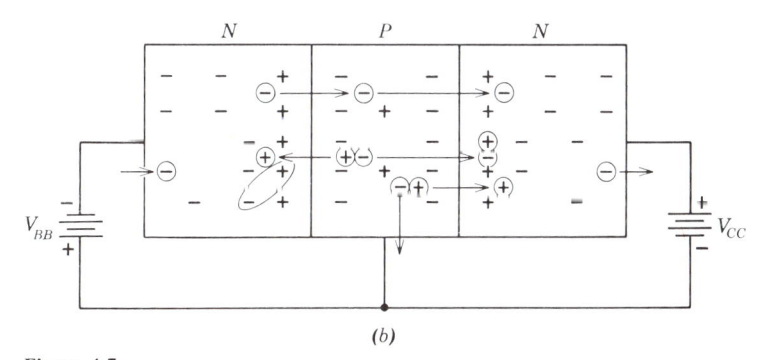

Figure 4-7
(a) Schematic diagram of a transistor and its bias sources. (b) Electron motion and internal generation of transistor currents.

V_{BB} will unbalance the equilibrium of electrons and holes, particularly those making up the potential barrier at the emitter-base junction. There will be a tendency to restore the balance between the number of electrons and holes in the emitter region. An extra electron entering from the bias source upsets this balance, and an electron must be removed from the emitter region. This result may be accomplished in two ways: another electron may transfer from the emitter region to the base region, or the electron may become "inactive" by recombination with a hole in the emitter region.

Regardless of which happens, the number of electrons in the emitter region will be restored and maintained in equilibrium. However, the number of holes in the emitter can be reduced, which upsets the balance of electrons and holes. Furthermore, an extra electron appearing in the base region continues a chain of events there.

Now let us turn our attention to the base region. When an extra electron arrives in the base, several reactions can occur: (1) another electron can go from the base to the collector, (2) the electron can recombine with a hole in the base, or (3) an electron can leave the base and return to the bias supply V_{BB}. Any of these reactions will restore the number of electrons that are active in the base, but recombination would reduce the number of holes to unbalance the equilibrium in the base region. Electron–hole recombination could trigger the generation of another electron–hole pair in the base, however, which would restore the number of holes there. An electron then can either enter the collector or leave at the external bias connection.

When an extra electron enters the collector from the base, a similar fate awaits it. It can recombine with a hole in the collector, triggering electron–hole pair generation, or it can force an electron to leave the collector at its connection to the bias supply V_{CC}. Either event restores the electron–hole balance in the collector, similar to the actions described for the base and emitter.

It should be remembered that electron motion anywhere in a closed path requires electron motion everywhere in that closed path. That is, electric current will pass in the circuit. Electrons going from the bias sources into the transistor regions constitute current in the transistor. Not only will there be emitter current, but also base current and collector current as indicated in Fig. 4-7.

Many electrons are available to enter the emitter from its bias source. By avoiding recombinations in the base, they will cause a continuous current in the collector-emitter circuit. Since the base region is small, there are few holes to permit electron–hole recombinations there. Base current will be small compared with collector current, typically 1/20 to 1/200 of the collector current. Emitter current is the sum of the base and collector currents, as it must be from Kirchhoff's law, so that collector current is almost equal to emitter current.

The magnitude of the emitter current, and consequently the collector current, is determined by the magnitude of the forward bias of the emitter-base diode. Since this bias controls the base current, which flows in the emitter-base diode, it

Sec. 4-6 Interaction of Internal Currents 167

also effectively controls the collector current. Thus, the amount of doping and the physical dimensions of the base region are parameters that affect transistor operation, controlling base current and its effect on collector current. Because of these effects, the transistor is often referred to as a *current amplifier* and is said to be a current-operated device.

Voltage and current relationships given in Fig. 4-8 are typical for a transistor of

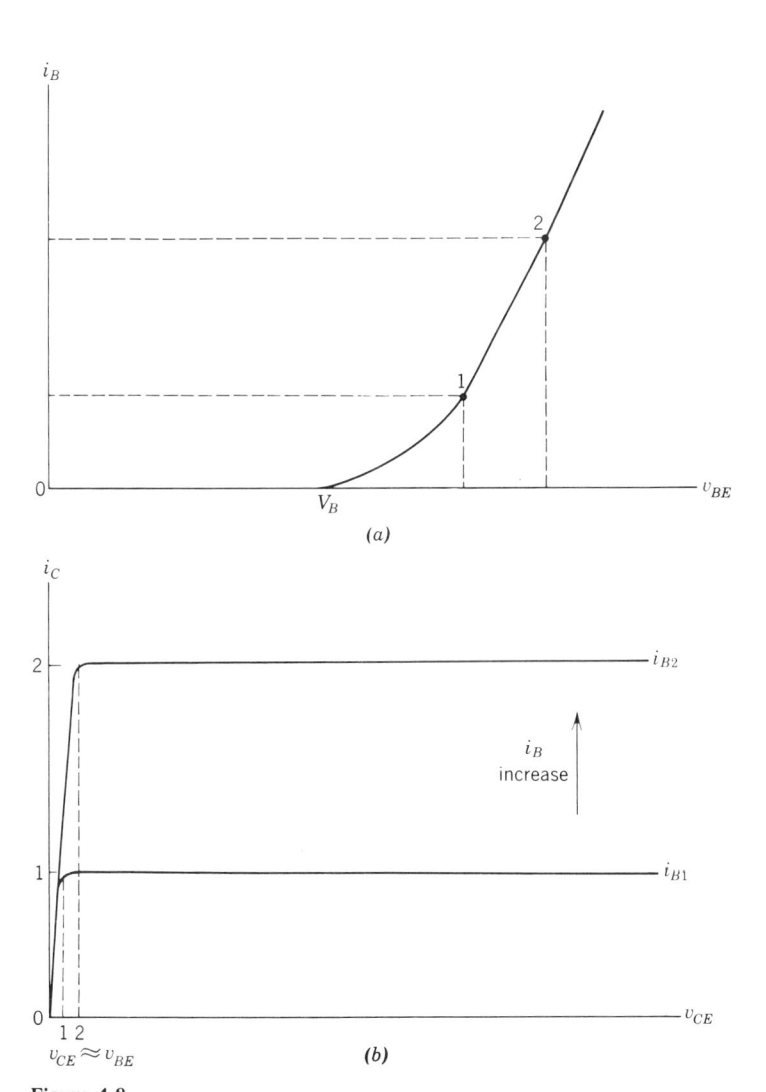

Figure 4-8
Voltage-current characteristic curves for *N-P-N* transistor. (*a*) For base-emitter junction. (*b*) For collector circuit.

the N-P-N type. Note the shape of the base-emitter curve in the upper graph, which describes the characteristic behavior of a forward-biased P-N junction, in this case the base-emitter junction. Once the barrier potential is exceeded by v_{BE}, base current i_B increases as expected of a diode.

Because of the current-control action of the transistor, collector current i_C also increases as the base current increases. This action is shown in the bottom graph. Note that i_C remains essentially constant after the base-collector P-N junction has achieved reverse bias. This condition will occur whenever $v_{CE} > v_{BE}$. As base current i_B increases, as from point 1 to point 2 in the graph, so does collector current i_C. Once the forward-biased B-E junction voltage exceeds the barrier potential and reverse bias is present on the B-C junction, the amount of base current determines the amount of collector current in the transistor.

Typical values of i_B may be 50 μA at point 2, controlling a collector current in the order of 5 mA.

A P-N-P transistor operates in a manner similar to that described for the N-P-N type. In each type, internal current carriers may be either electrons or holes, but external current consists solely of electrons in the connecting wires and bias sources. The only difference in circuit connections of external sources is the opposite polarities of emitter, base, and collector to provide the correct bias voltages for both P-N junctions.

4-7
Reverse-Bias Leakage Current

So far we have discussed internal currents of acceptor and donor charge carriers under conditions of forward and reverse bias. This description of internal currents is incomplete because of the effects of electron–hole pairs generated by thermal ionization. Such effects lead to additional carriers within the semiconductor and are particularly important in transistor operation because of the current amplification that occurs within the junction areas. Current that is caused by thermally generated electrons or holes in a reverse-biased junction is referred to as *leakage current*. Since the base-collector junction is reverse-biased, leakage current can contribute to the collector current.

Figure 4-9 illustrates carrier separation in a reverse-biased junction that leads to leakage current. An external battery provides reverse bias and separation of the electrons and holes in the depletion region, where they are thermally generated. In the figure, an electron on the left joins others in the N-region, and a hole on the right joins others in the P-region. Their migration will trigger electron motion in the external circuit to restore electrical equilibrium, thereby causing a reverse current in the junction. The other electron and hole recombine in the depletion region and do not contribute to current. Current that results is the

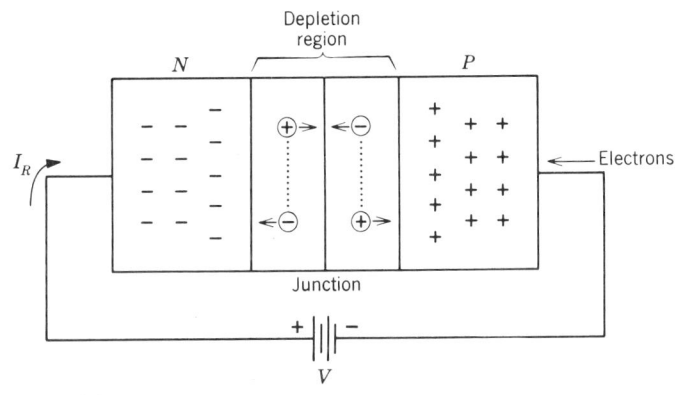

Figure 4-9
Thermally generated electron–hole pairs in the depletion region contribute to leakage current in a reverse-biased P-N junction.

leakage current of the reverse-biased junction. In the base-collector nomenclature, it is labeled I_{CBO}.

The amount of leakage current in a *P-N* junction increases exponentially with temperature. Electron–hole pairs in germanium will approximately double for each 10°C rise in temperature; their numbers triple in silicon for each 10°C rise. However, germanium diodes have about 1000 times the leakage current of silicon devices at 20°C. Since leakage current (reverse current) is undesirable in diodes, it is even more a problem in transistors because of current amplification. Silicon units have replaced germanium devices in most applications, and at competitive costs.

4-8
Common-Terminal Connections

There are three possible arrangements in which transistors can be connected in useful circuits. These are shown in Fig. 4-10. These have *common-emitter*, *common-base*, and *common-collector* connections, respectively, depending on which transistor terminal is connected to both bias supplies. Each of these circuit connections has specific advantages in some applications, but the common-emitter connection is by far the most widely used, partly because it is capable of more power gain than the others.

Correct bias connections for normal operation are easily remembered because *polarity of the emitter*, with respect to the base, *is determined first:* negative if the emitter semiconductor is *N*-type, and positive if it is *P*-type. The base is connected to the other side of the bias supply, either directly or through

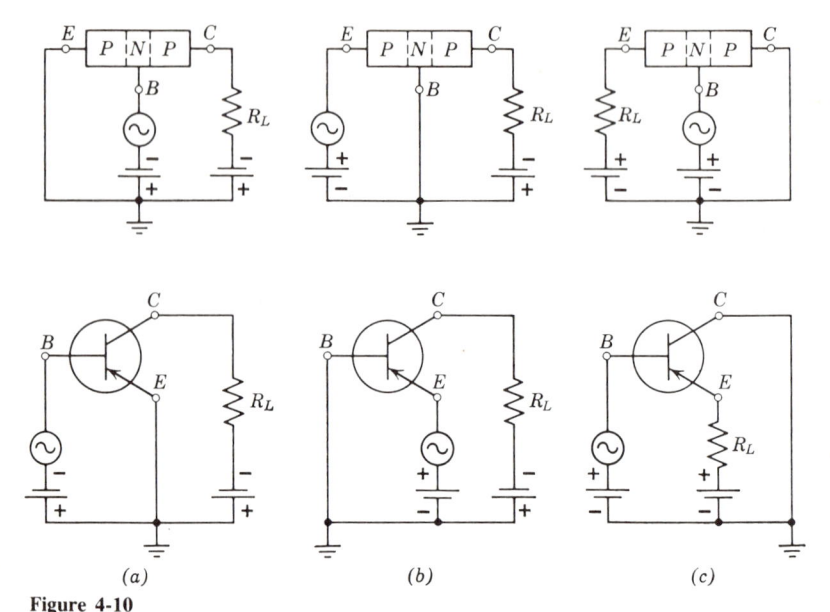

Figure 4-10
Physical connections and schematic diagrams of P-N-P transistors. (a) Common-emitter; input signal applied in base circuit. (b) Common-base; input signal applied in emitter circuit. (c) Common-collector; input signal applied in base circuit.

a signal source. Also, it should be easy to remember that the collector bias is always opposite in polarity to the emitter bias, because one of the internal P-N junctions must be forward biased and the other reverse biased.

The emitter-base junction is forward biased so that it offers *low impedance* to input signal currents which may then be sufficiently large to operate the transistor. Because of the relationship of the junction voltage and current in a forward-biased P-N diode, the emitter-base voltage will change very little for a wide variation in the base current. (It will be helpful to review Fig. 2-19 and associated description of the voltage–current behavior of silicon and germanium diodes.)

In a transistor that is correctly biased, *current* and *current changes* in the base-emitter diode will vary the current in the collector. This means that a transistor is essentially a current-operated device, although it is clear that current in the base circuit cannot operate the transistor unless a controlling voltage is applied. Many respected authors prefer to call a transistor a voltage-controlled device, and others prefer to think of it in terms of current control. Both are correct descriptive terms, of course, because of the necessary relationships between voltages and currents in electric circuit operation.

The emitter is always given forward bias, for reasons explained above, so it is *positive* when the emitter is made of P-type material. The collector is then biased negatively. The load resistor R_L is, of course, in the output circuit. Its voltage drop, its current, or its power is the output quantity desired. The ac component of either the voltage across R_L or the current through it is expected to be an enlarged reproduction of the input-signal voltage or current.

Note the arrow on the emitter electrode in each of the transistor symbols. Note also that the dc component of emitter current will flow from E toward B, that is, from emitter into base. The arrow is drawn in the conventional direction of current flow (not electron flow), and consequently the symbols in the three figures represent a P-N-P transistor. Without the arrow there would be no way of telling from the symbol which type of transistor is represented unless dc bias polarities were given. The dc bias polarities are not readily determined in the circuit diagrams of many transistor applications. We should surmise then (and we would be correct) that the arrow on the emitter electrode *points away from the base* in the symbol for an N-P-N transistor. These facts concerning transistor bias polarities and symbols should be committed to memory at once. A convenient way to identify the type of transistor represented by a symbol is to look at the emitter terminal. If the arrow points *towards the base*, it means that bias current is flowing from the outside *into* the emitter and from there toward the base and collector. This is what happens when the emitter is biased *positively*, and the transistor must therefore be the P-N-P type.

Of the three different ways to connect a transistor in a circuit, the common-emitter and the common-base connections are the most frequently used. There are various groups of characteristic curves showing relationships between two quantities, such as collector current and collector-to-base voltage (I_C vs. V_{CB}), for various constant values of a third quantity (I_E, emitter current). It is seen that the transistor input signal may be applied between base and emitter terminals (through a bias supply, of course) with either terminal grounded, and also between base and collector with the collector grounded. Observe, also, that the load resistor separates the third electrode from the grounded terminal in every case.

It is important that the impurity concentration be much larger in the emitter than in the base. To understand why, consider the P-N-P transistor (Fig. 4-8). It is desired that the collector, which is biased negatively, receive as many *holes* as possible. The emitter, which is strongly P-type with an abundance of free holes, will supply them and send them on their way toward the base. If the base has a much lower impurity concentration than the emitter, only relatively few of the holes will be neutralized, because of the *relative* scarcity of free electrons in the N material of the base. Thus the vast majority of the holes will go down the potential hill to the collector.

172 Junction Transistors Ch. 4

4-9
Characteristic Curves for Transistors

Manufacturers of transistors often supply graphs of the voltage and current relationships characteristic of any particular type of transistor. These graphs usually consist of a family of curves denoting the behavior over some region of voltage and current values within permissible ranges of operation. Operation outside these ranges may lead to damaging effects that could destroy the transistor. As an example of the curves supplied by a manufacturer, and for our discussion of transistor behavior, refer to Fig. 4-11. These curves describe the Texas Instruments 2N334 N-P-N transistor.

The transistor symbol for use in circuit diagrams is often oriented as shown in Fig. 4-11a, with the line representing the base drawn on the vertical and its external connection to the left. The emitter is drawn in the lower right part of the symbol. Note that the base and collector circuits contain variable voltage sources. As these sources are varied, the base current and the collector current will behave according to the graphs in Fig. 4-11b and c.

Let us see how these curves represent the operation of this transistor. When the source V_{BB} is zero, no base current will pass and there will be zero collector current, regardless of the magnitude of V_{CC}. This action is depicted by the line marked $I_B = 0$ along the bottom of the graph in Fig. 4-11c. However, as V_{BB} is increased, the base current will have a value given by the curve in (b), leading to larger amounts of collector current. For some constant value of I_B, there will be little increase in I_C even though collector voltage increases. However, I_C is many times the magnitude of I_B, a current amplification in the transistor. Once reverse bias of the collector-base junction has been established, in the region at the extreme left in Fig. 4-11c, collector current remains essentially constant with a constant base current controlling its magnitude. Increasing I_C with increasing V_{CE} shows about 10 per cent increase with each 100 per cent increase in V_{CE}.

It is interesting to note that constant collector current exists after V_{CE} is larger than V_{BE}. Reverse bias of the collector-base junction occurs only under this condition, which is necessary for transistor action.

The output characteristic curves in Fig. 4-11c constitute a family of curves for the indicated values of I_B. Other values of base current would yield a different family of curves, with the same general shape as these. For example, the curves for $I_B = 80$ and 90 μA would lie between those for 75 and 100 μA.

As the operating temperature of the transistor increases, the input curve would shift to the left in Fig. 4-11b. The V_{BE} junction voltage decreases about 25 mV for each 10°C increase in temperature.

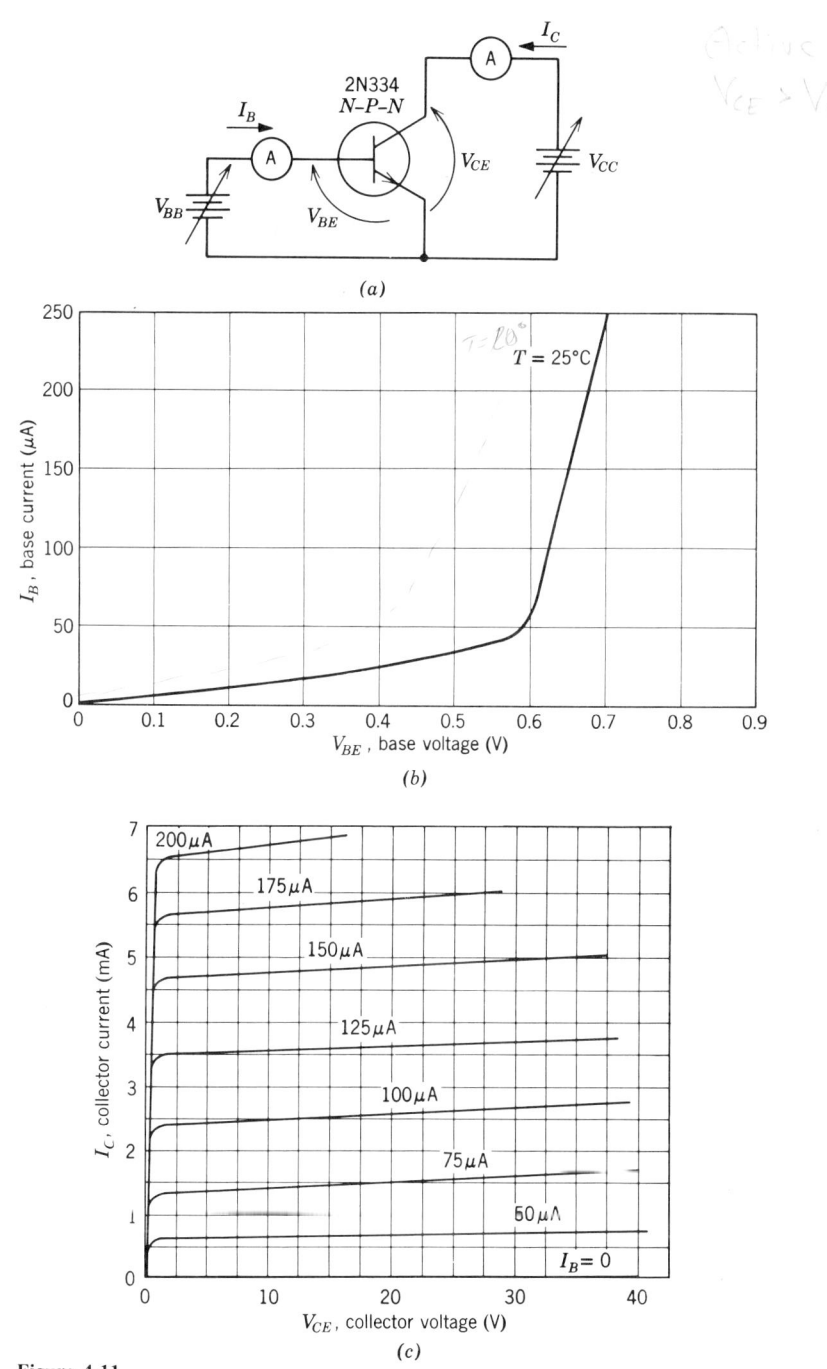

Figure 4-11
Circuit and curves of voltage and current relationships for Texas Instruments 2N334 transistor. (*a*) Schematic circuit diagram. (*b*) Input characteristic. (*c*) Output characteristics.

4-10
Letter Symbols for Transistor Circuits

Many voltages and currents in a circuit containing a transistor are important in its operation. In order to distinguish easily a particular voltage or current for discussion or to represent it in an equation, letter symbols have been adopted as symbols for these circuit parameters. Uppercase letters are used for quantities that are constant—for example, dc voltages and rms values of ac voltages and currents. Lowercase letters denote quantities that can vary in value, such as a sine wave of voltage or current.

Subscripts are also used to distinguish various circuit parameters that are symbolized by uppercase and lowercase letter symbols. Subscripts also can be either upper- or lowercase, again distinguishing constant and variable parameters. As an example, consider the two symbols for the voltage between base and emitter of a transistor: V_{be} and v_{BE}. The first one represents some constant value of voltage and the second represents a variable voltage. But their subscripts denote variable and constant quantities, an apparent contradiction according to the descriptions given above; how can this be? Evidently, V_{be} is the rms value of the ac voltage between base and emitter, whereas v_{BE} is a variable voltage somehow related to a constant voltage between base and emitter. Suppose we rewrite it as $v_{BE} = V_{BE} + v_{be}$. This expression shows that the voltage is the sum of dc and ac voltages between the base and emitter, and V_{be} is the rms value of v_{be}.

Other voltages and currents associated with a transistor circuit can be expressed in a similar manner. A few of them are introduced now, and others will be added as needed in later sections of the book. Once it is learned that uppercase and lowercase letters represent constants and variables, respec-

Table 4-1
Transistor Current and Voltage Symbols

I_B	Value of dc component of base current, its average value
i_b	Instantaneous value of ac component of base current
i_B	Instantaneous total base current (may be combined dc and ac components)
I_C	Value of dc component of collector current
i_C	Instantaneous total collector current
I_{CBO}	Reverse collector-to-base current when emitter is open-circuited
V_{CE}	Average voltage between collector and emitter
v_{ce}	Instantaneous value of ac component of collector-emitter voltage
v_{CE}	Instantaneous total collector-emitter voltage
V_{ce}	Rms value of ac component of collector-emitter voltage; rms of v_{ce}
V_{CC}	Supply voltage for collector circuit; may be battery or dc power supply
V_{BB}	Supply voltage for base circuit; may be battery or dc power supply

Sec. 4-11 Characteristic Data and Curves 175

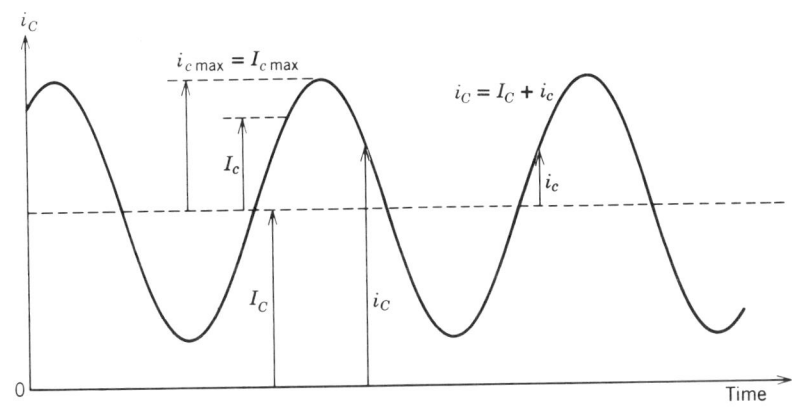

Figure 4-12
Sketch of collector current showing ac and dc components.

tively, there should be little difficulty in understanding their use in mathematical expressions.

A partial listing of voltage and current symbols that will be used for transistor circuits is given in Table 4-1 with an explanation for each. Figure 4-12 is labeled to show various current values that may be identified using this notation system for total collector current, which consists of a sine wave superimposed on the dc or average current.

Drill Problems

D4-3 The total base current of a transistor consists of a sinusoidal current having a peak amplitude of 50 μA superimposed on a dc current of 100 μA. Sketch the waveform of the base current for at least one cycle of the ac portion. Label the amplitudes of the components i_b, I_b, i_B, and I_B.

D4-4 Suppose the transistor in D4-3 has the characteristics given in Fig. 4-11c. Show the region of base current changes if the collector voltage is constant at 20 V.

D4-5 For conditions in D4-4, what will be the variation of collector current? Is its change also sinusoidal?

D4-6 For conditions in D4-4, what will be the variation of V_{BE}?

4-11
Characteristic Data and Curves; Alpha and Beta Factors

The common-emitter operation of a transistor may be described by a set of output characteristic curves as in Fig. 4-13. Notice that the collector current

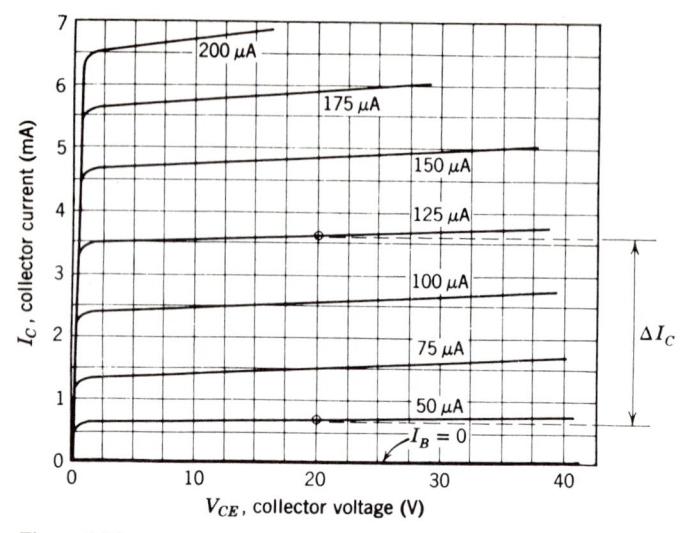

Figure 4-13
Common-emitter output characteristics for 2N334 transistor. (Courtesy Texas Instruments, Inc., Dallas, Tex.)

changes very little for substantial changes in collector-to-emitter voltage. The *ratio* of changes in collector current to changes in base current is the current-gain factor for the common-emitter connection. It is denoted by the Greek letter *beta* (β) and is defined as

$$\beta = \left[\frac{\Delta I_C}{\Delta I_B}\right]_{V_{CE} \text{constant}} \tag{4-1}$$

The base current is much smaller than the emitter current or collector current, because almost all of the emitter current passes to the collector; that is,

$$I_E - I_C = I_B \tag{4-2}$$

Connected in the common-emitter circuit, the base is on the input side, and the input signal will vary the base current. A corresponding change will take place in the collector current. The curves may be used to calculate a value for β:

$$\beta = \frac{(3.2 - 0.7)\text{mA}}{(125 - 50)\mu\text{A}} = \frac{2.5 \text{ mA}}{0.075 \text{ mA}} = 33.3$$

It can be seen that this current-gain factor is closely related to the spacing between constant base-current lines. The farther apart they are, the larger will be the value of β.

The behavior of a transistor may be studied by looking at the characteristic data and curves supplied by the manufacturer. Figure 4-14 shows a set of static

Sec. 4-11　　　　　　　　　　　　　　　　　　Characteristic Data and Curves　177

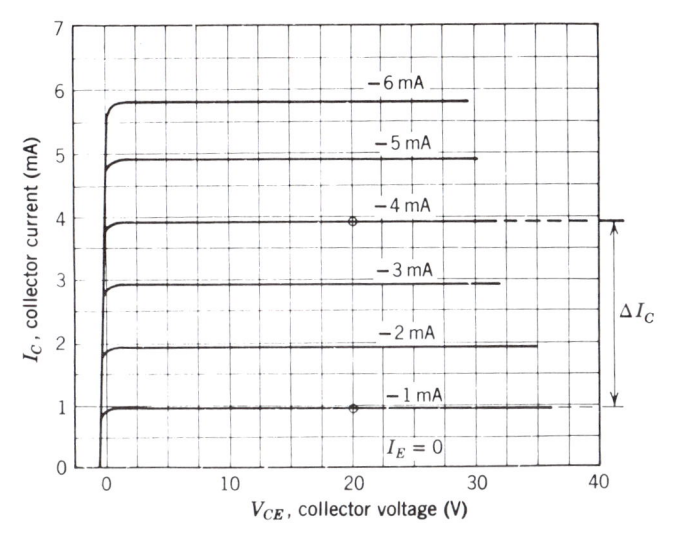

Figure 4-14
Common-base output characteristics of 2N334 transistor. (Courtesy Texas Instruments, Inc., Dallas, Tex.)

curves for the 2N334 transistor, a grown-junction silicon N-P-N type. These curves describe the operation of the transistor in a common-base connection. These *output characteristics* show how the collector current varies with the output voltage between the collector and base terminals for various values of emitter current. Again we are reminded that the transistor is a current-operated device.

For a constant emitter current, the collector current is practically unaffected by a substantial change in collector-to-base voltage. However, the *ratio* of a change in I_C to a change in I_E (at a constant V_{CB}) is an important factor. When this ratio is expressed as a *positive number*, it is called the *forward current gain* of the common-base connection and is designated by the Greek letter *alpha* (α).

$$\alpha = -\left[\frac{\text{change in } I_C}{\text{change in } I_E}\right]_{V_{CB} \text{ constant}} = -\left[\frac{\Delta I_C}{\Delta I_E}\right]_{V_{CB} \text{ constant}} \quad (4\text{-}3)$$

We may get an idea of the magnitude of this factor by calculating its value from the curves. At a constant collector-base voltage of 20 V, a change in I_E from -1 mA to -4 mA ($\Delta I_E = 3$ mA) causes a change in the collector current from 1 mA to 3.9 mA ($\Delta I_C = -2.9$ mA). The value of α is $-(-2.9)/3$, or 0.97. It is important to note that α is always less than unity, and it usually is greater than 0.9. The difference between the value of α and unity represents the magnitude of base current in a transistor, because the emitter current is the sum of base current and collector current. In general, the closer α is to unity, the better the transistor quality.

It should also be noted that the factor α applies only to the common-base connection of the transistor. The current gain for other common-terminal connections is defined in a different way.

The two factors α and β are closely related. One may be calculated from the other. From Eq. 4-2 we obtain

$$\Delta I_E - \Delta I_C = \Delta I_B$$

so that

$$\beta = \frac{\Delta I_C}{\Delta I_E - \Delta I_C} \tag{4-4}$$

The factor β can be expressed in terms of α by recognizing that the numerator and denominator may be divided by ΔI_E,

$$\beta = \frac{\Delta I_C / \Delta I_E}{\Delta I_E / \Delta I_E - \Delta I_C / \Delta I_E}$$

$$\beta = \frac{\alpha}{1 - \alpha} \tag{4-5}$$

Similarly, α may be expressed in terms of β by solving Eq. 4-5 for α:

$$\alpha = \frac{\beta}{1 + \beta} \tag{4-6}$$

These current-gain factors are sometimes referred to as *forward current ratio*, alluding to the ratio of input and output currents. They are also symbolized as h_f and h_F, where the subscript indicates whether ac or dc ratio is meant. In the case of common-emitter circuits, the forward current ratio may be expressed as h_{fe}. Here the subscript *fe* refers to "forward emitter." The quantities denoted by β and h_{fe} are the same for small changes in currents. Further consideration of the forward current ratio is given in later chapters, where amplifying properties of a transistor are treated in detail.

Drill Problems

D4-7 An *N-P-N* transistor is assumed to have linear collector characteristics; the family similar to Fig. 4-13 consists of parallel, horizontal lines that are equally spaced for similar increments of base current. Its beta value is 100. Sketch the family of curves for base current increments of 25 μA up to 200 μA.

D4-8 Using the same axes sketched in D4-7, show locations of base current increments for a transistor having a linear beta of 40.

D4-9 On the common-emitter output curves for the 2N334 given in Fig. 4-13, choose a collector-to-emitter voltage of 5 V and determine the value of β. Is its value constant (*a*) for all values of collector current? (*b*) for all values of collector-emitter voltage?

4-12
Some Important Characteristics and Maximum Ratings

Transistor manufacturers supply data sheets containing operating characteristics and maximum ratings that apply to a particular transistor. The information will usually include a description of the device, followed by sections on environmental tests, mechanical data (type of case, size, shape, etc.), absolute ratings, electrical characteristics, and typical characteristic curves. Application data and parameter information are usually included.

Before inserting a transistor in a circuit design, it is necessary to determine its ability to meet the specifications as to gain, power levels, etc., and to avoid using it where its absolute maximum ratings may be exceeded. These ratings, beyond which degradation or destruction of a transistor may occur, are established by the manufacturer. They are based on the semiconductor material, the manufacturing process, and internal physical construction. Since they represent the extreme capabilities of a transistor, they are not recommended for design conditions. A transistor will not necessarily withstand all maximum ratings simultaneously, so care is necessary in the design of transistor circuits. Transistors in a design that exceeds a maximum rating are more susceptible to destruction than are vacuum tubes.

The section of a data sheet that lists electrical characteristics will be referred to most often, because the limits to the electrical parameters that are important in a circuit design will be found there. Reverse-breakdown voltage, reverse-current values, internal capacitances, and amplification factors are among the parameters. Several of these are not important in vacuum-tube operation, but because of the diode construction of transistors, several diode parameters become important in their application.

As an example of the information supplied by a transistor manufacturer, the following is given for a Texas Instruments type 2N1302 Alloy-Junction Germanium Transistor:

Absolute Maximum Ratings (25°C)

Collector-base voltage	25 V
Emitter-base voltage	25 V
Collector current	300 mA
Total dissipation*	150 mW

* Derate 2.5 mW/°C above 25°C ambient.

180 Junction Transistors Ch. 4

Typical Electrical Characteristics (25°C)

I_{CBO}	Collector reverse current ($V_{CB} = 25$ V, $I_E = 0$)	3 μA
BV_{CBO}	Collector-base breakdown voltage ($I_C = 100$ μA)	25 V
h_{FE}	DC forward current ratio (in common-emitter circuit; ratio of collector and base currents)	50

This list is not exhaustive but is given as an example of the information supplied for a transistor. Neither is this type of transistor available only from Texas Instruments. The type number 2N1302 indicates that this transistor has been registered with the Joint Electron Device Engineering Council (JEDEC). The purpose of registration is to facilitate the purchase and distribution of semiconductor devices and to standardize electronic devices. Registration procedures are designed to ensure that devices differing from one another in their characteristics and performance are identified by different type numbers. Type numbers are assigned in numerical sequence as they are requested and approved. The numbers 1Nxxxx usually denote diodes or rectifiers, 2Nxxxx denote triode devices, and 3Nxxxx denote four-terminal (tetrode) construction. Any manufacturer that makes transistors that have the same characteristics as this one will number his product 2N1302.

4-13
Simple Transistor Circuit; The DC Load Line

We will use the circuit of Fig. 4-15a to study the behavior of a transistor that has both ac and dc currents and voltages. In this circuit the resistor R_B limits the current in the base-emitter diode that will be caused by V_{BB} and e_s acting in superposition. The resistor R_L is the *load* on the transistor in its collector circuit.

Since the base current i_B has both ac and dc components, the collector current i_C also has both ac and dc components. The base current and its variations will control the variations in collector current. Since i_C passes through R_L, the voltage drop across R_L will be related to the variations of i_C. This voltage will also exhibit variations in its waveform that conform to changes in i_C. Figure 4-15b shows the current and voltage variations in this circuit, assuming e_s is sinusoidal.

Variations in current and voltage can be related to the characteristic curves for the transistor. Output curves for the 2N334 operating in the common-emitter mode are given in Fig. 4-13. We will use this transistor to set up mathematical relations between the voltage and current parameters in the circuit of Fig. 4-15.

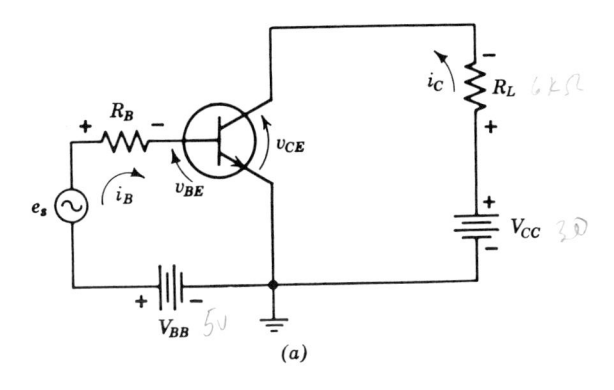

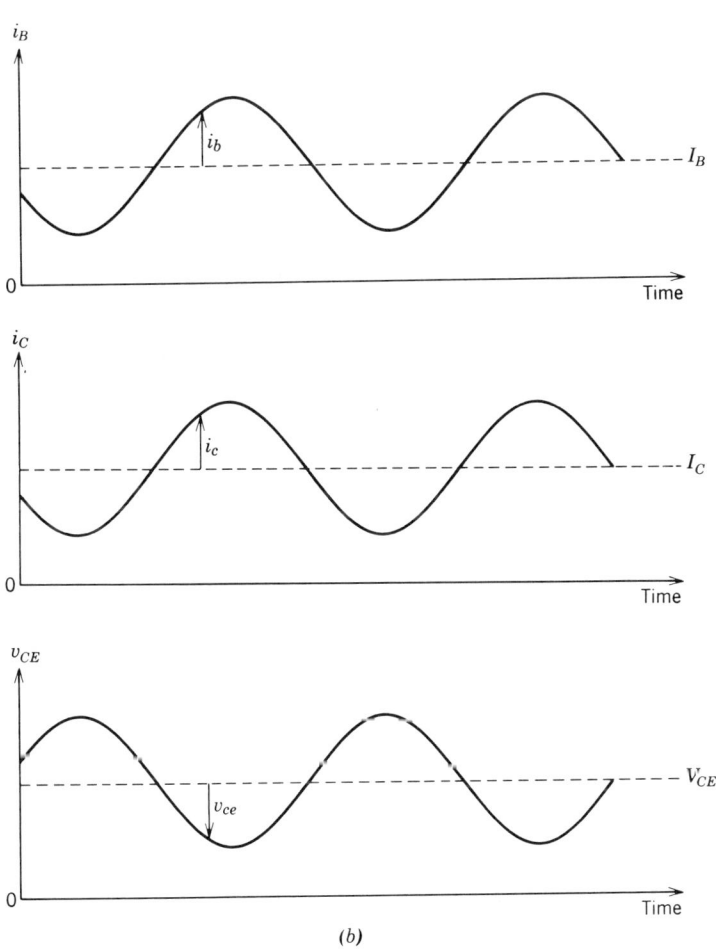

Figure 4-15
Common-emitter transistor circuit. (*a*) Schematic diagram. (*b*) Sketches of currents and voltages.

181

182 Junction Transistors

In order to analyze the circuit operation, we shall assume that R_L is 6 kΩ and the voltage sources, V_{CC} and V_{BB}, are 30 V and 5 V, respectively. These are typical values in this type of circuit. An examination of the output characteristic curves (V_{CE}–i_C) shows that changes in base current will cause simultaneous changes in collector voltage and current. Since the changes in collector current exist in the load (R_L), the voltage developed across the load will change. When collector current is zero, no voltage will be developed across R_L and the collector voltage will equal the source voltage V_{CC}. When collector current is flowing, conditions in the collector circuit may be expressed as

$$V_{CC} - v_{CE} = i_C R_L$$

This will be recognized as a statement of Kirchhoff's voltage law for the collector circuit. The expression may be rearranged in the form of a straight-line equation in terms of the two variables, i_C and v_{CE}:

$$i_C = v_{CE}\left(-\frac{1}{R_L}\right) + \frac{V_{CC}}{R_L} \qquad (4\text{-}7)$$

Equation 4-7 is called the "dc load line equation" for the collector circuit. When this equation is plotted on the output characteristic curves, as in Fig. 4-16, the resulting straight line is called the *dc load line* for the circuit. It may be drawn on these curves between $v_{CE} = V_{CC}$ (when $i_C = 0$) and $i_C = V_{CC}/R_L$ (when $v_{CE} = 0$). The intercepts on the axes in this case are 30 V and 5 mA. This load line represents all possible combinations of collector voltage and collector current that may exist in this particular circuit. Its location is determined by the value of load resistor (R_L) and the collector supply voltage (V_{CC}). Circuits containing other bias sources and load resistors will have their own unique dc load lines.

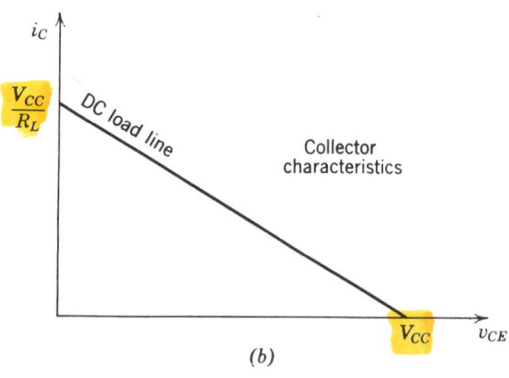

Figure 4-16
Location of dc load line on output characteristic curves.

Sec. 4-14   Steady-State Operation; The Q-Point   183

It can be seen from the input base-emitter curve given in Fig. 4-11b that the 5-V base bias supply (V_{BB}) is much too large for operation of this circuit. The base-emitter voltage, V_{BE}, should be in the range from about 0.6 to 0.7 V. This requires a voltage-dropping resistor, R_B, in the base circuit to limit the voltage to this range. What should be the value of R_B? Let us assume that the circuit is to operate about a nominal base current of 150 µA. Then the resistor should have a value given by

$$R_B = \frac{V_{BB} - V_{BE}}{I_B}$$

Taking the base voltage at the center of its range, the value of R_B in this circuit will be

$$R_B = \frac{(5.0 - 0.65)\text{V}}{150 \times 10^{-6}\text{A}} = 28 \text{ k}\Omega$$

Drill Problems

D4-10 Suppose it is desired to use a single bias source for both the collector and base circuits. The source V_{CC} can be used provided the base-dropping resistor is properly chosen. Sketch a circuit to show the connections that are made to use V_{CC} for biasing the base circuit, and calculate the value of R_B necessary to hold static base current at 100 µA.

D4-11 Determine the dc load lines for the circuit of Fig. 4-15 using a 2N334 transistor and resistor loads of 2 kΩ and 5 kΩ. Assume V_{CC} remains constant at 30 V.

D4-12 Assume that in D4-11 the static base current is 100 µA. For each of the load resistors, calculate or show the maximum amplitude of base current variations that may occur without distorting the output voltage. Assume the base current varies sinusoidally.

D4-13 Repeat D4-11 and D4-12 for $V_{CC} = 10$ V.

4-14
Steady-State Operation; The Q-Point

Using the circuit in Fig. 4-17, we will determine the roles of R_L and R_B in the operation of the circuit. Note that there are no variable signal sources in this circuit; only the dc bias conditions will exist. When transistor voltages and currents are not varying, the transistor circuit is said to be operating in its *steady-state* condition. The following discussion shows that values of bias voltage and circuit resistors can be selected to cause the transistor to operate in any desired region of its characteristics.

184 Junction Transistors   Ch. 4

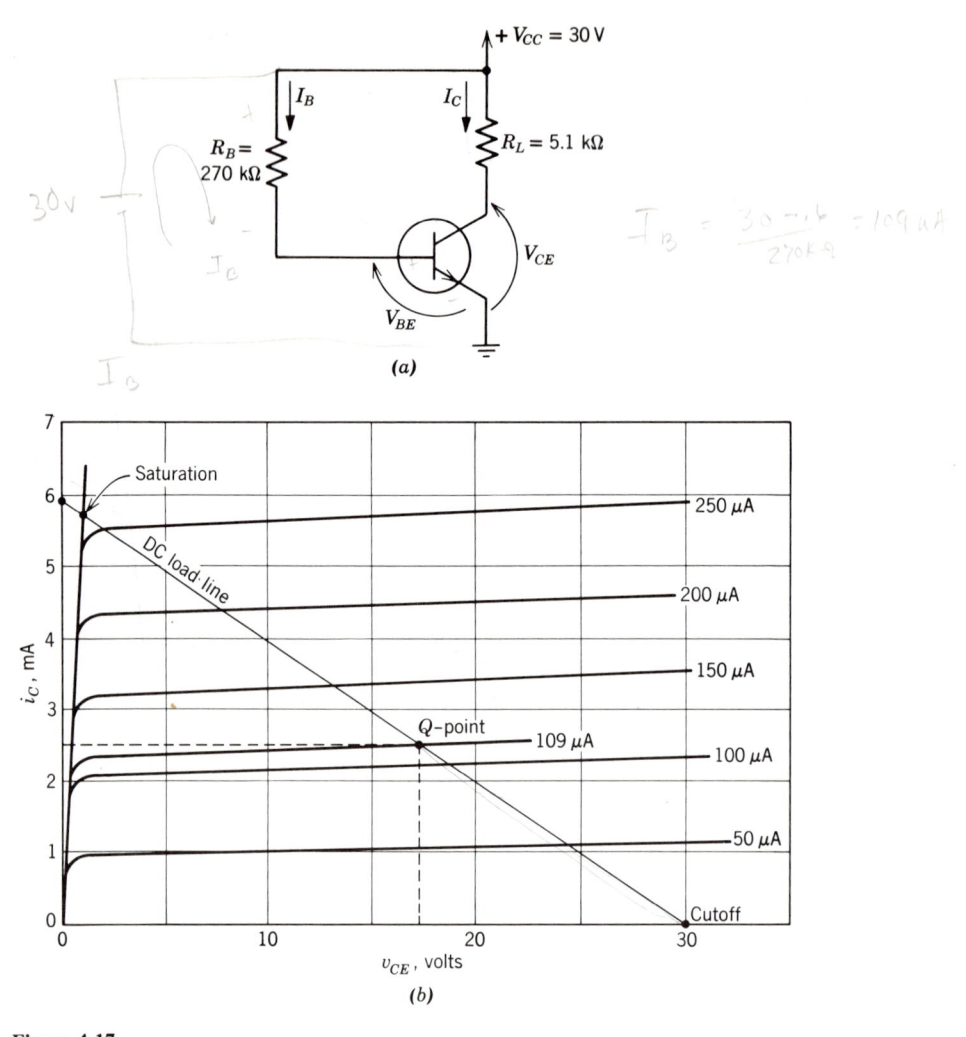

Figure 4-17
(a) Circuit containing biased transistor. (b) The emitter output characteristic curves for the transistor.

We have seen in the previous section that values of bias voltage and the base current-limiting resistor together determine the base current. For the circuit in Fig. 4-17, this relationship may be expressed as

$$\frac{V_{CC} - V_{BE}}{R_B} = I_B \tag{4-8}$$

For this circuit, assuming that the base-emitter voltage is approximately 0.6 V, the only possible value of base current is

$$\frac{(30 - 0.6)\text{V}}{270 \text{ k}\Omega} = 109 \ \mu\text{A}$$

Then we know that this transistor will have about 109 μA of base current. But we know nothing yet about the collector current and collector-emitter voltage.

In the previous section, it was shown that the dc load line describes all possible simultaneous values of collector current and collector-emitter voltage. Then, whichever point along the dc load line also lies on the curve for $I_B = 109 \ \mu$A will describe the point at which the transistor is operating in its steady-state condition (refer to Fig. 4-17b). The point of intersection is called the *Q-point* for this circuit, where Q refers to quiescent or steady-state operation.

Any two of the three parameters I_B, I_C, and V_{CE} will define the third and locate the Q-point on the characteristic curves. For the circuit of Fig. 4-17, I_B, I_C, and V_{CE} are approximately 109 μA, 2.5 mA, and 17.5 V, respectively. Of course, the location of the Q-point depends upon the values of R_B, R_L, and V_{CC}. Transistors with characteristic curves that permit operation over other ranges of current and voltage will have other Q-points. The possible locations of Q-points are virtually infinite, and the circuit designer may select the parameters that define the Q-point to meet functional requirements of the circuit.

4-15
Transistor Cutoff and Saturation

From a study of Fig. 4-17b, it is clear that sinusoidal variations of i_B about $I_B = 109 \ \mu$A would cause the transistor to operate along the dc load line. Sinusoidal changes in i_B would generate similar changes in i_C and v_{CE}, as in Fig. 4-15. All these changes occur about the Q-point as the zero reference for the resulting sine waves of current and voltage. If the amplitude of the sine variation of base current decreases, so will the extent of its positive and negative peak excursions. As its amplitude decreases to zero, the transistor circuit will continue to operate, but at the Q-point in a steady-state condition.

On the other hand, if the amplitude of the base current variation continues to increase, it can become large enough to force total base current to zero. When this condition occurs, the transistor is said to be *in cutoff*. In Fig. 4-17b, cutoff occurs whenever base current is forced to zero along the dc load line. Note that collector current is also zero, as it should be, since base current controls the amount of collector current in the transistor.

At the other extreme of base current variation, when total current is larger than the Q-point value, a magnitude will be reached at the upper end of the load line where no further increase in collector current is possible. When this

186 Junction Transistors   Ch. 4

condition exists, the transistor is said to be *in saturation*. Refer to Fig. 4-17b. Even though the base current may continue to increase its amplitude, no further increase in collector current is possible in this circuit. Thus, a transistor can be driven into both cutoff and saturation when the base-current changes along the load line are sufficiently large above and below the Q-point.

In the case of Fig. 4-17b, cutoff will exist whenever the base current decreases from the Q-point by about 109 μA. However, saturation will not occur until base current has increased to about 260 μA, an increase of 151 μA above the Q-point. Sometimes transistor circuits are purposely designed so that the transistor will be driven into cutoff and saturation. In these cases it may be desirable to have symmetrical variations about the Q-point. Then the center of the operating region would be chosen to locate the Q-point, in this case at about 130 μA. Placing the operating point at that value of i_B would require a smaller R_B, according to Eq. 4-8.

From the foregoing discussion and Fig. 4-15, it is noted that an *increase* in base current will cause a *decrease* in collector voltage. This result is the characteristic 180° phase difference between input signal and output voltage of a common-emitter circuit.

Drill Problems

D4-14 Output curves for the 2N1308 transistor are given in the Appendix. Design and calculate values for R_B and R_L to use this type of transistor in a circuit similar to that in Fig. 4-17. Use $V_{CC} = 5$ V and locate the Q-point at $I_B = 0.1$ mA and $V_{CE} = 2.5$ V.

D4-15 Using the design values obtained in D4-14, determine maximum and minimum amplitudes of i_b to cause both cutoff and saturation of the transistor. Assuming a sine wave for base current variations, which will occur first, cutoff or saturation?

4-16
Collector Dissipation

An important electrical rating of a transistor is its maximum permissible collector dissipation. This rating is usually expressed in milliwatts for transistors that are designed for handling small signals, and can be several watts for large units that are specifically constructed for operation at high power levels, for example, to drive a loudspeaker or a small motor.

Collector dissipation ratings refer to the amount of power being controlled by the transistor in its operation. They are indirectly a measure of heat that must be removed from the P-N junctions. Collector power in the common-emitter mode of operation is the mathematical product of i_C and v_{CE}. At the

Q-point, it is $I_C V_{CE}$. Since it is the product of collector current and collector-emitter voltage, collector dissipation can be plotted on the output characteristic curves, as in Fig. 4-18. Note that constant values of P_C result in hyperbolic curves.

Since the maximum permissible P_C is a rating assigned by the manufacturer for a given type of transistor, the circuit design using the transistor must not permit operation in a region where this rating will be exceeded. To do so could result in permanent damage to the transistor, resulting in an inoperative circuit and necessitating replacement of the transistor. Therefore, when selecting values of V_{CC}, R_L, and R_B (setting the Q-point and load line location) during design of a transistor circuit, it is important to avoid a collector dissipation that exceeds the rating of the transistor. In Fig. 4-18, the load line drawn across the P_C curves represents a circuit containing $V_{CC} = 40$ V and $R_L = 4$ kΩ. If the transistor can dissipate at least 100 mW, then safe operation can be achieved. However, it would be good practice to design the circuit for a dissipation that is smaller than the maximum permissible rating. We shall see in the next section that stability of the Q-point location cannot be completely assured, perhaps permitting destruction of the transistor if it is operated too close to its P_C rating.

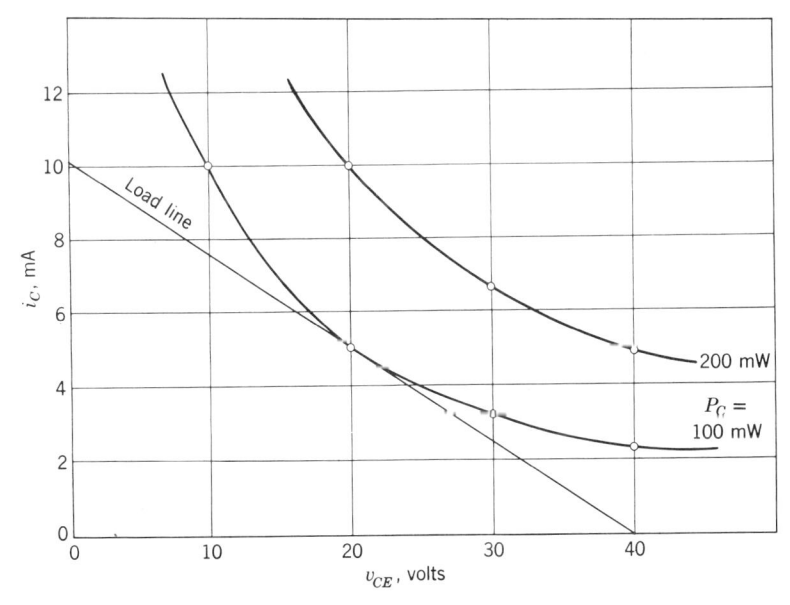

Figure 4-18
Collector dissipation curves plotted on output characteristics.

4-17
Effects of Temperature on the Q-Point

Several variable factors associated with the operating temperature of a transistor affect the location and stability of the Q-point. Interaction between these variables may shift the Q-point to an undesirable location on the characteristic curves of the transistor. For the best reproduction of an input signal without appreciable distortion, the Q-point should be located at approximately the center of the load line. Also, it should remain below the maximum dissipation level, to prevent self-destruction of the transistor because of internal heating. The location of the Q-point, and possible variations of it because of temperature effects, should remain in a region where cumulative self-heating is not possible. Self-heating that may result in the self-destruction of a transistor is referred to as *thermal runaway*.

Temperature variations affect certain characteristics of a transistor: forward current ratio (β), reverse collector current (I_{CBO}), and the quiescent base-emitter voltage (V_{BE}). When it is operated in the common-emitter mode, a transistor may be represented by the linear model of Fig. 4-19. In this diagram, the battery represents the quiescent value of base-emitter voltage, and the collector current includes both the reverse collector current caused by reverse bias of the collector-base junction and the current component caused by the forward current ratio.

The value of β can vary through a wide range of values from one transistor to another, even for those of the same type. For example, the 2N334 has a guaranteed β of 18 to 90, a 5-to-1 ratio. For a constant value of base current, the collector current at the Q-point can vary over a range of values in the ratio of 5 to 1 for transistors of the 2N334 type. Such variations in β can become quite a problem when it is necessary to replace a given transistor, unless some provision is made in the circuit to allow stabilization of the Q-point.

The curves of Fig. 4-20 show how the Q-point may be shifted when base

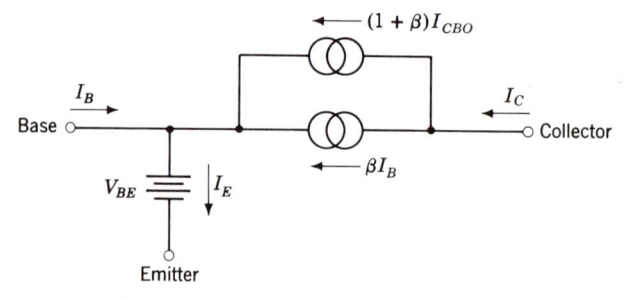

Figure 4-19
Model of common-emitter operation of a transistor.

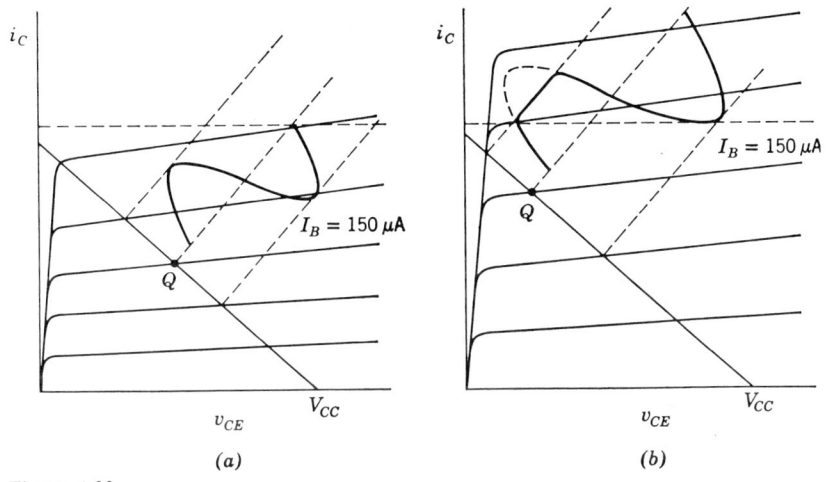

Figure 4-20
Shifted Q-point resulting from replacement with high-β device causes distortion of signals. (a) Low-β transistor. (b) High-β transistor.

current is held fixed, and a transistor having a low value of β is replaced with one having a much higher β. The value of β is also dependent on the operating temperature of the transistor, so that the Q-point may be shifted because of temperature effects, as well.

Increasing the operating temperature will increase the reverse collector current in a transistor. This component of the collector current is generated primarily by thermal ionization, although surface effects contribute a small fraction of the total. The component that occurs because of thermal ionization within the semiconductor crystal is usually predominant in shifting the Q-point because of temperature variations. The value of I_{CBO} approximately doubles for each 10°C increase in temperature.

In silicon transistors, the change in I_{CBO} with a change in temperature is usually not troublesome except at higher temperatures, because its value is small compared to the total collector current. In germanium transistors, however, thermal ionization may generate 1000 times as many current carriers as the same temperature would produce in silicon. The reverse collector current component due to thermal ionization may become large enough so that the total collector current is increased an appreciable amount. For a fixed base current, the Q-point will be shifted upward as in Fig. 4-20.

The value of the quiescent base-emitter voltage *decreases* about 2.5 mV/°C with *increases* in temperature. Since the base current is determined by this voltage in a fixed-base-current biasing arrangement

$$I_B = \frac{V_{BB} - V_{BE}}{R_B}$$

decreasing V_{BE} will increase I_B, and the Q-point will shift upward to higher collector current regions. In germanium transistors, V_{BE} is smaller than in comparable silicon units but can be important in either, depending on the base-bias source V_{BB}.

All of the factors considered, β, I_{CBO}, and V_{BE}, cause the Q-point to shift upward toward increasing collector current when the temperature of the transistor is increased. The operating temperature of a transistor is determined by two factors: (1) ambient conditions and (2) internal self-heating. Ambient conditions may be controlled externally, but the self-heating effects must be controlled internally, by controlling the variations of these three factors.

Self-heating occurs internally because of the power that is dissipated in the collector junction. It results in an increased temperature of the junction above the ambient temperature. The rate of dissipation of the heat will be determined by the rate of heating and the rate of cooling permitted by the physical conditions at the junction. The increase in junction temperature with increased power in a transistor results in the derating factors for transistors. For example, the 2N334 is derated at 1 mW/°C. Since its maximum collector dissipation rating is 150 mW, it cannot safely dissipate any power at 150°C and only 50 mW at 100°C. The derating necessary is dependent on the ability of the transistor to dissipate heat. Some high-power transistors are mounted on large metal bases of copper or aluminum, called *heat sinks*, and their outside surfaces may be formed as fins to radiate the heat more readily. In any event, self-heating must not be permitted at a rate that will be self-destructive.

The curves of Fig. 4-21 show the effect of a shift of the Q-point on the

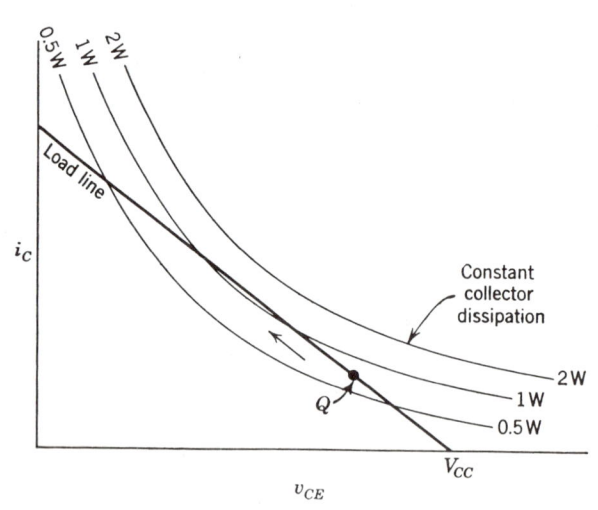

Figure 4-21
Effect of shifting Q-point location on collector dissipation.

collector dissipation. If the initial location of the Q-point is below the "knees" of the dissipation curves, then increases in operating temperature may shift the Q-point to higher collector currents and thus increase the operating temperature even more. If the rate of self-heating is greater than the cooling rate of the assembly, the effect will be cumulative and may lead to destruction of the transistor. However, if the initial location of the Q-point is above this region, increases in collector current because of the upward shifting of the Q-point will result in a lower dissipation.

We have looked at the possible variations that may occur in the operation of a transistor because of temperature variations. If any practical use is to be made of transistors in reliable circuits, some means must be provided to counteract internal parameter changes. This requires that external circuit components be used and connected in such a way that their actions will reduce the effects of parameter variations from one transistor to another and counteract the effects of temperature variations.

4-18
Q-Point Stabilization

There is some objection to using a single biasing resistor in the base circuit. The high external resistance of R_B tends to free the transistor for rather wide shifts in the Q-point location, resulting from the effects of temperature on the internal parameters.

In order to compare other biasing arrangements to the fixed-base-current method that used R_B alone, we will determine a *stability factor*, S, for transistor bias circuits. A measure of the stability of the dc bias conditions is the ratio of the change in collector current to the change in leakage current that caused it:

$$S = \frac{\Delta I_C}{\Delta I_{CBO}} \qquad (4\text{-}9)$$

where ΔI_C is the change in collector current caused by a change in the leakage current ΔI_{CBO}. A value of unity for S would be ideal, because this would indicate that collector current would change only by the amount of change in leakage current. Poor stability results when S is large, which is a consequence of current gain in a transistor that is biased. A circuit that tends to prevent changes in collector current or holds it to the same change as leakage current provides stability $S = 1$, the best that can be expected.

In the transistor model for common-emitter operation of a transistor given in Fig. 4-19, it is seen that total collector current may be expressed as

$$I_C = \beta I_B + (1 + \beta) I_{CBO} \qquad (4\text{-}10)$$

For changes in leakage current, the collector current will change:

$$\Delta I_C = \beta \Delta I_B + (1 + \beta)\Delta I_{CBO} \tag{4-11}$$

Dividing through by ΔI_{CBO} gives

$$S = \frac{\Delta I_C}{\Delta I_{CBO}} = \beta\left(\frac{\Delta I_B}{\Delta I_{CBO}}\right) + (1 + \beta)$$
$$= \beta(0) + (1 + \beta) = 1 + \beta \tag{4-12}$$

The stability factor is dependent upon the current gain of the transistor and will be greater for larger values of β. This means that the Q-point may shift over a wide range of values with changes in leakage current and for replacement transistors having a wide range of β values. For a transistor that has $\beta = 100$, the stability factor of the circuit in Fig. 4-19 will be 101, extremely large compared to the ideal value of unity.

The arrangement shown in Fig. 4-22 is often used to stabilize the operation of a transistor circuit so that the Q-point occurs in some desirable region of the characteristics. This scheme for base biasing overcomes the instability difficulties of temperature effects on internal parameters and variations of β between individual transistors. Even those of the same type, which may be used as replacements in the circuit, can have a wide range of β values.

A simpler form of this circuit is given in Fig. 4-23, in which the base circuit components are replaced by their Thévenin equivalent. The physical reason for improvement in bias stability may be explained in this way. If I_C tends to increase, perhaps because of an increase in leakage current generated by increasing temperature, the emitter current also increases. This action results in

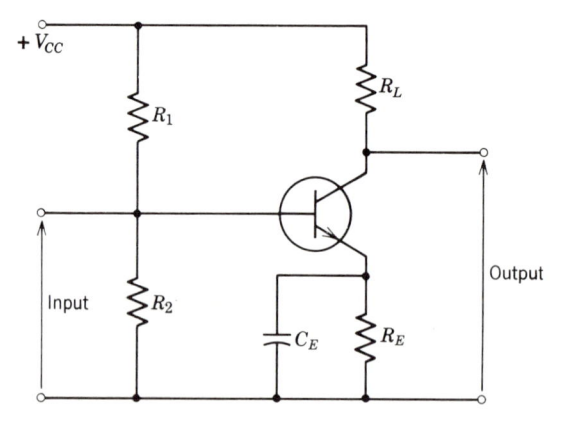

Figure 4-22
Bias network for stabilization of Q-point in common-emitter circuit.

Sec. 4-18　　　　　　　　　　　　　　　　　　Q-Point Stabilization　193

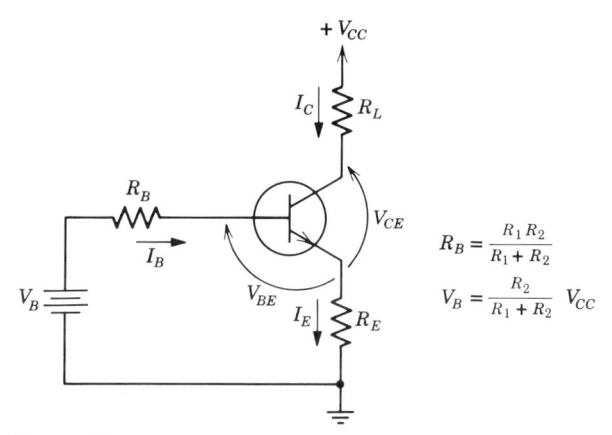

Figure 4-23
Equivalent dc circuit for Fig. 4-22.

a larger voltage drop across R_E, reducing both V_{BE} and I_B, and hence reducing I_C or holding its value lower than it would have been without R_E in the circuit.

Stability in this circuit results primarily from the presence of R_E in the emitter lead. Changes in emitter current, which is largely collector current, will change the emitter-base voltage. Any increase in I_{CBO} or βI_B in the collector circuit will decrease V_{BE}. A smaller voltage here reduces the quiescent base current, and the collector current is in turn reduced. Thus, operation of the circuit tends to maintain the Q-point at some preset location on the characteristics.

The stability of this bias circuit is given by

$$S = \frac{1+\beta}{1+\beta\left[\dfrac{R_E}{R_E+R_B}\right]} = \frac{1+\beta}{1+\beta\left[\dfrac{1}{1+R_B/R_E}\right]} \qquad (4\text{-}13)$$

Note that for $R_E = 0$, the value of S is the same as for fixed-base-current bias, $1 + \beta$. It will be seen also that for $R_B/R_E \ll 1$, S will approach the ideal value of 1. Of course, some value of S between these extremes will result from a given circuit design.

The capacitor shunting R_E in Fig. 4-22 bypasses ac signals around R_E. If there were no path for ac currents around R_E, the voltage developed across it would tend to reduce any input signal at the base, and a lower amplification would result in the collector circuit. This action is called *degenerative feedback* and is usually avoided in amplifying circuits. There are special circumstances in which it is purposely introduced for its advantages, but these are reserved for discussion in a subsequent chapter.

4-19
Locating the Q-Point

In the circuit of Fig. 4-23, Kirchhoff's voltage law for the collector circuit yields

$$V_{CC} - I_C R_L - (I_C + I_B)R_E - V_{CE} = 0$$

or

$$I_C = \frac{V_{CC} - I_B R_E}{R_L + R_E} + \left(-\frac{1}{R_L + R_E}\right) V_{CE} \qquad (4\text{-}14)$$

This expression is in the form of the dc load line discussed in previous sections. If we assume that $I_B R_E$ is small compared with V_{CC}, Eq. 4-14 will plot as a straight line on the I_C, V_{CE} output characteristics. The Q-point will be located somewhere along its length, depending upon the value of I_B in the circuit.

Kirchhoff's voltage law for the base circuit yields

$$V_B - I_B R_B - V_{BE} - (I_B + I_C)R_E = 0 \qquad (4\text{-}15)$$

After substituting the value of I_C from Eq. 4-14, an expression relating I_B and V_{CE} results:

$$V_{CE} = V_{CC} - (V_B - V_{BE})\left(1 + \frac{R_L}{R_E}\right) + I_B\left[R_B\left(1 + \frac{R_L}{R_E}\right) + R_L\right] \qquad (4\text{-}16)$$

Equation 4-16 is called the *bias line* for this circuit. For specific circuit components, the relationship of I_B and V_{CE} can be plotted on the output characteristics. The resulting curve will intersect the dc load line somewhere along its length. This intersection locates the Q-point, because only the values of I_B, V_{CE}, and I_C at the intersection can exist simultaneously in the circuit, according to Kirchhoff's voltage law.

Example. Determine the Q-point for a 2N334 transistor that is connected in a common-emitter circuit as in Fig. 4-22, in which $V_{CC} = 40$ V, $R_L = 7$ kΩ, $R_E = 1$ kΩ, $R_1 = 90$ kΩ, and $R_2 = 10$ kΩ. Calculate the stability factor for the circuit arrangement, and compare with the stability of a single base-bias resistor.

Solution: Using the equivalent circuit of Fig. 4-23 and assuming $V_{BE} = 0.6$ V for this silicon transistor,

$$V_B = \frac{10 \text{ k}\Omega}{(90 + 10)\text{k}\Omega}(40 \text{ V}) = 4 \text{ V} \qquad \text{and} \qquad R_B = \frac{90 \text{ k}\Omega(10 \text{ k}\Omega)}{(90 + 10)\text{k}\Omega} = 9 \text{ k}\Omega$$

Kirchhoff's voltage law applied in the collector and base circuits gives the two equations,

$$40 \text{ V} - I_C(7 \text{ k}\Omega) - I_C(1 \text{ k}\Omega) - I_B(1 \text{ k}\Omega) - V_{CE} = 0$$
$$4 \text{ V} - I_B(9 \text{ k}\Omega) - 0.6 \text{ V} - I_B(1 \text{ k}\Omega) - I_C(1 \text{ k}\Omega) = 0$$

Sec. 4-19 *Locating the Q-Point* 195

or
$$40 - 7I_C - I_C - I_B - V_{CE} = 0$$
$$4 - 9I_B - 0.6 - I_B - I_C = 0$$

where I_C and I_B are in milliamperes.

Eliminating I_C results in the relationship between I_B and V_{CE}:

$$V_{CE} = 12.8 + 79I_B \quad \text{(cf. Eq. 4-16)}$$

For certain values of I_B taken from plotted lines on the output curves in Fig. 4-24, the corresponding values of V_{CE} are calculated:

$$I_B(\mu A): \quad 75 \quad 100 \quad 125 \quad 150$$
$$V_{CE}(V): \quad 19.0 \quad 20.7 \quad 22.3 \quad 24.65$$

These corresponding values are graphed in Fig. 4-24 and labeled "bias line."

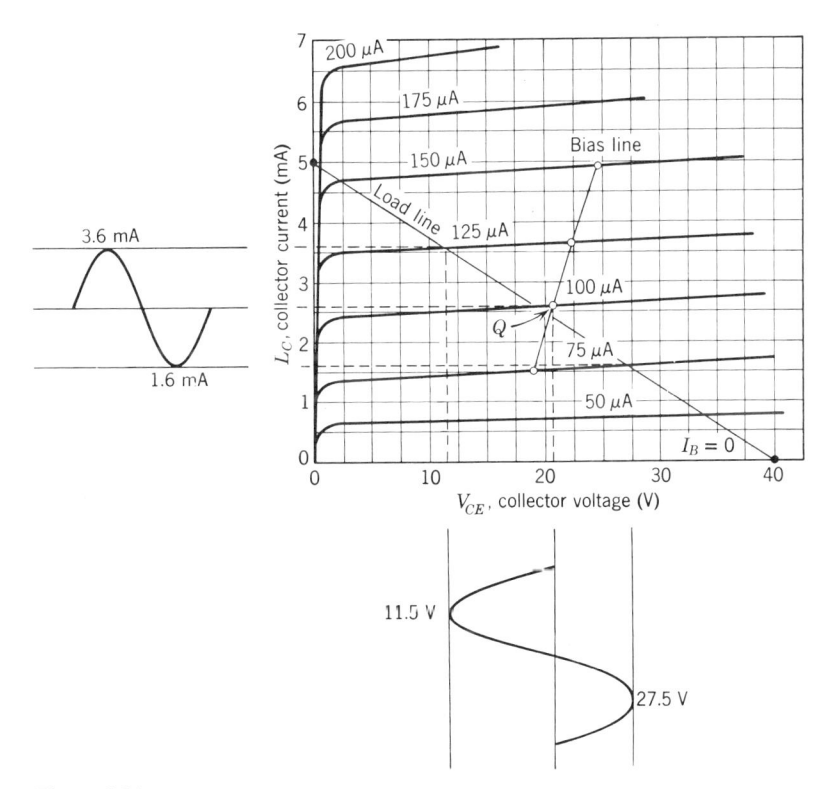

Figure 4-24
Common-emitter output characteristic curves for 2N334 transistor, showing locations of dc load line and bias line for circuit described in example.

The dc load line is also drawn between $V_{CC} = 40$ V and $I_C = 5$ mA. The Q-point occurs at their intersection, at about $I_B = 100\,\mu\text{A}$, $V_{CE} = 20.5$ V, and $I_C = 2.5$ mA.

The stability factor may be calculated from Eq. 4-13, assuming $\beta = 50$,

$$S = \frac{1+50}{1+50\left(\dfrac{1}{1+(9\,\text{k}\Omega/1\,\text{k}\Omega)}\right)} = \frac{51}{1+5} = 8.5$$

With a single base-bias resistor, S would equal 51, which is considerably larger. Even if the transistor were replaced by other similar units, with beta values in the range from 18 to 90 that is typical of the 2N334, the stability factor would range from only 6.5 to 9. Certainly this bias arrangement has reduced any effect of leakage current on the Q-point location, and has also stabilized the operation for a wide range of beta values.

4-20
Amplifying Action in a Transistor

The circuit of the foregoing example will be used to describe amplification in a common-emitter amplifier. A small varying potential applied in the input circuit, as shown in Fig. 4-22, will alter the base current in the transistor. These current variations will be reproduced as larger current changes in the collector. Collector current changes will generate a variable voltage across R_L, which may be taken as the output of the circuit. This means that *voltage* amplification may occur, as well as *current* amplification.

Base current variations will cause the transistor to shift its operation along the load line in Fig. 4-24, generating changes in both collector current and collector-emitter voltage. Suppose the base current varies sinusoidally about the Q-point, with $i_b = 25\,\mu\text{A}$ peak. Collector current will vary from about 1.6 to 3.6 mA, a peak value of 1 mA. The current amplification then is $1/0.025 = 40$.

Collector-emitter voltage varies from about 27.5 to 11.5 V, a peak value of 8 V. If a source of 1-V peak input signal caused the base current changes, then the voltage gain would be $8/1 = 8$. Power amplification, or power gain, is $40 \times 8 = 320$. These concepts of gain or amplification are developed in more detail in later chapters.

There is a voltage phase shift through this amplifier. When the input signal increases, it increases the base current and hence increases collector current, thereby decreasing collector-emitter voltage. The output load voltage goes into its negative half-cycle when the input signal is in its positive half-cycle. Conversely, decreases in signal voltage will increase the output, generating an output voltage that is 180° out of phase with the input signal.

4-21
Design Considerations

It should be clear that external circuit components and power supplies determine how a transistor behaves in a given circuit. Nothing inherent in the transistor can control its behavior, except for the limitations set by its characteristics and ratings. As long as it is operated within these limiting factors, its behavior is controlled by conditions external to the transistor. A circuit designer can make a complete circuit containing transistors perform as desired, limited only by the characteristics and ratings of the units chosen for the circuit.

It is not economical to hold manufacturing tolerances on ratings and characteristics closer than perhaps ±100 per cent of a specified value. Even if such control of the transistor's properties were possible within close tolerances, the temperature dependence of its internal parameters likely would negate any efforts to hold closer tolerances. Thus, the task of a circuit designer is clear: to select and interconnect circuit components in such a way that the circuit behaves reasonably well in spite of varying parameters and wide tolerances on transistor properties.

Proper circuit design usually hinges on making the circuit performance nearly independent of transistor parameters. Collector current and its variations are the principal parameters of concern in the design of a transistor circuit. As we have seen, variations in the base-emitter voltage, in leakage current, and in current gain affect collector current. Stability of the circuit about a selected Q-point is a major consideration of the circuit designer. Complex networks external to the transistor may be required to achieve it, even though they may be unnecessary for its basic function as an amplifier.

Questions

4-1   What is the origin of the word "transistor"?

4-2   What are the basic elements of a junction transistor? What are the approximate dimensions?

4-3   Draw from memory the symbol of the P-N-P and N-P-N transistors as they are represented in schematic diagrams.

4-4   Draw from memory an N-P-N transistor in a common-base circuit. Show bias voltages and label all parts.

4-5   Draw from memory a P-N-P transistor in a common-emitter circuit. Show bias voltages and label all parts.

4-6   With what polarity is the emitter normally biased? Why?

4-7   Why is a transistor classified as a *current-operated* device?

198 Junction Transistors   Ch. 4

4-8 What is meant by the statement that the Fermi level has a "fifty per cent probability of occupancy"?

4-9 What is the effect on the Fermi level when a *donor* impurity is added to a semiconductor?

4-10 What is the effect on the Fermi level when an *acceptor* impurity is added to a semiconductor?

4-11 What is an energy hill at a *P-N* junction?

4-12 Does forward bias across a *P-N* junction produce conduction by majority carriers or by minority carriers? Which way do they go across the junction?

4-13 Answer the preceding question for reverse-bias conditions.

4-14 In what direction do majority carriers in both regions move when reverse bias is applied?

4-15 Referring to the partial list of symbols in Section 4-10, tell what the following represent: v_{ce}, V_{CE}, I_e, i_E, v_{CE}, I_B.

4-16 Define current gain in a common-base transistor amplifier. What is the symbol?

4-17 Why can α never be greater than unity in a junction transistor? In fact, why can it not quite reach unity?

4-18 Define current gain in a common-emitter transistor amplifier.

4-19 Why does β have a large value compared with α?

4-20 Explain in a general way how amplification takes place through a transistor.

Problems

4-1 Choose a constant collector-to-base voltage of 5 V on the common-base characteristic curves for the 2N334 transistor (see Appendix) and determine the value of α. Is it constant for (*a*) all collector current, (*b*) all collector-base voltages?

4-2 On the common-emitter curves for the 2N1308 transistor (see Appendix), choose a collector-to-emitter voltage of 3 V and determine β. Is it constant for (*a*) all collector current, (*b*) all collector-emitter voltages?

4-3 Show that α and β are related by Eqs. 4-5 and 4-6. Calculate α in terms of β.

4-4 The output characteristics of a transistor that is used in a common-emitter amplifier can be approximated by horizontal lines with $i_C = 50\, i_B$. Sketch and dimension a set of these curves for i_B in increments of 50 μA from 0 to 250 μA. (*a*) The circuit uses a supply voltage $V_{CC} = 20$ V and

4-5 $R_L = 2 \text{ k}\Omega$. Construct a load line for the circuit on the characteristic curves. (b) Determine the static collector current and collector-emitter voltage necessary to operate the circuit at the midpoint of its load line. Estimate the value of base dropping resistor necessary to operate at this point.

4-5 The maximum permissible collector dissipation for the transistor of Problem 4-4 is given as 50 mW. Is this value exceeded in the circuit of Problem 4-4? (The collector dissipation is the product of collector current and collector voltage.)

4-6 It is of interest to know how the circuit of Problem 4-4 will behave if its supply voltage should decrease to 10 V. For this purpose, determine the effect of decreasing V_{CC} to 10 V, with other components in the circuit remaining the same.

4-7 An N-P-N silicon transistor is assumed to have linear output characteristics and a current amplification factor $\beta = 80$. It is used in the common-emitter connection and the circuit uses only one bias supply. The supply voltage $V_{CC} = 20$ V, $R_L = 3.3 \text{ k}\Omega$, and $R_B = 330 \text{ k}\Omega$. (a) Determine the load line and static operating point. (b) If the input signal is a sinusoidal current, determine the approximate amplitude of collector current that is permissible without serious waveform distortion in R_L. Is the limit set by collector current cutoff or saturation of the collector?

4-8 For the common-collector configuration, the short-circuit current gain may be approximated by β/α. Find an expression for this ratio in terms of (a) α only, (b) β only.

4-9 Using Fig. P4-9, sketch curves showing phase relations among e_s, i_B, i_C, and v_{CE}, for $e_s = E_m \sin \omega t$. What is the relative phase shift between e_s and v_{CE}?

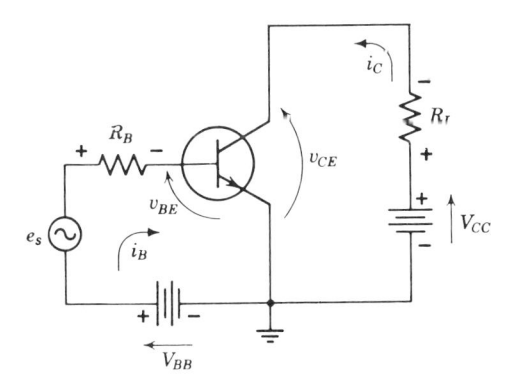

Figure P4-9

200   Junction Transistors   Ch. 4

4-10 A common-emitter amplifier using a 2N334 transistor has $V_{CC} = 30$ V and collector load resistor $R_L = 6$ kΩ. Plot the collector current i_C vs. base current i_B. Over what range can the curve be considered linear? What value of bias current in the base circuit would be most desirable?

4-11 (a) In the circuit of Fig. 4-15 discussed in Section 4-13, for no signal input, what power is being dissipated by the collector? (b) What power is supplied by the bias source?

4-12 The accompanying circuit in Fig. P4-12 uses a 2N334 transistor with $V_{CC} = 30$ V and $R_L = 6000$ Ω. Find R_f, assuming that static operating point is approximately at the center of the load line.

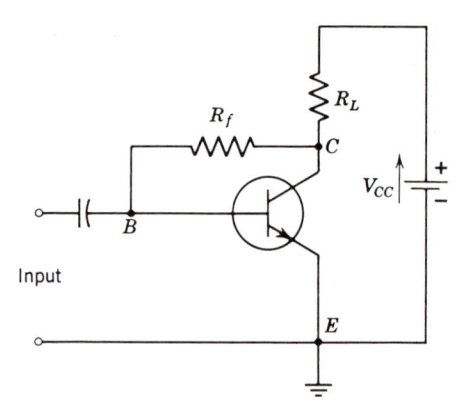

Figure P4-12

4-13 The internal potential barrier that results from the alignment of the P-side and N-side Fermi level is

$$V_0 = \frac{kT}{q_e} \ln \frac{n_e n_h}{n_i^2}$$

Compute V_0 if

$n_e = 1 \times 10^{16} =$ N-side doping
$n_h = 1 \times 10^{14} =$ P-side doping
$n_i^2 = 2.25 \times 10^{26} =$ intrinsic carrier density
$T = 300°$K $=$ room temperature
$k = 1.38 \times 10^{-23}$ J/°K $=$ Boltzmann's constant
$q_e = 1.602 \times 10^{-19}$ C $=$ electron charge

4-14 An N-P-N silicon transistor is assumed to have linear output characteristics and a current amplification factor $\beta = 100$. It is used in a common-emitter connection and uses a single bias supply for both collector and base circuits, $V_{CC} = 20$ V. The circuit is to be designed for a Q-point at $I_C = 1$ mA and $V_{CE} = 10$ V. Determine the required values for R_B and R_L to operate at this Q-point.

4-15 A P-N-P germanium transistor has a beta of 100 and a maximum permissible collector dissipation of 150 mW. Assuming it has a linear characteristic, design a common-emitter circuit that will develop a sinusoidal voltage of 5 V peak value across a load $R_L = 2$ kΩ. Use the smallest possible supply voltage V_{CC}. Determine the required value of R_B. (A sketch of the load line on i_C-v_{CE} coordinates may be helpful.)

4-16 An N-P-N silicon transistor having linear characteristics may have a value of β that lies between 30 and 120, depending on its method of manufacture, with a typical value being 80. (a) With a supply voltage $V_{CC} = 20$ V and $\beta = 80$, find the values of R_B and R_L that will place the Q-point at $I_C = 2$ mA and $V_{CE} = 4$ V. (b) Using the resistance values of part (a), determine the locations of Q-point when $\beta = 30$ and $\beta = 120$. What effect on circuit operation is noted?

4-17 A 2N334 transitor is used in a common-emitter connection with $V_{CC} = 30$ V and $R_L = 6$ kΩ. The base current varies sinusoidally between 50 and 150 μA. (a) What is the variation of collector current? (b) Calculate the current gain of the circuit.

4-18 A 2N404 P-N-P transistor is being used in a common-emitter circuit that has $V_{CC} = -10$ V and operates at a Q-point determined by $I_B = -0.15$ mA. Determine the smallest value of R_L that may be used without exceeding the maximum dissipation rating of 150 mW.

4-19 If the circuit described in Problem 4-18 uses the smallest permissible value of R_L and a single source for biasing the collector and base circuits, calculate the required value of R_B.

4-20 Using the curves in the Appendix, calculate the value of β at the point $V_{CE} = 2$ V, $I_B = 0.15$ mA, for each of the transistors 2N1302, 2N1304, 2N1306, and 2N1308. Compare these values with those listed for these types and discuss the differences, if any.

4-21 Repeat Problem 4-20 for the complementary types 2N1303 through 2N1309.

4-22 A 2N404 P-N-P transistor is to be used in free air at a temperature of 75°C. If it is used in a common-emitter circuit with $V_{CC} = -10$ V, what is the smallest permissible value of R_L that may be used without exceeding the dissipation rating for the device?

4-23 On a graph of common-emitter characteristics for the 2N334 transistor, plot maximum device dissipation curves for operation at 25°C, 100°C, and 150°C. Does the 2N334 have a *linear* derating factor for temperature?

4-24 The maximum ratings for a 2N334 transistor include $V_{CB} = 45$ V, $V_{BE} = 1$ V, and $I_C = 25$ mA. If this transistor is used in a common-emitter circuit in which $V_{CC} = 40$ V and $R_L = 2$ kΩ, will the device operate in excess of any of its maximum ratings? Assume operation at 25°C.

4-25 When the 2N334 transistor is used as a common-emitter amplifier it has $V_{CC} = +12$ V and a load resistance of 600 Ω. I_B is $+100$ µA. The base current is made to vary sinusoidally from $+50$ to $+150$ µA. Determine (*a*) the current gain, (*b*) the voltage gain, assuming v_{be} to be 0.1 V in amplitude, (*c*) the power gain.

4-26 A *P-N-P* transistor is connected in a common-emitter circuit with $V_{CC} = -10$ V, $R_L = 2$ kΩ, and base bias set by some R_B and V_{CC}. A sinusoidal voltage source drives the base of the transistor, and the output of the circuit is taken between the collector and circuit ground (the positive terminal of V_{CC}). (*a*) Sketch the circuit, showing the schematic of all components. (*b*) Sketch characteristics of the transistor, assuming it to have $\beta = 100$ and linear properties. (*c*) What value of base current will permit the input signal to have the largest excursion without distortion in the output? (*d*) What should be the value of R_B to set the operating point for largest input signal? (*e*) What should be the collector dissipation rating for the transistor to operate at this point?

4-27 A 2N334 transistor has $V_{CC} = 30$ V and $R_L = 5$ kΩ in a common-emitter circuit. Using V_{CC} to provide the base bias, what value of base resistor R_B is needed to set $V_{CE} = 20$ V? 10 V?

4-28 A 2N404 transistor is to be used in a common-emitter circuit having $V_{CC} = -10$ V. Choose values for R_L that will permit a maximum collector current variation for sinusoidal variations in base current, without exceeding any maximum ratings (*a*) when operating at 25°C; (*b*) when operating at 75°C.

4-29 For a 2N334 transistor connected in common-emitter mode, determine the maximum V_{CC} that may be used with a 6-kΩ collector load without exceeding the maximum collector dissipation of 150 mW at 25°C.

4-30 For the circuit in Fig. P4-30, determine: (*a*) graphical plot of i_C, v_{CE} output curves; (*b*) dc load line location, and sketch it on the graph obtained in (*a*); (*c*) bias line, and sketch it on the graph to locate the *Q*-point; and (*d*) the value of *S* for this circuit.

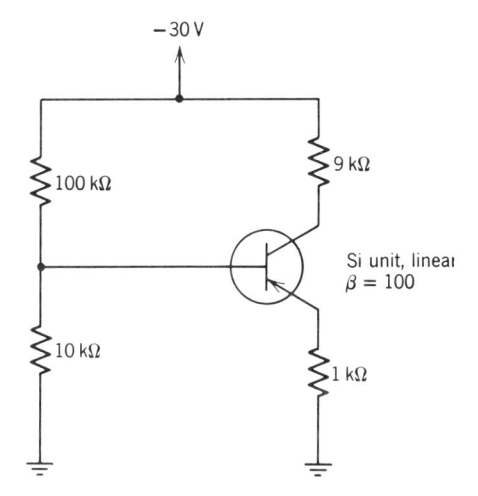

Figure P4-30

4-31 Using the circuit in Problem 4-30, replace the transistor with a similar unit with $\beta = 20$. Determine the values of base current, collector current, and collector-emitter voltage at the Q-point. Compare these values with those obtained in Problem 4-30c.

4-32 How much collector current is in the transistor of Fig. P4-32?

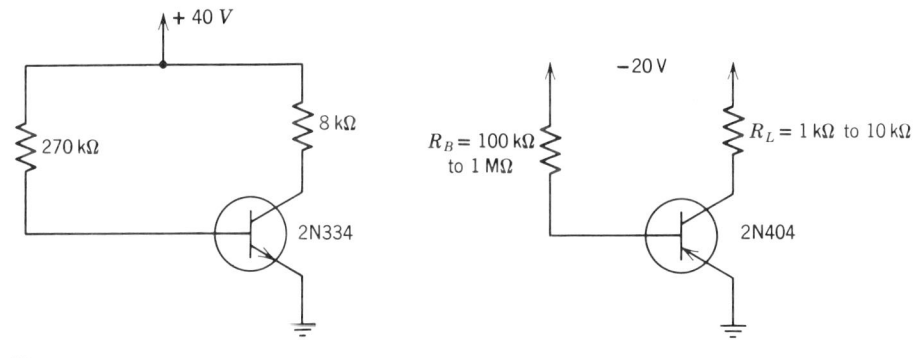

Figure P4-32

Figure P4-33

4-33 For the range of resistor values given in Fig. P4-33, determine from output characteristic curves the region of possible Q-point locations.

5
FIELD-EFFECT TRANSISTORS

This chapter continues the discussion of transistors to include unipolar devices, those that operate by the control of majority carriers only. This mechanism occurs because of the placement of P-N junctions in the physical construction of the devices. The field-effect transistor (FET) is a typical unipolar semiconductor device. Majority carrier control in the FET contrasts with the control of both majority and minority carriers in a junction transistor.

The FET has a very high input impedance and a high output impedance, which are favorable attributes in many circuit applications. It also can be used over a frequency range from low audio to the UHF spectrum. Special construction of the FET, called metal-oxide-semiconductor (MOS), permits the design of high-gain VHF–UHF amplifiers. This type of field-effect transistor is often labeled MOSFET and is sometimes referred to as IGFET, denoting insulated-gate field-effect transistor. Our discussions will reveal why these labels were chosen.

5-1
The Junction Field-Effect Transistor, JFET

There are two basic structures called field-effect transistors. Their control mechanisms are basically the same, but their internal structures and operating characteristics are decidedly different. First we will look at the junction FET. The block of P-type semiconductor shown in Fig. 5-1 has imbedded in it two

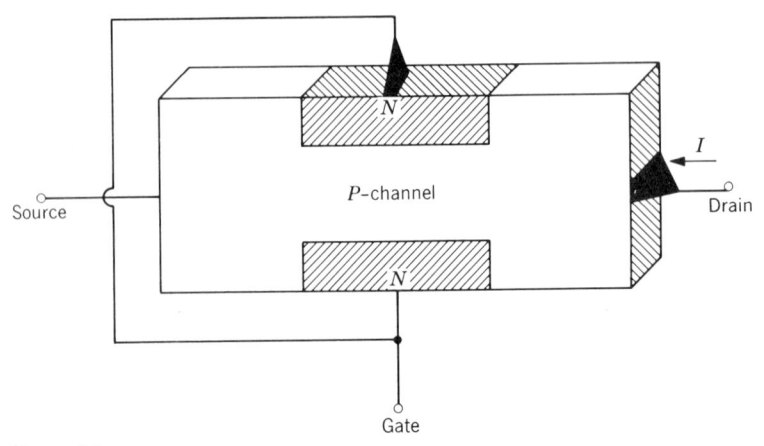

Figure 5-1
Basic structure of junction FET. Note that N-regions are electrically connected to gate.

regions of N-type semiconductor, a configuration first proposed by W. Shockley in 1952.*

Operating current of the FET passes between the connections marked *source* and *drain*. The semiconductor block between them is called the *channel*. Current control is obtained by applying a potential difference (electric field) between the source and the connection marked *gate*.

5-2
Depletion Region in the Channel

When the P-N junctions are formed in manufacture of the FET, they will exhibit the same carrier separation across the junction as discussed for other P-N junctions. A depletion region will be formed in the gate structure and in the channel, as shown in Fig. 5-2a. The center portion of the channel offers a predetermined resistance to current, dependent upon the material and the amount of impurity that is added to it. Normally this resistance is in the range between 100 and 2000 Ω. That portion of the channel that is occupied by the depletion region (void of carriers) has a very high resistance to current.

Applying an external voltage between the source and gate, as illustrated in Fig. 5-2b, will increase or decrease the depletion region depending upon its polarity. In turn, the depth of the channel between the gate connections will increase or decrease, varying the resistance of the channel. The larger the reverse-bias voltage, the greater the resistance. As the depletion region expands with increasing voltage, it will eventually reach the condition where

* W. Shockley, "A Unipolar Field-Effect Transistor," *Proc. I.R.E.*, **40**, 1365–76 (1952).

Sec. 5-3   *Symbols and Bias Polarities for Junction FET* 207

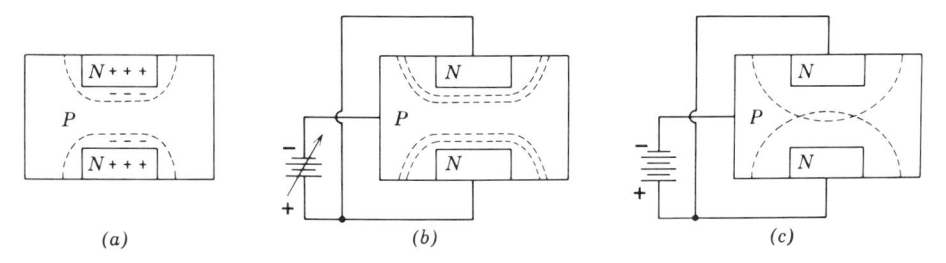

Figure 5-2
Depletion region in the channel of FET: (a) Depletion region about the P-N junctions. (b) Source-gate voltage varies the depletion region. (c) Depletion regions meet to cause "pinchoff."

both depletion regions meet, as in Fig. 5-2c. This condition is referred to as "pinchoff," alluding to the high resistance of the channel, which will effectively prevent current between the drain and source.

5-3
Symbols and Bias Polarities for Junction FET

Circuit diagrams containing JFETs use specific symbols that have been adopted for this purpose; the standard symbols are shown in Fig. 5-3. The only difference between the two schematic symbols is the direction of the arrow on the gate, which points in the direction that conventional forward current would flow in the P-N junctions, from P to N semiconductor. Since the channel semiconductor may be either P or N type, with a gate material of the other type, the direction of the arrow identifies the FET as P-channel or N-channel. A convenient memory aid is "N in" for the N-channel device.

A JFET acts as an electrical valve, working on the principle of a gate-source voltage controlling the drain current. Such current control occurs because of the changing resistance in the channel that results from the variations in its depletion region. A variable gate-source voltage controls the size of the depletion region and the depth of the channel. Drain current, then, depends

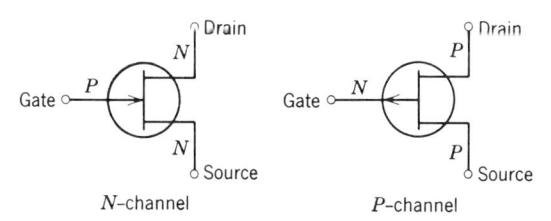

Figure 5-3
Symbols for N-channel and P-channel FETs.

208   Field-Effect Transistors   Ch. 5

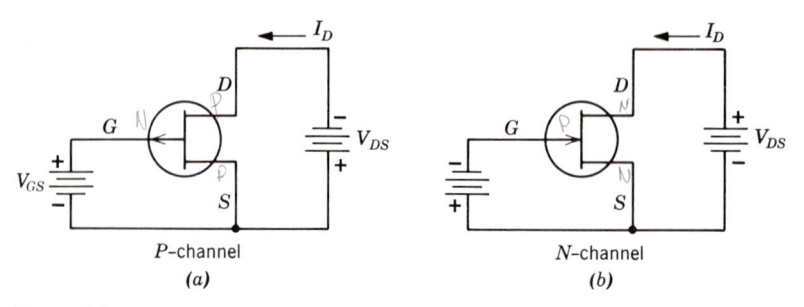

Figure 5-4
Bias polarities for operation of FET: (a) Biasing the P-channel type. (b) Biasing the N-channel type.

upon the gate-source voltage. It is also affected by the drain-source voltage below the *pinchoff voltage*, V_P, as we shall see.

In Fig. 5-4a, the P-channel FET is connected to two external batteries. The battery, V_{GS}, provides reverse bias for the gate-source P-N junction. A path for drain current I_D is provided through the battery in the drain-source circuit, V_{DS}. Figure 5-4b illustrates the biasing connections for the N-channel FET. Correct polarities of these biasing sources are simply reversed from those for the P-channel FET.

5-4
Static Characteristics of JFET

A family of curves may be determined from the operation of a JFET under the influence of applied dc operating conditions. The JFET, like other three-terminal devices, may be operated with any one of its terminals common to the biasing sources. To illustrate the effect of a control signal on the output current, Fig. 5-5a is the schematic diagram of a common-source circuit that yields the characteristic curves in Fig. 5-5b. This circuit connection of the JFET is analogous to common-emitter connection of a junction transistor. Variations in V_{GS} cause variations in I_D corresponding to changes in base current and collector current in the junction transistor. Note that the JFET current is controlled by a voltage.

Three regions of the curves are of interest. The normal operating region is in the center where the curves are essentially flat. This region represents a nearly constant-current source between drain and source terminals. The breakdown region at the right occurs because of reverse-bias breakdown of the drain-gate P-N junction and should be avoided in normal circuit operation of the JFET. Similarly, the nonlinear region at the left is not generally used.

Sec. 5-4   Static Characteristics of JFET   209

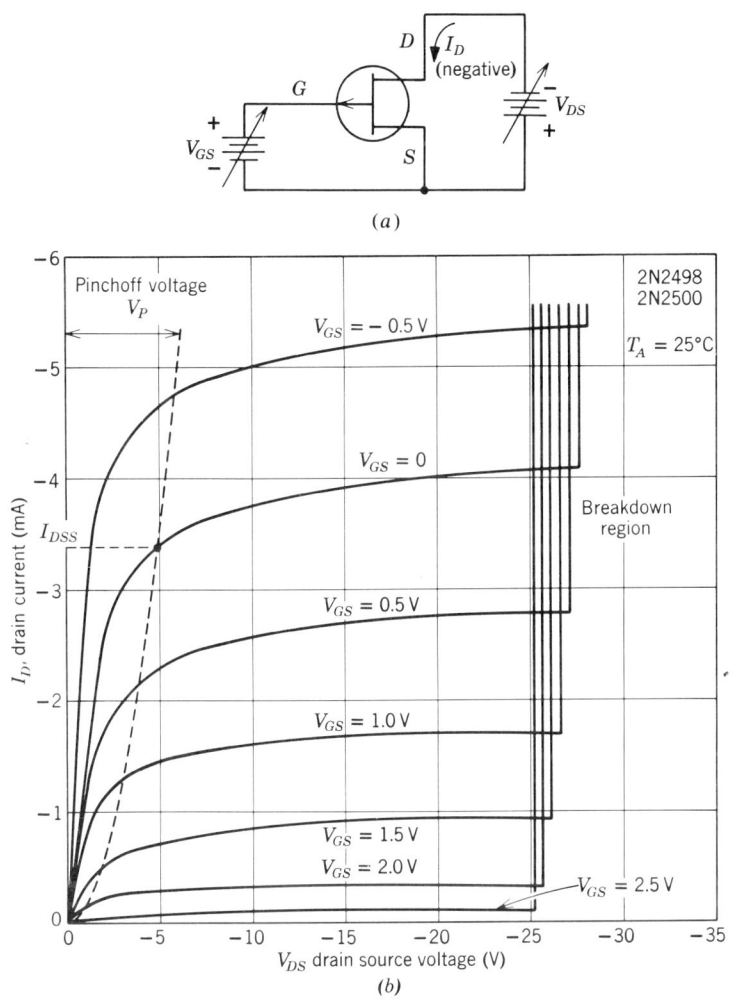

Figure 5-5
Static characteristics of P-channel JFET. (a) Common-source circuit. (b) Curves obtained from operation of the circuit. (Courtesy Texas Instruments, Inc., Dallas, Tex.)

The shape of these curves may be explained as follows. If an external voltage is applied between drain and source, V_{DS}, with zero gate voltage ($V_{GS} = 0$), drain current in the channel will set up reverse bias between the drain and gate. This bias is established by the resistive voltage drop along the channel. As V_{DS} increases, the depletion regions expand due to the increasing bias in the channel. There is an effective increase in channel resistance that prevents drain current from increasing in proportion to V_{DS}. The value of V_{DS} that causes drain current to be limited is called the "pinchoff voltage," V_P. Further increases in V_{DS} yield small increases in I_D.

Once the pinchoff region has been reached, I_D remains essentially constant until drain-gate avalanche occurs in the breakdown region. If a reverse voltage is applied to the gate, pinchoff will occur at a lower I_D value. This is so because the expansion of depletion regions caused by reverse bias on the gate adds to the reverse bias produced by I_D in the channel resistance. Maximum current for any V_{DS}, then, is reduced.

5-5
Electrical Ratings

An FET is subject to physical damage or destruction from excessive voltages and currents. Each unit that is distributed under a specific type number, such as the 2N2498, has been assigned electrical ratings that cannot be exceeded in any operating circuit in which it is used. Ratings include maximum currents, breakdown voltages, and maximum power dissipation capabilities. Specified parameters also are often selected by manufacturers to describe the performance of a particular type of FET.

An important parameter of an FET is the drain current at zero gate voltage, I_{DSS}. This current value is marked in Fig. 5-5. The first two subscripts denote the two terminals of interest, and the third letter indicates a short-circuit from the third terminal to the reference terminal. In this case, the current is between drain and source, and the gate is shorted to the source. Usually the quantity I_{DSS} is specified at a particular drain-source voltage, V_{DS}. Here, it is marked at $V_{DS} = -5$ V, the value of pinchoff voltage for $V_{GS} = 0$. Table 5-1 lists some important ratings and parameters for the 2N2498 P-channel FET.

Table 5-1
Parameters and Ratings for 2N2498 FET

Absolute maximum ratings, 25°C free-air temperature:

    Gate current   10 mA
    Dissipation   500 mW*

Electrical parameters:

BV_{DSS}	Drain-source breakdown voltage	20 V
BV_{GSS}	Gate-source breakdown voltage	20 V
I_{DSS}	Drain current ($V_{GS} = 0$, $V_{DS} = V_P$)	2 to 6 mA
$I_{D(off)}$	Pinchoff drain current	10 μA
	($V_{DS} = -15$ V, $V_{GS} = 6$ V)	
I_{GSS}	Gate leakage current ($V_{GS} = 10$ V, $V_{DS} = 0$)	0.01 μA

* Derate linearly to 175°C by 3.3 mW/°C.

Sec. 5-6　　　　　　　　　　　　　　　　　　　　Load Line for JFET Circuit　211

Drill Problem

D5-1　Obtain a manual that lists different types of FET units, and compare several *P*-channel and *N*-channel units with respect to maximum ratings and electrical parameters. Note also the different shapes and sizes of cases and the lead configurations for socket mounting in an operating circuit.

5-6
Load Line for JFET Circuit

We will analyze the behavior of the circuit whose schematic diagram is shown in Fig. 5-6. In this circuit, the gate-source voltage, v_{GS}, is the combina-

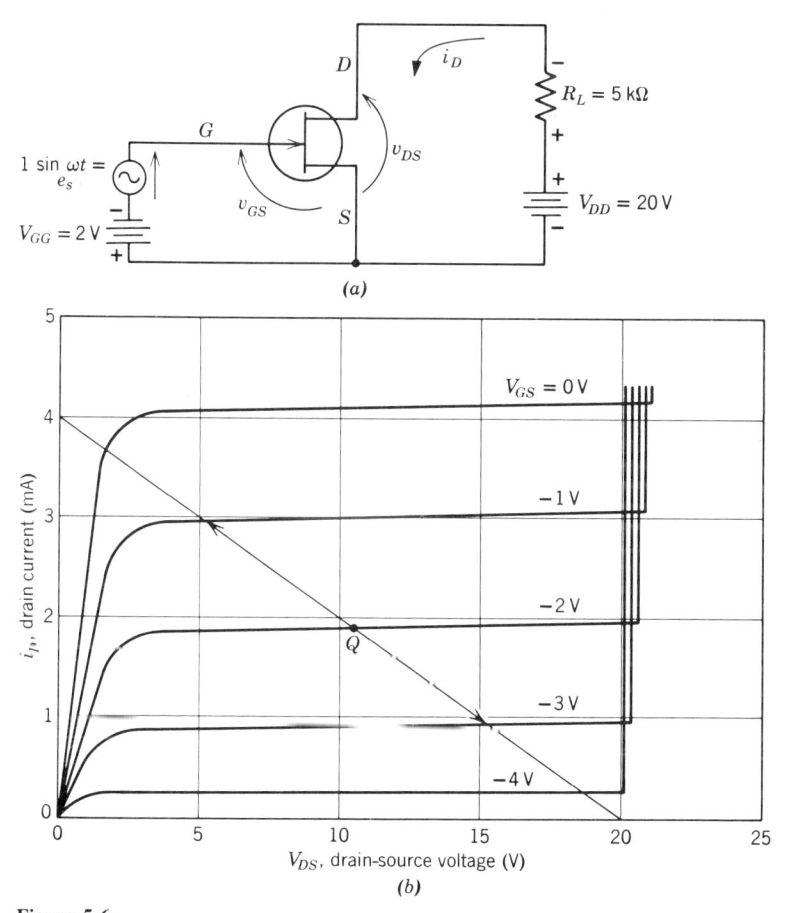

Figure 5-6
Operation of JFET along a load line. (*a*) Schematic diagram of common-source circuit. (*b*) Drain characteristics and location of load line.

212   Field-Effect Transistors   Ch. 5

tion of e_s and V_{GG} acting in superposition. Drain current, i_D, passes through the load resistor, R_L, in the drain circuit.

Kirchhoff's voltage law applied in the drain circuit may be expressed as

$$V_{DD} - i_D R_L - v_{DS} = 0$$

This expression may be rearranged in the form of a straight-line equation in terms of the two variables, i_D and v_{DS}:

$$i_D = v_{DS}\left(-\frac{1}{R_L}\right) + \frac{V_{DD}}{R_L} \qquad (5\text{-}1)$$

Equation 5-1 is the dc load line equation for the drain circuit. When this equation is plotted on the drain characteristic curves, as in Fig. 5-6, the resulting straight line defines the region in which the JFET will operate. For the voltage and component values given in the accompanying circuit diagram, the load line extends from $v_{DS} = 20$ V to $i_D = 4$ mA. The Q-point is located on the dc load line and on the line for $v_{GS} = -2$ V. As the source voltage, e_s, varies sinusoidally, the JFET will operate along the load line above and below the Q-point. For a 1-V variation in v_{GS}, the drain current and drain-source voltage will vary sinusoidally between 3 mA and 1 mA and from 5.5 V and 15 V, respectively. This region of operation is marked in Fig. 5-6.

Drill Problems

D5-2 Sketch the waveforms of i_D, v_{GS}, and v_{DS} for the circuit in Fig. 5-6. Note that there is a phase shift of 180 degrees from the gate to the drain in the voltage waveforms.

D5-3 Double the value of R_L in Fig. 5-6, and sketch the waveforms asked for in D5-2. Relate the changes in waveform to the operating region now occupied by the load line.

D5-4 Starting with the values given in the circuit of Fig. 5-6, reduce V_{DD} to 10 V and sketch the waveforms of current and voltage asked for in D5-2. Explain any changes or unusual features in the waveforms.

5-7
Self-Bias of JFET

The dc voltage source in the gate circuit (V_{GG}) is called the *bias* voltage source. Note that the negative terminal of the N-channel JFET in Fig. 5-6 is connected toward the gate. A resistor in series with the source terminal, as shown in Fig. 5-7, will provide a negative voltage between the source and gate because of the drain current passing through it. This arrangement provides

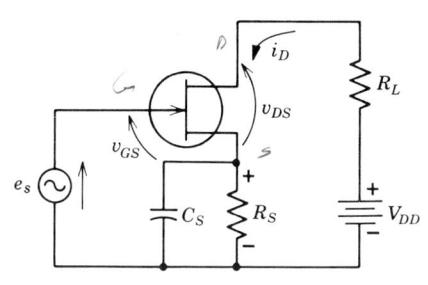

Figure 5-7 Self-bias network R_S-C_S is connected in source circuit of JFET.

self-bias for the JFET in the common-source circuit. It works quite well where drain current never decreases to zero.

Operation of the R_S-C_S combination to provide a negative gate-source voltage is easy to understand. The capacitor, C_S, is selected to have a small reactance compared with the ohmic value of R_S, at any frequency that may be a component of drain current, i_D. A small reactance provides a path for ac current around R_S, so that the dc current component of i_D will pass through R_S. Since only dc current is in the resistor, the voltage developed across it will be only dc. This dc voltage, developed in compliance with Ohm's law, provides self-bias for the JFET.

The gate of the JFET is seldom driven positive; the instantaneous gate-source voltage is usually negative. Therefore, the magnitude of the voltage developed across R_S should be larger than the maximum positive value of the input signal voltage, e_s. Bias voltage will be the required value of v_{GS} at some selected Q-point location.

5-8
Effect of Source Resistor on Q-Point

As an example of how to calculate values of R_S and C_S, we will use the circuit given in Fig. 5-7 and the drain characteristic curves of Fig. 5-8. For our calculations, we will assume these values for the circuit: $V_{DD} = 20$ V, $R_L = 4$ kΩ, $e_s = 2 \sin \omega t$. It will be desirable to operate this circuit at a Q-point defined by $v_{DS} = 10$ V and $v_{GS} = -2$ V.

When a resistor such as R_S is connected in the source circuit to generate self-bias voltage, the dc load line will have a slope determined by both R_L and R_S, because drain current passes through both of them:

$$i_D = v_{DS}\left(-\frac{1}{R_L + R_S}\right) + \frac{V_{DD}}{R_L + R_S} \qquad (5\text{-}2)$$

Thus, the dc load line will intercept the extremes on the v_{DS} axis at 20 V and on the i_D axis at a value determined by $(R_L + R_S)$. We have assumed 4 kΩ for R_L, but what will be the value of R_S? If the Q-point is to be located as stated above,

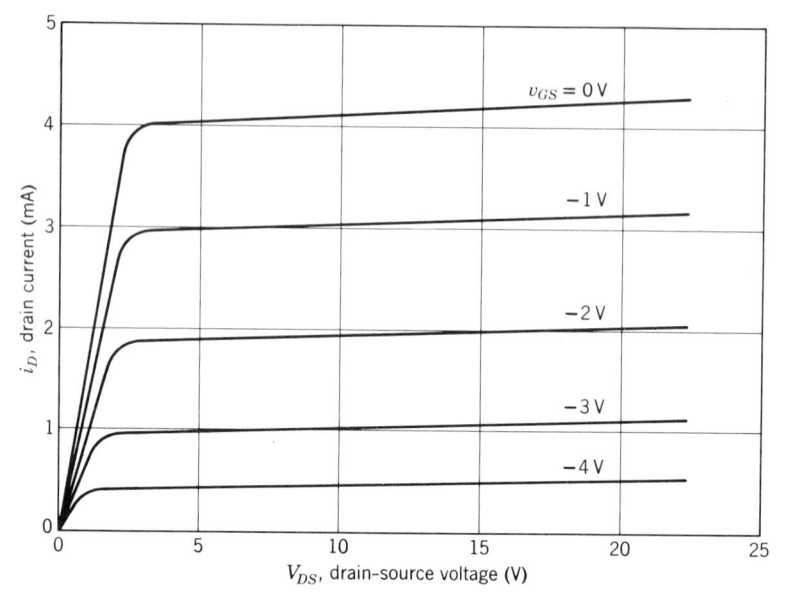

Figure 5-8
Typical drain characteristic curves for N-channel JFET.

then at that point i_D will be about 2 mA and v_{GS} will be about -2 V. Since the 2 mA will be in R_S, the required bias of -2 V will be developed across $R_S = 2\,\text{V}/2\,\text{mA} = 1\,\text{k}\Omega$. Now the total dc load in the drain circuit will be $R_L + R_S = (4 + 1)\text{k}\Omega = 5\,\text{k}\Omega$, and the load line can be drawn from 20 V to 4 mA on the two axes.

As a rule of thumb, the reactance of C_S should not be greater than one-tenth of R_S at the *lowest* frequency of e_s. If this condition is met, its reactance will be even smaller at higher frequencies. The smaller its reactance compared with the resistance it bypasses, the better it will pass ac currents around the resistor. This action permits the resistor to develop essentially a dc voltage for self-bias. In the case under study here, the reactance should be less than 1000/10 or 100 Ω at the lowest frequency present in the drain current. Let us assume that this frequency will be 20 Hz. The capacitance of C_S should then be such that

$$\frac{1}{\omega C_S} \leq 100\ \Omega$$

$$C_S \geq \frac{1}{2\pi \times 20 \times 100} = 79.5 \times 10^{-6}\,\text{F} = 79.5\ \mu\text{F}$$

A standard-size electrolytic capacitor of 100 μF would be selected. Its reactance would be slightly smaller than necessary to satisfy the "rule of thumb" stated above.

Drill Problems

D5-5 Draw the dc load line on the curves in Fig. 5-8 for the circuit conditions calculated in the preceding section, and locate the Q-point that results from selected circuit components.

D5-6 For the circuit discussed in the preceding section, determine the effect of the value of R_s on the location of the Q-point, when it is varied from 500 Ω to 5000 Ω.

D5-7 Can self-bias cause drain current cutoff? Why?

5-9 High Impedance DC Voltmeter Using FET

The field-effect-transistor (FET) has an extremely high input impedance and other characteristics usually attributed to a pentode vacuum tube. The voltmeter whose circuit is shown in Fig. 5-9 uses a single FET in its circuitry.

The input impedance of the voltmeter is determined by the string of resistors between the probe terminals. Since the overall accuracy is largely determined by these resistors, they should be stable and preferably have an accuracy of 1 per cent. To provide protection from transient overloads and stray ac voltages, the filter consisting of the 1-MΩ resistor and 0.02 μF capacitor is introduced at the input to the FET.

To analyze the circuit operation, assume that zero potential exists between points A and B, and that current is flowing in the FET and resistor R_1. Point C will then be negative with respect to point B. Adjustment of resistors R_2 and R_3 can make the potential at point D equal to that at point C, and the meter will indicate zero. When point C becomes more negative than point D, as when an input potential is connected between the probes at A and B, the meter will indicate the difference in potentials at C and D. Calibration of the meter according to the current change in the FET and resistor R_1 will permit the meter to read directly in volts to indicate the input potential applied to points A and B.

Switches S_1 and S_2 are ganged together to permit the selection of either positive or negative input potentials so that the meter will read upscale for either. The 5-kΩ potentiometer and diode associated with the meter are used to adjust the linearity of the calibration and to provide a shunt to limit the meter current to a value slightly greater than its full-scale rating.

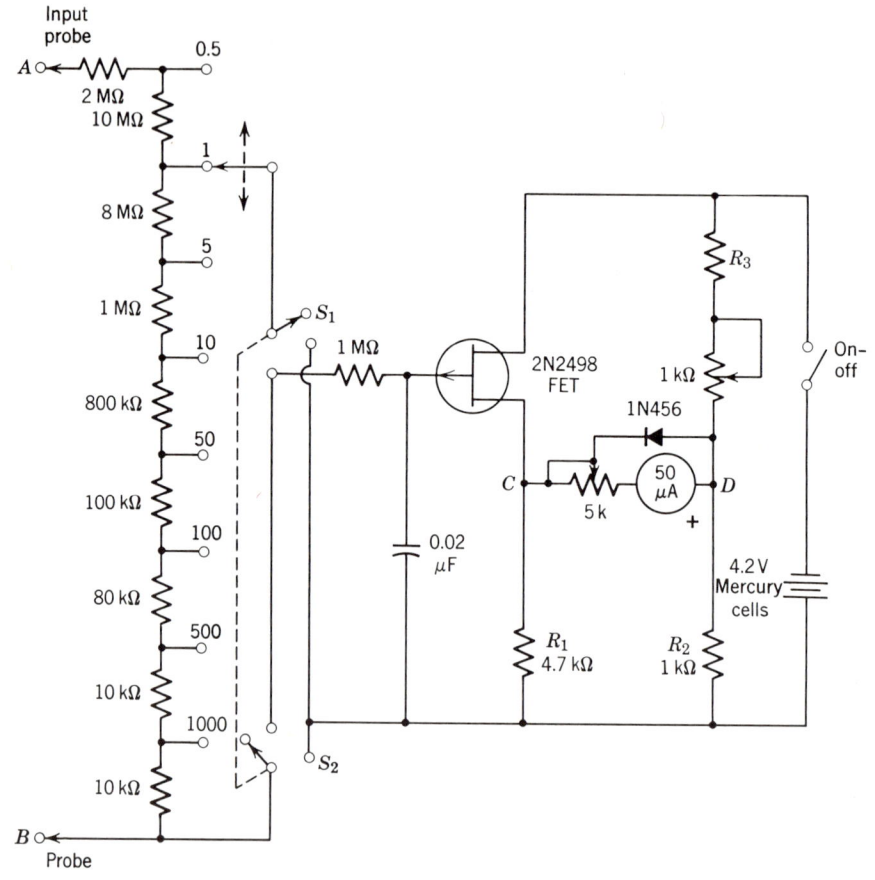

Figure 5-9
High impedance dc voltmeter using the FET. (Courtesy Texas Instruments Incorporated, Dallas, Tex.)

5-10
MOS Field-Effect Transistor (MOSFET)

The metal-oxide-semiconductor FET operates with a slightly different current control mechanism than the junction FET. A MOSFET is often referred to as an *induced channel* or *insulated gate* device. Why these names are descriptive can be learned from the construction of such devices, depicted in Fig. 5-10.

The basic construction of a MOSFET consists of a lightly doped or intrinsic slice of semiconductor with two separate regions of oppositely doped material diffused into the surface. In Fig. 5-10a, the host slice of P-type semiconductor

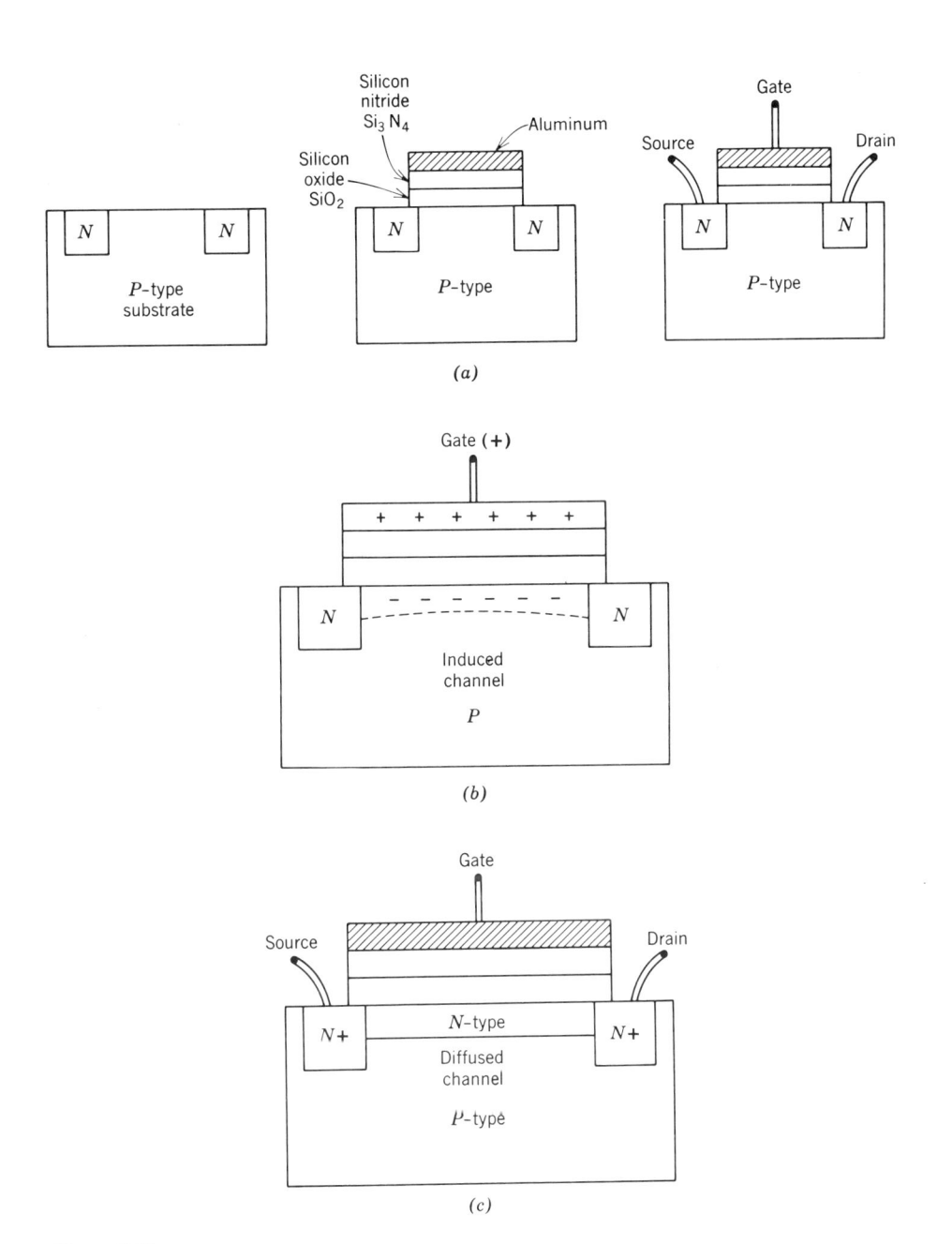

Figure 5-10
Construction of MOSFET. (a) Metal-oxide-semiconductor layering forms capacitor between gate and substrate. (b) Induced channel in enhancement-mode operation. (c) Diffused channel in enhancement-depletion-mode operation.

218 Field-Effect Transistors   Ch. 5

is called the *substrate*; the two regions of N-type semiconductor serve as drain and source for the MOSFET. During the construction the surface is covered by a layer of oxide and nitride, which will serve as an insulating dielectric between the channel and the gate. A layer of metal placed over the insulation becomes the gate connection. Connections to the doped regions are used for drain and source terminals.

In Fig. 5-10b, a positive potential is applied to the gate with respect to the substrate. Since the metal area of the gate forms a capacitor with the insulating layers and the substrate between source and drain, positive charges at the gate side of this capacitor will induce corresponding negative charges in the semiconductor. This action effectively generates a channel of N-type semiconductor between source and drain, an *induced channel*. As the gate keeps increasing its potential, the induced channel increases and reduces the resistance between drain and source. Any potential applied between drain and source will then cause a current to pass between them. In other words, drain current is "enhanced" by the gate potential. For this reason, the structure just described is called an enhancement-type MOSFET.

A depletion-type MOSFET can be constructed in a very similar fashion. Starting with the basic structure illustrated in Fig. 5-10a, the drain and source regions are heavily doped and separated by a moderately doped region between them. Figure 5-10c shows a diffused channel between drain and source. As before, as the gate potential is made more positive, drain current will be enhanced. However, as gate potential is made negative, a depletion region will form in the diffused channel. Drain current will decrease with decreasing gate voltage. This structure, therefore, is both enhancement-type and depletion-type MOSFET.

5-11
Symbols and Characteristics of MOSFET

The N-channel structure may be changed to P-channel by reversing the type of material making up the various regions. Of course, voltage polarities and current directions also reverse in going from N-channel to P-channel MOSFETs.

The symbols adopted for the enhancement mode and the enhancement-depletion mode types of MOSFET are given in Fig. 5-11. Note that the arrow points in the direction of conventional forward current in a P-N junction, from P to N. A memory aid for the direction of the arrow is the same as for the junction FET: "N in."

Characteristic curves for the JFET and the two types of MOSFET just described are similar, but they differ in some important respects. Figure 5-12 shows sketches of characteristic output curves for these three types of

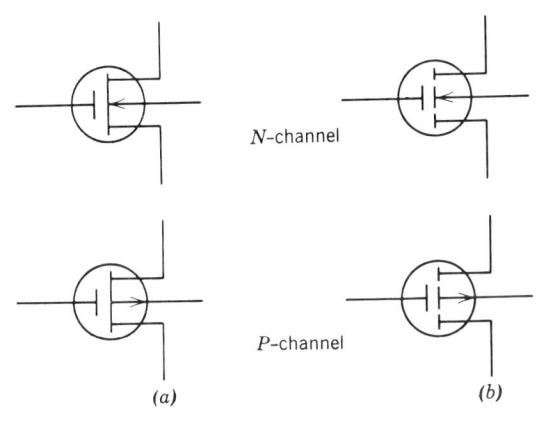

Figure 5-11
Schematic symbols for MOSFET. (*a*) Enhancement mode. (*b*) Enhancement-depletion mode.

transistors, each of which is an *N*-channel device. Those having *P*-channel operation would exhibit negative voltages and currents where these are positive.

5-12
Precautions Regarding MOSFET

Because of its physical structure and electrical properties, a MOSFET has a high impedance between the gate and source. In a typical application of a MOSFET, the gate-source circuit serves as the input connection for external signals. The impedance between the gate and source is typically a resistance of about $10^{14}\ \Omega$ at low frequencies. Such a magnitude of impedance offers attractive advantages in circuit applications that require minimum loading effects, but it also subjects the MOSFET to possible damage because of static electric charges.

A MOSFET may be packaged by a manufacturer to include a metal ring that is in contact with its external connecting leads. This ring serves as a short-circuiting connection between the gate and source to prevent the development of static charges. Possible damage to the transistor can be the result of static current in the picoampere range, which would generate 100 V across the 10^{14}-Ω impedance. To avoid damage to MOSFET units during handling it is necessary to observe certain precautions. Any protective packaging should be removed only after the MOSFET is installed in its circuit, or compensated by other connections of protective wiring while it is being handled. A short-circuit between gate and source may be sufficient precaution, but it is good practice to handle a MOSFET only by its case, not by its leads.

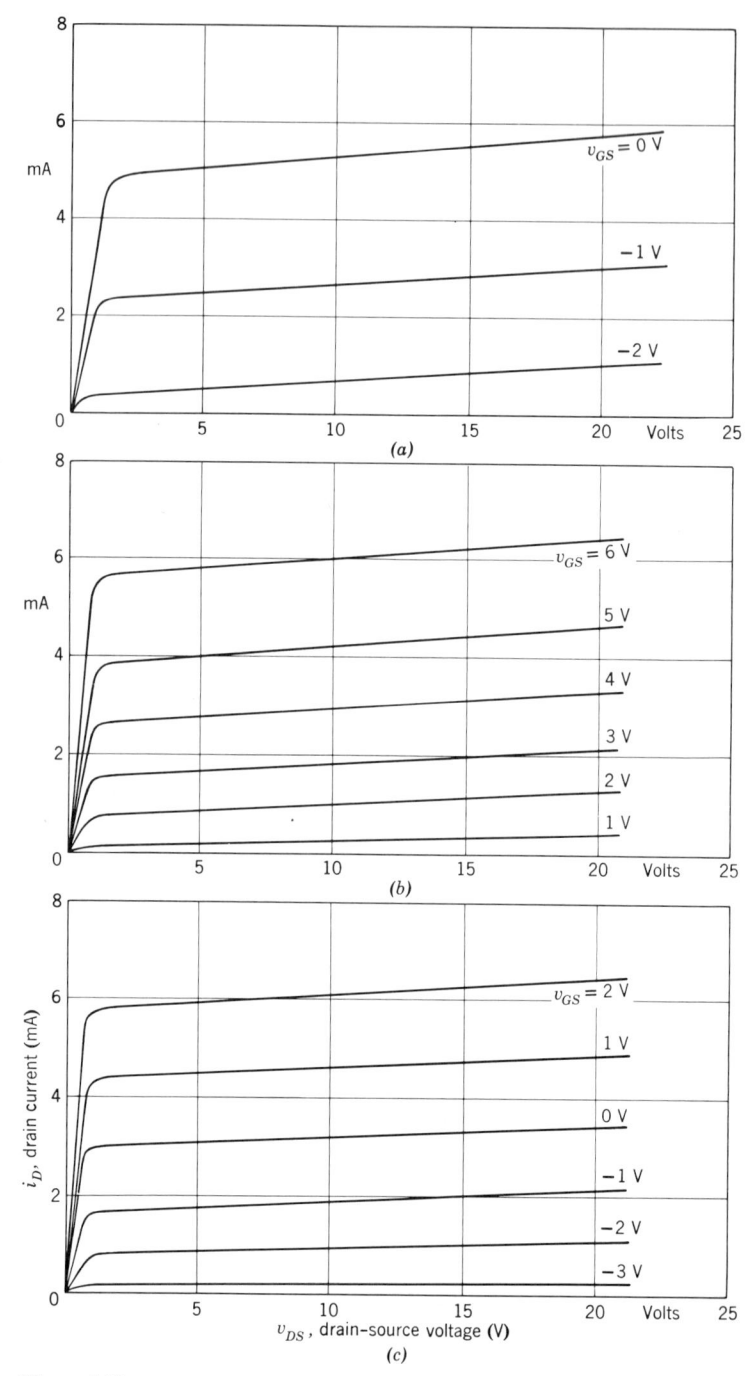

Figure 5-12
Output characteristic curves for N-channel FETs. (a) Depletion only—JFET. (b) Enhancement only—MOSFET. (c) Enhancement and depletion—MOSFET.

Questions

5-1 Why is a field-effect transistor referred to as a unipolar device?

5-2 Discuss briefly the control of channel current by changes in the depletion region.

5-3 What effect does "pinchoff" have on the channel current in a junction FET?

5-4 Sketch from memory the schematic symbol for a junction FET, and show the difference that denotes whether it is P-channel or N-channel.

5-5 What is "self-bias" of a JFET?

5-6 What are the basic differences between a JFET and a MOSFET?

5-7 Sketch the schematic symbol for a MOSFET, and show features that distinguish it as either P-channel or N-channel and as either enhancement or enhancement-depletion structure.

Problems

5-1 For the circuit and characteristics given in Fig. 5-6, V_{DD} is reduced to 10 V. (a) Sketch the dc load line and locate the Q-point. (b) What effect on circuit operation is noted?

5-2 Use the circuit and characteristics of Fig. 5-6. For $V_{DD} = 20$ V and $R_L = 5$ kΩ, determine a value of V_{GG} that will permit maximum e_s without severe waveform distortion in v_{DS}.

5-3 Design a self-biased FET amplifier, as in Fig. 5-7, using the characteristics in Fig. 5-8. Using $V_{DD} = 20$ V and $I_{DQ} = 2$ mA, calculate values for R_L and R_S.

5-4 For the circuit design in Problem 5-3, calculate a value of C_S that is appropriate for an audio amplifier.

6
VACUUM TUBES

The vacuum diode discussed in Chapter 2 was shown to have two active electrodes, the cathode and the anode. Current passing through the diode was controlled by the voltage between anode and cathode. Other vacuum tubes, such as *triode, tetrode,* and *pentode,* have additional active electrodes within their enclosing envelopes. As their names imply, a triode has three active electrodes, the tetrode four, and the pentode five. These electrodes provide additional means of controlling the current passing between anode and cathode.

High-vacuum tubes serve as important elements in circuits designed to perform several different functions in electronic applications: (a) rectification, changing alternating current to direct current; (b) oscillation, the production of electric voltages and currents having desirable frequencies and waveforms; (c) amplification, the strengthening of voltages and currents; (d) modulation, the process of causing an electric voltage or current, such as a radio signal, to change its waveform in accordance with variations in sound or in another kind of signal; (e) demodulation or detection, a kind of rectification devised to recover from a modulated signal a voltage variation similar to the one that was used to generate the modulation; (f) electron-beam production, as in the picture tube of a television receiver or an X-ray tube; (g) photoelectric action, the change of light energy into electric current as in the phototube or electric eye.

224 *Vacuum Tubes* Ch. 6

(a)

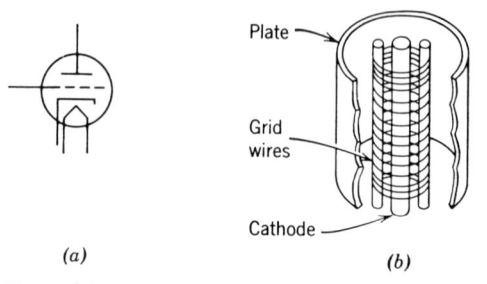

Figure 6-1
(a) Symbol of a high-vacuum triode. (b) Sketch of electrode configuration in a triode.

6-1
Physical Characteristics of the Triode

The symbol that represents a high-vacuum triode tube, and a simplified vertical cross section of the electrode configuration, are shown in Fig. 6-1. The cathode structure is the same as that described for thermionic diodes in Chapter 2. It is represented here as a cylinder. The grid consists of a helix of fine wire coaxial with, and mounted close to, the cathode. The plate is a metallic cylinder also coaxial with the cathode but several times farther away from it than is the grid. In most triodes the cylindrical plate surrounds the other elements of the tube, although in some tubes it is not a circular cylinder.

6-2
Function of the Grid

The grid was introduced by Dr. Lee DeForest in the year 1906. It heralded an era of rapid development in vacuum-tube applications and performance. The ability to vary plate current by means of very small variations in grid-to-cathode voltage made possible the strengthening of radio signals, which was the first application of vacuum-tube amplification. Other kinds of tube performance, including the generation of powerful high-frequency currents by vacuum-tube oscillators, the variation (or modulation) of vacuum-tube plate currents in accordance with frequencies and amplitudes of sound waves of the human voice or musical instruments, and the separation of voice frequencies from inaudible frequencies in very weak radio waves (detection), soon were discovered. All these basic operations were accomplished with the triode.

The grid controls the flow of electrons in the space between the cathode and the anode. The potential on the grid wires determines in a large measure the nature of the potential distribution within the tube. Figure 6-2 shows how the potential within the tube is distributed when the grid is negative with respect to the cathode and when the cathode is cold. The straight-line sections, a, show

Sec. 6-2   *Function of the Grid*   225

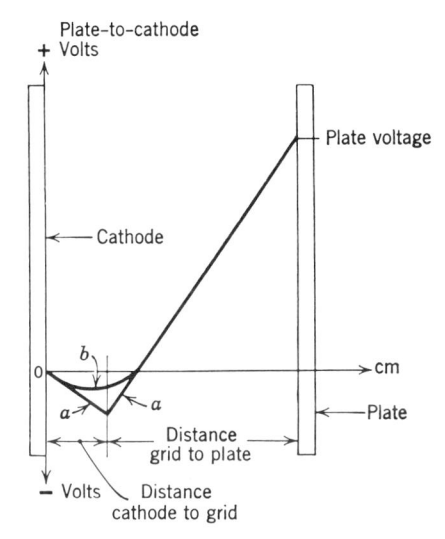

Figure 6-2
Potential distribution in vacuum triode tube. No space charge.

the potential distribution along a straight line from cathode to plate through a grid wire; the curved section, *b*, shows a potential distribution along a straight line midway between the grid wires.

When the cathode is heated to normal emitting temperature, an abundant supply of electrons appears immediately in front of the cathode. They are "boiled out" of the cathode, so to speak, and many of them leave it with sufficient velocity to carry them against the opposing forces of the electric field, so that they pass between the grid wires and on to the plate. If the plate potential is not high enough to produce a continually rising potential from cathode to plate so that all the electrons proceed immediately toward the plate, many of them reenter the cathode. When this situation prevails, there is a definite dip in the potential-distribution curve along the lines between grid wires, like the *b* segment of the line in Fig. 6-3. The amount of accumulated negative space charge (electrons) in front of the cathode determines how many *electrons per second* may pass on to the plate. The plate current that flows in this kind of situation is said to be *space-charge-limited*. It must be remembered that the potential of the grid wires is very influential in determining the nature of the potential distribution in the grid-cathode region. In normal operation the plate current of a triode is space-charge-limited.

The rate of flow of electrons to the plate is controlled by the potential of the grid. The electrons that leave the space-charge region and pass through the grid plane cause the potential of the space charge to rise, since negative charges are being removed. The result is that the repelling action of the space charge against electrons trying to enter the space-charge region from the cathode is decreased, and so more electrons enter to take the places of those that have left and passed on to the plate.

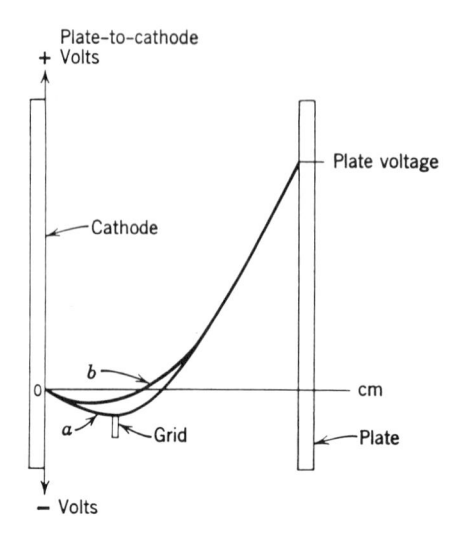

Figure 6-3
Potential distribution in a triode (the current is space-charge-limited): (*a*) through a grid wire, (*b*) midway between two grid wires.

It may then be said that although the plate current of a high-vacuum tube is space-charge-limited, the potential of the grid with respect to the cathode is the controlling factor in determining plate current when the plate voltage and cathode temperature are kept constant. The space charge limits the number of electrons per second that permanently leave the cathode. The potentials in the space between cathode and anode are not sufficiently high to produce an electric field strong enough to send all of the electrons immediately to the plate as fast as they leave the cathode.

6-3
Types of Triode Tubes

High-vacuum triodes are classified generally as either voltage amplifiers or power amplifiers. Voltage-amplifier triodes carry small current, seldom more than 8 or 10 mA and usually less. Their function is to cause a current change in a load resistance, which produces a relatively large voltage variation across the resistance. The current change is caused by very small changes in the voltage between the grid and cathode of the tube.

Twin triodes are used where two stages of amplification or two other separate triode functions are required. Figure 6-4 shows terminal connections for the 6F8G. The tube has double heaters in parallel connected to pins 2 and 7.

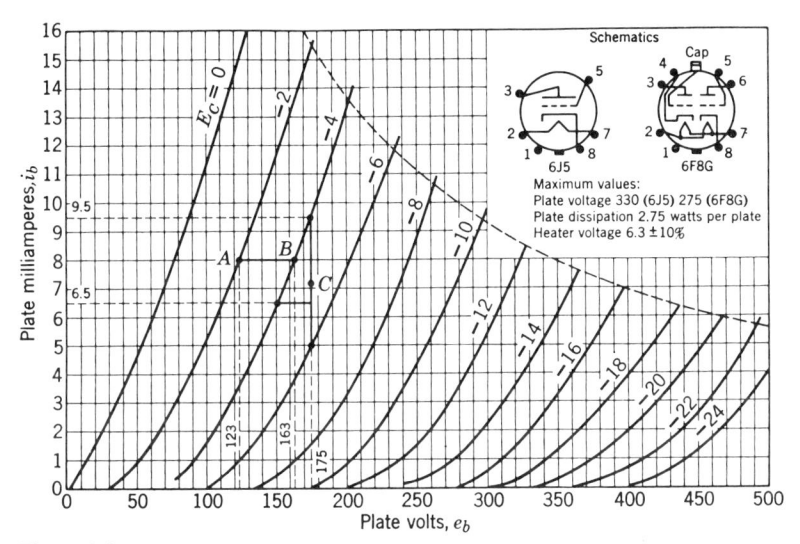

Figure 6-4
Average plate characteristics of 6J5, 12J5, 6F8G vacuum triodes.

6-4
Plate-Characteristic Curves of a Triode

Inasmuch as plate current in a high-vacuum triode depends on the potential of the grid with respect to the cathode (hereafter called either grid potential or grid voltage), it is natural to inquire what happens when the grid potential is kept constant and the plate potential is varied. A "family" of curves that give this information is shown in Fig. 6-4. Such a group of curves is required for the study of tube operation because all three quantities—plate voltage, plate current, and grid voltage—change in normal operation of the tube.

Consider a definite condition of operation such that the grid potential is -4 V and the plate potential is 163 V. The plate current for these potentials is 8 mA. If the grid is held at 4 V negative with respect to the cathode while a positive potential of 163 V is applied to the plate, the plate current will remain at 8 mA. This situation is called a static condition because no potentials or currents are changing. No current flows in the *grid circuit* of a high-vacuum tube if the grid is negative with respect to the cathode.

The static resistance of this tube for the static condition just described is

$$R_{dc} = \frac{\text{plate volts}}{\text{plate amperes}} = \frac{163}{0.008} = 20{,}375 \; \Omega$$

It is readily understood that there will be a static resistance corresponding to every point on the plate-characteristic curves. Any point not located on one of

228 *Vacuum Tubes* Ch. 6

the characteristic curves given in Fig. 6-4 will lie on a curve representing the proper grid potential that corresponds to the plate current and voltage for that point. That is, we can readily see that curves may be determined for many grid-voltage values between those represented by any two adjacent curves on the chart.

In Fig. 6-4, the grid voltage curve for $E_c = 0$ V is very similar to the forward voltage-current curve for a diode tube. Of course, it should be, because the grid has no effect when its voltage with respect to the cathode is zero and the triode operates as a diode. The curves for increasing negative grid voltage also resemble the diode characteristic but are shifted toward positive plate voltages.

6-5
Letter Symbols for Vacuum Tubes and Circuits

Many voltages and currents in a circuit containing a vacuum tube are important in its operation. In order to distinguish easily a particular voltage or current for discussion or to represent it in an equation, letter symbols are used according to the following legends for various parameters that we shall use in the remainder of the book.

CONTROL GRID

e_c	Instantaneous total voltage
i_c	Instantaneous total current
E_c	Average or quiescent voltage
I_c	Average or quiescent current
e_g	Instantaneous value of alternating-voltage component
i_g	Instantaneous value of alternating-current component
E_g	Effective value of alternating-voltage component
I_g	Effective value of alternating-current component
E_{cc}	Supply voltage
C_g	Grid capacitance
e_{c1}	Instantaneous total control-grid voltage
e_{c2}	Instantaneous total screen-grid voltage, and so forth (Subscripts 1, 2, 3, etc., are used to specify the grids.)

CATHODE

E_f	Voltage impressed
I_f	Current
E_{ff}	Voltage of supply
I_s	Saturation current
C_k	Capacitance (bypass)
R_k	Resistor

PLATE

e_b	Instantaneous total voltage
i_b	Instantaneous total current
E_b	Average or quiescent voltage
I_b	Average or quiescent current
e_p	Instantaneous value of alternating-voltage component
i_p	Instantaneous value of alternating-current component
E_p	Effective value of alternating-voltage component
I_p	Effective value of alternating-current component
E_{bb}	Supply voltage
P_p	Plate dissipation
r_p	Dynamic plate resistance

MUTUAL

C_{gp}	Grid–plate capacitance
C_{gk}	Cathode–grid capacitance
C_{pk}	Plate–cathode capacitance
C_{gh}	Grid–heater capacitance
C_{ph}	Plate–heater capacitance
g_m	Transconductance
μ	Amplification factor
$E_{gm}, I_{gm}, E_{pm}, I_{pm}$	Maximum instantaneous values of the alternating components
R_b	Direct-current resistance of plate load
r_b	Alternating-current resistance of plate load
R_g	Direct-current resistance of grid load
C	Blocking capacitor
L_g	Grid-inductance coil
L_p	Plate-inductance coil

6-6
Three Dynamic Factors

The analysis of vacuum-tube operation requires that certain tube factors, generally called *parameters*, be known. These parameters, the amplification factor, dynamic plate resistance, and transconductance, tell how two of the three important quantities—plate voltage, grid voltage, and plate current—are related while the third is kept constant. Their values may be determined from the plate-characteristic curves.

The *amplification factor* of a vacuum tube may be defined as *the ratio of a small change in plate voltage to the change in grid voltage required to restore the plate current to the value it had before the plate voltage was changed.* If the grid potential is changed slightly and if it is possible to change the plate voltage at the same time at such a rate that the plate current is prevented from changing, the rate of change of plate voltage with respect to grid voltage would be another way of defining the amplification factor of the tube.

Stated more exactly the amplification factor is the *ratio* of the *rate of change of plate voltage with respect to time* to the required *rate of change of grid voltage with respect to time*, such that the *plate current of the tube is constant.* Assuming that small changes in the two voltages occur in the same short time interval, we say that the rate of change of plate voltage with respect to time is $\Delta e_b / \Delta t$ and the rate of change of grid voltage with respect to time is $\Delta e_c / \Delta t$. Δe_b represents a *very small* change in plate voltage that takes place in a *very small* time interval, Δt. The amplification factor (μ) is, then,

$$\mu = \frac{\Delta e_b / \Delta t}{\Delta e_c / \Delta t} \quad (i_b \text{ constant})$$

$$\mu = \frac{\Delta e_b}{\Delta e_c} \quad (i_b \text{ constant}) \tag{6-1}$$

There should be a minus sign on one of the voltage terms here, because the voltages change in opposite directions. That is, if Δe_b is positive, Δe_c must be negative in order to restore the plate current to its original value, and viceversa. The amplification factor μ is a positive number, so it is the *absolute value* of the ratio given in Eq. 6-1.

The amplification factor tells *how many times more effective the grid voltage is than the plate voltage in controlling the plate current.* For example, the amplification factor of the type 6J5 tube, as given in the tube manual, is 20. This means that a change of 1 V in grid potential will require a change of 20 V in plate potential in the opposite direction to keep the plate current from changing. This does not mean that when this tube is used in an amplifier circuit it will multiply the voltage twenty times, but it does give some information about how much voltage amplification may be expected. Tubes with high amplification factors are chosen for voltage-amplifier work. It will be seen later that pentode tubes, which have three grids instead of one, possess more advantages than triodes for voltage-amplification service, and higher amplification factor is one of these advantages.

The amplification factor of a tube may be determined from measurements made on the plate characteristics. Its value depends on where the "operating point" is located. There are certain regions, however, over which the amplification factor is nearly constant. Observe the graphical construction at point A in Fig. 6-4. At a current of 8 mA, a change in grid voltage from -2 to -4 V requires a change in plate voltage from 123 to 163 V to keep the current constant.

$$\mu = -\frac{\Delta e_b}{\Delta e_c} = -\frac{163 - 123}{(-4) - (-2)} = -\frac{40}{-2} = 20 \tag{6-2}$$

The *dynamic plate resistance*, which is often called the ac plate resistance, is a tube parameter defined as the *rate of change of plate voltage with respect to plate current, grid voltage being kept constant*. That is, with constant potential on the grid, a small change in plate voltage will cause a small change in plate current, and the ratio of those two changes is the dynamic plate resistance.

The graphical construction at point B in Fig. 6-4 shows how to determine graphically the dynamic plate resistance of the tube. At point B, the characteristic curve has a certain slope. The slope is defined as the slope of a straight line drawn tangent to the curve at the point. The slope of the tangent line is the ratio of the altitude to the base of a right triangle that has the tangent line for its hypotenuse.

The slope of the curve is given by $\Delta i_b / \Delta e_b$. The dynamic plate resistance is the reciprocal of the slope of the curve at the point; thus

$$r_p = \frac{\Delta e_b}{\Delta i_b} \qquad (e_c \text{ remaining constant}) \tag{6-3}$$

The construction at point B shows that

$$r_p = \frac{175 - 150}{0.0095 - 0.0065} = \frac{25}{0.003} = 8333 \ \Omega$$

The dynamic plate resistance is a very important tube factor. It is used, as will be shown later, in setting up an equivalent circuit (representing the tube and other circuit parts) that is useful in performance calculations. The dynamic plate resistance is present only while the plate current is changing. Electronic and radio circuits accomplish nothing unless tube currents and voltage change.

It is useful to know how effective the grid of a high-vacuum tube is in *controlling plate current*. This effectiveness is expressed by the third important tube factor, the *transconductance*, which is represented by the symbol g_m. Conductance is defined as the reciprocal of resistance, and, since resistance may be calculated by dividing voltage by current (Ohm's law), conductance is given by current divided by voltage. The "trans" part of the word transconductance denotes a transition through the tube from one side (the input, or grid side) to the other (the output, or plate side). In short, the transconductance of a tube is the *ratio of a slight change in plate current to the slight change in grid voltage that caused it, the plate voltage being held constant while the changes take place*.

That is,

$$g_m = \frac{\Delta i_b}{\Delta e_c} \qquad (\text{plate voltage remaining constant})$$

The transconductance (often called *mutual conductance*, indicating that it is mutual to the grid and plate circuits) is determined graphically in Fig. 6-4 at point C. Here Δe_c was chosen to be -2 V, and the corresponding i_b turned out to be 4.5 mA.

$$g_m = \frac{\Delta i_b}{\Delta e_c} = \frac{0.0095 - 0.0050}{(-4)-(-6)} = \frac{0.0045}{2} = 0.00225 \text{ mho} = 2250 \text{ } \mu\text{mho}$$

It may be seen algebraically that the amplification factor is equal to the transconductance times the dynamic plate resistance.

$$g_m r_p = \frac{\Delta i_b}{\Delta e_c} \times \frac{\Delta e_b}{\Delta i_b} = \frac{\Delta e_b}{\Delta e_c} = \mu$$

That is,

$$\mu = g_m r_p \tag{6-4}$$

The numerical values obtained above for these parameters do not check exactly, principally because they must all be determined at the same point on the plate-characteristic curves instead of at three separate points A, B, and C, as was done here. Three separate points were chosen in the explanation above in order to prevent complications on the diagram. Another source of error is the accuracy to which the curves can be read. The values determined agree fairly well with the values given in the tube manual. It must be kept in mind that the location of the point on the chart influences all three numerical values. Usually the dynamic plate resistance is called simply plate resistance. It is so listed in tube manuals.

The effect of the grid voltage on plate current can be expressed by an equation that includes the amplification factor

$$i_b = k(e_b + \mu e_c)^n$$

where n is about 1.5 and k is an empirical constant. This expression is a modified form of Child's law.

Drill Problems

D6-1 Determine the three dynamic parameters of a 12AU7A triode at the operating point $e_b = 200$ V, $e_c = -6$ V; at $e_b = 100$ V, $e_c = -4$ V.

D6-2 Assuming the modified Child's law for a certain triode in which the plate current is 10 mA when $e_b = 200$ V and $e_c = -6$ V, calculate the approximate plate current at $e_b = 100$ V and $e_c = -2$ V. Assume that the amplification factor remains constant and equal to 12 over this range.

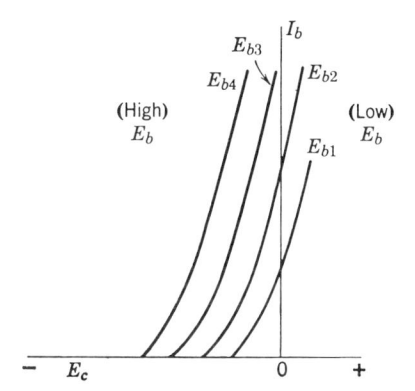

Figure 6-5
Static-transfer characteristics of a triode.

6-7
Grid-Plate Transfer Characteristics of a Triode

A useful set of curves is shown in Fig. 6-5. As indicated, the plate voltage is constant for each curve. The points that determine a curve may be obtained from the curves of Fig. 6-4 by drawing a vertical line at each desired plate-voltage value (this is equivalent to holding the voltage constant) and obtaining a plate-current reading at each value of grid voltage desired. A plot of a set of data, for each plate-voltage value chosen, gives the curves of Fig. 6-5. A set of curves of this type will be used later to help describe what happens in the important process called amplification.

Drill Problem

D6-3 Determine the three dynamic parameters of a 6AU6 pentode at the operating point $e_b = 200$ V, $e_c = -3$ V; at $e_b = 50$ V, $e_c = -1$ V. (*Hint:* The transfer curve may be helpful.)

6-8
Grid Bias

A triode tube in a simple electronic circuit is shown in Fig. 6-6. Although plate current is usually supplied by a rectifier and filter, a battery is shown here for convenience. Imagine first that the terminals 1 and 2 in the grid circuit are wired together instead of being supplied with an alternating voltage (E_1). Then the steady value of plate current may be set at any desired value from zero to the maximum the tube can carry by merely adjusting the sliding contact on the voltage-dropping resistor across the bias-supply battery that has a fixed voltage (E_{cc}). The dc voltage (E_c) in the grid circuit of an electron tube is called the *bias*

234   Vacuum Tubes   Ch. 6

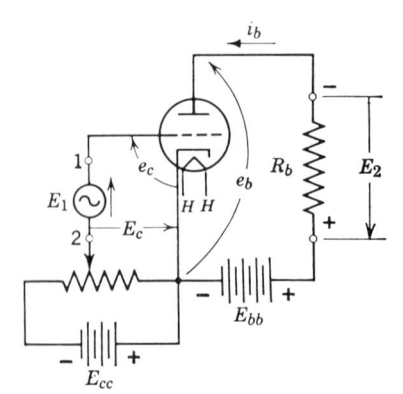

Figure 6-6
Simple triode circuit.

voltage. Often it is referred to merely as the *bias*. E_1, when replaced, will be in series with E_c.

6-9
Grid Bias Arrangements

Ordinarily a battery is used to supply grid bias only in laboratory exercises or in test work. Bias batteries are not used in practice; they are expensive, require too much space, and wear out too quickly.

A resistor in series with the cathode of a tube will provide a negative voltage for use as grid bias. The circuit of Fig. 6-7 shows the arrangement called *self-bias*. This circuit works well in applications where the plate current never decreases to zero. Its operation is easy to understand.

The cathode capacitor C_k serves to bypass the ac components of plate current around R_k so that its voltage drop is a steady dc voltage. It will be explained later that although the tube current flows *downward only* in this circuit, it continually changes in *instantaneous value*. This means that it has ac

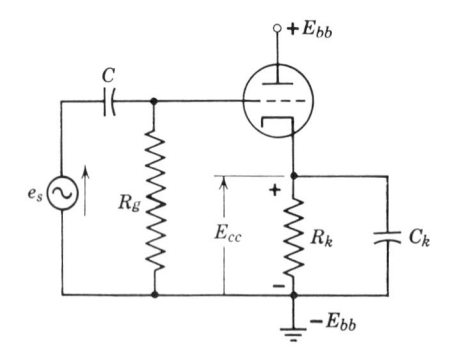

Figure 6-7
Self-bias by cathode resistor.

Sec. 6-10   Grid-Leak Bias   235

components superimposed upon a steady dc component. Only the dc component flows through R_k, making the cathode *positive* in polarity with respect to ground. It is important to note here, then, that the voltage drop across R_k makes the grid negative with respect to the cathode. The input signal voltage, e_g, is *generally* kept small enough so that its instantaneous maximum value *never makes the grid positive with respect to the cathode*. However, in certain applications the grid does go positive with respect to the cathode. This situation will be discussed later.

Drill Problems

D6-4 A triode type 12AU7A is used in the circuit of Fig. 6-6. Determine the value of R_b if $E_1 = 0$ V, $E_c = 4$ V, and $e_b = 150$ V. What magnitude of supply voltage E_{bb} can be used?

D6-5 The cathode resistor R_k in Fig. 6-7 has the value 1 kΩ. If the capacitive reactance of C_k is to be always less than or equal to one-tenth of R_k, what is the lowest frequency permissible in the plate-current variations?

6-10
Grid-Leak Bias

It is possible to provide negative grid-bias voltage by connecting a resistor to the grid terminal of the tube and bypassing the ac signal around it with a capacitor as in Fig. 6-8. This is done in high-frequency amplifiers, oscillators, television circuits, and radio receivers.

Some electrons in a vacuum tube land on the grid when it is not made negative with respect to the cathode by connecting it to a negative point, as has already been described.

The very small electron current entering the grid circuit will produce a dc voltage across the series *grid-leak* resistor connected to the grid terminal of the tube. This electron flow is interpreted as a *positive current* flowing toward the

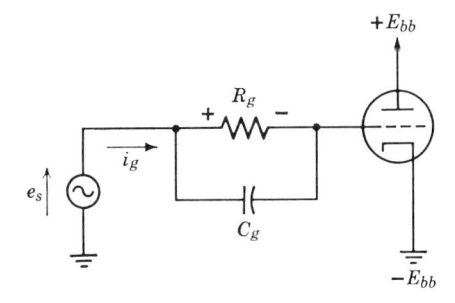

Figure 6-8 Grid-leak bias circuit.

grid, which makes the grid end of the resistor negative with respect to its other end. The parallel capacitor is thus charged to the same polarity, and it will discharge slightly during negative half-cycles of the signal voltage on the grid, thus maintaining the bias voltage across the resistor at a practically constant level. Electrons not used in charging the capacitor back up during positive half-cycles of signal voltage, pass down through the input circuit to the cathode of the tube, and are neutralized by positive charges. The cathode is usually grounded when grid-leak bias is used. Grid-leak resistance values are of the order of 0.5 MΩ or more.

6-11
The Triode in a Circuit

A plate-circuit resistor (R_b) is connected in series with the plate of the tube in Fig. 6-6. This is an important circuit element because the *ac component* of the voltage E_2 across this resistor is the *output voltage* of this simple amplifier stage. To make this more meaningful, let us consider that an ac generator or a transformer winding is applying a small alternating voltage E_1 to terminals 1 and 2. It will be seen later, in detail, how the tube causes a variation of current in R_b to such an extent that the *variations* in the voltage E_2 across R_b will be *an enlarged reproduction of the input voltage E_1*. It will also be shown that the variations in E_2 will be displaced 180 degrees in phase with respect to those of E_1.

The vast majority of vacuum electronic tubes operate with their grids always negative with respect to their cathodes. It will be recalled that the plate-characteristic curves of the 6J5 tube (Fig. 6-4) did not include a condition of positive grid voltage. In Fig. 6-6, however, the input voltage E_1 is alternating in nature and so it has positive and negative half-cycles. Therefore, in order to prevent the grid from going positive at any time, the negative bias voltage (E_c) should be sufficiently large. A graph of the voltages in the grid circuit is shown in Fig. 6-9a. Hereafter in this text, the alternating component of grid-cathode voltage will be denoted by the symbol e_g or E_g, which is standard notation, instead of by E_1 as is done in Fig. 6-6.

It is a very important matter to learn that *the alternating component of plate current (i_p) is in phase with the alternating component of grid voltage* in any high-vacuum tube with resistance load in which only the control-grid voltage is forced to vary. Figure 6-9b shows the alternating component of plate current (i_p) superimposed upon the direct component of plate current (I_b). The direct component of plate current flows in the tube even without the input voltage E_g applied. The plate current flowing when there is no varying component in the input voltage is denoted by the symbol I_{bo}. Note that the instantaneous total plate current is represented by the symbol i_b. Figure 6-9c shows how the voltage between the plate and cathode of the tube varies with time while the

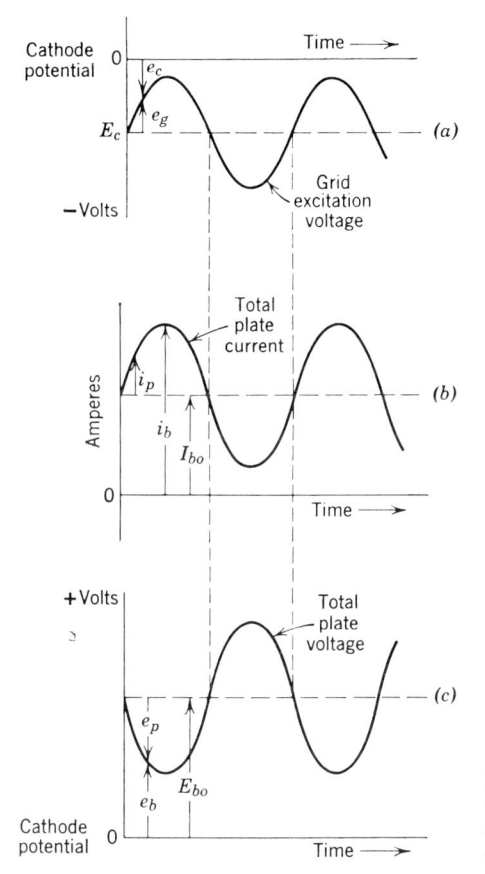

Sec. 6-12   Simple Circuit Equations   237

Figure 6-9
(a) Voltages in the grid circuit of Fig. 6-6.
(b) Currents in the plate circuit of Fig. 6-6.
(c) Voltages in the plate circuit of Fig. 6-6.

performance just described above is going on. The voltage E_b is the average value (dc component) of plate-to-cathode voltage; e_b is the instantaneous total value of plate-to-cathode voltage; and e_p is the instantaneous value of the varying component of plate-to-cathode voltage. It will be seen later that e_p shows up entirely across the load resistor (R_b) because the alternating component of plate current does not encounter enough resistance in going through the battery to cause the battery-terminal voltage to have a varying component.

6-12
Simple Circuit Equations

An examination of the circuit of Fig. 6-6 and a familiarity with the symbols for current and voltage components will permit the setting up of mathematical relationships among those quantities.

The instantaneous total plate current will be the sum of the steady dc component and the instantaneous value of the varying component:

$$i_b = I_b + i_p \tag{6-5}$$

Using Kirchhoff's voltage law, we see that in the plate circuit, E_{bb} is a voltage rise above cathode potential, E_2 is a voltage fall below the positive battery potential, and e_b (plate to cathode) is a voltage fall back to cathode potential. Hence

$$E_{bb} - E_2 - e_b = 0$$

That is, the output voltage is

$$E_2 = E_{bb} - e_b \tag{6-6}$$

but E_2 is also equal to $i_b R_b$, so that

$$E_{bb} - e_b = i_b R_b \tag{6-7}$$

In the grid circuit we also start at the cathode, go down the voltage drop (E_c), up the voltage rise E_1, now called e_g, and down the amount e_c to get back to cathode level. Hence

$$-E_c + e_g - e_c = 0$$

or the instantaneous total grid-to-cathode voltage is

$$e_c = -E_c + e_g \tag{6-8}$$

Although $i_p R_b$ is the instantaneous ac component of voltage drop across the plate-circuit resistor (R_b), there is a constant dc voltage drop across it equal to the product of the dc component of current and the resistance. Letting E_{bo} represent the dc component of plate-to-cathode voltage, we have

$$E_{bb} = E_{bo} + I_{bo} R_b \tag{6-9}$$

The amount of negative bias voltage provided by the cathode-bias arrangement is easily calculated. The dc component of tube current and the ohms value of R_k obey Ohm's law. In a triode tube the cathode resistor current is equal to the dc plate current. In a tube with a screen grid and a control grid, the cathode current is the sum of the plate current and the screen-grid current. The latter is usually much smaller than the plate current, but both flow through R_k.

Assume that a grid bias voltage $E_C = -4\,\text{V}$ is needed in a particular application where the dc component of plate current, I_b, is 6 mA. The required value of cathode-bias resistance is

$$R_k = \frac{4}{0.006} = 667\,\Omega$$

It is desirable that the bypass capacitor have a reactance not greater than one-tenth the resistance of R_k at the *lowest* frequency at which the tube is to operate. Assume this to be 50 Hz. The capacitance of C_k should then be such that

$$\frac{1}{\omega C_k} = \frac{667}{10}$$

$$C_k = \frac{10}{2\pi \times 50 \times 667} = 47.7 \times 10^{-6} \text{ F} = 47.7 \, \mu\text{F}$$

A standard-size electrolytic capcitor of 50 μF would be chosen. The voltage drop across it would be evaluated as the product of the current in R_k and the resistance:

$$E_c = 0.006 \times 667 = 4 \text{ V}$$

yielding a bias voltage of -4 V.

6-13
The Screen-Grid Tetrode and the Pentode

The screen-grid tube was developed through research in the tube design directed toward the elimination of difficulties that arise when triodes are used at high frequencies. The capacitance between the control grid and the plate of a triode allows a small part of the plate-circuit energy to be fed back, within the tube, to the grid circuit. This starts surges of current, known as oscillations, in both plate and grid circuits and puts an end to the amplification process.

The presence of a mesh of fine wire between the control grid and the plate appreciably reduces the grid-plate capacitance. This mesh is the screen grid. It usually is connected to the cathode through a capacitor of appropriate size to prevent ac current from flowing through the power supply. The control-grid-plate capacitance can be further reduced by bringing the control-grid connecting wire to a metal cap at the top of the tube envelope.

The screen-grid tetrode was found to have other desirable characteristics not possessed by the triode. The electrostatic field between the screen grid and the cathode, especially in the control-grid region, is affected only slightly by variations in plate voltage in the tube's normal range of operation. On the other hand, the control grid is just as effective as it is in the triode tube in determining the potential distribution of the electrostatic field. This means that the effectiveness of the grid voltage as compared to that of the plate voltage in determining plate current is much greater in the tetrode than in the triode, that is, the amplification factor of the screen-grid tetrode is much larger.

The plate resistance is much higher in a tetrode than in a triode of comparable size, because Δe_b must be substantially larger to produce a given Δi_b at constant E_c than in the case of the triode. The transconductance may be made high also by appropriate design and construction.

There is a tendency for electrons that are knocked out of the plate of a screen-grid tetrode by the electrons forming the plate current to pass to the screen. This is objectionable when the ac component of voltage in the plate circuit (superimposed on the dc plate-supply voltage, of course) is high. Under these conditions the plate voltage dips to low values and the secondary-emission electrons from the plate pass to the screen in large numbers. Then distortion becomes excessive and oscillations are likely to start.

To overcome this effect, a third grid, called the *suppressor grid*, was installed just in front of the plate, and the tube became a pentode (Fig. 6-10). The suppressor is usually connected to the cathode of the tube and is thus at full negative potential with respect to the plate. It forces secondary electrons back into the plate while offering practically no obstruction to the passage of the main stream of electrons from the cathode to the plate.

The pentode tube has a higher amplification factor and plate resistance than the tetrode. The suppressor grid isolates the plate still more from the control grid and the cathode. As a result, the capacitance between the control grid and the plate is even less than in the tetrode. An example of the reduction in this capacitance is seen when the grid-plate capacitance of the 6J5 triode, which is 3.4 pF, is compared to that of the 6SJ7 pentode, which is 0.005 pF, a reduction ratio of 680 to 1.

The suppressor grid is able to cause the return of secondary electrons to the plate even when the plate voltage is low. This is because those electrons are emitted with comparatively low velocities and the opposing electric field between the plate and suppressor is sufficiently strong to prevent those electrons from passing back through the suppressor-grid mesh.

The greatly reduced capacitance between plate and control grid in pentodes makes possible their use in amplifiers at high frequencies where triodes would not be practical. Oscillations produced by feedback of energy through the tube from the plate circuit to the grid circuit would cause the triode to quit amplifying. Tetrodes are seldom used now except in special applications. A set of plate characteristics of a voltage-amplifier pentode is shown in Fig. 6-11.

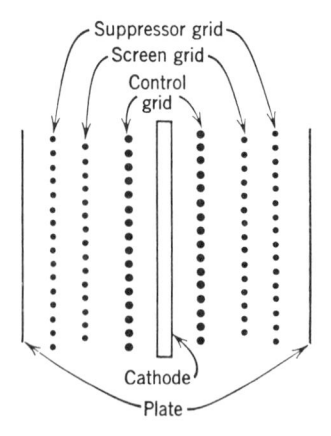

Figure 6-10
Elements of a pentode tube.

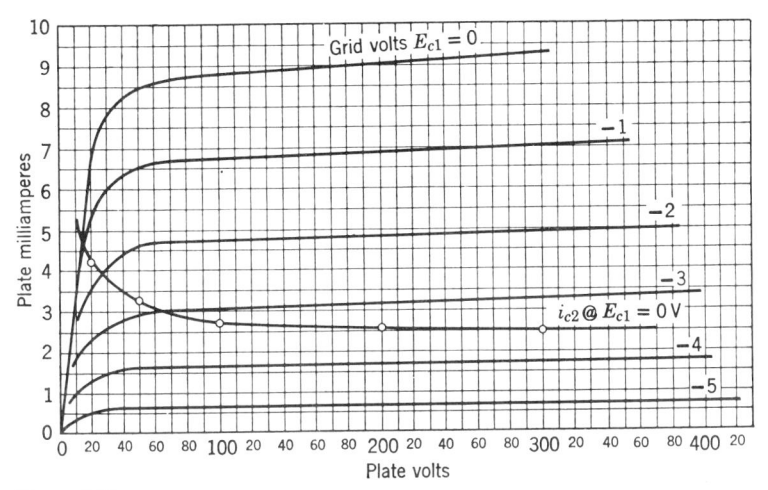

Sec. 6-13   The Screen-Grid Tetrode and the Pentode   241

Figure 6-11
Average plate characteristics of type 6SJ7 pentode. $E_{c2} = 100$ V, $E_{c3} = 0$ V.

The plate characteristics are shown for a screen-grid voltage of 100 V, which is a typical operating potential for the screen. Superimposed on the plate characteristics is a curve showing the variation of screen-grid current with increasing plate voltage, when the control grid is at 0 V. Screen current will be smaller for higher (more negative) bias values. Although it is small, it must be taken into account if the pentode is self-biased, as explained earlier.

The screen-grid potential may be obtained from the plate supply (E_{bb}) when the screen voltage is to be equal to or smaller than E_{bb}. The screen may be connected directly to E_{bb} when the design requires operation at this voltage. If the screen voltage is to be less than E_{bb}, a voltage-dropping resistor such as R_s in Fig. 6-12 is required.

The screen current can be used to reduce the screen-to-cathode voltage to the design value. For example, if $E_{bb} = 250$ V and the screen voltage is to be 100 V, then R_s must drop 150 V across it. Then the value of R_s can be calculated,

$$R_s = \frac{150 \text{ V}}{\text{screen current, A}}$$

Suppose the circuit is operating at a bias voltage of -4 V and the screen current is 1.5 mA. The value of R_s will be 100 kΩ.

In an operating circuit the control-grid potential will be varying, and the screen current will vary in the same manner. In order to avoid variations in the screen voltage, a large capacitor (C_s) can be connected between the screen grid and the negative terminal of E_{bb}, as shown in Fig. 6-12. The reactance of C_s is

242   Vacuum Tubes   Ch. 6

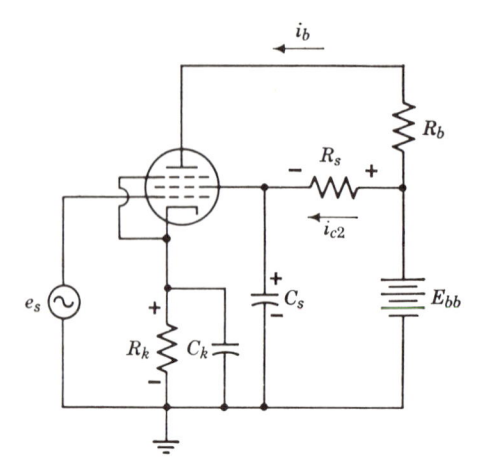

Figure 6-12
Pentode circuit supplying screen-grid voltage from plate supply through bias network R_s-C_s.

usually chosen to be one-tenth or less of R_s at the *lowest* frequency of current variations. The capacitor C_s provides a low-reactance path for ac around R_s, so that the screen voltage will remain practically constant at 100 V as required.

Drill Problem

D6-6 A voltage amplifier is to be designed using a 6SJ7 pentode in the circuit of Fig. 6-12. A power supply of 250 V is available. Based on certain requirements for the circuit, the plate load will be 50 kΩ and cathode bias will be provided by 1 kΩ shunted by 20 μF. Determine the required values for the screen-bias network to hold $E_{c2} = 100$ V. (*Hint:* Determine the operating point, and consult a tube manual for the magnitude of screen current flowing at that point.)

6-14
Effect of Cathode-Biasing Resistor on Q-Point

When a resistor R_k, bypassed for alternating current by a capacitor, is used in the cathode branch to provide grid-bias voltage, the *dc load line* is drawn with slope $-1/(R_b + R_k)$. The quiescent point Q will be located on the *bias line* at its intersection with the dc load line. The quiescent grid voltage will, in most cases, have a value such that the Q-point will not lie on one of the constant-grid-voltage curves (E_c) of the static characteristics, as published by a tube manufacturer.

A self-biased triode and its average-plate-characteristic curves are shown in Fig. 6-13. The dc load line is drawn for the 20,000-Ω load in the conventional manner. The bias line is drawn through points P_1 and P_2, which are located in

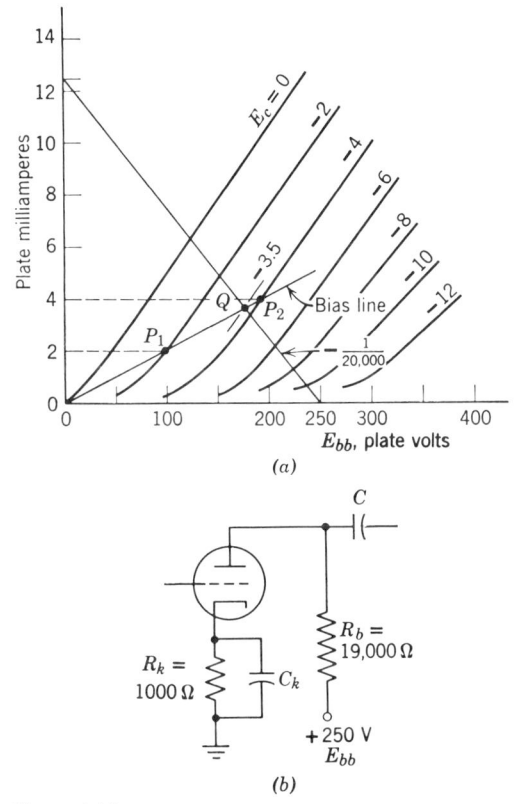

Figure 6-13
Location of Q-point for cathode-bias operation. (a) Bias line construction. (b) Plate circuit; cathode-bias operation.

the following manner: Without a signal on the grid, a 2-V self-bias requires a direct current of $2/1000 = 2$ mA; a 4-V self-bias requires $4/1000 = 4$ mA, and so on. Of course, zero self-bias requires zero current. The bias line is drawn through the origin and points P_1, P_2. The Q-point must lie on the dc load line as well as on the bias line; therefore, their intersection locates the Q-point for cathode-bias operation. In this example, the Q-point falls on the static line at $E_c = -3.5$ V.

Drill Problems

D6-7 In the circuit associated with Fig. 6-13, show the effect of the value of R_k on the location of the Q-point, when it is varied from 500 Ω to 5000 Ω.

D6-8 Can cathode bias cut off plate current in the circuit? Why?

6-15
Graphical Analysis of Triode-Tube Performance

Let us first examine only the right side of Fig. 6-14. A load resistor (R_b = 40,000 Ω) is in the plate circuit, and so the slope of the load line is determined by $-1/R_b = -1/40,000$.

The load line for this circuit passes through the 320-V point on the voltage axis, and the point $e_b = 0$, $i_b = 0.008$ A (8 mA) on the current axis, obtained from Eq. 6-7.

The grid bias is $E_c = -6$ V. Observe that if no other grid voltage than this existed, the plate current would be $I_{bo} = 3.8$ mA, and the plate-to-cathode voltage would be $E_{bo} = 168$ V. This point on the chart where the load line intersects the grid-bias voltage curve is called the *quiescent point* (Q). The quiescent plate current (I_{bo}), the quiescent plate-to-cathode voltage (E_{bo}), and the grid-bias (quiescent grid-to-cathode voltage) are given by the coordinates of this point. The voltage drop across the resistor is

$$I_{bo}R_b = E_{bb} - E_b = 320 - 168 = 152 \text{ V}$$

as obtained from the chart. This checks with the product of R_b and the plate current as read from the chart; thus,

$$I_{bo}R_b = 0.0038 \times 40,000 = 152 \text{ V}$$

Now let us examine the upper left-hand portion of Fig. 6-14. The static-transfer characteristics are shown for convenient values of constant plate voltage. Referring to the curve for $E_b = 80$ V, it is easy to see how three points were obtained from the plate characteristics to use in plotting the dynamic-transfer characteristic curve. A vertical line at 80 V would show that when $E_c = 0$, $i_b = 8$ mA; when $E_c = -2$, $i_b = 3.3$ mA; and when $E_c = -4$, $i_b = 0.4$ mA. That is how the static-transfer characteristics may be plotted when only the plate characteristics are available. However, data for plotting the static-transfer characteristics may be obtained directly by placing the tube in a test circuit arranged so that grid and plate voltages may be controlled. Readings are taken of plate current and grid bias at fixed values of plate voltage.

The static-transfer characteristics are useful in locating points that determine the dynamic-transfer characteristic. We know that when the grid voltage of the tube is varied, the variation in plate current is accompanied by a variation in plate-to-cathode voltage. The dynamic-transfer characteristic relates the instantaneous values of grid voltage and plate current.

The value of the load resistance plays an important part in determining the location and shape of the dynamic-transfer characteristic, since it fixes the slope of the load line. For every instantaneous value of the plate current, there is a corresponding value of instantaneous plate voltage, and these pairs of current and voltage values are coordinates of points that lie on the load line.

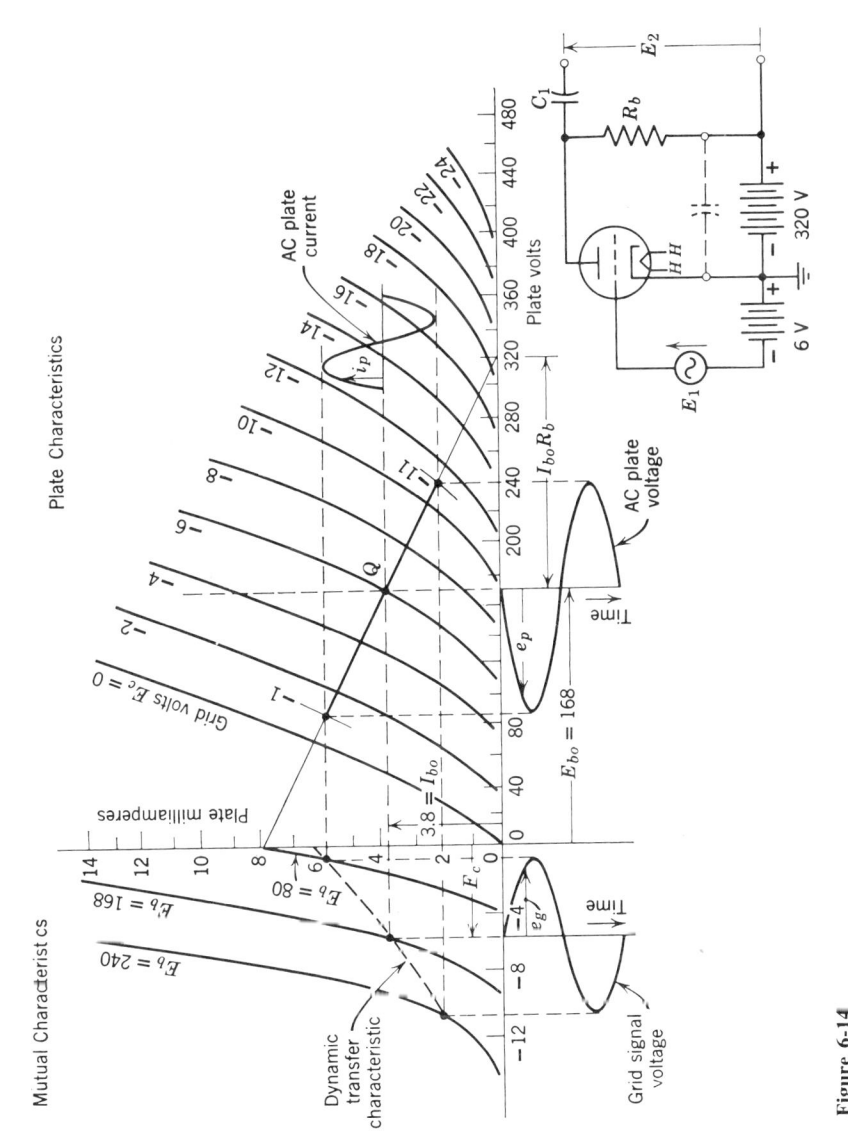

Figure 6-14
Showing dynamic operation of triode and graphical determination of voltage amplification.

Since each of these points corresponds to an instantaneous grid-voltage value, it can be seen that there are points in the static-transfer characteristic chart that correspond to points on the load line. Corresponding points are connected in Fig. 6-14 by dashed lines, to show how to locate the dynamic-transfer characteristic.

Locations of points on the dynamic-transfer characteristic may also be determined by calculation. After the static-transfer characteristics (I_b against E_c at constant values of plate voltage) have been plotted and the quiescent point (Q) determined, one may use the equation

$$e_b = E_{bb} - i_b R_b$$

For example, at the instant the plate voltage is 80 V, i_b will be

$$(320 - 80)/40{,}000 = 0.006 \text{ A}$$

Again, when $e_b = 240$ V, i_b is equal to

$$(320 - 240)/40{,}000 = 0.002 \text{ A}$$

and so on.

Figure 6-15 shows how the shape of the dynamic-transfer characteristic determines the waveform of the plate current flowing in the tube and the load, and therefore the faithfulness of the amplification effected by the circuit. The grid-bias value and the amplitude of the signal voltage are also very influential in determining plate-current waveform.

The grid-bias voltage, E_c, is -6 V, and the ac input voltage E_g has been chosen to be a sine-wave voltage of 5 V maximum value. This is shown superimposed upon the grid bias in the lower left-hand region of Fig. 6-14. As the ac voltage increases in the positive direction, the plate current increases

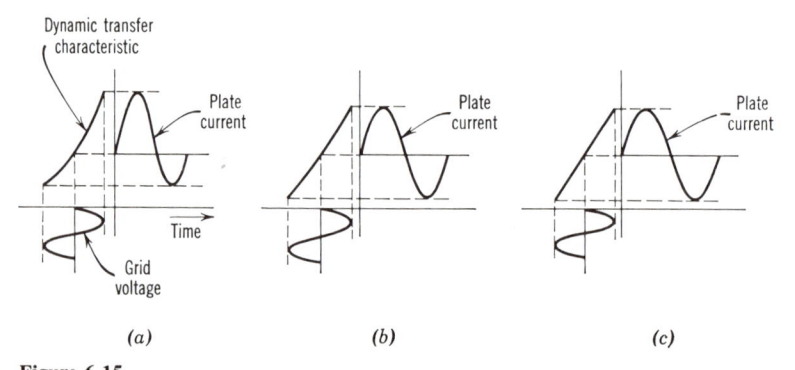

Figure 6-15
Effect of shape of dynamic-transfer characteristic on waveform of plate current in a vacuum tube. (*a*) Much curvature, much distortion; (*b*) little curvature, little distortion; (*c*) no curvature (straight line), no distortion.

Sec. 6-15   *Graphical Analysis of Triode-Tube Performance*   247

and the *instantaneous operating point moves up the load line*. At the highest position of the point, the instantaneous grid-to-cathode voltage, e_c, is -1 V, the instantaneous plate current is 5.8 mA, and the instantaneous plate voltage is 87 V.

The ac voltage then decreases, and, while this is going on, the *instantaneous operating point* moves *down* the load line through Q and on down to its lowest position. The lowest position is reached when the grid is most negative (-11 V); this occurs, obviously, at the negative peak of the ac input voltage. At this position we have the maximum instantaneous plate voltage, which is 238 V. The ac input voltage continues and comes up to zero from its negative value of -5 V, and so the plate current rises with the total instantaneous grid voltage, e_c, until the quiescent value, 3.8 mA, is again reached. All the variations just described occur during every cycle of the applied voltage.

Note the ac component of the plate voltage superimposed upon the dc, or average, value (E_{bo}) 168 V. Its variation is such that its instantaneous value (e_p) goes from zero (corresponding to 168 V) up to 70 V in the positive direction (238 − 168), and down 82 V in the negative direction. Because the dynamic-transfer characteristic *is not a straight line*, the output voltage is not a perfect reproduction of the sine-wave form of the input voltage. It is very important to observe that the ac component of plate voltage is 180 *degrees out of phase* with the ac component of grid voltage. This is a natural result, because when the ac component of grid voltage is at its positive maximum the instantaneous plate current is maximum and the $i_b R_b$ drop through the plate-circuit resistor (R_b) is maximum. Thus, at that instant the plate-to-cathode voltage is a minimum and the ac component of the plate voltage is at its negative maximum. When two ac voltages of the same frequency are so timed that one has its negative maximum value at the same instant the other has its positive maximum value, they are 180 degrees out of phase.

At the instant the ac component of grid voltage is at its greatest negative value, the plate current is a minimum and the $i_b R_b$ drop is a minimum. At that instant the plate voltage on the tube is a maximum; this means that the ac component of the plate voltage is at its positive maximum.

A few words must be said here about the output voltage (E_2) being the ac component (e_p) of the plate-to-cathode voltage of the tube. Referring to the circuit diagram in Fig. 6-14, we see that the load resistor (R_b) is in series with the plate-supply battery and that this series circuit is connected between the plate and cathode of the tube. Properly, e_p should be divided so that part of it shows up across R_b as E_2 and the remainder across the battery. Actually, the battery has such low resistance to the flow of plate current that the part of e_p that exists across the battery is negligible. Hence all of e_p shows up across R_b because it is considered to be connected between plate and cathode *when only ac voltage is involved.*

We may find a capacitor connected across a plate battery, and usually there is one in parallel in a power supply filter, which bypasses the ac components of plate current, offering very low impedance to them and thus preventing ac variations in the plate-supply voltage (E_{bb}).

Thus we see that the tube receives a voltage of 5 V maximum value and puts out a voltage of about 76 V maximum value across its load resistor. This is a *voltage amplification* of 76/5 = 15.2.

Questions

6-1 Describe the physical form of the electrodes in a conventional vacuum triode. What are the advantages of cylindrical construction?

6-2 Explain how the grid controls the flow of electrons in a vacuum triode.

6-3 What is meant by the potential distribution in a vacuum tube? How is the potential distribution affected by making the grid more negative with respect to the plate?

6-4 Consider a vacuum triode with proper dc voltages on its electrodes. What conditions inside it result in a dip in the potential-distribution curve just off the cathode? What will be the effect of increasing only the plate potential? Of making the grid still more negative?

6-5 Plate current flows in a vacuum triode with normal dc voltages on its electrodes. Why is it called space-charge-limited current? Why are some electrons that are emitted from the cathode unable to reach the plate? Where do they go?

6-6 What are the meanings of the symbols E_c, e_g, i_g, I_g, R_k, E_b, e_p, i_p, I_b, r_p, R_b?

6-7 Draw from memory a plate-characteristic curve of a triode, labeling all important lines. Then draw a few more, including one that goes through the origin. What quantity is constant on each of the curves?

6-8 Give the definition, in words, of each of the following tube parameters: (*a*) amplification factor, (*b*) dynamic plate resistance, (*c*) transconductance.

6-9 Write the symbolic equations that define each of the three tube parameters. How is g_m related to r_p and μ? Check the units.

6-10 What is a grid-plate transfer characteristic? What quantity is constant when a static transfer characteristic is drawn?

6-11 How does a power amplifier tube differ from a voltage amplifier tube in respect of currents?

6-12 Name three ways of obtaining grid bias for a triode.

6-13 Why is a capacitor used with cathode bias? with screen bias?

6-14 How is the ohms value of a cathode-biasing resistor determined? of a screen resistor?

6-15 How is the capacitance of a cathode bypass capacitor calculated? Why is its value determined for the *lowest* operating frequency?

6-16 Refer to the simple triode circuit of Fig. 6-6. The *ac component* of grid-to-cathode voltage is properly denoted by e_g. (*a*) At what instant is the plate current a maximum? Answer in terms of possible value of e_g. What is true about e_c at that instant? (*b*) At the instant of maximum plate current, what is known about i_b, i_p, e_b, e_p? (*c*) Since the instantaneous value of total plate current is a maximum when E_g is at its *positive* maximum, why is e_b at its *minimum* at that instant? (*d*) Consider that E_p may be measured with a very high-resistance voltmeter across R_b. Explain why E_p is 180° out of phase with respect to E_g. The ac voltage across E_{bb} is entirely negligible because its impedance is essentially zero at moderate frequencies.

6-17 Refer to the circuit of a triode amplifier, and explain in words how voltage amplification takes place.

6-18 Explain how to draw the dc load line on a set of average plate characteristics.

6-19 What determines the location of the quiescent point in a triode circuit?

6-20 What would be the effect on voltage amplification if the value of R_b were increased? decreased?

6-21 What is the purpose of the capacitor C_k in self-bias? of the resistor R_k?

Problems

6-1 A vacuum triode operating under space-charge-limited current conditions has a 3-V dip in front of its cathode. Calculate the normal component of velocity with which electrons must leave the cathode in order to reach the lowest point of the dip.

6-2 An electron that was just able to pass the lowest point of the dip in Problem 6-1 goes on to the plate, which is at +100 V with respect to the cathode. (*a*) With how many electronvolts of energy will it strike the plate? (*b*) With what velocity, in meters per second, will it strike the plate?

6-3 In a simple voltage-amplifier circuit like Fig. 6-6, the maximum and minimum instantaneous values of e_b are 150 V and 90 V, respectively. (*a*) What is the rms value of e_p? (*b*) What is the average (dc) value of plate-to-cathode potential? (*c*) If $R_b = 5000\ \Omega$ what is the rms value of

250 Vacuum Tubes   Ch. 6

i_p? (d) If E_{bb} = 250 V, calculate the dc component of plate current, using Eq. 6-9. Should this current be obtainable from E_b/R_b? Why? Check it.

6-4 On a set of plate characteristics for a 12AU7A tube, determine graphically values of μ, g_m, and r_p in the neighborhood of E_b = 170 V, E_c = 0 V. Check $\mu = g_m r_p$.

6-5 Draw several static transfer characteristics for the 12AU7A tube, using the average plate characteristics.

6-6 Plot a set of curves showing plate voltage on the ordinate and grid voltage on the abscissa for several values of constant plate current in the 12AU7A tube. Measure $\Delta e_b/\Delta e_c$. What parameter is represented by the slope of these curves?

6-7 What is the transconductance of a tube that has an amplification factor of 12 and a dynamic plate resistance of 5000 Ω?

6-8 A tube in a circuit like that of Fig. 6-6 is biased at −10 V. What is the largest rms value of ac input voltage that may be used without driving the grid positive?

6-9 Cathode bias is produced by a resistor R_k = 800 Ω carrying 6 mA of current. What is the bias voltage? Calculate the required size of the bypass capacitor for a minimum frequency of 159 Hz.

6-10 What should be the resistance of a bleeder section on a power supply so that it will provide −5 V of bias when the current is 60 mA before the bias connection is made to the tube?

6-11 Using the characteristic of the 2A3 triode, determine the amplification factor, plate resistance, and transconductance for a plate supply voltage of 200 V and a grid voltage of −35 V.

6-12 The characteristics of a certain triode may be approximated by the expression $i_b = 10^{-5}(155 e_c + 8 e_b)$. Determine the values of μ, r_p, and g_m.

6-13 Draw the dynamic transfer characteristic for a 12AU7A tube for E_{bb} = 250 V and plate load R_L = 2500 Ω.

6-14 The plate current in a triode is 8 mA when E_{bb} = 225 V and E_{cc} = −14 V. Assuming μ as approximately constant and equal to 10, what should be the approximate plate current if E_{bb} = 300 V and E_{cc} = −25 V?

6-15 The plate resistance of a triode is 8000 Ω and the transconductance is 3000 μmho. If the plate voltage is increased by 40 V, what is the increase in plate current, assuming the grid voltage is maintained constant?

6-16 A certain triode has the following tube coefficients:

$$\mu = 40, \quad g_m = 2000 \, \mu\text{mho}, \quad r_p = 20{,}000$$

If the plate current is to go up 0.2 mA when the grid voltage is increased

1 V in the negative direction, in what direction, and by how much, must the plate voltage be changed?

6-17 The plate current of a certain pentode connected as a triode can be expressed approximately by the equation

$$i_b = 41(e_b + 10e_c)^{1.41} \times 10^{-6} \text{ A}$$

What is the amplification factor?

6-18 A 6J5 triode operates into a 20-kΩ plate load and is powered by a 200-V dc source in the plate circuit. A 6-V battery is used to provide fixed bias in the cathode circuit. (a) Draw from memory the schematic diagram of the circuit. (b) Determine the values of plate current, plate voltage, and grid bias when the grid circuit has no ac signal applied; i.e., the *operating point*.

6-19 In the circuit shown in Fig. P6-19, tube 2 is to operate at $i_b = 1$ mA, $e_c = -6$ V. Determine the values of i_b, e_b, and e_c for tube 1 and the necessary value for R_k. (*Hint:* The plate current of both tubes flows through R_k and produces the −6 V bias for both of them.)

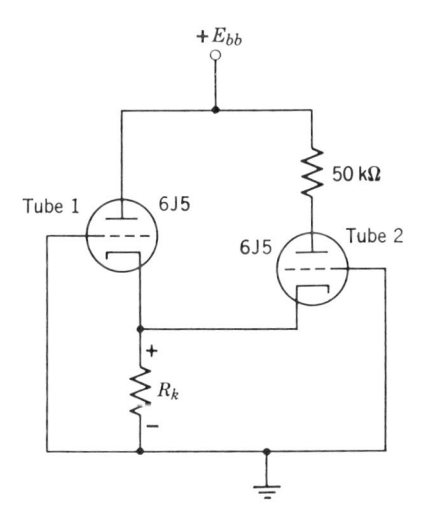

Figure P6-19

6-20 The grid circuit of a 6J5 triode uses a fixed bias of −4 V. The plate circuit contains a 25-kΩ load and is supplied from a 250-V supply. Calculate the range of plate supply voltage required to hold plate current constant for plate loads varying from 10 kΩ to 100 kΩ. (*Hint:* The plate supply voltage equals the sum of load voltage and plate voltage, e_b, at all times.)

252   Vacuum Tubes   Ch. 6

6-21 For a signal $e_s = 20 \sin(\omega t + \phi)$ in the plate circuit as shown in Fig. P6-21, determine the peak-to-peak variation of the plate current. (*Hint:* The grid bias will remain constant.)

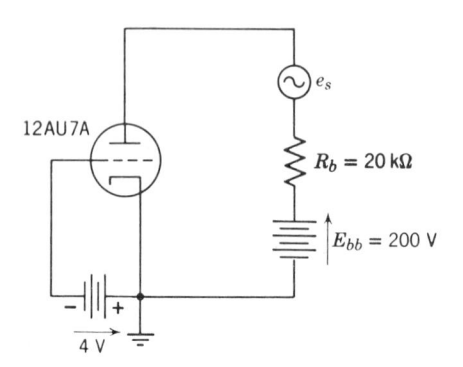

Figure P6-21

6-22 One section of a 12AU7A triode is used in the circuit of Fig. 6-6. Supply voltage E_{bb} is 200 V, and $R_b = 2$ kΩ. What value of grid bias is required to make the quiescent plate voltage $= 100$ V?

6-23 A 12AU7A triode is used in the circuit of Fig. 6-6. Determine the necessary value of R_b for $E_1 = 0$ V, $E_c = 0$ V, $e_b = 200$ V, $E_{bb} = 300$ V.

6-24 The cathode resistor R_k in Fig. 6-7 has the value 1.2 kΩ and $C_k = 10$ μF. If the reactance of C_k is to be always less than or equal to one-tenth of R_k, what is the lowest permissible frequency of variation in plate current?

6-25 A triode has $g_m = 4000$ μmho and $r_p = 12$ kΩ. (*a*) What change in plate current will occur for a grid voltage change from -2 to -6 V? (Assume e_b remains at 150 V.) (*b*) How much change in plate voltage is required to return plate current to its original value, with e_c remaining at -6 V? (State the magnitude and direction of the change in plate voltage.)

6-26 Using the plate characteristics for the 6SN7GT triode, (*a*) plot the static transfer characteristic for $E_b = 200$ V, and (*b*) plot the dynamic transfer curve for $E_{bb} = 200$ V and $R_b = 20$ kΩ.

6-27 State the purpose of each of these circuit elements in Fig. 6-12: R_b, R_k, R_s, C_k, C_s, E_{bb}.

6-28 A 6SJ7 pentode is operated with $E_{bb} = 300$ V and $R_b = 30$ kΩ. The screen voltage is set at 100 V and the suppressor grid is connected to the cathode. Fixed bias holds the quiescent control grid potential at -2 V. For a signal $e_s = 2 \sin \omega t$ in the grid circuit, determine the variation of plate current and the variation of plate voltage. Make graphical plots of the time variations of e_s, i_b, and e_b. Label the graph with numerical values at significant points, in a manner similar to the symbolic markings in Fig. 6-9.

6-29 In the circuit of Fig. P6-29, if the grid is made 1 V more positive than normal relative to the cathode, the voltage across R_k is 15 V larger than normal. What value of e_s is required to cause this to happen?

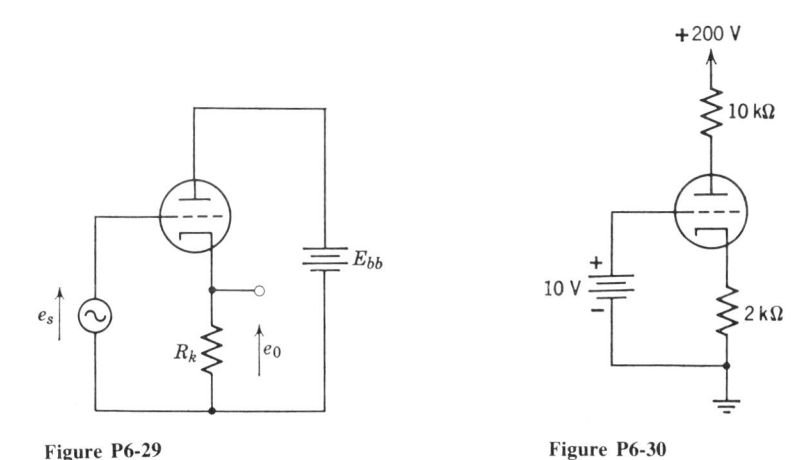

Figure P6-29

Figure P6-30

6-30 In the circuit shown in Fig. P6-30, it is known that plate current is related to grid voltage by the expression $i_b = (30 + 2e_c)$, where i_b is in milliamperes and e_c is in volts. How much plate current flows in this circuit?

6-31 Determine numerical values for each of these quantities in Fig. P6-31: I_b, E_b, E_c, e_p, i_p, e_g.

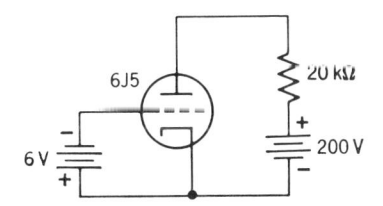

Figure P6-31

7
SMALL-SIGNAL EQUIVALENT CIRCUITS

We have discussed the operation of transistors and vacuum tubes from a dc viewpoint, with some attention to ac variations of voltages and currents in the circuits containing them. Now we will determine models for their behavior that may be used for analysis of circuit operation when the ac signals are of small amplitudes.

Small-signal operation is said to occur whenever voltage and current changes are within a region of linear characteristics of the transistor or vacuum tube being used. Within such a region, the device parameters may be considered constant in value, permitting the use of a circuit model containing these parameters and the use of linear circuit analysis methods introduced in Chapter 1 for general ac circuits.

The key to small-signal analysis is an equivalent circuit for the transistor or vacuum tube, which will be developed in this chapter. It is the combination of device parameters and other ac circuit components that will best approximate the behavior of the circuit in the region of operation. Once the ac equivalent circuit for small-signal operation can be established, operation of the circuit may be analyzed by Thévenin and Norton theorems, loop current methods, and other approaches discussed in Chapter 1. As always, Ohm's and Kirchhoff's laws of electric circuits will govern our analysis.

7-1
The AC Load Line

The circuit in Fig. 7-1 is the familiar common-emitter voltage amplifier we have seen in earlier chapters. The significant changes include capacitors in the

256 Small-Signal Equivalent Circuits Ch. 7

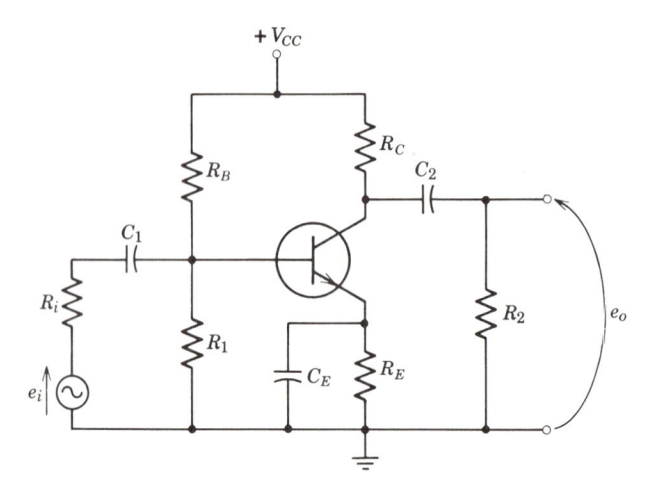

Figure 7-1
Common-emitter transistor circuit, which includes ac signal coupling capacitors.

input and output circuits. These capacitors are used to *couple* ac voltages and currents and to *block* dc components. Circuit design will often include capacitors whose reactances are small compared with the resistor values in the circuit. Capacitor C_2, for example, may have negligible reactance compared with resistor R_2, which may well be the bias network for a succeeding stage of amplification. Then as far as ac signals are concerned, resistors R_C and R_2 are in parallel.

An equivalent circuit for ac operation of the circuit is given in Fig. 7-2. All dc bias sources have been replaced by short-circuits (their ac equivalents), and capacitors are represented by ac short-circuits. The block labeled "transistor ac equivalent" contains elements that will be determined in subsequent sections of this chapter.

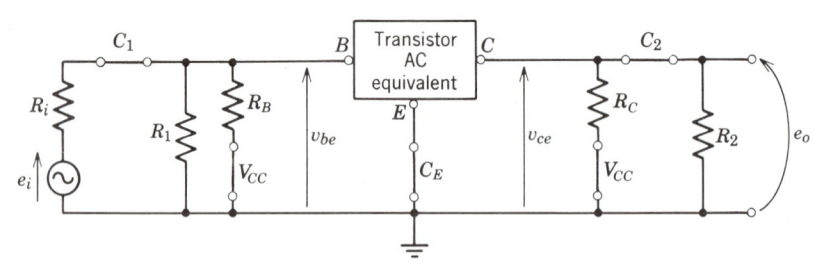

Figure 7-2
Equivalent circuit for network in Fig. 7-1 for ac operation.

Sec. 7-1   The AC Load Line   257

You will recall that the dc conditions in the circuit of Fig. 7-1 determine the location of a Q-point and the region of operation about it for dc considerations. However, when ac circuit components are included through the addition of coupling capacitors, transistor behavior will be altered. ==The bias and Q-point location will still be determined by dc conditions==, but variations in voltages and currents will be governed by the ac impedances and circuit paths. An *ac load line* can be constructed on the characteristic curves for the transistor being used.

==The ac load line will pass through the Q-point==, since the transistor operates there when ac signals pass through zero amplitudes. Its slope is determined by the equivalent ac load in the collector circuit. In the circuit of Fig. 7-3, the ac collector load is the parallel combination of R_3 and R_2. The dc collector load is R_1 in series with R_2.

We shall design the circuit to have an ac load resistance of 2000 Ω. This means that R_2 must be larger than 2000 Ω since it is in parallel with R_3. Choosing $R_2 = 3000$ Ω and $R_3 = 6000$ Ω would make the load resistance 2000 Ω. The ac load line would fall as shown in Fig. 7-4 after the dc load line has been drawn for 3000 Ω, allowing 100 Ω for R_1. The determination of the locations of the load lines will now be explained.

The quiescent point is selected at 20 μA base current. The dc load line passes through the quiescent point and has a slope of −1/3100. It may be drawn as a straight line having this slope and passing through the quiescent point. It will intercept the collector-current axis and collector-emitter voltage axis according to the equation for the load line:

$$i_C = v_{CE}\left(-\frac{1}{R_{DC}}\right) + \frac{V_{CC}}{R_{DC}}$$

where V_{CC} is the voltage intercept at $i_C = 0$, and i_C is the current intercept at $v_{CE} = 0$. Current and voltage values at Q are: $i_C = -0.8$ mA, $v_{CE} = -2.25$ V.

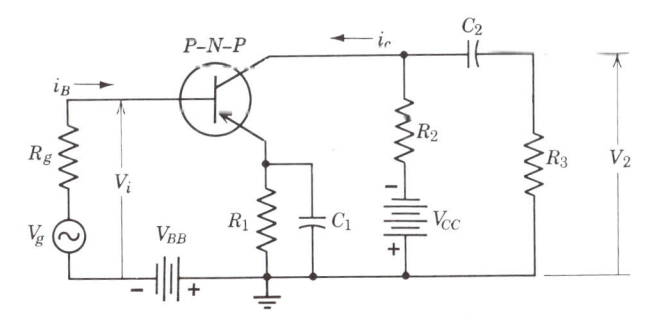

Figure 7-3
Transistor circuit having different ac and dc collector loads.

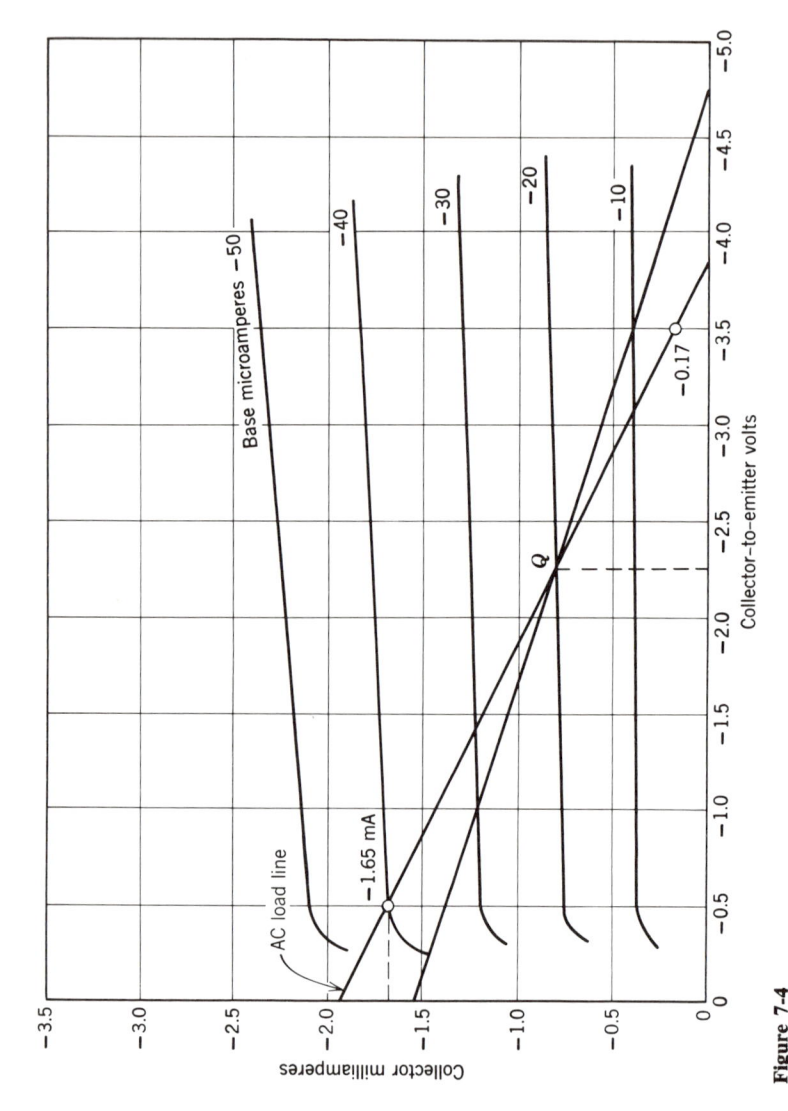

Figure 7-4
Common-emitter characteristic curves of a transistor, Q at -2.25 V, -0.8 mA.

Sec. 7-1   The AC Load Line   259

Using these values in the equation we may obtain V_{CC}, the voltage intercept. This value is -4.75 V. The current intercept is $V_{CC}/R_{DC} = -1.525$ mA.

The ac load line also passes through Q and is a straight line having a slope of $-1/2000$. Using the point-slope form of the equation for the straight line

$$-\frac{1}{R_{AC}} = \frac{(i_C - i_Q)}{(0 - v_Q)}$$

the current intercept will be -1.925 mA.

Changes in collector-emitter voltage will result from changes in base current along the ac load line. It can be seen that the base current changes caused by an input signal produce large changes of collector-emitter voltage. Assume that the input signal causes a sinusoidal variation in base current corresponding to collector-emitter voltage variation from -0.5 to -3.5 V. The corresponding change in base current is from $-40\ \mu$A to $-4\ \mu$A, as a close approximation. The change in collector current, meanwhile, is from -1.65 mA to -0.17 mA.

The gains will now be computed from the graphical data. The ratio of the ac collector current to the ac base current is

$$A_i = \frac{i_C}{i_B} = \frac{-1.65 - (-0.17)}{[-40 - (-4)] \times 10^{-3}} = 41.2$$

The voltage gain involves the use of the input-signal voltage value. The input resistance of this transistor with common-emitter connections is given as 1200 Ω. To obtain a change in base current of 36 μA (40 − 4), the required amplitude of the sine-wave input signal can be computed.

$$\Delta V_i = R_i \times \Delta i_B = 1200 \times 36 \times 10^{-6}$$
$$\Delta V_i = 43.2\ \text{mV}$$

The voltage gain is then

$$A_v = \frac{\Delta v_{CE}}{\Delta V_i} = \frac{-0.5 - (-3.5)}{0 - 0.0432} = -69.5$$

The negative sign means there is a 180-degree phase shift in voltage through the amplifier. The power gain is the product of the current and voltage gains

$$A_p = A_i \times A_v = 41.2 \times 69.5 = 2860$$

This is a gain of 34.6 dB.

Although the curves used are plotted on small graph paper and the readings can be only close approximations at best, the results are well within acceptable limits. It will be found in Chapter 8 that the common-emitter transistor amplifier has gains with these orders of magnitude.

260 Small-Signal Equivalent Circuits   Ch. 7

7-2
Theory of Hybrid Parameters

The hybrid parameters, or *h-parameters* as they are commonly called, are gaining favor with manufacturers for use in specifications for transistors. It is helpful in learning about them to describe their use in setting up equivalent circuits for general, linear, four-terminal networks. Then they may be extended for use in describing the behavior of transistors in their small-signal mode of operation.

We have learned that a linear four-terminal network of any configuration between a generator and a load can be represented by a new, simple series circuit derived by applying Thévenin's theorem or by a new, simple parallel circuit obtained by applying Norton's theorem.

The *hybrid* equivalent circuit is the most widely used in the analysis of transistor circuits. It is termed *hybrid* because it contains both *admittance* and *impedance* parameters. The ease of measurement of these parameters, which are the *h-parameters* of the circuit, has contributed to the widespread use of the equivalent circuit.

A set of *h*-parameters can be derived for any *black box* network having linear elements and two input and two output terminals. Consider the situation in Fig. 7-5, where the external currents and voltages at the terminals of the black box network are measurable quantities. In general, the output current is a function of the output voltage and input current; the input voltage can be written as a function of input current and output voltage:

$$v_1 = f(i_1, v_2) \qquad (7\text{-}1)$$

$$i_2 = g(i_1, v_2) \qquad (7\text{-}2)$$

This is merely another way of saying that any two of the input and output quantities may be expressed in terms of the other two quantities.

Expressing these two equations in total differential form yields:

$$dv_1 = \left.\frac{dv_1}{di_1}\right|_{v_2\,\text{constant}} di_1 + \left.\frac{dv_1}{dv_2}\right|_{i_1\,\text{constant}} dv_2 \qquad (7\text{-}3)$$

$$di_2 = \left.\frac{di_2}{di_1}\right|_{v_2\,\text{constant}} di_1 + \left.\frac{di_2}{dv_2}\right|_{i_1\,\text{constant}} dv_2 \qquad (7\text{-}4)$$

Now if the differential changes in voltages and currents are taken as single-

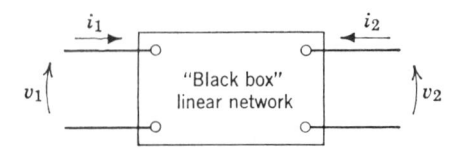

Figure 7-5
Four-terminal "black box" network.

frequency sinusoidal variations, these equations may be written in terms of phasors to represent the varying quantities:

$$V_1 = \left.\frac{dv_1}{di_1}\right|_{v_2 \text{ constant}} I_1 + \left.\frac{dv_1}{dv_2}\right|_{i_1 \text{ constant}} V_2 \tag{7-5}$$

$$I_2 = \left.\frac{di_2}{di_1}\right|_{v_2 \text{ constant}} I_1 + \left.\frac{di_2}{dv_2}\right|_{i_1 \text{ constant}} V_2 \tag{7-6}$$

The coefficients of I_1 and V_2 are the h-parameters of the network. They are usually defined as follows:

$$V_1 = h_{11} I_1 + h_{12} V_2 \tag{7-7}$$

$$I_2 = h_{21} I_1 + h_{22} V_2 \tag{7-8}$$

Since Eqs. 7-7 and 7-8 apply to the black-box network, an equivalent circuit may be drawn that fits these equations. The equivalent circuit is given in Fig. 7-6. It can be seen that the input circuit (Eq. 7-7) is represented by Thévenin's equivalent circuit, and the output circuit is a Norton equivalent of Eq. 7-8. The input circuit equation is the sum of voltages existing there, whereas the output circuit equation is a nodal sum of currents in the output side of the network.

Each of the h-parameters may be determined from measurements in the network, when the input and output voltages and currents are sinusoidal, according to this tabulation derived from Eqs. 7-7 and 7-8:

$$h_{11} = \left.\frac{V_1}{I_1}\right|_{V_2=0} \quad \Omega \tag{7-9}$$

$$h_{12} = \left.\frac{V_1}{V_2}\right|_{I_1=0} \quad \text{(dimensionless)} \tag{7-10}$$

$$h_{21} = \left.\frac{I_2}{I_1}\right|_{V_2=0} \quad \text{(dimensionless)} \tag{7-11}$$

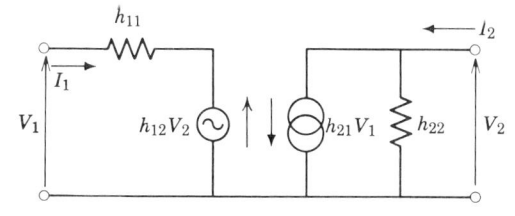

$$h_{22} = \left.\frac{I_2}{V_2}\right|_{I_1=0} \quad \text{mhos} \tag{7-12}$$

Figure 7-6
Equivalent circuit for "black box" network, showing placement of h-parameters.

262 Small-Signal Equivalent Circuits Ch. 7

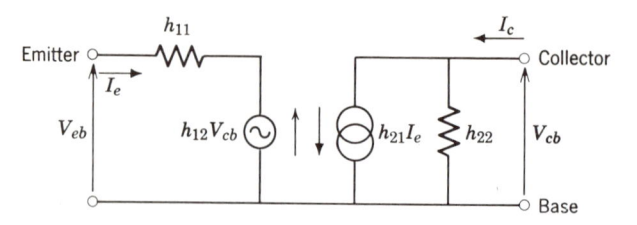

Figure 7-7
h-Parameter equivalent circuit for transistor in common-base circuit.

Figure 7-7 is the circuit for a transistor operating in the common-base configuration. Note that the input and output voltages and currents can be related to those of the black-box network by this set of equations:

$$V_{eb} = h_{11}I_e + h_{12}V_{cb}$$
$$I_c = h_{21}I_e + h_{22}V_{cb}$$
(7-13)

The values of the common-base h-parameters may be measured for the transistor by using the definitions and restrictions of Eqs. 7-9 through 7-12. Since they are defined for sinusoidal quantities, the measurements will be made using the small-signal operation of the transistor, for which linearity may be assumed in the network.

7-3
Transistor Equivalent Circuit Using h-Parameters

We are now equipped to draw the equivalent circuit of a transistor for dynamic operation employing h-parameters. The grounded-base circuit is discussed first.

Figure 7-8 shows the actual and equivalent circuits. The impedance h_{11} is put

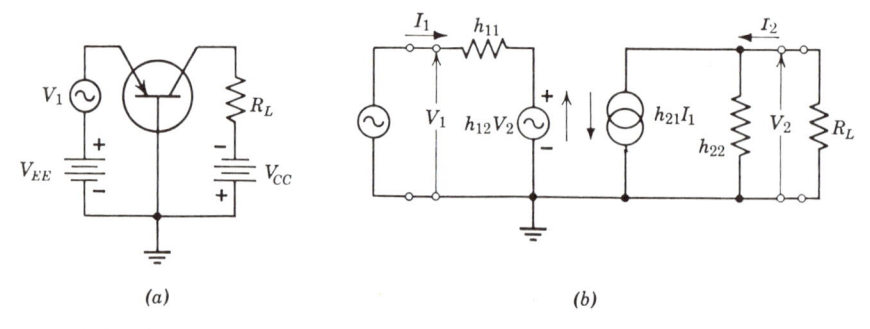

Figure 7-8
Common-base amplifier. (a) Actual circuit. (b) Equivalent circuit.

Sec. 7-4   Mathematical Analysis of Transistor Amplifier   263

in place of its equivalent in the circuit of the "black box." By definition, h_{22} is a *conductance*. The element labeled h_{22} is a resistance whose value is $1/h_{22}$; when V_2 is multiplied by the h_{22} value, a *current* is obtained, since this multiplication actually amounts to *dividing* V_2 by $1/h_{22}$.

The constant-current generator in the output part of the equivalent circuit for the black box can be replaced by $h_{21}I_1$. Thus the equivalent-circuit nomenclature is established.

We are now prepared to write the operating equations for the equivalent circuit of the *grounded-base transistor* amplifier.

7-4
Mathematical Analysis of Transistor Amplifier

Refer to the equivalent circuit of the common-base transistor in Fig. 7-8. The input section lends itself to the writing of a loop-voltage, or *mesh*, equation. The output section lends itself to the writing of a current-node or nodal equation.

$$V_1 = h_{11}I_1 + h_{12}V_2 \qquad (7\text{-}14)$$

$$I_2 = h_{21}I_1 + h_{22}V_2 \qquad (7\text{-}15)$$

where, as explained in Section 7-2, h_{11} is a resistance and h_{22} is a conductance.

Consider the quantities that are known in these equations. The h-parameters are given for the transistor by the manufacturer. Typical values for a specific transistor are as follows:

Input resistance, output circuit shorted:

$$h_{11} = 40\ \Omega$$

Reverse voltage transfer ratio, input circuit open:

$$h_{12} = 3.23 \times 10^{-4}$$

Forward current transfer ratio, output circuit shorted:

$$h_{21} = -0.95$$

Output conductance, input circuit open:

$$h_{22} = 1.4\ \mu\text{mho}$$

The remaining known quantities are the applied voltage, V_1, and the load current, I_2. Evidently, the simultaneous equations may be solved for the unknowns, I_1 and V_2. This is a simple task, and the results are:

$$V_2 = \frac{h_{11}I_2 - h_{21}V_1}{h_{11}h_{22} - h_{12}h_{21}} \, V \tag{7-16}$$

$$I_1 = \frac{h_{22}V_1 - h_{12}I_2}{h_{11}h_{22} - h_{12}h_{21}} \, A \tag{7-17}$$

The voltage gain, A_v, is the ratio of V_2 to V_1. We can get an expression for V_2 in terms of only the *h*-parameters, V_1, and the load resistance.

$$I_2 = -V_2/R_L \tag{7-18}$$

Substituting this in Eq. 7-16,

$$V_2 = \frac{-h_{11}\dfrac{V_2}{R_L} - h_{21}V_1}{h_{11}h_{22} - h_{12}h_{21}}$$

Calling the denominator a constant D to save tedious writing, we can rewrite the equation:

$$V_2 D = -h_{11}\frac{V_2}{R_L} - h_{21}V_1$$

$$V_2\left(D + \frac{h_{11}}{R_L}\right) = -h_{21}V_1 \tag{7-19}$$

$$V_2 = \frac{-h_{21}V_1}{D + \dfrac{h_{11}}{R_L}} = -\frac{h_{21}V_1}{h_{11}h_{22} - h_{12}h_{21} + \dfrac{h_{11}}{R_L}}$$

The voltage gain, V_2/V_1, is then

$$A_v = \frac{-h_{21}}{h_{11}\left(h_{22} + \dfrac{1}{R_L}\right) - h_{12}h_{21}} \tag{7-20}$$

Checking the dimensions, A_v should be dimensionless (volts/volts).

The numerator h_{21} is the forward-current transfer ratio and therefore dimensionless. The denominator terms will be written in dimensional form:

$$\text{ohms}\left(\frac{1}{\text{ohms}} + \frac{1}{\text{ohms}}\right) - \frac{\text{volts}}{\text{volts}} \times \frac{\text{amperes}}{\text{amperes}}$$

Thus the denominator is also dimensionless.

The *h*-parameters supplied by the manufacturer of the transistor hold for small-signal operation *with a specific Q-point*.

This analysis is intended to present the general procedure for setting up the equivalent circuit and writing equations for finding the important quantities involved in transistor amplifier behavior. In the next chapter we shall present a more detailed analysis of the three types of transistor amplifier—common base, common emitter, and common collector—using the equivalent circuit for each case.

7-5
Manufacturer Designation of h-Parameters

The subscripts on the h-parameters used in the preceding sections are not generally used by transistor manufacturers. Instead, they use a system of letter subscripts to designate which terminal of the transistor is common in a circuit. For example, h_{11} is designated as h_{ib} in the common-base circuit connection and as h_{ie} in the common-emitter circuit. The first letter refers to the *input* parameter and the second letter designates the common terminal. The h-parameters are labeled as follows, depending on the common terminal of the circuit in which the transistor is being used:

h_{11}: h_{ib}, h_{ie}, h_{ic} h_{12}: h_{rb}, h_{re}, h_{rc}
h_{21}: h_{fb}, h_{fe}, h_{fc} h_{22}: h_{ob}, h_{oe}, h_{oc}

The first letter of the double-subscript system has a special meaning. The i refers to the *input* impedance; r designates the *reverse* voltage feedback ratio; f refers to the *forward* current ratio; o is used to designate the *output* admittance.

7-6
Calculation of h-Parameters from Manufacturer's Data

The data supplied by a transistor manufacturer may include values of the h-parameters for a specific Q-point. A circuit in which the transistor is to operate may require a different Q-point. The values of h-parameters may be quite different at this operating point. Before making an analysis of the circuit behavior by the methods of the next chapter, it will be necessary to calculate the values of h-parameters that apply to the circuit operating point.

As an example of the necessary calculations, we will use the following data supplied for the 2N334 transistor:

Common-Base Design Characteristics, at 25°C

h_{ib}	Input impedance	($V_{CB} = 5$ V, $I_E = -1$ mA)	56 Ω
h_{ob}	Output admittance	($V_{CB} = 5$ V, $I_E = -1$ mA)	0.5 μmho
h_{rb}	Feedback voltage ratio	($V_{CB} = 5$ V, $I_E = -1$ mA)	350×10^{-6}
h_{fb}	Current transfer ratio	($V_{CB} = 5$ V, $I_E = -1$ mA)	-0.97

266   Small-Signal Equivalent Circuits   Ch. 7

In addition to these numerical values of the common-base characteristics, charts are provided to permit their calculations for other operating points. The given values apply only to the condition $V_{CB} = 5$ V, $I_E = -1$ mA. For other emitter-current conditions, the chart of Fig. 7-9 may be used.

It will be noted that the quantity $1 + h_{fb}$ is plotted instead of h_{fb}. The reason for this is clear. Since the parameter h_{fb} is nearly unity and changes very little for different emitter currents, the small changes would not be apparent in the chart of Fig. 7-9. However, the quantity $1 + h_{fb}$ may change greatly with only a small change in h_{fb}. For example, suppose h_{fb} varies from -0.97 to -0.94. The change is only 0.03; compared with the value of h_{fb} it is small. The change in $1 + h_{fb}$ is from 0.03 to 0.06, a ratio of 2 to 1.

An important result of plotting the variations of the h-parameters is their wide range of possible values with emitter currents. Over the range of currents

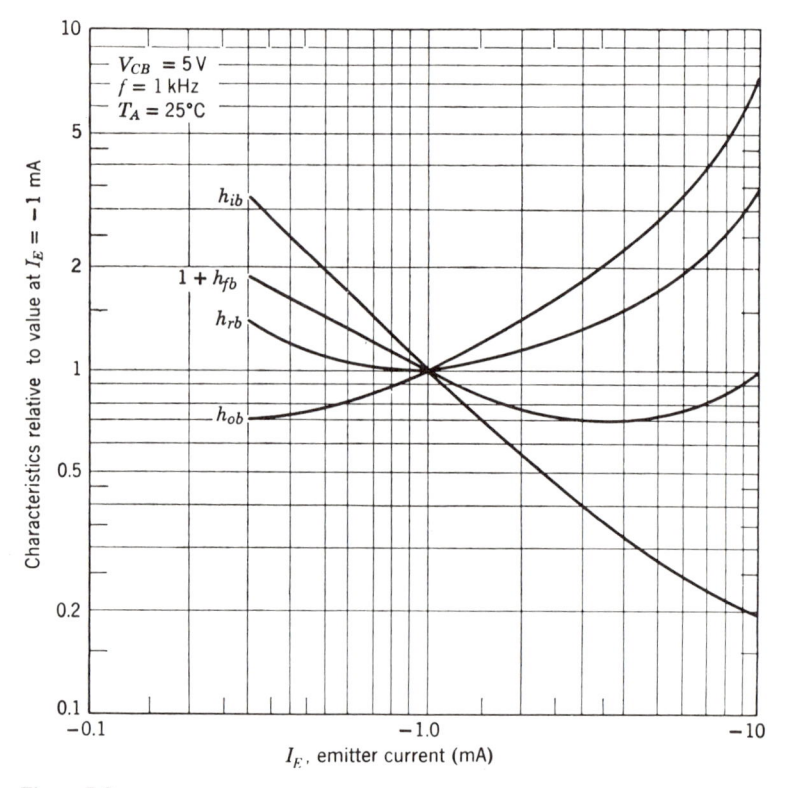

Figure 7-9
Common-base characteristics vs. emitter current 2N334 transistor; h-parameter variations with changes in I_E.

Sec. 7-7   *Formulas for Calculating h-Parameters*   267

from -0.3 to -10 mA, they may change their values by an order of magnitude or more. Therefore, it becomes important to determine the *h*-parameters for the particular *Q*-point being used in a circuit.

7-7
Formulas for Calculating *h*-Parameters

The manufacturer data sheets usually list the common-base characteristics and *h*-parameters. However, a transistor is more often used in the common-

Table 7-1
Conversion Equations for *h*-Parameters (Numerical Values Are Typical at 25°C)

Symbol	Common-Base	Common-Emitter	Common-Collector
h_{ib}	30–80 Ω	$\dfrac{h_{ie}}{1+h_{fe}}$	$-\dfrac{h_{ic}}{h_{fc}}$
h_{rb}	$5\text{–}200 \times 10^{-6}$	$\dfrac{h_{ie}h_{oe}}{1+h_{fe}} - h_{re}$	$h_{rc} - 1 - \dfrac{h_{ic}h_{oc}}{h_{fc}}$
h_{fb}	-0.90 to -0.99	$-\dfrac{h_{fe}}{1+h_{fe}}$	$-\dfrac{1+h_{fc}}{h_{fc}}$
h_{ob}	0.5 to 1.5 μmho	$\dfrac{h_{oe}}{1+h_{fe}}$	$-\dfrac{h_{oc}}{h_{fc}}$
h_{ie}	$\dfrac{h_{ib}}{1+h_{fb}}$	600–2000 Ω	h_{ic}
h_{re}	$\dfrac{h_{ib}h_{ob}}{1+h_{fb}} - h_{rb}$	3×10^{-4}	$1 - h_{rc}$
h_{fe}	$-\dfrac{h_{fb}}{1+h_{fb}}$	10 to 100	$-(1+h_{fc})$
h_{oe}	$\dfrac{h_{ob}}{1+h_{fb}}$	20 μmho	h_{oc}
h_{ic}	$\dfrac{h_{ib}}{1+h_{fb}}$	h_{ie}	600–2000 Ω
h_{rc}	1	$1 - h_{re} \approx 1$	1
h_{fc}	$-\dfrac{1}{1+h_{fb}}$	$-(1+h_{fe})$	-10 to -100
h_{oc}	$\dfrac{h_{ob}}{1+h_{fb}}$	h_{oe}	20 μmho

emitter mode and occasionally will be connected in the common-collector configuration. In order to analyze circuit behavior using the h-parameter equivalent circuit for the transistor, it is necessary to calculate the values of the h-parameters for these circuit modes, in addition to any variations because of the Q-point conditions. The formulas of Table 7-1 are useful in making the necessary calculations.

An example will show how to use these formulas. Suppose it is desired to determine the values of h-parameters for a common-emitter circuit that uses a 2N334 transistor. The common-base parameters are listed above and will be used in the equations of Table 7-1 to calculate the values of the common-emitter parameters. For the purpose of this example, it is assumed that the Q-point of the circuit will be fixed at the same conditions in the two circuits.

The values of the common-emitter parameters may be calculated from their common-base values by using the equations listed under common-base in Table 7-1. The value of h_{ie} is calculated:

$$h_{ie} = \frac{h_{ib}}{1 + h_{fb}} = \frac{56\,\Omega}{1 - 0.97} = \frac{56}{0.03}\,\Omega$$
$$= 1870\,\Omega$$

Similarly, the value of h_{fe} is calculated as

$$h_{fe} = -\frac{h_{fb}}{1 + h_{fb}} = -\frac{-0.97}{1 - 0.97} = \frac{0.97}{0.03} = 32.3$$

The other parameters are calculated in a similar manner. Here again we see the factor $1 + h_{fb}$, which is plotted in Fig. 7-9

Drill Problems

D7-1 Using the equations of Table 7-1 and Fig. 7-9, determine the values of common-emitter h-parameters for emitter current $= -5$ mA, $V_{CB} = 5$ V, for the 2N334 transistor.

D7-2 Calculate the voltage gain of a transistor whose common-base h-parameter values are $h_{11} = 50\,\Omega$, $h_{22} = 0.5\,\mu$mho, $h_{12} = 200 \times 10^{-6}$, $h_{21} = -0.98$, and which is driving a load resistance of $1000\,\Omega$.

7-8
Actual and Equivalent Circuits of Vacuum-Tube Amplifier

The circuit of a simple amplifier is shown in Fig. 7-10a. The heater circuit for the cathode has been omitted, since it serves merely to provide thermionic emission at the cathode and does not enter otherwise into the circuit operation.

Sec. 7-8　　　　　　　　*Actual and Equivalent Circuits of Vacuum-Tube Amplifier*　269

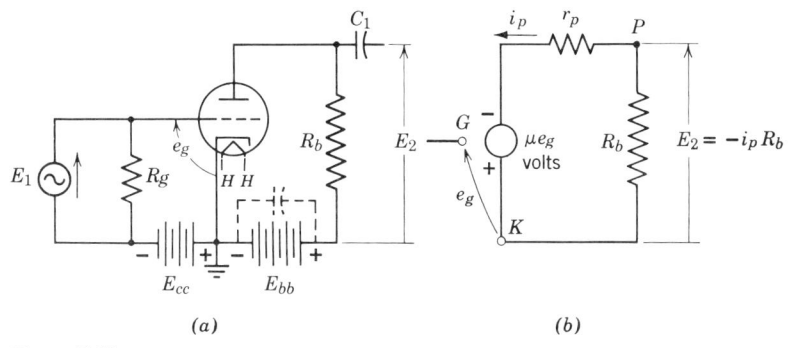

Figure 7-10
(a) Simple amplifier circuit. (b) Its equivalent circuit.

The grid-circuit resistor, R_g, usually has a high ohmic value, perhaps 500,000 Ω, and serves to accept the signal voltage and to connect the negative-bias voltage to the grid. The varying input voltage, E_1, is placed on R_g and so is "superimposed" upon the steady dc grid-bias voltage supply E_{cc}. Thus E_1 causes the grid to be, alternately, first less and then more negative with respect to the cathode than the dc bias-voltage value. It is very important that no grid current should flow, and therefore the grid should not be allowed to go positive with respect to the cathode.

If grid currents were to flow in a small-signal voltage-amplifier circuit, several detrimental effects would result. The grid of a vacuum-tube voltage amplifier is not designed to carry more than an extremely small current. Such grids are easily damaged. When grid current flows, the ac component is supplied by the input circuit, which usually cannot function properly when "loaded" in this way. Distortion of the input-voltage waveform results; this means that the output voltage of the amplifier will be distorted.

Although the ac component of the grid-to-cathode voltage, e_g, is across R_g and equal to E_1, there is no ac voltage loss in the bias battery, so it is correct to consider that e_g exists at the grid-cathode terminals of the tube.

The plate-load resistor, R_b carries the plate current at all times. Since the plate current will vary because of variations in grid-to-cathode voltage, E_2 will be a varying voltage with a larger amplitude than E_1.

Figure 7-10b shows how the actual circuit may be depicted by a simple series circuit in which the tube is represented by a generator having an induced voltage equal to μe_g. It is important to observe that *no dc voltages appear in the equivalent circuit*. Only *varying voltages* are used in working with the equivalent circuit.

The *coupling capacitor* C_1 is large enough to provide negligible reactance at the operating frequencies. Furthermore, the ac voltage drop across the power supply battery is also negligible. A bypass capacitor across E_{bb} could be used to

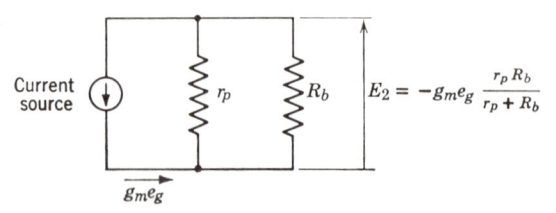

Figure 7-11
Circuit equivalent to Fig. 7-10a; constant-current source.

Current source, r_p, R_b, $E_2 = -g_m e_g \dfrac{r_p R_b}{r_p + R_b}$, $g_m e_g$

ensure zero ac voltage drop, if necessary. This means that the ac output voltage across R_b is available where E_2 is shown. It has the same value between the right-hand plate of C_1 and ground.

The equivalent circuit, Fig. 7-10b, is derived by applying Thévenin's theorem. The open-circuit ac voltage (R_b disconnected) is the plate-to-cathode voltage e_b, which is equal to $-\mu e_g$. The impedance looking back toward the tube, with R_b disconnected, is the internal ac resistance of the tube, denoted by r_p. The polarity of the equivalent generator is as indicated, because i_p will be in its *positive* half-cycle and flowing *into the plate* while e_g is in its *positive* half-cycle since the two are in phase. It is worth mentioning here that, when e_g is in its *negative* half-cycle, i_p will also be in its negative half-cycle *even though current is still flowing into the plate* because i_b *is always positive*.

Another type of equivalent circuit for Fig. 7-10a may also be set up with the help of Norton's theorem. It employs a *constant-current* source, with the current given by $g_m e_g$.

A short-circuit across R_b will carry ac current limited only by r_p:

$$i_p = \frac{\mu e_g}{r_p} = g_m e_g$$

Looking back toward the tube and determining the impedance to be put in parallel with the load, as required by Norton's theorem, we see only r_p. The equivalent circuit, using a constant-current source shown in Fig. 7-11, is more often used in connection with pentode tubes than with triodes, because its plate resistance is usually much larger than any practical R_b used with it. The pentode, it will be remembered, acts almost as a constant-current source.

7-9
Simple Amplifier Equations

The equivalent circuit of Fig. 7-10b represents the amplifier stage for the purpose of mathematical analysis. It is convenient to assume that the voltages and currents are of sine-wave form and to use effective values in the analysis. The symbols for the effective values of the ac components of grid voltage, plate voltage, and plate current are E_g, E_p, and I_p, respectively, as given in Fig. 7-12.

Sec. 7-10   AC Load Line for Vacuum Triode   271

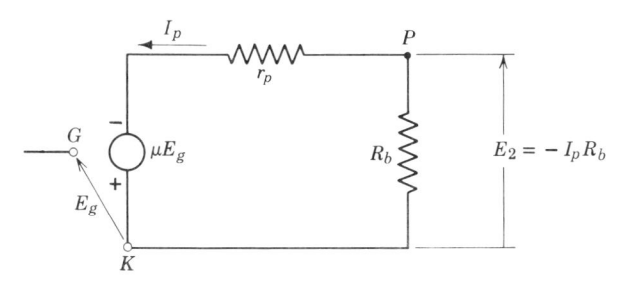

Figure 7-12
Equivalent circuit of Fig. 7-10b.

It is seen in Fig. 7-12 that

$$I_p = \frac{\mu E_g}{r_p + R_b} \qquad (7\text{-}21)$$

Considering E_2 to be the effective value of the voltage across R_b,

$$E_2 = -I_p R_b = \frac{-R_b}{r_p + R_b} \times \mu E_g \qquad (7\text{-}22)$$

In the original circuit of Fig. 7-10a we call E_1 the effective value of the ac component of grid-to-cathode voltage. Then $E_1 = E_g$, and we find that the ratio of the output voltage to the input voltage is, from Eq. 7-22,

$$\frac{E_2}{E_1} = \frac{-\mu R_b}{r_p + R_b} = A_v \qquad (7\text{-}23)$$

This is called the *voltage amplification* of the amplifier, and it is sometimes denoted by A_v. The voltage amplification is, then, the ratio of the effective value of the alternating output voltage to the effective value of the alternating input voltage. The negative sign indicates the 180-degree phase shift between E_g and E_2.

7-10
AC Load Line for Vacuum Triode

It usually happens that the ac output voltage E_2 is applied to a circuit element, such as a resistor, which must be taken into account in determining the effective *ac load* on the amplifier. Figure 7-13 shows such an arrangement. The capacitor C_1 is large enough that its reactance is negligible compared with R_2 at the frequency of signal E_1. Resistor R_2 could be the input bias network for a succeeding stage of amplification. It provides a load for the ac voltage variations in the plate circuit.

272 Small-Signal Equivalent Circuits   Ch. 7

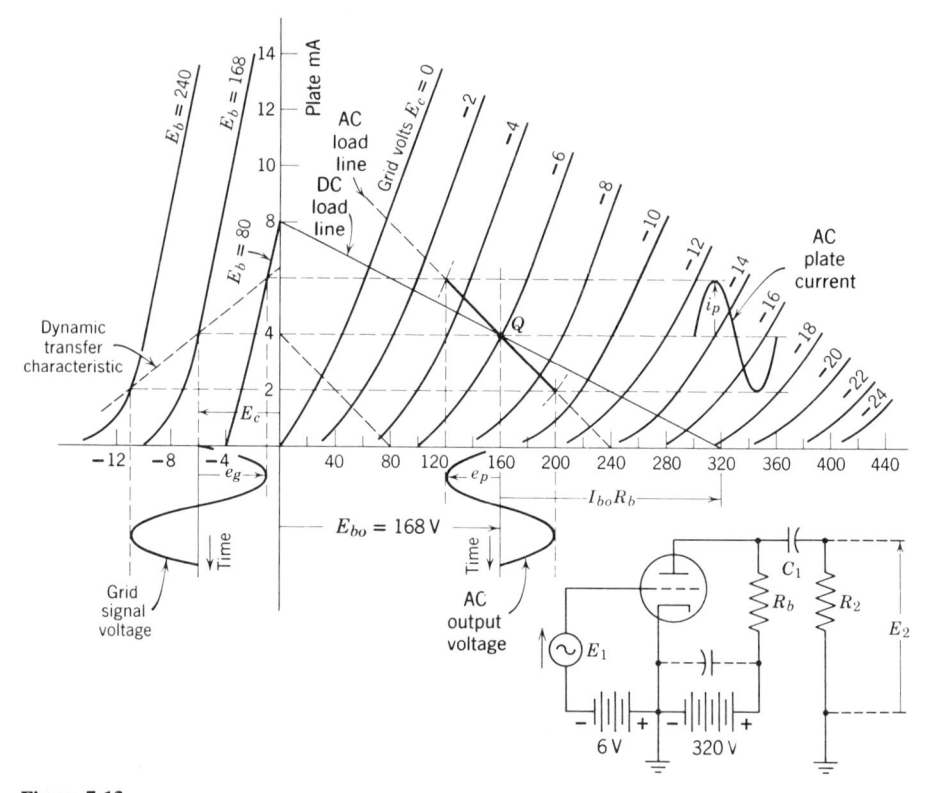

Figure 7-13
Graphical analysis for circuit shown; ac load line differs from dc load line because of C_1-R_2.

Graphical analysis of the circuit in Fig. 7-13 shows that the dc and ac load lines pass through Q, and their slopes will differ by the difference between the dc load (R_b) and the ac load ($R_b \| R_2$). The addition of R_2 by capacitive coupling did not change the dc conditions, but the ac current in the tube will be larger. Some of the ac current will pass through C_1 and R_2.

Drill Problems

D7-3 In the circuit of Fig. 7-13, (a) determine the end points of the dc load line when $R_b = 20\,\text{k}\Omega$. (b) What is the slope of the load line on the tube characteristic curves? (c) What will be its slope if the plate-supply voltage is reduced to $200\,\text{V}$? (d) Determine the ac load line for $R_2 = 1\,\text{M}\Omega$ and $R_b = 20\,\text{k}\Omega$, $E_{bb} = 320\,\text{V}$.

D7-4 In the circuit shown (Fig. D7-4), the plate is loaded with an *output transformer* that supplies power to the 8-Ω load in its secondary

Sec. 7-11 Output Power 273

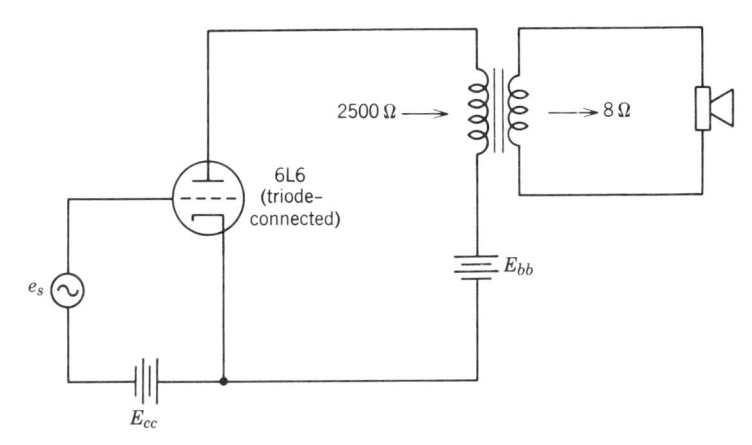

Figure D7-4

winding. (This is typical of a circuit used to drive a loudspeaker.) The primary winding has a resistance of 100 Ω. The turns ratio of the transformer is such that the 8-Ω load is *coupled* into the primary winding as an effective ac resistance of 2500 Ω. On a set of characteristic curves for the tube (see Appendix), choose a value for E_{bb} and locate a Q-point so that a 30-V peak-to-peak ac grid-voltage variation may be used without appreciable distortion of the output waveform. (*a*) Locate a suitable Q-point on the characteristics. (*b*) Sketch the dc and ac load lines. (*c*) What will be the expected variation of plate voltage?

7-11
Output Power

The tube does not actually enlarge the input voltage. It does, however, make possible the creation of a larger voltage that has the same frequency and practically the same waveform as the input voltage. With proper load resistance, grid bias, and amplitude of input voltage (often called signal voltage), the amplifier stage will produce an output voltage that is practically an exact reproduction (although larger) of the input voltage, even though its waveform differs very much from sine-wave form. The complicated waveforms produced in a microphone circuit by musical tones and speech may be reproduced faithfully (with high fidelity) by voltage amplifiers.

Although the purpose of a voltage amplifier is to produce a large voltage output, the plate-circuit resistor, commonly called the load resistor, must dissipate power. The equation for the ac power in the load resistor, called power output and denoted by P_o, is

274 Small-Signal Equivalent Circuits   Ch. 7

$$P_o = I_p^2 R_b = \frac{\mu^2 E_1^2 R_b}{(r_p + R_b)^2} \tag{7-24}$$

It is seen that the *output power involves only the ac component of plate current*. The dc component of plate current also flows through the load resistor in this circuit and contributes to the power that the resistor must dissipate, but not to the useful power output.

The smaller the ohms value of R_b, the less will be the dc power loss. R_b may be replaced by an inductance with low dc resistance but high reactance to reduce the alternating current through it as much as possible; this amounts to establishing as high an ac voltage buildup across it as possible. If such an inductance is used, the *dc load line* will be almost directly vertical.

An illustration involving a dc load line with a very steep slope is given in the study of a power amplifier with a transformer-coupled load in Drill Problem D7-4. The resistance of the transformer primary winding is usually very small compared to the ac resistance "coupled in" from the load.

Operation of power amplifiers is discussed in Chapter 9. We shall study these circuits after further development of amplifier behavior in the next chapter.

7-12
Variation of Vacuum Tube Parameters

Just as the *h*-parameters of a transistor vary depending upon the location of the chosen Q-point, so are there variations in the values of vacuum-tube parameters. Figure 7-14 depicts the range of variations in the parameters of the 6SN7 double triode tube. Plate characteristic curves for this tube type are given in the Appendix.

Manufacturers' manuals may list the parameter values for some specific Q-point location. It should be understood that radically different values may occur for other regions of the tube characteristics, as shown in Fig. 7-14. For making calculations of the tube's performance in a given amplifier circuit, using the analytical procedure of Section 7-9, parameter values that are correct for the chosen Q-point must be used. Of the three values graphed in Fig. 7-14, note that the amplification factor has the most nearly constant value for a wide range of Q-point locations. The other two variables exhibit a tendency toward rapid changes in value, particularly at low plate voltages.

7-13
Circuit Models for JFET and MOSFET

JFET and MOSFET units may be modeled for low-frequency operation by the ac equivalent circuit in Fig. 7-15. Note the open circuit between gate and source, indicating the high impedance present there. Gate-source input resis-

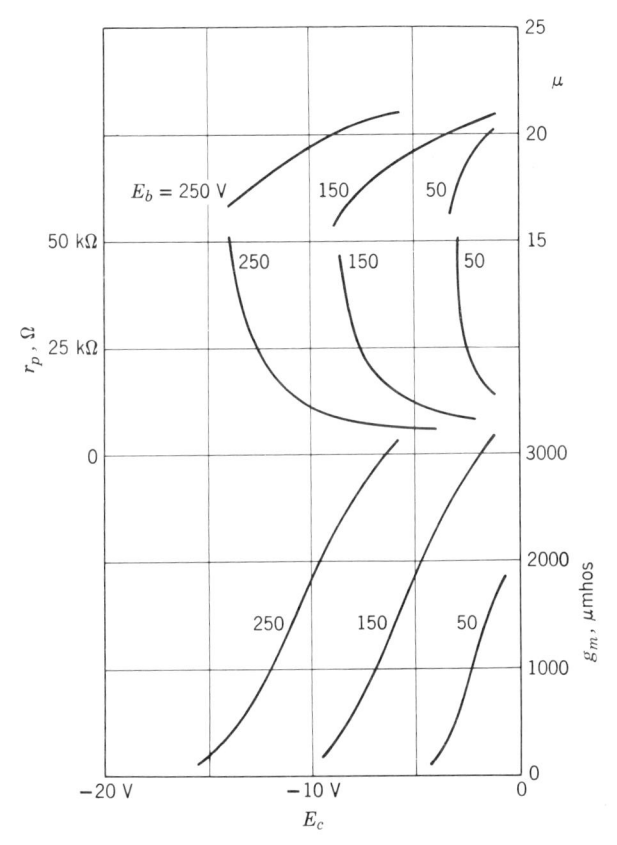

Figure 7-14
Variations of parameter values in the 6SN7 triode for different combinations of plate voltage and grid voltage.

tance (at dc) is typically 10^8 to 10^{10} Ω in a JFET and is usually higher in MOSFET units, from 10^{10} to 10^{14} Ω. These values are so large that the input from gate to source can be considered an open circuit.

This model is similar to the Norton equivalent that was developed for the vacuum triode. The FET transconductance, y_{fs}, relates changes in v_{GS} and i_D:

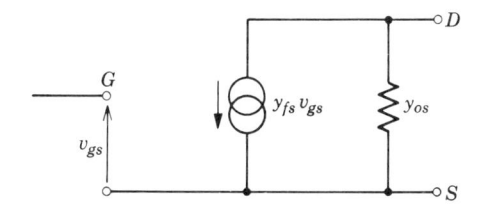

Figure 7-15
Small-signal ac equivalent circuit for FET.

$$y_{fs} = \left.\frac{\Delta i_D}{\Delta v_{GS}}\right|_{\text{constant } v_{DS}}$$

The symbol y_{fs} indicates that the relationship is the forward, small-signal admittance of the FET. This parameter has values in some range that may be identified by a manufacturer and published as a characteristic of a particular type of FET. For example, electrical characteristics published for the 2N3796 include values for y_{fs} from 900 to 1800 μmhos. This range means that a 1-V change in gate-source voltage can generate a drain current change from 0.9 to 1.8 mA.

The other parameter in the model, y_{os}, is the output admittance for small-signal operation. It is the ratio of changes in i_D and v_{DS}:

$$y_{os} = \left.\frac{\Delta i_D}{\Delta v_{DS}}\right|_{\text{constant } v_{GS}}$$

Typical values for the 2N3796 are in the range from 12 to 24 μmhos.

Questions

7-1 Explain how to sketch an ac load line on collector characteristic curves for a transistor voltage amplifier.

7-2 What effect does a coupled ac load have on the location of the Q-point?

7-3 Why is the h-parameter equivalent circuit for a transistor said to be a hybrid circuit?

7-4 The parameter h_{11} is in the input circuit of the transistor equivalent. In a common-emitter mode, what does h_{11} represent on the input characteristic curves?

7-5 Explain the units of each of the four h-parameters.

7-6 Refer to the circuit of a common-emitter amplifier and explain how voltage amplification takes place.

7-7 When is the ac load different from the dc load on a voltage amplifier?

7-8 Sketch from memory the h-parameter equivalent circuit for the three common-mode circuits.

7-9 Why should there be no grid current in a small-signal triode amplifier?

7-10 Draw from memory the Thévenin equivalent circuit of a vacuum-tube amplifier and label all components.

7-11 Why is there a phase shift of 180° from input to output in a triode voltage amplifier?

7-12 In a triode voltage amplifier circuit, what would be the effect of increasing R_b? a larger r_p? a larger amplification factor?

Problems

7-1 Calculate the theoretical voltage amplification of a transistor having the following parameters in common-emitter circuit:

$$h_{11} = 2720 \ \Omega \qquad h_{21} = 55$$
$$h_{12} = 3.12 \times 10^{-4} \qquad h_{22} = 14 \ \mu\text{mho}$$
$$R_L = 0.5 \ \text{M}\Omega$$

How many decibels gain is this?

7-2 A certain transistor may be considered to have a linear beta of 100. It is connected in the common-emitter mode as a voltage amplifier, such as the circuit of Fig. 7-1. Assume these values for circuit components: $V_{CC} = 14$ V, $R_B = 100$ kΩ, $R_1 = 10$ kΩ, $R_C = 4$ kΩ, $R_E = 200$ Ω, $R_2 = 9$ kΩ, and capacitors are large enough to have negligible reactances. (a) Sketch collector characteristics for the transistor over a region large enough to plot dc and ac load lines for the circuit. (b) Locate and sketch both load lines on the characteristics. (c) What current gain can be expected from this circuit?

7-3 Sketch a small-signal equivalent circuit for the circuit in Fig. 7-3.

7-4 Determine the common-base h-parameters for a 2N1308 transistor when $V_{CB} = 5$ V and $I_E = 10$ mA, using the curves in the Appendix.

7-5 Calculate the values of common-emitter parameters for the 2N1308 transistor at the operating point specified in Problem 7-4.

7-6 For the circuit given in Fig. P7-6, draw the dc load line and the ac load line.

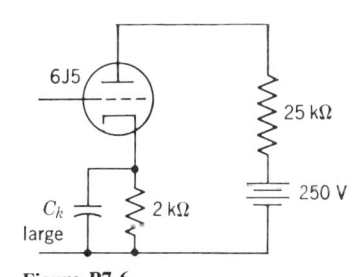

Figure P7-6

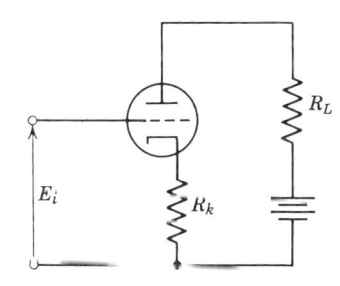

Figure P7-7

7-7 For the circuit shown in Fig. P7-7, $\mu = 15$, $r_p = 10$ kΩ, $R_L = 6000$ Ω, $R_k = 1500$ Ω. Draw Thévenin's equivalent circuit and solve for I_p, the rms ac plate current, if $E_i = 1/\underline{0°}$ V.

7-8 A 2-kHz signal of 2 V peak to peak is applied to the input of the circuit in Problem 7-6. What is the ac output?

278 Small-Signal Equivalent Circuits   Ch. 7

7-9 In the circuit of Fig. P7-9, $E_s = 1/\underline{0°}$ V, $\mu = 20$, and $r_p = 7000\ \Omega$. The plate load R_b is 5 kΩ, $R_1 = 1$ kΩ, $R_2 = 1$ kΩ, and C is large enough so that R_1 is adequately bypassed. Solve for the ac output, E_o.

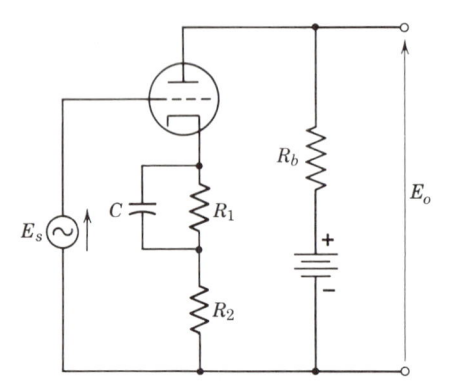

Figure P7-9

7-10 For the circuit of Fig. P7-10, calculate an expression for the voltage gain $A_v = E_o/E_s$ in terms of tube coefficients and circuit impedances only. Assume identical tubes and small capacitive reactance compared with R_g.

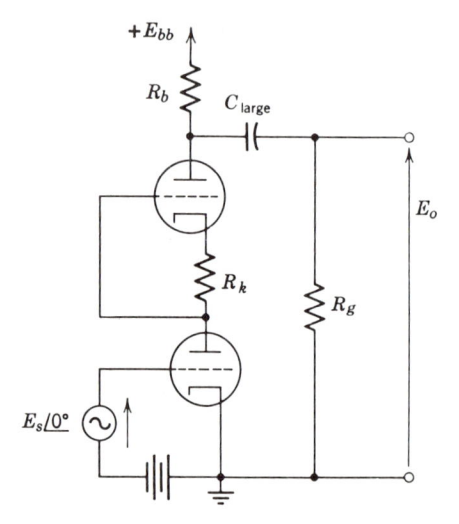

Figure P7-10

8
ANALYSIS OF AMPLIFIER OPERATION

This chapter presents some fundamental properties of electronic circuits that operate as amplifiers. The performance of single-stage circuits containing transistors and vacuum tubes as the amplifying components will be studied for ac operation under small-signal conditions. Our analysis of amplifier operation will take into account the effects of external resistors and capacitors as well as the device parameters. The discussion is confined, for the most part, to the audio range of frequencies, although some attention is given to high-frequency amplifiers.

Equivalent circuits developed in Chapter 7 are applicable, assuming that the circuit under study is properly biased to establish a suitable operating point where device parameters can be considered constants over the range of voltage and current variations that exist. It will also be assumed that device parameters are known or available for the Q-point at which small-signal operation takes place. Calculations of voltage gain, current gain, and power gain will be made for the three common-mode connections of transistors, and input and output impedances will be found for the circuits. Further calculations involve similar properties of amplifier circuits that employ vacuum tubes.

8-1
The Basic Single-Stage Amplifier

For our analysis of amplifier performance, we refer to the basic circuit in Fig. 8-1. A single-stage amplifier is energized by a source of ac voltage or current and provides a replica of this input signal in some output load.

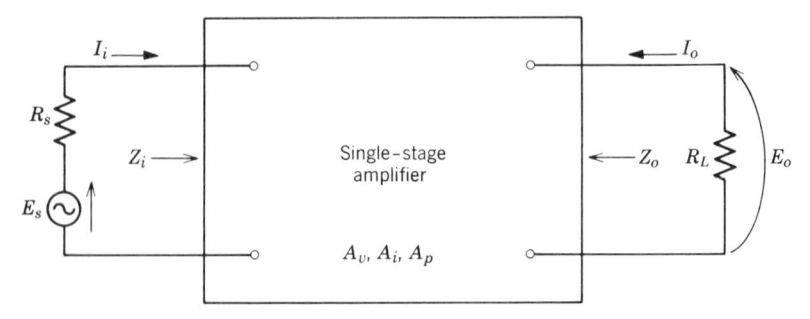

Figure 8-1
Basic configuration of a single-stage amplifier, including input signal source and output load.

Of interest are the voltage gain, current gain, and power gain from input to output, and the input and output impedances that affect circuit performance. The three gains, introduced in Chapter 7, are defined here:

Voltage gain: A_v = load voltage/input voltage = E_o/E_s

Current gain: A_i = load current/input current = I_o/I_i

Power gain: A_p = output power/input power = P_o/P_i

In all these expressions it is assumed that only ac quantities are being considered. Since the voltages and currents are symbolized by capital letters, they evidently are rms values of single-frequency sinusoidal signals. Therefore, our methods of analyzing ac circuits using Ohm's and Kirchhoff's laws and phasor algebra can be used.

In addition to voltage gain, current gain, and power gain, input impedance and output impedance are of prime importance. Only pure resistance loading will be studied; this means that these impedances will turn out to be pure resistances in the frequency range of operation. Reactive loads on tube and transistor amplifiers cause the load line to become an ellipse and introduce such extensive complications that mathematical analysis is far beyond the scope of this book. At very high frequencies, inductance and capacitance values show up in tubes and transistors to add further complications.

Input resistance of a transistor amplifier is simply defined as the input voltage divided by the input current *when the load is connected. Output resistance* is defined as the output voltage divided by the output current *when the supply source is connected*; it is the resistance presented to the output terminals *by the amplifier circuit when the source is connected.* Of course, the load resistance is not included. However, the load resistance should be selected to match the output resistance of the stage for maximum power to the load.

The transistor amplifier is essentially a power amplifier. It is, as previously emphasized, a current-actuated device. This means that its control (input)

Sec. 8-2　　　　　　　　　　　　　　　　　　　　　Two-Stage Voltage Amplifier　281

circuit takes power, and the predominant objective in the circuit design is to obtain maximum power amplification consistent with other desirable features of performance such as minimum distortion, long life, and stability.

8-2
Two-Stage Voltage Amplifier

The elementary amplifier circuits studied in Chapter 7 produced an output voltage across a load resistor in the plate circuit of the tube or the collector circuit of the transistor. It often happens that this voltage is not sufficiently large to perform the task desired, and so further amplification is required.

Figure 8-2 shows two-stage voltage amplifiers, each of which has resistance-capacitance coupling between its stages. C_1 and R_{g2} form the coupling network for the tube amplifier and C_1, R_{b2} for the transistor amplifier. In each case, C_1 serves to block the dc potential of the first stage from reaching the input circuit

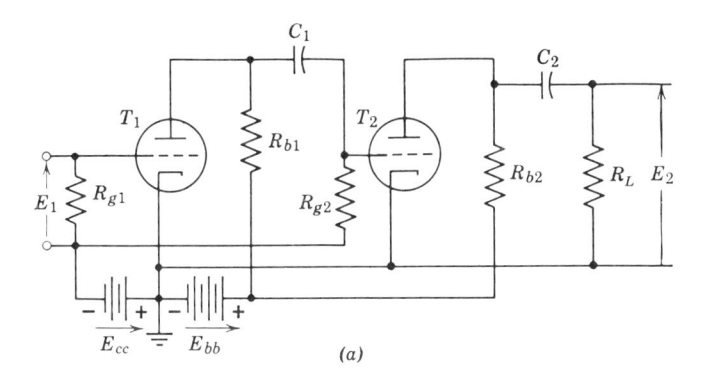

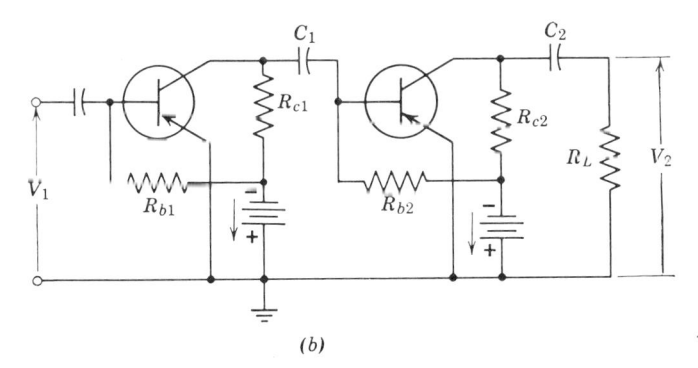

Figure 8-2
Two-stage resistance-capacitance coupled voltage amplifiers. (a) Vacuum-tube amplifier. (b) Common-emitter transistor amplifier.

282 *Analysis of Amplifier Operation*   Ch. 8

of the next stage, while allowing the amplified ac voltage to "drive" the second stage. C_1 is made large enough in capacitance that the ac voltage loss across it is negligible in the range of working frequencies of the amplifier.

8-3
Operating Conditions; Classes of Amplifiers

Through selection of Q-point location (bias conditions) and amplitudes of signals applied to the input of an amplifier, it is possible to cause current in the transistor or tube during the complete cycle of input signal or only during a fraction of it. The letters A, B, and C are used to designate operation under different bias conditions. Bias conditions are effective in establishing operation over different time periods of the input cycle, and are identified as either Class A, B, or C conditions.

Although the following description applies to operation of vacuum tubes, it can be correlated with similar conditions in transistor circuits. Different classes of amplifiers are considered separately.

Class A. Figure 8-3 shows the nature of operation called Class A_1. "Class A" means that plate current flows during the complete grid-voltage cycle, and the subscript 1 means that grid current does not flow. The grid is not driven positive.

If the bias (E_c) were not so large, the grid would go positive, grid current would flow, and the operation would be Class A_2. This type of operation is not desirable, because the flow of grid current usually prevents proper performance of the source supplying the grid-input voltage, and the advantage gained from larger plate-current amplitude is offset by larger losses within the tube itself.

Class AB. When the grid bias has a value that will permit current cutoff for less than a half-cycle, the operation is Class AB. If the grid does not go

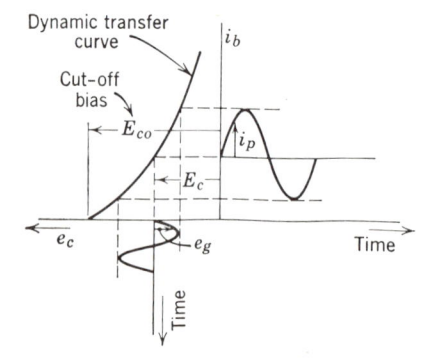

Figure 8-3
Class A_1 operation. Plate current flows during complete cycle. Grid does not go positive.

positive, the operation is Class AB$_1$; if it does, grid current flows and the operation is Class AB$_2$.

It is seen in Fig. 8-4 that the plate-current wave is far from sinusoidal in shape. This *distortion* prohibits the use of Class AB operation in an amplifier stage employing a single tube. It will be found later that in a push-pull amplifier, a type that employs two tubes in a special circuit in a single stage, Class AB operation is practical and desirable. Indeed, satisfactory performance is possible with a push-pull stage in which plate current flows alternately in the two tubes for only a half-cycle at a time.

Class B. Although Class B operation is confined mostly to the push-pull type of amplifier, it is used in special applications with one tube. One-tube amplifier operation is often called single-ended operation to distinguish it from push-pull operation. In Class B the grid bias is set at the current cutoff point. The grid-input voltage is necessarily quite large and usually is made to drive the grid a few volts positive, producing Class B$_2$ operation (Fig. 8-5). Class B is mostly confined to power amplification in push-pull stages.

Class C. When power amplification at only one frequency or over a narrow band of frequencies is desired, we may use Class C in either a single-ended or a push-pull stage. In Fig. 8-6 it is seen that plate current flows during appreciably less than a half-cycle. The bias of a Class C amplifier is set anywhere between approximately 1.5 and 2 or more times cutoff bias. The plate current flows only in "spurts," and therefore the distortion is great. Class C amplifiers usually are driven with sinusoidal grid voltages. The use of a parallel-resonant circuit in the plate circuit results in the production, across its terminals, of a practically pure sine wave of voltage having large amplitude. The result is that there is no distortion problem when the grid voltage is of only one frequency. Class C amplifiers are usually confined to high-frequency power work. They have greater efficiency than the other classes.

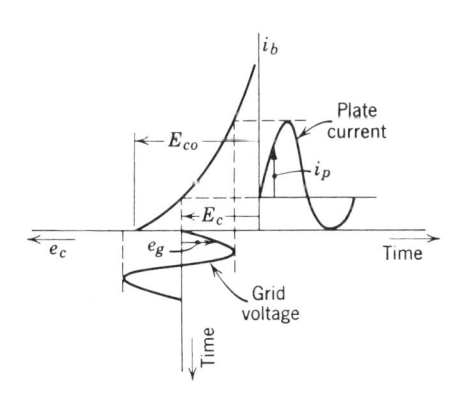

Figure 8-4
Class AB$_1$ operation.

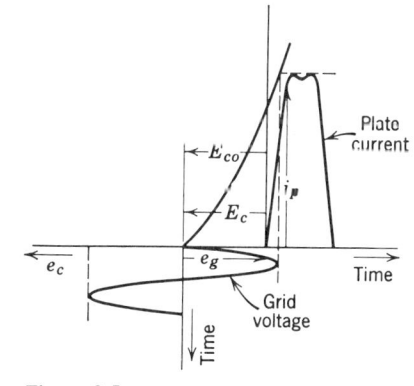

Figure 8-5
Class B$_2$ operation.

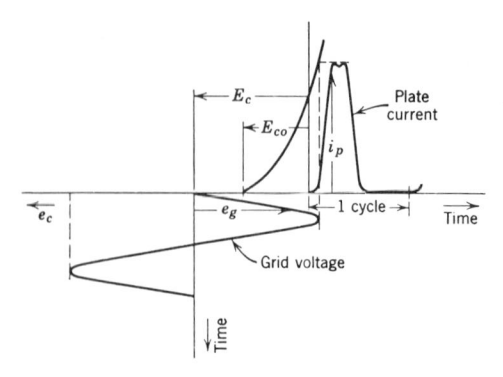

Figure 8-6
Class C operation.

Comparisons of some general properties of amplifiers operated as classified are shown in Table 8-1.

Table 8-1
General Properties of Triode Amplifiers

Class	Bias (Approx.)	Plate-Current Flow	Approx. Relative Efficiency of Triode Power Amplification	Distortion
A_1	1/2 cutoff	Full cycle	Low (15–20%)	Minimum
A_2	1/2 cutoff	Full cycle	Low; seldom used	—
AB_1	2/3 cutoff	Less than full cycle	Appreciably better than A_1; still relatively low (25–30%)	Low
AB_2	2/3 cutoff	Less than full cycle	Better than A_1 (25–35%)	Low
B_1	Cutoff	One-half cycle	Not often used, but in PP less than with Class B_2	—
B_2	Cutoff	One-half cycle	Good (40–60% in PP)	Low in PP, high in single tube
C	1.5–2.5 times cutoff	120–160°	High (60–85%)	Low at single frequency

8-4
Distortion in Amplifiers

The performance of voltage amplifiers throughout the range of audible frequencies, from about 40 to 15,000 Hz, is an important and interesting topic. If a microphone is placed against a rotating machine, it will develop a weak voltage due to the machine's vibration. The frequency of this voltage may be low or high, depending on how fast the machine rotates or vibrates. It is usually desirable to amplify this voltage so that an indicating instrument may be activated or some other registration effected. An ordinary motor may run at 1750 rpm. This is approximately 29 rev/s. The signal picked up with the microphone would in all probability have a frequency very near to 29 Hz. A voltage amplifier supplied with this signal should be capable of amplifying it without much distortion.

Speech is easily understood—even when coming from a loudspeaker driven by amplifiers incapable of good reproduction above 2500 or 3000 Hz—and music has a good sound if a frequency band from 100 to 4000 Hz is employed. However, the handling of overtones that enrich musical sounds requires that the frequency range of the amplifier be much greater. The amplification should be constant from approximately 40 to 15,000 Hz, for audible response by the average human ear.

Frequency distortion is present in an amplifier when it does not amplify signal voltages of all frequencies by the same amount. Consider, for example, a musical sound that produces in a microphone a voltage wave composed of only two frequencies. One of the components (the fundamental) has a frequency of 500 Hz and a peak value of 1 mV; the other has a frequency of 1000 Hz and a peak value of 0.5 mV. If the fundamental is amplified tenfold to 10 mV, the 1000-Hz component should also be multiplied tenfold to 5 mV. If, instead, the components are multiplied by different amounts, the amplifier has frequency distortion.

Phase distortion is another undesirable characteristic of an amplifier. Consider a voltage made up of a 500-Hz component and a 1500-Hz component that happens to be displaced, say, 10 degrees lagging from the fundamental. At the output of the amplifier, the 1500-Hz component should still be lagging 10 degrees behind the 500-Hz component. If it does not, the amplifier has phase distortion.

Unless phase distortion is severe, it is not very noticeable in the amplification of sound. In high-frequency work it is very noticeable, and in some applications, such as television, it cannot be tolerated.

Nonlinear distortion exists when an amplifier puts out frequencies that are not present in the input signal. Curvature of the dynamic-transfer characteristic results in nonlinear distortion. The generation of a sound harmonic by the tube, due to curvature of the dynamic-transfer characteristics, results from large signal excursions. This is discussed in more detail in Chapter 9.

Proper choice of operating voltages and load impedances will help to minimize this type of distortion. Larger load impedance tends to straighten the dynamic-transfer characteristic, reducing its curvature. Operating voltages may be chosen so that the amplifier operates over the most linear part of the characteristic.

Drill Problems

D8-1 The signal input to an amplifier is expressed as $e_{in} = 10 \sin \phi - 3 \cos 3\phi$. Its output signal may be expressed as $e_{out} = 100 \sin \phi - 20 \cos 3\phi$. Is the signal distorted in passing through the amplifier? If so, what type of distortion is it?

D8-2 Show an example of a signal that is *phase distorted* in passing through an amplifier.

D8-3 The input signal to an amplifier is $e_{in} = 10 \sin \phi - 5 \cos 2\phi$. The output signal is $e_{out} = 100 \sin \phi - 30 \cos (2\phi - 30°)$. What type(s) of distortion is introduced by this amplifier?

D8-4 The output signal from an amplifier may be expressed in terms of its input signal as $e_{out} = K e_{in}^2$. For an input signal of $e_{in} = a \sin \phi - b \sin \theta$, calculate the output and determine whether distortion has been introduced; if so, what type of distortion is it?

8-5
Frequency Response of *R-C* Coupled Amplifiers

An ideal amplifier could handle any signal that might be applied to its input, providing an output that is an exact replica. However, practical circuits cannot behave as the ideal; a practical amplifier will contain circuit elements that unavoidably cause its performance to deteriorate in certain ranges of frequency. The *R-C* coupled amplifier (Fig. 8-2) will amplify signals equally well over some frequency range and not so well in other ranges.

An *R-C* coupled *audio amplifier* is designed to process signals over a frequency range that can be detected by the human ear. The ratio of output voltage to input voltage (voltage amplification, A_v) over this range will plot as shown in Fig. 8-7. The region of the curve that is flat defines the *midband*

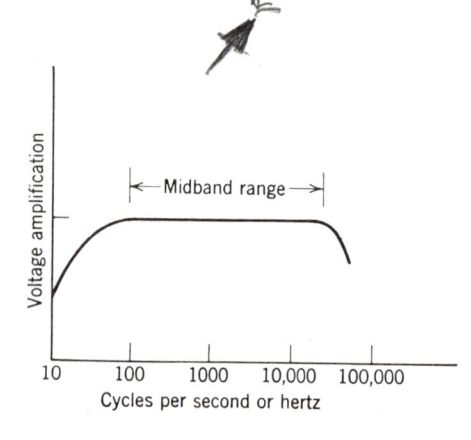

Figure 8-7
Frequency-response curve of a resistance-capacitance coupled amplifier, designed to amplify audio signals.

frequency range for the amplifier. The other regions are referred to as the *low-frequency* and *high-frequency* ranges, respectively, in which amplification is less than in the midband.

Amplification reduction in the low-frequency range is caused by increasing reactance in coupling capacitors. Ideally, their reactance should offer as little opposition as possible to ac currents. At midband and high frequencies this result is obtained, but as signal frequencies drop, capacitive reactance values become large enough to claim a portion of the amplified voltage. Thus, there is less voltage at the output terminals, and amplification will be smaller.

Amplification will be reduced at higher frequencies because of parasitic capacitance effects in a practical circuit. Such effects may arise because of wires being close together but having different potentials, and from similar proximity of current-carrying wires and a metal chassis on which the amplifier is built. Internal characteristics of a transistor or tube may also exhibit capacitance effects that will appear in amplifier operation as shunts across resistor loads and input circuits. As the signal frequency increases, their reactances reduce sufficiently to provide current paths around the load or input components. This results in a reduction in signal voltages and hence a reduction in amplification in the high-frequency range.

The frequency response of an amplifier depends upon several factors. Parameters that limit the internal operation of transistors and tubes combine with external circuit components in determining how a given amplifier will function over some frequency range. We shall analyze the behavior of R-C coupled amplifiers over three separate ranges of frequency described above. We shall look first at operation within the midband range and then extend the discussion to include low-frequency and high-frequency ranges.

8-6
Analysis of Common-Emitter Amplifier

An equivalent circuit for small-signal analysis of the frequency response of a common-emitter amplifier is given in Fig. 8-8. Bias resistors are included in the generator and load resistances, and capacitors are excluded for reasons explained above. The analysis will begin with a look at operation in the midband frequency range, in which only resistive circuit components are involved.

At the node a, the current relations are

$$h_{fe}I_1 + h_{oe}V_2 + V_2/R_L = 0 \tag{8-1}$$

from which we obtain

$$V_2 = \frac{-h_{fe}R_L I_1}{1 + h_{oe}R_L} \tag{8-2}$$

288 Analysis of Amplifier Operation   Ch. 8

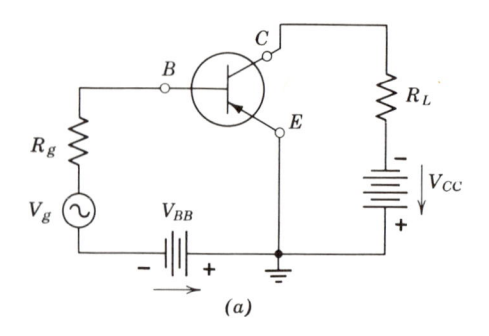

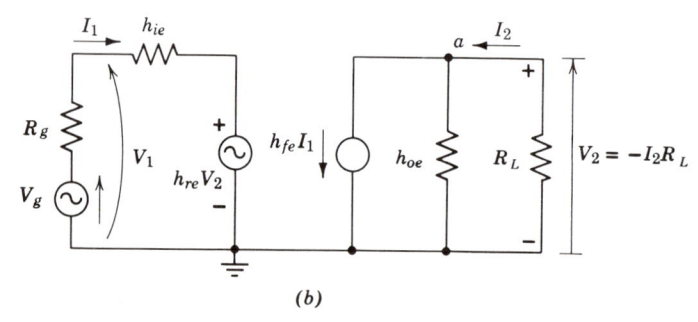

Figure 8-8
Circuits of common-emitter transistor amplifier. (a) Actual circuit. (b) Equivalent circuit.

The output current is $-V_2/R_L$,

$$I_2 = \frac{h_{fe}I_1}{1 + h_{oe}R_L} \tag{8-3}$$

The input voltage is

$$V_1 = h_{ie}I_1 + h_{re}V_2 \tag{8-4}$$

Using Eq. 8-2,

$$V_1 = h_{ie}I_1 - \frac{h_{re}h_{fe}R_LI_1}{1 + h_{oe}R_L} \tag{8-5}$$

$$V_1 = \frac{I_1[h_{ie}(1 + h_{oe}R_L) - h_{re}h_{fe}R_L]}{1 + h_{oe}R_L} \tag{8-6}$$

The *current gain*, A_{ie}, as before, is I_2/I_1, which is obtained from Eq. 8-3:

$$A_{ie} = \frac{I_2}{I_1} = \frac{h_{fe}}{1 + h_{oe}R_L} \tag{8-7}$$

Sec. 8-6   *Analysis of Common-Emitter Amplifier*

Note that if R_L becomes zero (short circuit), the current amplification becomes equal to h_{fe}, the forward-current transfer ratio.

The *voltage gain*, A_{ve}, is obtained from Eqs. 8-2 and 8-7:

$$A_{ve} = \frac{V_2}{V_1} = \frac{-h_{fe}R_L}{h_{ie}(1 + h_{oe}R_L) - h_{re}h_{fe}R_L} \tag{8-8}$$

Since this ratio is a negative number, this means that V_2 is 180 degrees out of phase with respect to V_1. To explain it in another way, as V_1 increases in the positive direction, so do the currents I_1 and I_2. While this is happening, V_2 is becoming a larger *negative* value. Since there is a phase reversal, A_{ve} is a negative number. The negative sign may be omitted in deriving the following power formula.

The *power gain*, A_{pe}, is obtained in equation form by multiplying A_{ve} by A_{ie}:

$$A_{pe} = \frac{(h_{fe})^2 R_L}{h_{ie}(1 + h_{oe}R_L)^2 - h_{re}h_{fe}R_L(1 + h_{oe}R_L)} \tag{8-9}$$

Input resistance of the common-emitter circuit is easily calculated from Eq. 8-6:

$$R_{ie} = \frac{V_1}{I_1} = \frac{h_{ie}(1 + h_{oe}R_L) - h_{re}h_{fe}R_L}{1 + h_{oe}R_L} = \frac{h_{ie} + DR_L}{1 + h_{oe}R_L} \tag{8-10}$$

This result shows the dependence of input resistance on the value of the load resistance, an unexpected relationship. The factor D is defined in Section 7-4.

To determine an expression for the *output resistance* of the common-emitter amplifier, we need to write the circuit equations for the equivalent circuit. The input generator is left connected, but R_L is disconnected for the calculation.

Using node a, we write the current equation,

$$I_2 - h_{oe}V_2 - h_{fe}I_1 = 0 \tag{8-11}$$

and for the input circuit,

$$h_{re}V_2 + I_1(h_{ie} + R_g) = 0 \tag{8-12}$$

Rearranging these equations,

$$h_{oe}V_2 + h_{fe}I_1 = I_2$$
$$h_{re}V_2 + (h_{ie} + R_g)I_1 = 0$$

Solving for V_2 in terms of I_2 using determinants,

$$V_2 = \frac{\begin{vmatrix} I_2 & h_{fe} \\ 0 & h_{ie} + R_g \end{vmatrix}}{\begin{vmatrix} h_{oe} & h_{fe} \\ h_{re} & h_{ie} + R_g \end{vmatrix}} = \frac{(h_{ie} + R_g)I_2}{(h_{ie} + R_g)h_{oe} - h_{re}h_{fe}} \tag{8-13}$$

The output resistance is the ratio of V_2 and I_2, which may be obtained from Eq. 8-13,

$$R_{oe} = \frac{V_2}{I_2} = \frac{h_{ie} + R_g}{(h_{ie} + R_g)h_{oe} - h_{re}h_{fe}} = \frac{h_{ie} + R_g}{h_{oe}R_g + D} \qquad (8\text{-}14)$$

The resistance of the generator, whether it is part of a signal source or the impedance looking back into an amplifier stage preceding this one, is significant in the value of output resistance of this amplifier.

Using the equations derived in this section, we can calculate the gains and resistance values of an amplifier for the midband frequency range. For known values of circuit resistors and h-parameters, we may calculate currents and voltages present in the small-signal circuit. For any particular transistor the h-parameters are obtainable from manufacturer data, or they may be measured in the laboratory. It would be a waste of time to memorize these equations. It is sufficient to understand their derivations and to know the meanings of the symbols used.

8-7
Comparison of the Three Connections

The expressions for gains and resistances seem rather cumbersome, but they are easy to evaluate. The expressions are similar for each of the three common connections and only require that the correct h-parameters be used. Table 8-2 lists the expressions that are used to evaluate the various gains and resistances.

Table 8-2
Equations for Calculating Gains and Resistances

Current gain	$A_{i-} = \dfrac{h_{f-}}{h_{o-}R_L + 1}$
Voltage gain	$A_{v-} = \dfrac{-h_{f-}R_L}{h_{i-} + DR_L}$
Power gain	$A_{p-} = A_{v-}A_{i-}$
Input resistance	$R_{i-} = \dfrac{h_{i-} + DR_L}{h_{o-}R_L + 1}$
Output resistance	$R_{o-} = \dfrac{R_g + h_{i-}}{h_{o-}R_g + D}$

The factor D represents $(h_{i-}h_{o-} - h_{r-}h_{f-})$, which may be calculated separately and used as a multiplier in several of the expressions.

Sec. 8-7 *Comparison of the Three Connections* 291

As an example of the calculations necessary to evaluate these properties of transistor amplifiers operating in the three basic connections, the h-parameters for the 2N334 transistor shown in Table 8-3 will be used. The common-base h-parameters are furnished by the manufacturer, and the remaining h-parameters have been calculated using the expressions of Table 7-1.

Table 8-3
h-Parameters of 2N334 N-P-N Transistor

Common-Base h-Parameters
$h_{ib} = 56\,\Omega$, $\quad h_{ob} = 0.5\,\mu\text{mho}$, $\quad h_{rb} = 350 \times 10^{-6}$, $\quad h_{fb} = -0.97$
Common-Emitter h-Parameters
$h_{ie} = 1870\,\Omega$, $\quad h_{oe} = 16.7\,\mu\text{mho}$, $\quad h_{re} = 9.2 \times 10^{-4}$, $\quad h_{fe} = 32.3$
Common-Collector h-Parameters
$h_{ic} = 1870\,\Omega$, $\quad h_{oc} = 16.7\,\mu\text{mho}$, $\quad h_{rc} = 1$, $\quad h_{fc} = -33.3$

Example 8-1. For the purposes of these calculations, we will assume that the source and load resistors are equal and have a value of $1000\,\Omega$. We will calculate and compare the numerical values of voltage gain, current gain, power gain, input resistance, and output resistance for each of the three basic connections of a transistor amplifier. The equivalent circuit of Fig. 8-8 is used for the analysis.

First, we calculate the factor D given in Table 8-2.

Common-base:

$$\begin{aligned}
D_b &= h_{ib}h_{ob} - h_{rb}h_{fb} = (56)(0.5 \times 10^{-6}) - (350 \times 10^{-6})(-0.97) \\
&= (28 + 339.5) \times 10^{-6} \\
&= 367.5 \times 10^{-6}
\end{aligned}$$

Common-emitter:

$$\begin{aligned}
D_e &= (1870)(16.7 \times 10^{-6}) - (920 \times 10^{-6})(32.3) \\
&= (3120 - 2970) \times 10^{-6} \\
&= 150 \times 10^{-6}
\end{aligned}$$

Common-collector:

$$\begin{aligned}
D_c &= (1870)(16.7 \times 10^{-6}) - (1)(-33.3) \\
&= 33.303 \\
&\doteq 33.3
\end{aligned}$$

Next, we will evaluate the *current gain* for each of the three connections.

Common-base:

$$A_{ib} = \frac{h_{fb}}{h_{ob}R_L + 1} = \frac{-0.97}{(0.5 \times 10^{-6})(10^3) + 1}$$
$$\doteq -0.97$$

Common-emitter:

$$A_{ie} = \frac{32.3}{(16.7 \times 10^{-6})(10^3) + 1}$$
$$\doteq 32.3$$

Common-collector:

$$A_{ic} = \frac{-33.3}{(16.7 \times 10^{-6})(10^3) + 1}$$
$$\doteq -33.3$$

The *voltage gain* may be calculated using the equation of Table 8-2.

Common-base:

$$A_{vb} = \frac{-h_{fb}R_L}{h_{ib} + DR_L} = \frac{-(-0.97)(10^3)}{56 + (377.5 \times 10^{-6})(10^3)}$$
$$= 17.2$$

Common-emitter:

$$A_{ve} = \frac{-(32.3)(10^3)}{1870 + (150 \times 10^{-6})(10^3)}$$
$$= -17.3$$

Common-collector:

$$A_{vc} = \frac{-(-33.3)(10^3)}{1870 + (33.3)(10^3)}$$
$$= 0.95$$

The *power gain* for each basic connection may be calculated.

Common-base:

$$A_{pb} = A_{vb}A_{ib} = 17.2 \times 0.97$$
$$= 16.68$$

Common-emitter:

$$A_{pe} = 17.3 \times 32.3$$
$$= 558$$

Common-collector:

$$A_{pc} = 0.95 \times 33.3$$
$$= 31.6$$

Sec. 8-7   Comparison of the Three Connections   293

The *input resistance* for each of the basic connections is found in a similar manner, using the equation of Table 8-2.

Common-base:

$$R_{ib} = \frac{56 + (377.5 \times 10^{-6})(10^3)}{(0.5 \times 10^{-6})(10^3) + 1} = 56.3 \, \Omega$$

Common-emitter:

$$R_{ie} = \frac{1870 + (150 \times 10^{-6})(10^3)}{(16.7 \times 10^{-6})(10^3) + 1} = 1870 \, \Omega$$

Common-collector:

$$R_{ic} = \frac{1870 + (33.3)(10^3)}{(16.7 \times 10^{-6})(10^3) + 1} \doteq 34.6 \, \text{k}\Omega$$

The values of *output resistance* may be evaluated in the same way. The resulting numerical values are listed in Table 8-4 for comparison of the three common connections.

Table 8-4
Comparison of Transistor Performance in the Three Basic Common-Terminal Connections

Type of Connection	R_i	R_o	A_v	A_i	A_p
Common-base	56.3 Ω	1.2 MΩ	17.2	−0.97	16.68
Common-emitter	1870 Ω	170 kΩ	−17.3	32.3	558
Common-collector	34.6 kΩ	86.5 Ω	0.95	−33.3	31.6

In general, transistor amplifiers (except those using the FET) have very low input resistance compared with the input resistance of low-frequency vacuum-tube amplifiers. The output impedance, on the other hand, is usually high, with the exception of the common-collector circuit. This connection has a high input resistance and a very low output resistance.

The values obtained for the voltage gains show that the common-emitter amplifier is the only one through which a 180-degree phase reversal occurs. This is shown by the minus sign associated with the calculated value of voltage gain. This circuit is the one most frequently used in transistor amplifiers, because it has the largest power gain and its voltage and current gains are quite favorable. The values of input and output resistances are also favorable for practical operation of the circuit. The same transistor was used in the example

294 Analysis of Amplifier Operation Ch. 8

solely for the purpose of comparing the performances of the three circuit connections. It should be clear that another type would yield entirely different numerical values. However, the methods used here for evaluating the circuit performance would be valid for any type of transistor and for any source and load resistors.

In practical circuits it is often necessary to match the impedances of a circuit and its load and to match the circuit to its source generator. As indicated in Table 8-4, the common-base amplifier could match a low-resistance driver to a high-resistance load, while the common-collector amplifier could match a high-resistance driver to a low-resistance load. The latter connection is often used for the primary purpose of obtaining its input and output resistance-matching properties. It will be seen in later discussions that the cathode-follower vacuum-tube amplifier has the same property.

Another important property of the common-emitter amplifier circuit has a practical value. Only one supply battery is needed to drive the collector and base circuits, as was discussed in connection with biasing methods. The other two basic connections need two separate supplies, a distinct disadvantage in practical designs where costs of components are significant.

8-8
Analysis of Small-Signal Two-Stage Transistor Amplifier

Two common-emitter amplifiers connected in cascade are shown in Fig. 8-9. The parameters of the transistors are listed below the circuit diagram.

Parameters of Transistors

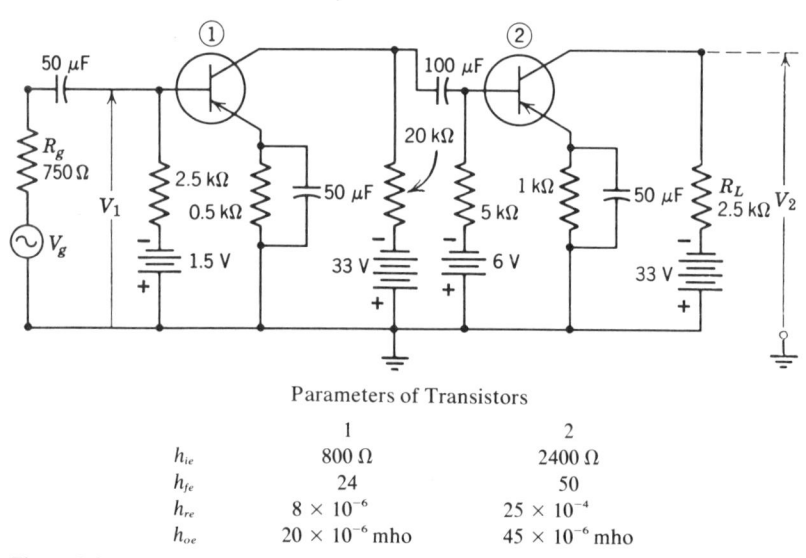

	1	2
h_{ie}	800 Ω	2400 Ω
h_{fe}	24	50
h_{re}	8×10^{-6}	25×10^{-4}
h_{oe}	20×10^{-6} mho	45×10^{-6} mho

Figure 8-9
Two-stage small-signal transistor amplifier.

Sec. 8-8   *Analysis of Small-Signal Two-Stage Transistor Amplifier*   295

The performance will be analyzed by first drawing the equivalent circuit and then applying the equations developed in this chapter to one stage at a time. Although there are other kinds of equivalent circuits such as the T or π type, we shall again use the loop and nodal forms with which we are now familiar.

The complete equivalent circuit is shown in Fig. 8-10. It is advantageous to think of the parts inside the circles as active coupling networks having obtainable input and output impedances. We have determined such quantities before and they will be useful in this analysis. We desire to determine current gain and power gain.

Since the load resistance is known, we should begin there and work toward the input end. The equations already derived will be used.

$$A_{i2} = \frac{I_L}{I_3} = \frac{h_{fe}}{1 + h_{oe}R_L}$$

$$= \frac{50}{1 + (45 \times 10^{-6})(2.5 \times 10^3)} = 44.4$$

$$R_{i2} = h_{ie} - \frac{h_{re}h_{fe}R_L}{1 + h_{oe}R_L}$$

$$= 2400 - \frac{(2.5 \times 10^{-4})(50)(25 \times 10^3)}{1 + 0.1125} = 2120 \ \Omega$$

Observe that the current I_2 entering the second amplifier circuit is divided among 20 kΩ, 5 kΩ, and R_{i2} in parallel. The equivalent resistance of these three is R_e kΩ, such that

$$\frac{1}{R_e} = \frac{1}{20} + \frac{1}{5} + \frac{1}{2.12} = 0.72$$

$$R_e = 1.4 \ \text{k}\Omega$$

The voltage drop across each parallel branch is then $-1400I_2$, and the current input to the second stage is I_3, where

$$I_3 = \frac{-1400I_2}{2120} = -0.66I_2$$

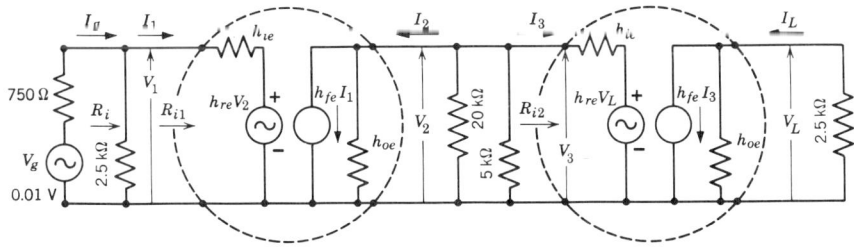

Figure 8-10
Equivalent circuit of amplifier in Fig. 8-9.

so that

$$\frac{I_3}{I_2} = -0.66$$

Going now to the first stage, we see that R_e is the load (R_{L1}) on this stage. The current gain through the first stage and its input resistance are now found as they were above for the second stage.

$$A_{i1} = \frac{I_2}{I_1} = \frac{h_{fe}}{1 + h_{oe}R_{L1}} = \frac{24}{1 + (20 \times 10^{-6})(1.4 \times 10^3)} = 23.3$$

$$R_{i1} = h_{ie} - \frac{h_{re}h_{fe}R_{L1}}{1 + h_{oe}R_{L1}} = 800 - \frac{8 \times 10^{-4}(24)(1.4 \times 10^3)}{1 + (20 \times 10^{-6})(1.4 \times 10^3)} = 773 \, \Omega$$

The 2.5-kΩ resistor that biases the input of the first stage is paralleled by R_{i1}. Consequently, not all of the generator current enters the transistor. This cuts down the current gain of the whole amplifier insofar as the ratio of input and output currents is concerned. This two-branch parallel circuit will divide I_g, so that

$$I_1 = \frac{2500 I_g}{2500 + 773} = 0.76 I_g$$

The current gain ahead of the first stage is

$$\frac{I_1}{I_g} = 0.76$$

The input resistance to the amplifier, as presented to the generator terminals, is

$$R_i = \frac{2500 R_{i1}}{2500 + R_{i1}} = \frac{2500(773)}{2500 + 773} = 590 \, \Omega$$

The *overall current gain*, A_{iT}, is now determined by multiplying all the current gains together:

$$A_{iT} = \frac{I_L}{I_g} = \frac{I_L}{I_3} \times \frac{I_3}{I_2} \times \frac{I_2}{I_1} \times \frac{I_1}{I_g} = 44.4 \times -0.66 \times 23.3 \times 0.76 = -519$$

The *overall power gain*, A_{pT}, can be expressed in terms of the power gain and the terminal resistances. Since power gain in pure resistance circuits is the ratio of output volt-amperes to input volt-amperes, it is also equal to the ratio of output I^2R and input I^2R:

$$A_p = \frac{V_2 I_2}{V_1 I_1} = \frac{I_2^2 R_2}{I_1^2 R_1} = (A_i)^2 \frac{R_2}{R_1}$$

The total power gain through this amplifier is then

$$A_{pT} = (A_{iT})^2 \frac{R_L}{R_i} = (-519)^2 \times \frac{2500}{590} = 1,141,360$$

The equivalent decibel gain is 60.6 dB.

Sec. 8-9   *Analysis of R-C Coupled Tube Amplifier*

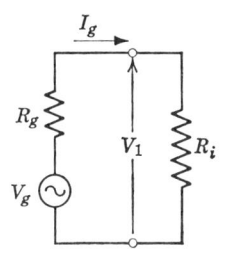

Figure 8-11
Circuit for determining power-input equation.

Output power is obtained from the input power multiplied by the overall power gain. The generator circuit is represented in Fig. 8-11. The input power, P_i, is $V_1 I_g$ and $I_g = V_1/R_i$. Therefore,

$$P_i = \frac{V_1^2}{R_i}$$

By voltage-divider action,

$$V_1 = \frac{R_i}{R_g + R_i} V_g$$

so that

$$P_i = \frac{R_i V_g^2}{(R_g + R_i)^2}$$

Using values already determined, including $V_g = 0.01$ V,

$$P_i = \frac{590(0.01)^2}{(750 + 590)^2} = 3.28 \times 10^{-8} \text{ W}$$

The output power, P_o, then is

$$P_o = P_i \times A_{pT} = 3.28 \times 10^{-8} \times 1.141 \times 10^6 = 0.0374 \text{ W}$$
$$= 37.4 \text{ mW}$$

The foregoing analysis illustrates the application of node and loop equations to the problem of evaluating the performance characteristics of a transistor amplifier. Because the stages are not isolated from each other, the load on one influences the load on the preceding one. Starting the analysis at the final output end makes it possible to determine the load impedances presented to preceding stages in easy, logical steps.

8-9
Analysis of *R-C* Coupled Tube Amplifier

The single-stage *R-C* coupled voltage amplifier circuit to be analyzed is shown in Fig. 8-12. Note that the second resistor (R_2) coupled by the capacitor

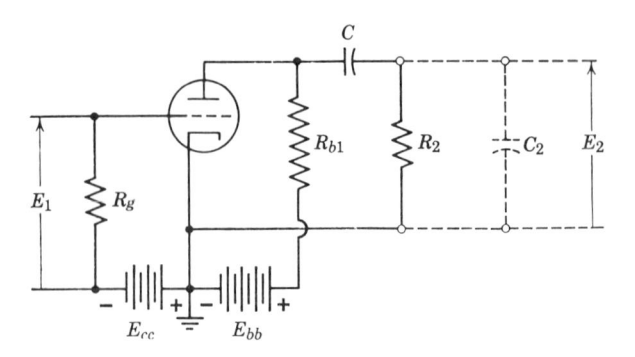

Figure 8-12
One-stage resistance-capacitance coupled amplifier.

(C) may be the grid-circuit resistor (R_{g2}) of a second stage or it may represent some other kind of load. The imaginary capacitor (C_2) is not actually connected in the circuit, but it represents a real capacitive effect caused by wires being close to each other and to the metal chassis on which the amplifier is built. It also represents the capacitive effect of the second tube if a second stage is present.

The drop in amplification in the high-frequency range is due to the effect of the output capacitance represented by C_2. At high frequencies the reactance of the coupling capacitor (C) is negligible because its capacitance is relatively large. On the other hand, C_2 is of the order of 10 or 20 pF, and its reactance becomes low enough so that it acts more and more like a short circuit across the output as the frequency of the input and output voltages increases above the midband values.

The drop in amplification in the low-frequency range is largely due to the action of C. It should offer as little opposition as possible to current flow through R_2. At high and midband frequencies its reactance is nearly zero, and so for the purpose of analysis its effect may be ignored. But at low frequencies its reactance becomes great enough to cause it to claim a portion of the amplified voltage developed across the plate resistor (R_{b1}), and so there is less voltage at the output terminals, i.e., across R_2.

The equivalent circuit for analysis of midband operation is shown in Fig. 8-13. Since C_2 and C have negligible reactances at midband frequencies, they are omitted. We shall now obtain a mathematical expression by means of which the voltage amplification may be calculated.

Let r_b represent the equivalent resistance of R_{b1} and R_2 in parallel. This (r_b) may be called the ac load resistance, since ac components of plate current flow through both R_2 and R_{b1}. Incidentally, in this type of amplifier the dc component of plate current flows through only R_{b1}. From earlier work on

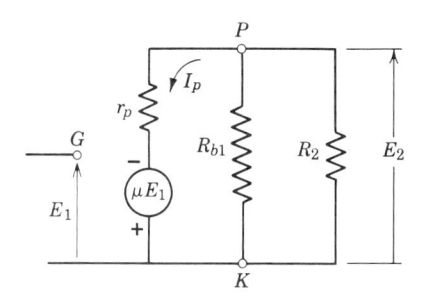

Sec. 8-9 Analysis of R-C Coupled Tube Amplifier 299

Figure 8-13
Equivalent circuit of R-C coupled amplifier in the midband range.

parallel resistances we see that

$$\frac{1}{r_b} = \frac{1}{R_{b1}} + \frac{1}{R_2} \tag{8-15}$$

Using r_b in place of R_{b1} and R_2, we find that the voltage amplification is

$$A_v = \frac{E_2}{E_1} = \frac{-I_p r_b}{E_1} = -\frac{\mu E_1}{r_b + r_p} \times \frac{r_b}{E_1} = -\frac{\mu r_b}{r_b + r_p} \tag{8-16}$$

I_p is expressed as a negative quantity because E_2 is negative whenever E_1 is positive. Their ratio must be negative. Dividing numerator and denominator by r_b,

$$A_v = -\frac{\mu}{1 + r_p/r_b}$$

Dividing now by r_p, and using Eq. 8-15,

$$A_v = -\frac{\mu/r_p}{1/r_p + 1/r_b} = -\frac{\mu/r_p}{1/r_p + 1/R_{b1} + 1/R_2} \tag{8-17}$$

That is, the voltage amplification in the midband range, which we shall denote by A_{vm}, is given by

$$A_{vm} = -\frac{g_m}{1/r_p + 1/R_{b1} + 1/R_2} = -g_m R_e \tag{8-18}$$

in which R_e is the equivalent resistance of r_p, R_{b1}, and R_2 in parallel. None of the quantities in Eq. 8-18 depends on frequency. Therefore, the amplification in the midband range does not change with frequency. When the frequency is low enough to make X_c no longer negligible, or high enough to make X_{c2} no longer so high that its current drain may be neglected, the limit of the midband range of frequencies has been passed.

Let us examine Eq. 8-18 to see what contributes to high-voltage amplification. Evidently the tube should have high transconductance and a high tube-amplification factor. Increasing either R_2 or R_{b1} (or both) helps to decrease the denominator of Eq. 8-18, and thus increases the voltage amplification.

There are practical limits to how high these resistance values may be made. Tube design and manufacture produce g_m and μ as high as practicable. R_{b1} is limited by the $I_b R_{b1}$ drop, which causes the plate-to-cathode voltage to be small or the required plate-supply voltage (E_{bb}) to be too high. If R_2 represents R_{g2}, the grid-circuit resistor of the following circuit may not be larger than a certain value recommended for the tube by the tube manufacturer, because electrons tend to become "trapped on the grid" instead of "leaking off to the cathode," and then they adversely affect tube operation by altering the bias value.

Example 8-2. Calculate the midband voltage amplification of the amplifier of Fig. 8-12 in which a 6SF5 tube is used having $g_m = 1500$ μmho; $r_p = 66{,}000$ Ω; and in which $R_{b1} = 250{,}000$ Ω and $R_2 = 500{,}000$ Ω. $C_c = 0.01$ μF.

Solution: From Eq. 8-18,

$$A_{vm} = -\frac{1500 \times 10^{-6}}{1/250{,}000 + 1/500{,}000 + 1/66{,}000}$$

First multiply the numerator and the denominator by 10^6:

$$A_{vm} = -\frac{1500}{4 + 2 + 15.2} = -\frac{1500}{21.2} = -70.7$$

This is a theoretical value and is probably a little higher than that which can actually be obtained. The minus sign is the result of the division of E_2 by E_1 and of E_2 being opposite in phase to E_1.

Drill Problem

D8-5 A 6SF5 triode is used in the circuit of Fig. 8-12 with $E_{bb} = 300$ V and $E_{cc} = -1.5$ V. The plate load R_{b1} is 150 kΩ and R_2 is 470 kΩ. (*a*) Determine dc and ac load lines, assuming operation in the midband frequency range. (*b*) Estimate from the ac load line the voltage gain for an input signal having a 2-V swing peak to peak. (*c*) Calculate the voltage gain using Eq. 8-18 and compare with the result of (*b*). (Use $\mu = 100$, $g_m = 1000$ μmho, $r_p = 66{,}000$ Ω.)

Low-Frequency Range

In analyzing the low-frequency performance of an amplifier stage that is coupled to the next stage by means of a capacitor, the reactance of the coupling capacitor is a matter of primary importance. As the frequency decreases, $X_C = 1/2\pi f C$ becomes larger. The ac component of voltage across the load resistor, being applied to the series combination of C and R_2 (Fig. 8-12), is then

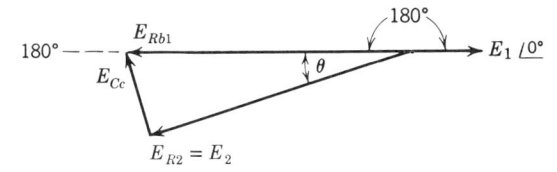

Sec. 8-9 *Analysis of R-C Coupled Tube Amplifier* 301

Figure 8-14
Phasor diagram for *R-C* coupled voltage amplifier, with resistive load, operated at frequency below midband range. The output voltage E_{R2} leads its midband frequency position by an angle θ.

divided between these two circuit elements. That is, part of the ac voltage available for output appears across C and cannot be utilized as output voltage. Furthermore, the actual output voltage across R_2 is no longer 180 degrees out of phase with the input voltage E_1. It leads the 180-degree phase position by an angle θ, as shown in Fig. 8-14. For a given amplifier, the value of the phase-shift angle θ depends on the frequency of operation. Figure 8-15 shows how it varies with the product of frequency, the capacitance of the coupling capacitor, and the equivalent resistance R_{el} defined in terms of the resistances in the equivalent circuit for *low-frequency operation*. The voltage amplification at any particular frequency below the midband range will be a certain fraction of the midband value, depending on the frequency. This fraction is shown by the voltage-amplification curve in Fig. 8-15.

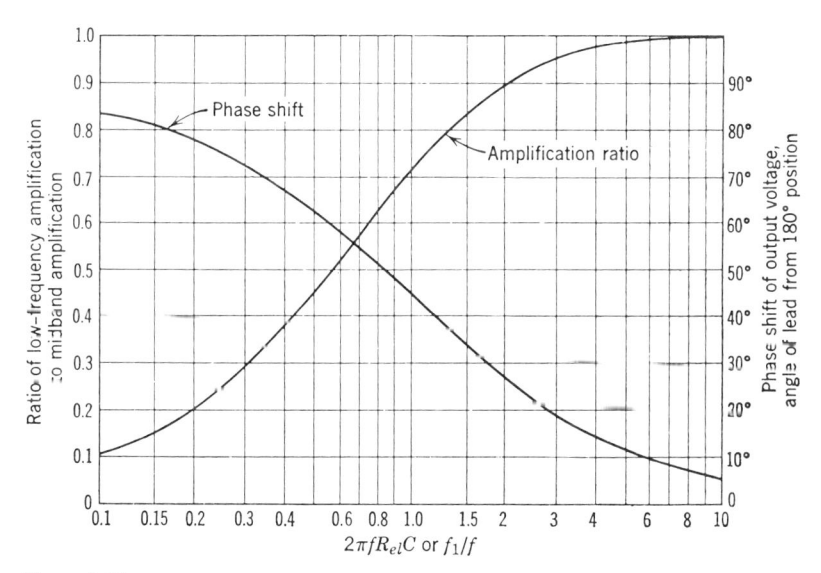

Figure 8-15
Curves showing amplification ratio and phase shift in low-frequency range. Resistance-capacitance coupled amplifier.

302 Analysis of Amplifier Operation Ch. 8

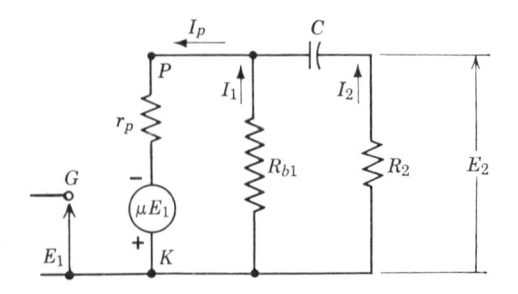

Figure 8-16
Equivalent-circuit diagram of a resistance-capacitance coupled amplifier operating at frequencies below the midband range.

The equivalent circuit for low-frequency operation is shown in Fig. 8-16. Note that C_2 does not appear. As stated before, it is very small and its reactance is extremely high except at high frequencies. Because of this, its shunting (bypassing) effect is negligible in the midband and low-frequency ranges. The symbol R_2 is used instead of R_{g2} for the sake of consistency with the midband-analysis equation. R_{g2} symbolizes the grid-current resistors of a second amplifier stage, which would not be part of the circuit of a single-stage amplifier. R_2, however, may be used in general to represent any resistive load fed by the amplifier. The symbols for current and voltage represent effective values of ac components. It should be reemphasized that the ac component of grid-to-cathode voltage is used in the equivalent circuit. The symbol for the effective value of the ac component of grid-to-cathode voltage is E_g, but we are using E_1 here because $E_1 = E_g$ since there is no other ac voltage in the grid circuit that could make E_g different from E_1.

In Fig. 8-16 we see that, starting at the top of the equivalent generator and going counterclockwise, the summation of voltages around the left-hand loop gives

$$+\mu E_1 - I_1 R_{b1} - I_p r_p = 0 \tag{8-19}$$

Using the other loop we obtain, going clockwise,

$$-I_1 R_{b1} + I_2 R_2 + I_2 X_C = 0 \tag{8-20}$$

But

$$I_p = I_1 + I_2 \quad \text{and} \quad X_C = \frac{1}{j\omega C} = -\frac{j}{\omega C}$$

where $\omega = 2\pi f$. Substituting these relations into Eqs. 8-19 and 8-20, and rearranging terms:

$$I_1 R_{b1} + (I_1 + I_2) r_p = \mu E_1 \tag{8-21}$$

Sec. 8-9　　　　　　　　　　　　　　　*Analysis of R-C Coupled Tube Amplifier*　303

$$I_1 R_{b1} - I_2(R_2 - j/\omega C) = 0 \tag{8-22}$$

$$I_1(R_{b1} + r_p) + I_2 r_p = \mu E_1 \tag{8-23}$$

The output voltage (E_2) is $-I_2 R_2$. The minus sign is used because at the instant the currents flow as indicated, the top of R_2 is negative with respect to the bottom (ac ground connection of actual circuit) and at that instant E_1 is positive with respect to ground. Hence E_2 is in its range of negative values with respect to E_1. If X_C were zero, E_2 would be directed exactly opposite to E_1 in the phasor diagram.

Using $E_2 = -I_2 R_2$, we obtain the expression for the voltage amplification in the low-frequency range:

$$A_{vl} = \frac{E_2}{E_1} = -\frac{I_2 R_2}{E_1} \tag{8-24}$$

The minus sign indicates 180-degree phase shift through the tube. Let us now solve Eq. 8-22 and 8-23 for I_2. From Eq. 8-21,

$$I_1 = \frac{\mu E_1 - I_2 r_p}{R_{b1} + r_p} \tag{8-25}$$

Substituting this into Eq. 8-22,

$$\frac{\mu E_1 R_{b1} - I_2 r_p R_{b1}}{R_{b1} + r_p} - I_2\left(R_2 - \frac{j}{\omega C}\right) = 0 \tag{8-26}$$

Multiplying both sides by ($R_{b1} + r_p$),

$$\mu E_1 R_{b1} - I_2 r_p R_{b1} - I_2(R_{b1} + r_p)\left(R_2 - \frac{j}{\omega C}\right) = 0$$

$$\mu E_1 R_{b1} - I_2 r_p R_{b1} - I_2\left(R_{b1} R_2 + r_p R_2 - j\frac{R_{b1} + r_p}{\omega C}\right) = 0 \tag{8-27}$$

$$I_2 = \frac{\mu R_{b1} E_1}{r_p R_{b1} + R_{b1} R_2 + r_p R_2 - \dfrac{j(R_{b1} + r_p)}{\omega C}} \tag{8-28}$$

Using Eq. 8-24, the voltage amplification in the low-frequency range is

$$A_{vl} = \frac{-\mu R_{b1} R_2}{r_p R_{b1} + R_{b1} R_2 + r_p R_2 - \dfrac{j(R_{b1} + r_p)}{\omega C}} \tag{8-29}$$

Equation 8-29 may be used to compute the voltage amplification at any value of frequency in the low-frequency range. Inasmuch as the midband amplification is a constant and rather easy to calculate, the amplification in the low- and high-frequency ranges may be expressed in terms of A_{vm}.

304 *Analysis of Amplifier Operation* Ch. 8

Since $\mu = g_m r_p$,

$$A_{vl} = \frac{-g_m r_p R_{b1} R_2}{r_p R_{b1} + R_{b1} R_2 + r_p R_2 - \frac{j(R_{b1} + r_p)}{\omega C}} \quad (8\text{-}30)$$

Dividing numerator and denominator by $(r_p R_{b1} + R_{b1} R_2 + r_p R_2)$,

$$A_{vl} = -\frac{g_m \dfrac{r_p R_{b1} R_2}{r_p R_{b1} + R_{b1} R_2 + r_p R_2}}{1 - \dfrac{j(R_{b1} + r_p)}{\omega C (r_p R_{b1} + R_{b1} R_2 + r_p R_2)}} \quad (8\text{-}31)$$

Referring to Eq. 8-18 and converting it to an equivalent form,

$$A_{vm} = -g_m \frac{1}{\dfrac{1}{r_p} + \dfrac{1}{R_{b1}} + \dfrac{1}{R_2}}$$

$$= -g_m \frac{r_p R_{b1} R_2}{r_p R_{b1} + R_{b1} R_2 + r_p R_2} = -g_m R_e \quad (8\text{-}32)$$

in which R_e represents the parallel resistance combination shown in two forms. Equation 8-31 may then be written

$$A_{vl} = \frac{-g_m R_e}{1 - \dfrac{j}{\omega C R_{el}}} = \frac{A_{vm}}{1 - \dfrac{j}{\omega C R_{el}}} \quad (8\text{-}33)$$

in which R_{el} is an equivalent resistance for the low-range computation:

$$R_{el} = \frac{r_p R_{b1} + R_{b1} R_2 + r_p R_2}{R_{b1} + r_p} \quad (8\text{-}34)$$

Since every quantity on the right-hand side of Eq. 8-33 is constant except ω, it is only necessary to substitute the value of the frequency and to solve for the amplification *after A_{vm} has been determined*. Note that if the frequency is large enough to make the j term of the denominator very small compared to unity, the amplification becomes the midband value, as it should. The ratio of A_{vl} to A_{vm} is therefore

$$\frac{A_{vl}}{A_{vm}} = \frac{1}{1 - \dfrac{j}{\omega C R_{el}}} \quad (8\text{-}35)$$

The frequency at which the denominator of Eq. 8-35 becomes $(1 - j1)$ will be useful as a reference to which all other frequencies may be compared. It is seen that this frequency can be determined from

$$1 = \frac{1}{\omega C R_{el}} = \frac{1}{2\pi f_1 C R_{el}}$$

where f_1 is the frequency that makes the expression equal to unity. Solving for the frequency f_1 yields

$$f_1 = \frac{1}{2\pi CR_{el}}$$

Substituting this value into Eq. 8-35 yields

$$\frac{A_{vl}}{A_{vm}} = \frac{1}{1 - j(f_1/f)} \qquad (8\text{-}35a)$$

The frequency f_1 is often referred to as the *low-frequency half-power frequency*. At this frequency the output voltage of an amplifier is reduced from its midband amplitude by an amount that represents only half the power in the load. Let us see that this is so. Since the amplification is a measure of the output voltage and the output voltage squared is proportional to the load power, then we may use Eq. 8-35a to calculate load power at different frequencies. We will calculate the frequency f at which the power ratio is one-half in the following manner:

$$\frac{1}{2} = \frac{1^2}{[1 - j(f_1/f)]^2} = \frac{A_{vl}^2}{A_{vm}^2}$$

or

$$2 = 1^2 + \frac{f_1^2}{f^2}$$

This expression is true when $f = f_1$.

It is convenient to use a curve giving the value of this ratio for all values of ωCR_{el} or f_1/f. Reich* has shown a set of amplification-ratio and phase-shift curves that are convenient for determining quite accurately the voltage amplification in the low-frequency and high-frequency ranges. Figure 8-15 shows the curves for the low-frequency range. The constant R_{el} and the frequency in the low range at which the amplification is desired being known, it is necessary only to compute the product $2\pi fR_{el}C$ and find its value on the horizontal scale. The point directly above, on the amplification curve, has an ordinate on the relative-amplification scale denoting the ratio of the amplification in question to the midband amplification, which is known from Eq. 8-18.

Example 8-3. An example of the determination of amplification in the low range follows. The circuit of Example 8-2 will be used again. Assume that the voltage amplification at 60 Hz is desired. The equivalent resistance for low-range determinations is obtained by using Eq. 8-34:

---

* H. J. Reich, *Theory and Application of Electron Tubes*, McGraw-Hill, New York, 1944.

$$R_{el} = \frac{66{,}000 \times 250{,}000 + 250{,}000 \times 500{,}000 + 66{,}000 \times 500{,}000}{250{,}000 + 66{,}000}$$

$$= \frac{10^6(16{,}500 + 125{,}000 + 33{,}000)}{10^6(0.25 + 0.066)}$$

$$= \frac{174{,}500}{0.316} = 552{,}200$$

$$\omega R_{el} C = 2\pi \times 60 \times 552{,}200 \times 0.01 \times 10^{-6} = 2.08 = f_1/f$$

The curve of Fig. 8-15 shows the ratio A_{vl}/A_{vm} to be 0.9. Since $A_{vm} = 70.7$, from Example 8-2 above,

$$A_{vl} = 0.9 \times 70.7 = 63.6$$

The phase shift, as read from the other curve of Fig. 8-15, is found to be 26 degrees beyond the 180-degree position, making 206° between E_1 and E_2. To determine A_{vl} directly, use Eq. 8-29 ($\mu = 100$ as given in the tube manual):

$$A_{vt} = -\frac{100 \times 0.25 \times 0.5 \times 10^{12}}{10^{12}(0.066 \times 0.25 + 0.25 \times 0.5 + 0.066 \times 0.5) - \dfrac{j10^6(0.25 + 0.066)}{2\pi \times 60 \times 0.01 \times 10^{-6}}}$$

$$= -\frac{12.5}{0.0165 + 0.125 + 0.033 - j0.316/3.77}$$

$$= -\frac{12.5}{0.1745 - j0.0839}$$

$$= -\frac{12.5}{0.1745 - j0.0839} \times \frac{0.1745 + j0.0839}{0.1745 + j0.0839}$$

$$A_{vl} = -\frac{2.18 + j1.05}{0.0305 + 0.00706} = -57 - j28 = 63.6\ \underline{/206.2°}$$

The angle of lead, θ, with reference to the 180-degree phase position is seen to be 26.2° (Fig. 8-17). The relative positions of the phasors verify the general phasor diagram shown earlier in Fig. 8-14.

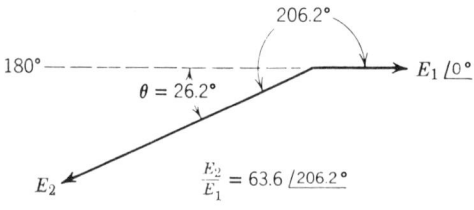

Figure 8-17
Phasor diagram showing phase shift at low frequency of output voltage (E_2) with respect to its midband 180° position. Shift is in the leading direction.

Sec. 8-9   Analysis of R-C Coupled Tube Amplifier   307

High-Frequency Range

Attention has been called to the presence of capacitance at the output terminals of an amplifier. Plate-to-cathode capacitance of the tube, capacitance between plate-circuit wiring, and the input capacitance of the following stage or other connected load all contribute to the bypassing of high-frequency energy. The combined capacitance of these effects is represented by C_2 in the equivalent-circuit diagram shown in Fig. 8-18. The current I_2 flows through an equivalent impedance (Z_2), which is given by the equation

$$\frac{1}{Z_2} = \frac{1}{R_2} + \frac{1}{X_{C_2}} \tag{8-36}$$

Since

$$\frac{1}{X_{C_2}} = \frac{1}{1/j\omega C_2}$$

$$\frac{1}{Z_2} = \frac{1}{R_2} + j\omega C_2 \tag{8-37}$$

The current I_p flows through a total equivalent impedance made up of R_{b1} in parallel with Z_2. Calling this total equivalent impedance Z_e, we have

$$\frac{1}{Z_e} = \frac{1}{R_{b1}} + \frac{1}{R_2} + j\omega C_2 \tag{8-38}$$

The plate current (I_p) is then given by

$$I_p = \frac{\mu E_1}{r_p + Z_e} \tag{8-39}$$

in which Z_e is a complex number. The output voltage (E_2) is equal to I_p multiplied by the total equivalent impedance (Z_e) through which it flows.

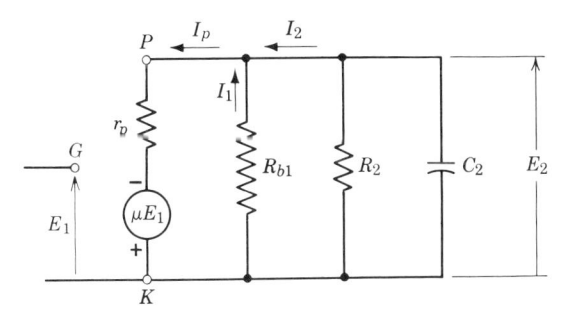

Figure 8-18
Equivalent-circuit diagram of a resistance-capacitance coupled amplifier operating at frequencies above the midband range.

308 Analysis of Amplifier Operation   Ch. 8

$$E_2 = -I_p Z_e = \frac{-\mu E_1 Z_e}{r_p + Z_e} \tag{8-40}$$

and the voltage amplification in the high-frequency range, A_{vh}, is

$$A_{vh} = \frac{E_2}{E_1} = \frac{-\mu Z_e}{r_p + Z_e} \tag{8-41}$$

Now insert the expression for Z_e in order to make this equation useful. Let us first divide the numerator and denominator by $r_p Z_e$:

$$A_{vh} = \frac{-\mu/r_p}{\dfrac{1}{Z_e} + \dfrac{1}{r_p}} \tag{8-42}$$

Using Eq. 8-38,

$$A_{vh} = \frac{-\mu/r_p}{\dfrac{1}{R_{b1}} + \dfrac{1}{R_2} + \dfrac{1}{r_p} + j\omega C_2} = \frac{-g_m}{\dfrac{1}{R_{b1}} + \dfrac{1}{R_2} + \dfrac{1}{r_p} + j\omega C_2} \tag{8-43}$$

Observe that this equation is very similar to Eq. 8-18, which is for the midband amplification. Indeed, if the frequency is so low that the value of $j\omega C_2$ is negligible, it becomes identical with Eq. 8-18, as it should. Equation 8-18 allows us to write Eq. 8-43 in the form

$$A_{vh} = \frac{-g_m}{(1/R_e) + j\omega C_2} = \frac{-g_m R_e}{1 + j\omega C_2 R_e} \tag{8-44}$$

And, from Eq. 8-32,

$$A_{vh} = \frac{A_{vm}}{1 + j\omega C_2 R_e} \tag{8-45}$$

$$\frac{A_{vh}}{A_{vm}} = \frac{1}{1 + j\omega C_2 R_e} \tag{8-46}$$

We may calculate a frequency in the high-frequency operating region at which the load power is one-half its value in the midband region. This frequency is often referred to as the *upper half-power frequency*. We will label it f_2, and calculate its value in the manner used for the lower half-power frequency:

$$1 = 2\pi f_2 C_2 R_e$$

or

$$f_2 = \frac{1}{2\pi C_2 R_e}$$

Sec. 8-9   *Analysis of R-C Coupled Tube Amplifier*   309

Substituting this value into Eq. 8-46 yields

$$\frac{A_{vh}}{A_{vm}} = \frac{1}{1 + j(f/f_2)} \qquad (8\text{-}46a)$$

To show that the load power is one-half its midband value at $f = f_2$, square both sides of Eq. 8-46a and set equal to $1/2$:

$$\frac{1}{2} = \frac{1^2}{[1 + j(f/f_2)]^2} = \frac{A_{vh}^2}{A_{vm}^2}$$

or

$$2 = 1^2 + \frac{f^2}{f_2^2}$$

This expression will be true only when $f = f_2$. Therefore the load power is one-half of its midband value when the driving frequency is f_2.

The range of frequencies between f_1 and f_2 is called the *bandwidth* of the amplifier. This concept is presented in some detail in Chapter 1. It can be used as a *figure of merit* when comparing the performance of one circuit design with another. It is a measure of the range of driving frequencies that may be amplified by the circuit in order to maintain load power between its full value and one-half that amount. Further, it will be seen that within the bandwidth the phase shift of the amplifier will remain within $\pm 45°$ of the 180-degree phase shift at midband frequencies.

Equation 8-46 suggests that a curve of A_{vh}/A_{vm} plotted against $\omega C_2 R_e$ or f/f_2 would be very useful in determining A_{vh} after A_{vm} is known. Such a curve, and one for phase shift, is shown in Fig. 8-19.

Example 8-4. Now determine the voltage amplification of the circuit in Example 8-2 at a frequency in the high range. Assume that the total output capacitance C_2 is 50 pF and that the frequency in question is 50,000 Hz.

Solution: Assuming that we do not know the amplification in the midband range, let us use Eq. 8-43.

$$A_{vh} = \frac{-g_m}{\dfrac{1}{R_{b1}} + \dfrac{1}{R_2} + \dfrac{1}{r_p} + j\omega C_2}$$

$$= -\frac{1500 \times 10^{-6}}{10^{-6}\left(\dfrac{1}{0.25} + \dfrac{1}{0.5} + \dfrac{1}{0.066}\right) + j 2\pi \times 50{,}000 \times 50 \times 10^{-12}}$$

$$= -\frac{1500}{4 + 2 + 15.3 + j15.71} = -\frac{1500}{21.3 + j15.71}$$

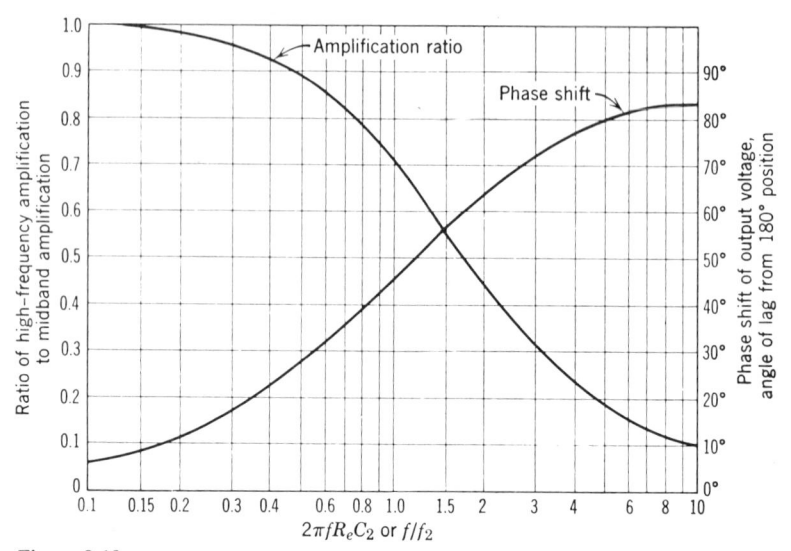

Figure 8-19
Curves showing amplification ratio and phase shift in high-frequency range. Resistance-capacitance coupled amplifier.

$$= -\frac{1500}{21.3 + j15.71} \times \frac{21.3 - j15.71}{21.3 - j15.71} = -\frac{31,800 - j23,565}{449.44 + 246.80}$$

$$A_{vh} = -45.8 + j33.8 = 56.9/143.6°$$

We shall verify this result by the shorter method of solution that employs the universal curves of Fig. 8-19. The mid-frequency-band amplification was found to be 70.7. Solving for the value to be used on the horizontal scale, we obtain

$$\omega C_2 R_e = 2\pi \times 50,000 \times 50 \times 10^{-12} \times \frac{10^6}{21.3} = 0.742 = \frac{f}{f_2}$$

$R_e = 10^6/21.3$ was determined in the solution for A_{vm} in Example 8-2. At this value (0.742) on the horizontal axis of Fig. 8-19, we find the amplification ratio to be 0.80, so that $A_{vh} = 0.80 \times A_{vm} = 0.80 \times 70.7 = 56.5$. The phase shift is seen to be 36° lagging the midband 180-degree position, or an advance in phase of 144°.

Drill Problems

D8-6 Using the curves of Figs. 8-15 and 8-19, sketch the relative voltage gain of an amplifier that has a bandwidth extending from 100 Hz to 100 kHz.

D8-7 Show that the gain of an R-C coupled amplifier is "down 3 dB" from its midband gain at f_1 and f_2. (*Hint:* Start with Eqs. 8-35 and 8-46 and calculate the absolute magnitudes; then express in terms of decibels.)

8-10
Analysis of Transistor-Amplifier Operation Outside Midband

Transistor-amplifier operation in the low-frequency range, where the coupling capacitor in Fig. 8-9 no longer has negligible reactance, is analyzed in the same manner as for vacuum tubes. However, the coupling network between stages has a slightly more complicated form because the electrode currents must pass through "internal resistances" in the two adjacent stages.

For low-frequency operation, the equivalent circuit of the interstage network is shown in Fig. 8-20. R_{o1} is the output resistance of the first stage, and R_{i2} is the input resistance of the second stage. The output voltage of the first stage must be V_2 in the analysis previously presented, so here V must be of proper value to produce V_2 across R_{b2} and R_{i2} in the midband *when C_1 is not present.* Therefore, this current will be

$$I_m = V_2/R_b$$

To determine the voltage V,

$$V = I_m(R_a + R_b)$$

and, combining these two equations,

$$V_2 = V\frac{R_b}{R_a + R_b} \qquad V = \frac{R_a + R_b}{R_b}V_2$$

Now consider Fig. 8-20b, where V is applied:

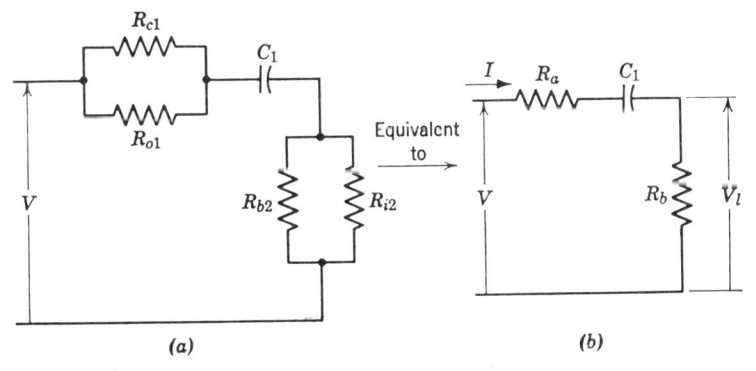

Figure 8-20
Equivalent interstage coupling network of transistor amplifier in Fig. 8-9.

312 Analysis of Amplifier Operation   Ch. 8

$$I = \frac{V}{R_a + R_b - j/\omega C_1}$$

The low-frequency output voltage is

$$V_l = \frac{VR_b}{R_a + R_b - j/\omega C_1}$$

and

$$\frac{V_l}{V_2} = \frac{R_a + R_b}{R_a + R_b - \dfrac{j}{\omega C_1}} = \frac{1}{1 - \dfrac{j}{\omega C_1 (R_a + R_b)}}$$

This is the same form as Eq. 8-35 for the vacuum-tube amplifier and has the same meaning, since $A_{vl} = V_l/V_i$ and $A_{vm} = V_2/V_i$, where V_i is the input-signal voltage.

In the high-frequency range, where C_1 is neglected but the shunting capacitance C_s is appreciable, there are four resistances in parallel: R_{o1}, R_{C1}, R_{b2}, and R_{i2}. As for the tube, the total shunt capacitance is made up of the output capacitance of the transistor and the capacitance of the circuit wiring. The value of output capacitance is supplied by the transistor manufacturer. Using the equivalent circuit in a procedure similar to that for the low-frequency range, the ratio of output voltage (or gain) in the high range to output voltage (or gain) in the middle range turns out to be

$$\frac{V_h}{V_2} = \frac{1}{1 + j\omega C_T R_{eq}}$$

where C_T is the total shunting capacitance and R_{eq} is the parallel equivalent of the four resistors listed above. This is the same form as Eq. 8-46 for operation of a vacuum-tube amplifier in its high-frequency region. It has the same meaning, and the transistor amplifier performance may be determined from Figs. 8-15 and 8-19 in the same way.

8-11
Improving Amplifier Response at Low Frequencies

The gain of an amplifier falls off as the operating frequency is reduced in the low-frequency range. Increase in the capacitance of the coupling capacitor to the upper limit imposed by space, cost, or other conditions is the first step in improving low-frequency response.

Another effective way to maintain the voltage gain at, or nearer to, the midfrequency value while the frequency is decreasing in the low range is to use the "decoupling" arrangement of C_d and R_d in parallel shown in Fig. 8-21. At the low end of the *midband range* of frequencies, the effect of R_d as an ac load

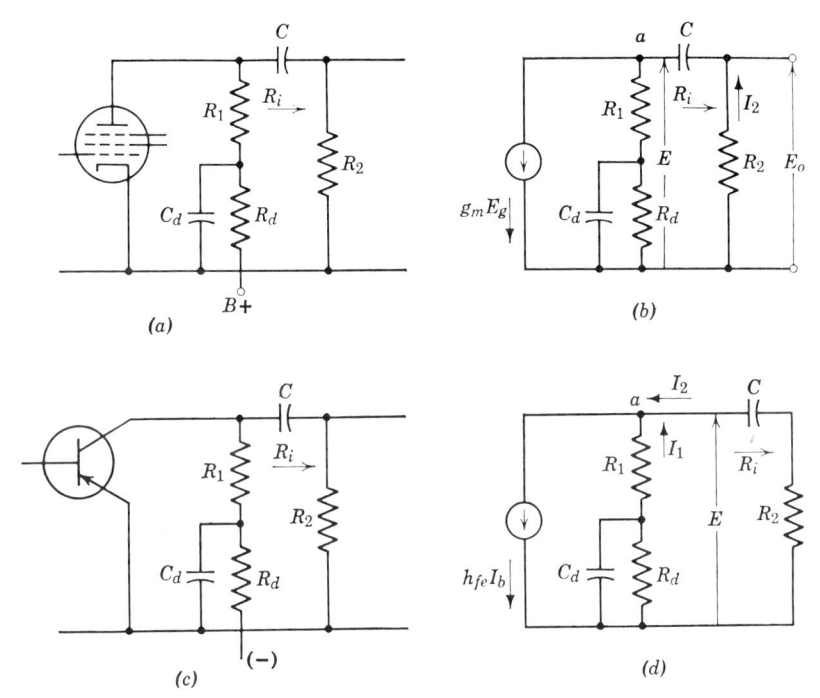

Figure 8-21
Low-frequency compensation arrangement for vacuum-tube and transistor amplifiers.

resistor is canceled by the low reactance of C_d. But as the frequency *decreases*, the impedance of the parallel circuit *increases* and the ac voltage drop across the load impedance of the amplifier (this is the output voltage) holds up very well. The effect is as if the ac load line were tilted to become more nearly horizontal, thus increasing the ac voltage swing, E, and offsetting the growing voltage drop across C, as far as the output voltage E_o or transistor current I_2 is concerned.

A pentode tube is shown because it will be used instead of a triode if the objective is constant voltage gain over as wide a frequency range as possible. Pentodes perform the same functions as triodes, and in some respects they do them better.

The ac plate resistance of the tube and the internal-collector resistance of the transistor are not shown because they are large compared with $(R_1 + R_d)$. It is desired that the output voltage $I_2 R_2$ in the tube circuit and I_2 in the transistor circuit be independent of frequency. This is asking a lot of the transistor circuit, because the input resistance of a second stage is so low. In order to make the reactance of C very small compared to R_{ie}, a capacitance of many microfarads must be used.

314   Analysis of Amplifier Operation   Ch. 8

The current equation at node a in the transistor circuit is

$$h_{fe}I_b = -\frac{E}{R_i - \dfrac{j}{\omega C}} - \frac{E}{R_1 + \dfrac{R_d(-j/\omega C_d)}{R_d - j/\omega C_d}} \qquad (8\text{-}47)$$

and

$$E = -I_2\left(R_i - \frac{j}{\omega C}\right) \qquad (8\text{-}48)$$

From these two equations, I_b, the base current of the transistor stage, can be expressed in terms of I_2, the input current to the next stage or load. The algebra is cumbersome, although the assumption that $R_d = 1/\omega C_d$ helps. Eliminating E gives the result

$$h_{fe}I_b = \left[\frac{\omega^2(R_i + R_1)CC_d - j\omega(C + C_d)}{\omega^2 R_1 CC_d - j\omega C}\right]I_2 \qquad (8\text{-}49)$$

By means of a procedure in calculus, an expression that is the *rate of change* of I_2 with respect to frequency can be obtained. Since it is desirable that I_2 *does not change* with frequency, this *rate of change* is ideally *zero*. Setting the rate-of-change expression equal to zero leads algebraically to the following circuit requirement:

$$R_1(C + C_d) = (R_1 + R_i)C$$

Canceling $R_1 C$ on both sides gives

$$R_1 C_d = R_i C \qquad (8\text{-}50)$$

This says that the time constants of the compensating circuit and the coupling circuit should be equal. Note that although R_i is shown as a coupling resistor in the transistor circuit, it really includes the input resistance to the next stage. For the tube, the input impedance to the next stage will be practically infinite at such low frequencies unless the grid goes positive.

The analysis for the pentode amplifier will be the same as for the transistor, as can be seen from the fact that their equivalent circuits are identical. The current I_2 flows through the grid-leak resistor of the next stage or through an adequately large load resistance. Another beneficial effect is derived from this compensation circuit, in that making the time constants equal (Eq. 8-50) results in zero phase shift in the coupling circuit at the output of this stage in the low-frequency region.

8-12
Bandwidth Reduction in Cascaded Stages

When two or more stages of amplification are interconnected so that the output of one becomes the input of the next, they are said to be in *cascade* connection. If we denote the lower half-power frequency of one stage as f_1, the relative amplification of *n identical stages connected in cascade* is

$$(A_{rel})^n = \left(\frac{1}{[1 + (f_1/f)^2]^{1/2}}\right)^n = \frac{1}{[1 + (f_1/f)^2]^{n/2}} \qquad (8\text{-}51)$$

The half-power frequency for the composite amplification is also determined where the relative gain is reduced to $1/\sqrt{2}$ of its maximum value. This requires that the denominator of Eq. 8-51 equal $\sqrt{2}$

$$\left[1 + \left(\frac{f_1}{f}\right)^2\right]^{n/2} = \sqrt{2}$$

This expression may be rearranged as

$$\frac{f_1}{f} = \sqrt{2^{1/n} - 1} \qquad (8\text{-}52)$$

A similar analysis of the upper half-power frequency, f_2, yields a similar expression for the bandwidth reduction

$$\frac{f}{f_2} = \sqrt{2^{1/n} - 1} \qquad (8\text{-}53)$$

Table 8-5
Bandwidth Reduction Factors

$n =$	1	2	3	4	5	6
$\sqrt{2^{1/n} - 1} =$	1.000	0.643	0.510	0.435	0.387	0.350

The bandwidth reduction factor is given in Table 8-5 for several values of n. It is seen that if an overall bandwidth is necessary for an amplifier, each stage of a cascaded circuit must have a materially greater individual bandwidth. For example, three identical stages will require that each stage have a bandwidth that is approximately twice the desired bandwidth for the whole system.

8-13
Intercommunication System Using Vacuum Tubes

Private communication systems that provide talking facilities between offices and to remotely located points have come into general use. Many of them are electronically operated and consist of an audio-frequency amplifier with a loudspeaker and a suitable switching arrangement. The speaker, usually a permanent magnet type, serves as a microphone as well as a receiving device. The units are usually provided with a talk–listen switch that must be depressed when the operator is speaking and released when he wishes to listen. However, some master stations are provided with a privacy earphone that may be connected into the circuit by raising the switch to a stable position. When this is done the conversation is permitted in the same manner as with a telephone.

Figure 8-22 is a schematic diagram of a simple master unit, without the privacy earphone feature, that may be used with from one to five substations. If the substation shown were called No. 1, it would be connected X to X, R to R, and *No.* to 1 of the master circuit. Then a second substation could be called No. 2 and it would be connected X to X, R to R, and *No.* to 2 of the master circuit. A three-conductor cable would be used for making the connections. A description of the operation will now be given.

Suppose that the master station is to call a substation, in this case No. 1. The selector switch is turned to 1 and the talk–listen switch is depressed. The sound waves of the voice strike the diaphragm of the loudspeaker, which vibrates and *moves the attached voice coil in the magnetic field of its permanent magnet.* The induced voltage sends very weak current through the primary of the input transformer. Thus a voltage is applied, by transformer action, to the grid of T_1. It is amplified and then applied to the grid of T_2, its strength being regulated by the volume control.

The *output-transformer secondary winding* delivers the strong audio power to the second section of the talk–listen switch (2 in dotted position), where it is passed on to the selector switch and contact No. 1. From there it goes to the substation speaker *No.* terminal, through the voice coil, and then back on line R to ground, thence to the output-transformer secondary, completing the circuit.

If points A and B are connected by the nonprivate jumper, as indicated by the dashed line, the person at the substation need only answer without pushing the switch in line X. Tracing his reply circuit from the grounded R terminal, we go through the substation speaker to the *No.* terminal, to selector switch 1, through the second and third sections of the talk–listen switch (which is in the up position for listening), and through the primary of the input transformer, and also through the 47-Ω resistor, to ground. The amplifier thus receives the voice signal and delivers it from the output transformer to the master speaker.

If private operation is desired (this means that the master station amplifier cannot reproduce sounds originating at any substation unless the substation talk switch is depressed), the jumper between points A and B is removed and connected between points B and C instead. With this arrangement, calling by the master is accomplished over the same circuits as described above. The substation's answer comes through the substation switch on line X and goes through the input-transformer primary to ground and back to the substation on R. The 5-Ω resistor prevents the B–C jumper from short-circuiting the substation speaker.

It is important that the person at the master station not be disturbed by sounds from any substation when communication is not desired. This is

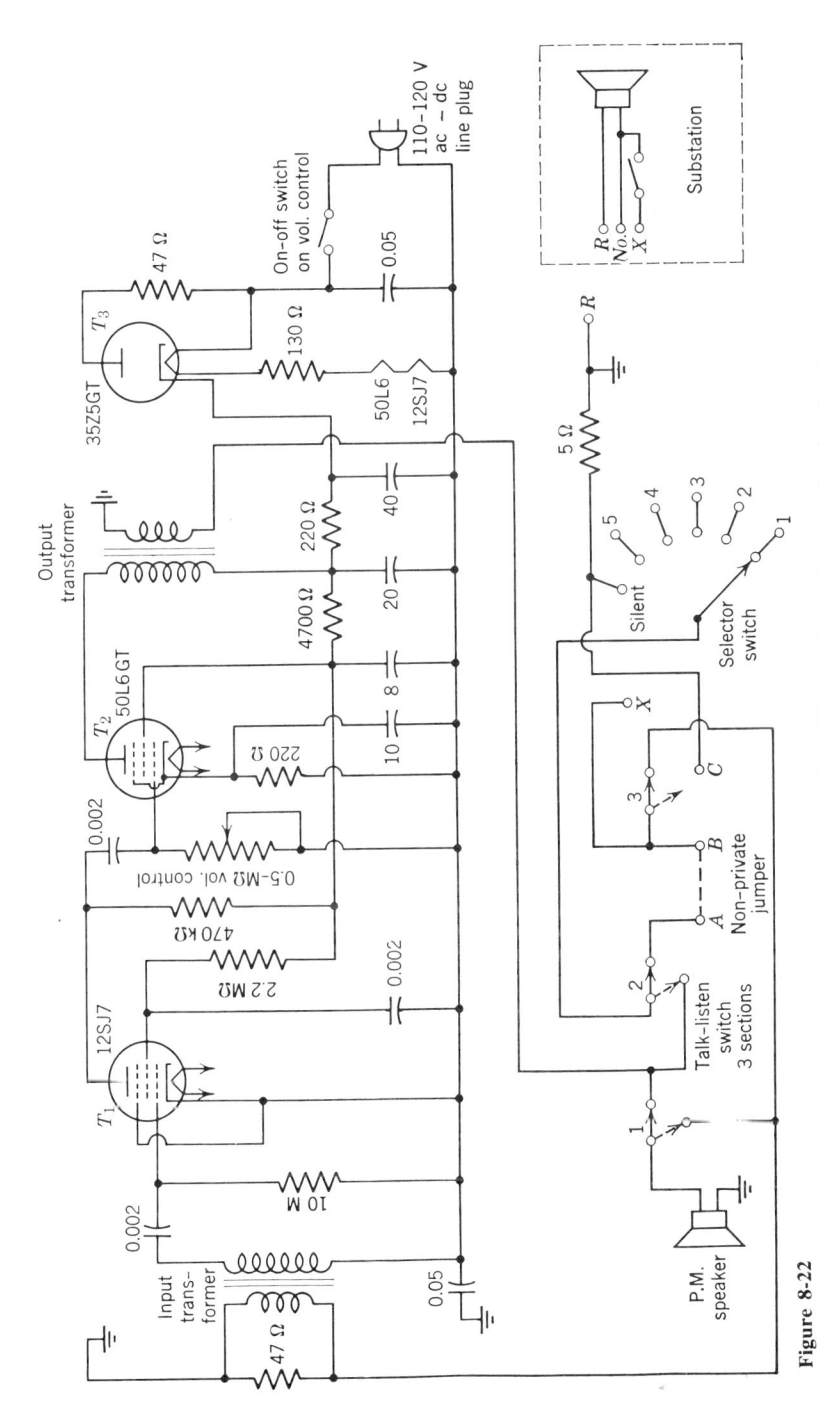

Figure 8-22
Schematic diagram of intercommunication system master station. All capacitor ratings are in microfarads.

automatically achieved with the connection arranged for private operation, because the substation switch must be closed before sound is produced in the master station. But, in *nonprivate operation*, the master operator will hear sound from the substation to which the selector switch is connected. To eliminate all sound, the selector switch must be placed in the *silent* position, which is the contact beyond No. 5 in the diagram. With this setting the master station will receive from any substation whose talk switch is depressed. In fact, it is always necessary to depress any substation switch to which the selector switch is not connected if the master station is to be called.

There is a provision for the master station to call all stations at once. The selector-switch position for this feature is not shown in the diagram. Lines running from all five terminals on the selector switch come to separate, closely spaced points but remain separate until the switch is moved to that position, in which event they are all tied together and receive power in parallel.

One advantage of the nonprivate connection is that we may answer at a substation from a great distance away from the unit. This means the operator need not leave his working place to come to the station to press the talk switch.

8-14
Intercommunication System Using Transistors

Figure 8-23 is a complete circuit of an intercommunication system, designed by Radio Corporation of America, using transistors and operating from a 12-V dc power supply. For ac operation a filtered power supply operating on 115 V, 60 cycles, and using crystal diodes or selenium rectifiers is recommended.

When switch S_1 is pressed to *talk*, Sp 1 serves as a microphone and sends the signal through capacitor C_1 to the first amplifier stage. The output stage delivers audio power through capacitor C_9 and switch S_2 to the speaker Sp 2. The person at the substation presses the switch to the *talk* position to answer, and uses the same circuit to deliver audio power to Sp 1, which then serves as a speaker.

By means of a multiple-contact switch with suitable connections, the audio power output through C_9 can be directed to other substations, as is done in Fig. 8-22.

Questions

8-1 What are the functions of the coupling resistor and capacitor in an R-C coupled amplifier? Must they have large or small values, and why?

8-2 Compare the four classes of amplifiers described in regard to current flow and amount of grid bias.

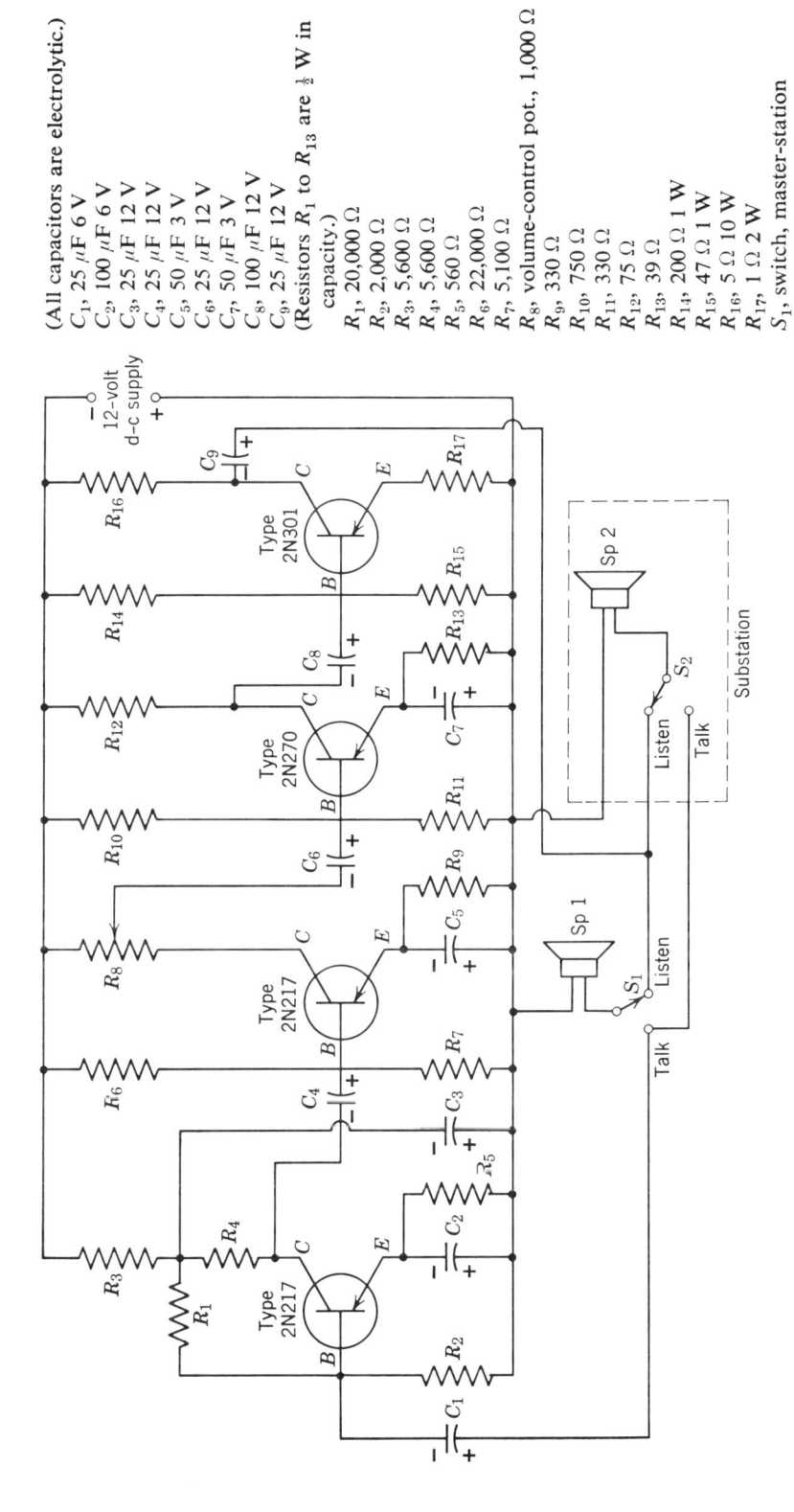

(All capacitors are electrolytic.)
C_1, 25 μF 6 V
C_2, 100 μF 6 V
C_3, 25 μF 12 V
C_4, 25 μF 12 V
C_5, 50 μF 3 V
C_6, 25 μF 12 V
C_7, 50 μF 3 V
C_8, 100 μF 12 V
C_9, 25 μF 12 V

(Resistors R_1 to R_{13} are ½ W in capacity.)
R_1, 20,000 Ω
R_2, 2,000 Ω
R_3, 5,600 Ω
R_4, 5,600 Ω
R_5, 560 Ω
R_6, 22,000 Ω
R_7, 5,100 Ω
R_8, volume-control pot, 1,000 Ω
R_9, 330 Ω
R_{10}, 750 Ω
R_{11}, 330 Ω
R_{12}, 75 Ω
R_{13}, 39 Ω
R_{14}, 200 Ω 1 W
R_{15}, 47 Ω 1 W
R_{16}, 5 Ω 10 W
R_{17}, 1 Ω 2 W
S_1, switch, master-station
S_2, switch, substation
SP 1, Speaker, master-station, 12 Ω 1 W
SP 2, Speaker, substation, 12 Ω 1 W

Figure 8-23
Transistorized intercommunication system. (Courtesy Radio Corporation of America.)

320 Analysis of Amplifier Operation   Ch. 8

8-3 Name and identify the types of distortion that may be present in an amplifier.

8-4 Draw from memory a curve showing the variation of voltage amplification with frequency of an R-C coupled amplifier. What are some of the causes of drop in amplification in the low- and high-frequency ranges?

8-5 Explain why there is a 180-degree phase shift in voltage through a common-emitter amplifier stage.

8-6 Draw the equivalent circuit that represents a voltage amplifier stage in the midband range.

8-7 Explain how to use the "universal" curves (Fig. 8-15) showing voltage gain and phase shift as a function of frequency and circuit constants.

8-8 What difference, if any, is there in the methods of using the "universal" curves for amplifier operation in the high-frequency and low-frequency ranges?

8-9 Why is a transistor amplifier regarded as a *current-controlled* device? Compare its most useful amplifying function with that of the vacuum-tube amplifiers studied thus far.

8-10 Define the following for a transistor amplifier stage: (*a*) input resistance, (*b*) output resistance, (*c*) voltage gain, (*d*) current gain, (*e*) power gain.

8-11 It is desired to compare the various gains and impedances of transistor amplifiers classified as common-base, common-emitter, and common-collector. Make a table and designate the gains and input and output impedances as high, medium, or low. Which are negative? In which is there phase shift?

8-12 Name the principal advantages of the common-emitter connection of transistors in amplifiers.

8-13 Why is the overall bandwidth of cascaded stages of an amplifier smaller than the bandwidth of the individual stages?

Problems

8-1 An R-C coupled amplifier stage and its triode tube have the following constants: $R_{b1} = 0.25$ MΩ, $R_2 = 0.5$ MΩ, $C = 0.005$ μF, $r_p = 7700$ Ω, $g_m = 2600$ μmho. Determine the voltage amplification and the angle by which the output voltage leads the input voltage (*a*) in the midband range, (*b*) at 100 Hz. Do this by direct computation and check your results by using the curve of Fig. 8-15.

8-2 The amplifier of Problem 8-1 has a total shunting capacitance of 1200 pF. Calculate the voltage gain at 15,920 Hz. Check your result by using the curve of Fig. 8-19.

8-3 Calculate the frequencies at the half-power points for the amplifier of Problems 8-1 and 8-2. For the low range $R_{el} = X_C$ and for the high range $R_e = X_{C2}$.

8-4 A 6SF5 triode is used in a preamplifier with cathode bias. $E_{bb} = 300$ V, $R_b = 0.147$ MΩ, $R_g = 0.1$ MΩ (next stage), $r_p = 66,000$ Ω, $g_m = 1500 \times 10^{-6}$ mho, $C = 0.02$ μF. (a) Draw the dc and ac load lines, choosing a Q-point that will give linear operation. Then calculate the bias resistance. (b) Using a signal input of 0.5 V peak, determine the peak value of output voltage and the voltage gain in the midband range.

8-5 A manufacturer lists the following data for using a high-μ triode in a small-signal voltage amplifier.

E_{bb} (V)	R_b (MΩ)	R_g (MΩ)	R_k (Ω)	C (μF)	g_m (μmho)	r_p (Ω)
180	0.25	0.5	2150	0.006	1325	53,000
300	0.50	1.0	2980	0.003	1325	53,000

(a) Calculate the midband voltage gain in each type of operation. (b) Calculate the rms signal-voltage input for 39 V output when E_{bb} is 180 V, and for 48 V output when E_{bb} is 300 V.

8-6 Assume that the total output capacitance of the amplifier of Problem 8-4 is 75 pF. (a) Determine the voltage gain at 40,000 Hz with and without C_2. (b) At what frequency above midband is the gain down 3 dB?

8-7 Compute the gain of the amplifier of Problem 8-4 at 80 Hz. What is the low-frequency half-power point?

8-8 An amplifier stage using a 6SJ7 tube operates with $E_{bb} = 250$ V, bias -3 V, $R_b = 30$ kΩ, $r_p = 1.25$ MΩ, $R_2 = 30$ kΩ. (a) Draw the dc and ac load lines. (b) Find the maximum and minimum instantaneous values of plate to cathode voltage for a grid signal with 1-V peak. (c) Find the amplitude of the plate current swing and the rms value of plate current, assuming sine-wave form. Calculate the power that must be dissipated by the plate: $P_p = E_{bo}I_{bo}$.

8-9 A pentode tube is used in a small-signal amplifier circuit with constants given below. (a) Calculate the voltage gain in the midband. (b) What are the upper and lower half-power frequencies? (c) How much signal voltage is needed to produce 100 V rms across R_{g2}?

$$r_p = 1.0 \text{ MΩ}, \quad g_m = 4 \times 10^{-3} \text{ mho}, \quad R_b = 0.25 \text{ MΩ},$$
$$R_{g2} = 0.25 \text{ MΩ}, \quad C = 0.0032 \text{ μF}, \quad C_T = 159 \text{ pF}$$

8-10 A two-stage amplifier uses a 6SN7 twin triode with circuit elements as listed and identified with reference to Fig. 8-2. $R_{g1} = 0.10$ MΩ, $R_{b1} = 0.1$ MΩ, $C_1 = 0.014$ μF, $C_{w1} = 500$ pF (due to wiring), $R_{g2} = 0.22$ MΩ, $R_{b2} = 0.22$ MΩ, $C_2 = 0.0065$ μF, $C_{w2} = 500$ pF, $R_L = 0.47$ MΩ. Tube constants are: $g_m = 2600$ μmho, $r_p = 7700$ Ω, $C_{gp} = 4$ pF, $C_{gk} = 2.4$ pF, $C_{pk} = 0.7$ pF. (a) What is the overall midband gain? (b) How much input signal is needed for an output voltage of 100 V rms from the second stage? (c) At what frequency above the midband range is the output voltage of the first stage only 0.707 of the midfrequency output? (d) Calculate, for an input signal of 0.1 V rms, the output of the second stage at the frequency determined for part (c).

8-11 After working Problem 8-10, determine the frequency above the midband range at which the gain of the *whole amplifier* is down 3 dB below the midband gain of the unit as a whole.

8-12 A transistor has the following parameters for common-emitter, base-input connection, when $V_{CE} = -1.3$ V and the dc collector-current is -0.3 mA: $h_{ie} = 4800$ Ω, $h_{re} = 9.1 \times 10^{-4}$, $h_{fe} = -45$, $h_{oe} = 12.4 \times 10^{-6}$ mho. Calculate the input resistance R_i and the output resistance R_o. Assume the driving generator resistance $= 500$ Ω. Calculate the current gain and power gain with $R_L = 4700$ Ω.

8-13 A two-stage transistor amplifier is shown in Fig. P8-13. The parameters are listed below it. (a) Calculate the stage gain at the midband frequencies. (b) Calculate the input and output resistances of the transistors. (c) Calculate V_i/V_2, the stage gain at 15.92 Hz. (d) Calculate V_h/V_2, the stage gain at 15,920 Hz (assume a total shunting capacitance of 0.003 μF), C_{CE} + shunt output capacitance $= 100 \times 10^{-12}$ F (or 100 pF).

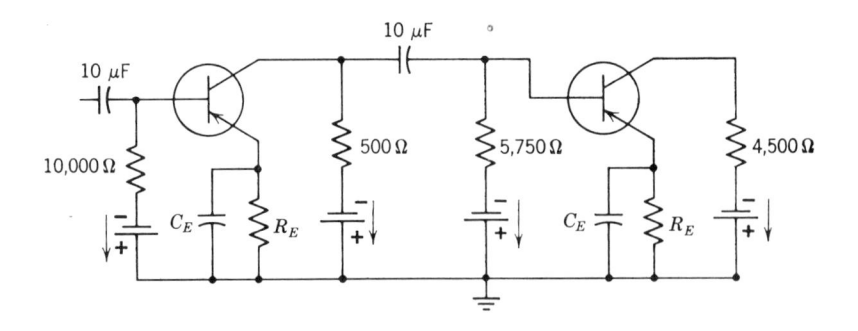

Figure P8-13
$h_{ie} = 2000$ Ω; $h_{fe} = 42$; $h_{re} = 4.5 \times 10^{-4}$; $h_{oe} = 28 \times 10^{-6}$ mho; $C_w = 15$ pF = wiring capacitance, $C_{CE} = 42$ pF = output capacitance. Assume that C_E has zero reactance.

8-14 Solve Problem 8-13 after replacing the transistors with two having the following parameters: $h_{ie} = 1667\,\Omega$, $h_{fe} = -44$, $h_{re} = 4.95 \times 10^{-4}$, $h_{oe} = 22.8 \times 10^{-6}$ mho, $C_{CE} = 40$ pF. Shunt output capacitance is 60 pF.

8-15 (a) In the amplifier circuit of Fig. P8-15, what is the voltage gain if the effect of C is neglected? Assume $\mu = 20$, $r_p = 10{,}000\,\Omega$. (b) What is the lower half-power frequency?

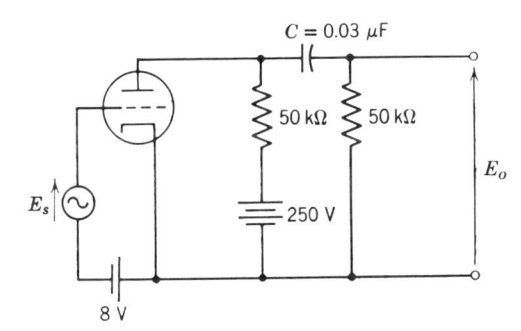

Figure P8-15

8-16 An R-C coupled amplifier consists of a single stage using a 6J5 tube. For the circuit, the following values apply: $E_{bb} = 250$ V, $R_b = 50{,}000\,\Omega$, $R_g = 500{,}000\,\Omega$, $R_k = 3900\,\Omega$ and is well bypassed. The coupling capacitor is $0.01\,\mu\text{F}$; stray capacitance $= 12$ pF. Assume $r_p = 1400\,\Omega$ and $\mu = 20$. Determine the input admittance at the upper half-power frequency.

8-17 The circuit of an amplifier is shown in Fig. P8-17, and dc supply voltages have been omitted for convenience. Calculate the voltage gain E_o/E_s.

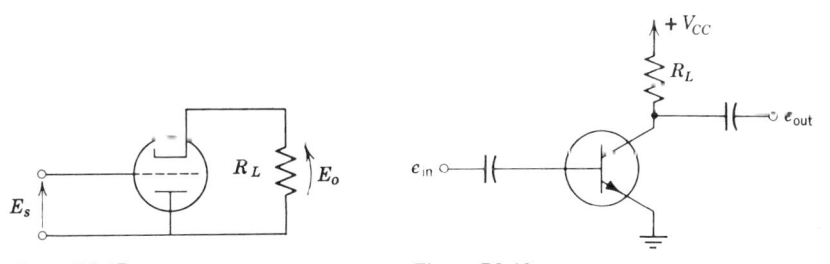

Figure P8-17

Figure P8-18

8-18 Sketch the small-signal equivalent for the circuit shown in Fig. P8-18 using h-parameters for the transistor.

324 Analysis of Amplifier Operation   Ch. 8

8-19 An amplifier produces an output voltage that is related to an input voltage by the expression $e_{out} = ke_{in}^2$. Does the amplifier produce distortion of the input signal? What type of distortion will it produce for $e_{in} = 10 \sin \omega t$?

8-20 An R-C coupled, single-stage voltage amplifier has $f_1 = 100$ Hz and $f_2 = 100$ kHz. Its response is flat in the middle-frequency region of operation. Sketch and label curves showing its relative amplification and phase shift throughout the frequency spectrum, to include f_1 and f_2. Express the relative amplification as a ratio compared to the midband gain, and in dB, at f_1 and f_2.

8-21 Using the h-parameters listed in the Appendix for the 2N404 transistor, calculate and construct a table of values similar to Table 8-4, using $R_L = 1$ kΩ and $R_g = 1$ kΩ.

8-22 By making calculations of input resistance, determine the effect of R_L in a common-emitter circuit. Sketch your results (R_i vs. R_L) on semilogarithmic graph paper, for a 2N404 transistor operating at $V_{CE} = -6$ V, $I_C = -1$ mA.

8-23 Calculate the effect of R_g on output resistance for a 2N404 operating at $V_{CE} = -6$ V, $I_C = -1$ mA. Sketch the results (R_o vs. R_g) on semilogarithmic graph paper.

8-24 A common-emitter uses the 2N334 at an operating point $V_{CE} = 12$ V, $I_C = 2.5$ mA. (*a*) Determine the values of h-parameters, assuming V_{CB} is constant at 5 V. (*b*) Calculate voltage and current gains, assuming $V_{CC} = 25$ V.

8-25 For the circuit conditions of Problem 8-24, calculate the effect of R_L on voltage gain and on current gain, as its value varies from zero to infinity (from short-circuit to open-circuit conditions).

8-26 Using the values of h-parameters found in part (*a*) of Problem 8-24, calculate the required value of C so that the common-base amplifier of Fig. P8-26 will have a lower half-power frequency of 100 Hz.

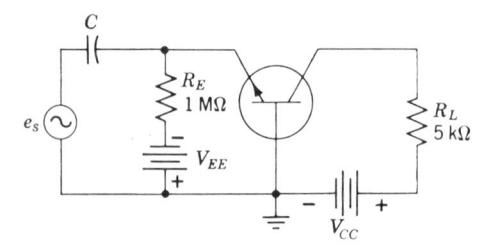

Figure P8-26

8-27 The transistor of Fig. P8-27 has these parameters: $h_{ie} = 2\ k\Omega$, $h_{fe} = 300$, $h_{re} = 2 \times 10^{-4}$, $h_{oe} = 30 \times 10^{-6}$ mho. Calculate the value of coupling capacitor C needed to reduce the gain by 3 dB at 100 Hz.

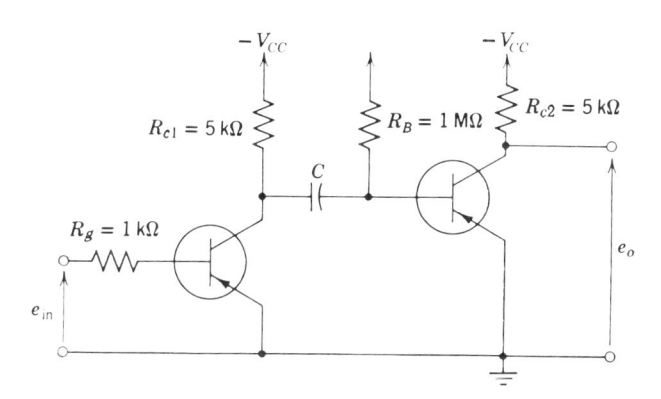

Figure P8-27

8-28 For each of the circuits in Fig. P8-28, $\mu = 20$, $r_p = 10\ k\Omega$, $E_s = 2/\underline{0°}$ V, $V = 1/\underline{0°}$ V, and operation may be considered linear. (*a*) Draw and label Thévenin's equivalent circuit. (*b*) Calculate I_p. (*c*) Determine the impedance presented to the source V. (Consider that $E_s = 0$ for this calculation.)

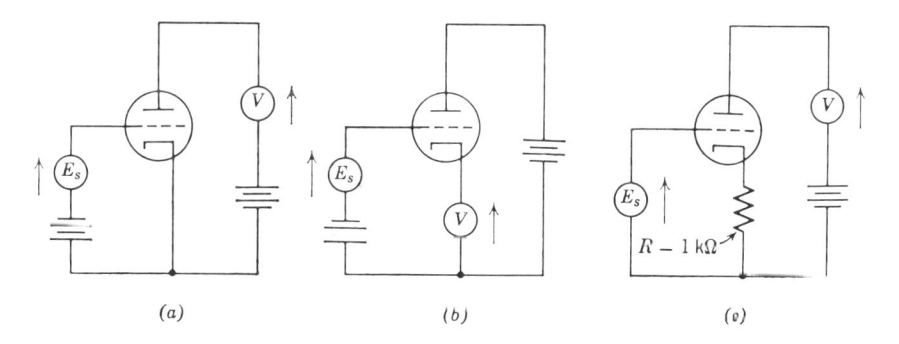

Figure P8-28

8-29 Three identical stages of an amplifier are connected in cascade. The amplifier is to have a bandwidth covering the audio frequency range, from 100 Hz to 20 kHz. What must be the bandwidth of the individual stages?

9
POWER AMPLIFIERS

The amplification and control of power at low and high frequencies involve many of the principles already studied. The location of operating regions along load lines plays a major role in the analysis of performance. The signals are large and extend over the audio-frequency range in amplifiers of speech, music, and many kinds of control signal. At the higher frequencies used in radio and TV transmitters, the frequency range of operation of power amplifiers is very narrow; in fact, it would be ideal in some cases if operation were at a single frequency rather than in a narrow band. This can be appreciated when it is learned that a parallel-resonant circuit is often used to obtain high selectivity of frequencies, resistive impedance, and maximum signal between stages and at the output of the amplifier.

Power tubes are built to have large current capacities, low values of plate resistance, and good power-handling capabilities. Triodes, pentodes, and beam-power tubes are used. *Power transistors* are built to have good heat dissipation capabilities and are often required to be attached to special shapes and sizes of metal called *heat sinks*.

We shall discuss several concepts that become important in the operation of power amplifiers but are usually negligible in voltage amplifiers. The large signals introduce some effects of nonlinearities in tube and transistor characteristics, such as distortion and variations of parameters that affect amplification. We shall learn that some classes of operation are more suited to particular load applications than others.

328 Power Amplifiers   Ch. 9

9-1
Single-Stage Power Amplifier

The simple one-tube and one-transistor amplifier circuits we have studied in previous chapters may be used as power amplifiers, provided the active device is capable of handling the voltages and currents necessary and the load is properly chosen to receive power that may be generated. The circuit shown in Fig. 9-1 is the basic configuration of a single-stage power amplifier.

The component labeled *load* in the diagram may be a single resistor in some cases, but usually the loads required are relatively small resistances that require an impedance-matching transformer between the active device and the load. Three general types of loading may be encountered in power amplifiers: (a) direct connection, (b) transformer-coupled connection, and (c) shunt, or L-C coupling. These are shown in Fig. 9-2.

Power-amplifier operation usually requires a large driving voltage or current, which may be the output of a tube or transistor voltage amplifier. If the

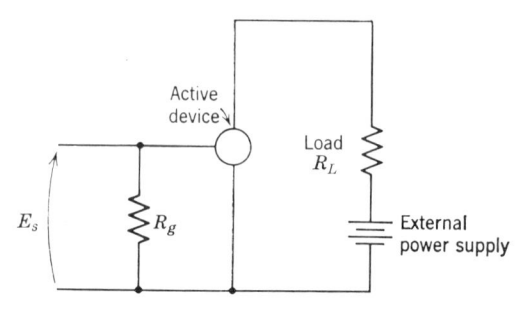

Figure 9-1
Basic power amplifier; active device may be tube or transistor in electronic applications.

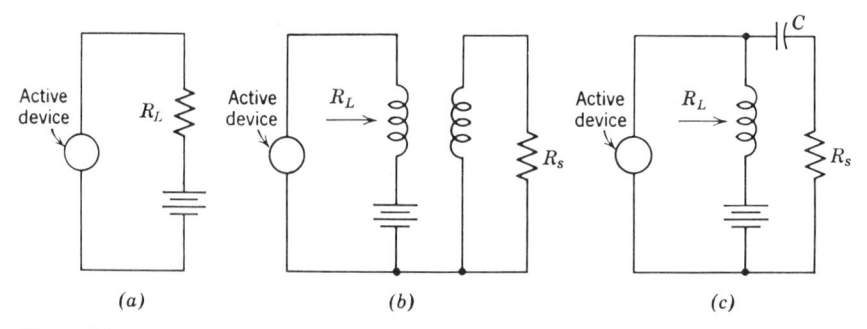

Figure 9-2
Types of load coupling in power amplifiers. (a) Direct connection. (b) Transformer coupling. (c) Shunt L-C coupling.

operation of the circuit in Fig. 9-1 is limited to Class A_1, the only current taken from the driving source is the very small amount taken by R_g. Thus, the power delivered by the driving source may be very small. However, the power delivered to the load may be many times larger, because the load current is being controlled by the active device. The load power is derived from the external power supply, rather than from the driving source. Similarly, for a transistor as the active device, base current may be very small but may control a much larger collector current in the load. In this way, power amplification occurs in the circuit; it is a *power amplifier*.

9-2
AC Load of a Power Amplifier

A power amplifier primarily controls power. It causes sufficiently large current variations in the load circuit that substantial ac power is present in the load. This output power is controlled by the signal coming from the driving source, and the power delivered from this source may be very small. The output power is generated in the load, designated R_L in the circuit of Fig. 9-1. If R_L were merely a resistor, the output power would be dissipated as heat. If, on the other hand, R_L represented the effective ac resistance of a control circuit that required ac power for its operation, or of a loudspeaker, the output power would do something useful in addition to disappearing as heat. The important idea to be gained is that R_L *represents an ac load*.

We are interested in the ac component of current in R_L, because only that component of the total current is useful in the production of output power. It is desirable, then, to minimize the resistance path for dc components of current, because the power developed in the dc paths is lost as far as useful power in the load is concerned. This may be done by employing the shunt-feed circuit (Fig. 9-2c) or by using a transformer to couple the load device to the output circuit (Fig. 9-2b).

A transistor or triode tube performs most efficiently as a power amplifier when its output-circuit impedance is a pure resistance ($Z_L = R_L$) that is roughly one-half to two times the value of the dynamic internal resistance. When the load is chosen to be within this range of values, maximum power is developed in the load and reasonably low distortion (about 5 per cent) results. The load will often be a device that has a low value of ac resistance—for example, a loudspeaker that has an 8-Ω voice coil. The 8-Ω ac load cannot be connected directly into the circuit for maximum power, because the dynamic resistance of the active device will be many times this value. The actual load must be coupled into the circuit in such a way that an *effective ac load* is present in the circuit. Either shunt coupling or transformer coupling may be used to make the load "look like" a much larger value than the actual load.

330 Power Amplifiers Ch. 9

We will now study these two concepts separately. We are interested in knowing that the dynamic resistance must be matched reasonably well to an effective load and how transformer coupling achieves this result.

It will be recalled that Class A operation of a single-stage triode amplifier yields ac plate current that may be expressed as

$$I_p = \frac{\mu E_g}{r_p + R_L}$$

Then the power developed in R_L is

$$P = I_p^2 R_L = \frac{\mu^2 E_g^2 R_L}{(r_p + R_L)^2}$$

Now let us define a *power sensitivity* as the ratio of output power and the square of input voltage

$$\text{Power sensitivity} = \frac{P}{E_g^2}$$

Then the sensitivity may be written as

$$\text{Power sensitivity} = \frac{\mu^2 R_L}{(r_p + R_L)^2}$$

$$= \frac{(\mu^2/r_p^2) R_L}{(1 + R_L/r_p)^2} = \frac{\mu g_m (R_L/r_p)}{(1 + R_L/r_p)^2} \quad (9\text{-}1)$$

Equation 9-1 may be used to determine the ratio of the load resistance to plate resistance that will permit the maximum output power to be developed. The factor in Eq. 9-1 that contains this ratio is plotted in Fig. 9-3 as a function of the ratio. It will be noted that the output power remains essentially constant (between 100 and 88 per cent) for the load varying within the range $\tfrac{1}{2} r_p$ to $2 r_p$.

Suppose it is desired to drive an 8-Ω loudspeaker with a power amplifier having resistance 2 kΩ and to operate the amplifier so that maximum power can be delivered to the 8-Ω load. What should be the properties of the transformer

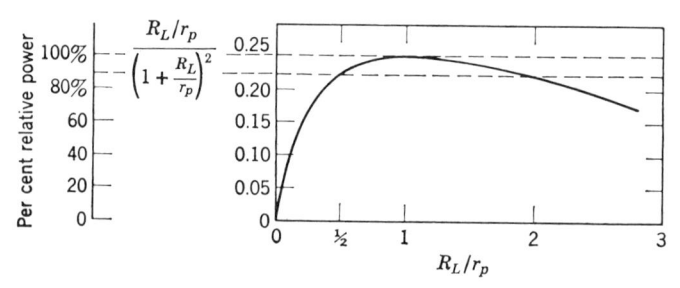

Figure 9-3
Power sensitivity related to R_L/r_p ratio.

Sec. 9-2   AC Load of a Power Amplifier   331

to be used as the impedance-matching device? Consider the output circuit in Fig. 9-4. The 8-Ω load is designated R_s and the load, R_L. If the transformer can be considered almost ideal, i.e., if the ac power supplied to the primary winding is transformed to the secondary without significant losses, then we may write

$$E_p I_p = E_s I_s$$

For an ideal transformer, the primary and secondary voltage and current values will be related by the number of turns in the primary and secondary windings as

$$\frac{n_p}{n_s} = \frac{E_p}{E_s} = \frac{I_s}{I_p}$$

The secondary load may be expressed as the ratio of secondary voltage to secondary current, and the primary load may be written as the ratio of primary voltage to primary current:

$$R_L = \frac{E_p}{I_p} \quad \text{and} \quad R_s = \frac{E_s}{I_s}$$

Then the load, R_L, may be written in terms of the transformer turns ratio and the secondary load:

$$R_L = \left(\frac{n_p}{n_s}\right)^2 R_s \tag{9-2}$$

In order to make the 8-Ω loudspeaker appear as 2 kΩ to the amplifier, the transformer turns ratio must be

$$\frac{n_p}{n_s} = \left(\frac{2000}{8}\right)^{1/2} = 15.8$$

Therefore a transformer having a turns ratio of 15.8 would be used.

Now we see the reason for using a so-called *output transformer* in the plate circuit of a vacuum tube power amplifier. For similar reasons, a transistor power amplifier may use a transformer to couple a small load into its output

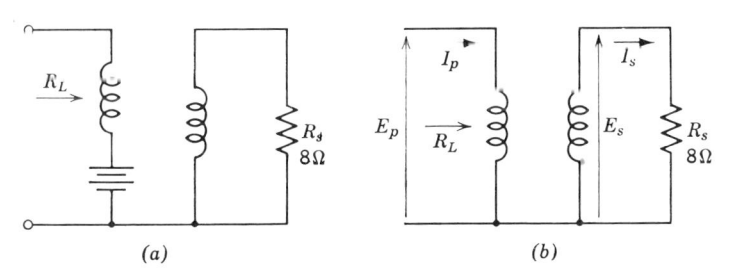

(a)   (b)

Figure 9-4
Analysis of transformer coupling. (*a*) Output circuit. (*b*) Equivalent circuit.

circuit. Coupling transformers are also used between amplifier stages operating at radio frequencies, and the primary and secondary windings are parts of *tuned circuits* to permit efficient operation over a narrow range of frequencies. The fundamental reason for using transformer coupling is to match two dissimilar impedances.

9-3
Transistor Power Amplifiers

The power that may be delivered to a load in a transistor power amplifier depends on several factors. These include: (1) maximum current ratings, (2) maximum breakdown voltage ratings, (3) permissible distortion, and (4) maximum permissible dissipation. The last factor is influenced by temperature and often requires special care in designing the amplifier circuit to avoid exceeding the maximum dissipation rating when temperature effects may become important.

It is extremely important to maintain bias current as nearly constant as possible. A shift of the Q-point may cause the collector dissipation to rise, thereby further raising the temperature and resulting in thermal runaway and eventual destruction of the transistor. The saturation current I_{co} that flows when emitter current is zero is greatly affected by temperature and can cause a shift in the Q-point if some provision is not made to counteract its effect on the dissipation.

There are various methods of stabilizing collector current against effects of temperature variations. One of them is illustrated in Fig. 9-5, where the R_E-C_E

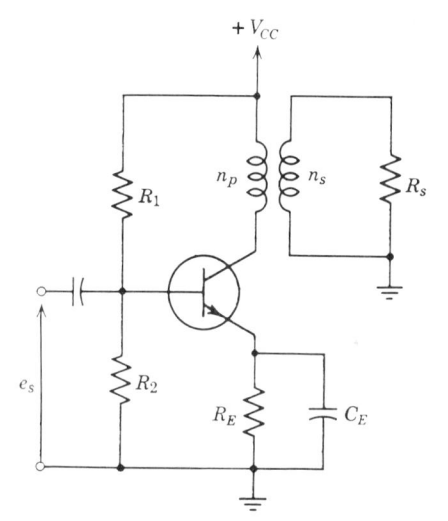

Figure 9-5
Transformer-coupled power amplifier.

Sec. 9-3   *Transistor Power Amplifiers*   333

combination in the emitter circuit reduces the base voltage and base current for increases in total collector current. The effect is to maintain the Q-point below dissipation values that may destroy the transistor. This circuit was described in some detail in the chapter on transistors.

The design of a power amplifier is concerned primarily with obtaining the required power in the load, using the smallest size and weight possible, while maintaining a suitably low level of distortion. This requires the high efficiency of Class B or C, but because of its high distortion Class C is practically limited to tuned loads, where its inherent filtering provides freedom from distortion. Classes A and B are generally used for audio and low rf applications. Class B is usually employed as a *push-pull* arrangement, requiring two transistors and a center-tapped transformer for coupling the load. Even in Class A operation, an output transformer is generally required to match the load to the transistor.

The procedures for analysis of amplifier operation developed for tube amplifiers are fully applicable to transistor amplifiers. It is more important in transistor amplifiers, however, to assure that maximum ratings are not exceeded in their operation. Transistors are more prone to destruction than are tubes. The additional circuit components needed in transistor circuits will be described in the following sections, as the various classes of operation are presented.

Example. The circuit of Fig. 9-5 uses a transistor that has a maximum $v_{CE} = 40$ V and maximum permissible collector current of 300 mA. The circuit is to be designed for maximum output power at the smallest possible input signal. Assuming a linear characteristic for the transistor, sketch the region of operation that is permitted and determine the effective ac load resistance and the Q-point. If $R_s = 8\ \Omega$, what is the required turns ratio of the coupling transformer?

Solution: The linear $i_C - v_{CE}$ characteristics are shown in Fig. 9-6. The

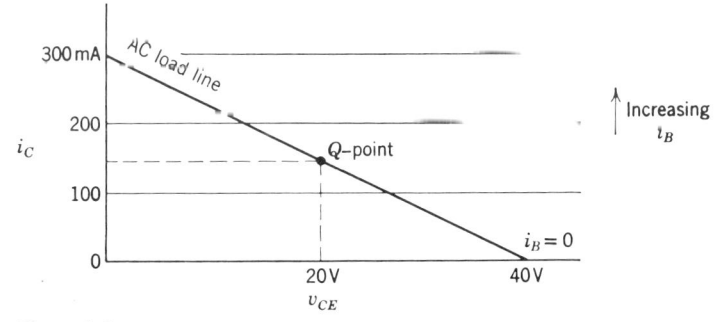

Figure 9-6
Idealized common-emitter characteristics of a transistor.

maximum voltage that can appear across the transistor is 40 V, occurring at $i_C = 0$. The maximum permissible collector current is the other end of the ac load line.

Maximum power will be developed in the ac load when the input signal permits operation from one end of the load line to the other. This requires that the Q-point be located at the midpoint of the load line, at $i_C = 150$ mA and $v_{CE} = 20$ V.

The ac load may be calculated from the slope of the load line

$$R'_L = \frac{40}{300} \times 10^3 = 133 \ \Omega$$

The required turns ratio is calculated:

$$\frac{n_p}{n_s} = \left(\frac{133}{8}\right)^{1/2} = 4.08$$

9-4
Output Power and Efficiency

From the signal variations shown in Fig. 9-6, the ac power developed across the transformer primary can be calculated to be

$$P_o = V_{ce}I_c = \frac{V_{ce\ max}I_{c\ max}}{2} = \frac{(v_{CE\ max} - v_{CE\ min})(i_{C\ max} - i_{C\ min})}{8} \quad (9\text{-}3)$$

The output power may also be computed by using the ac load resistance

$$P_o = I_c^2 R'_L \quad (9\text{-}4)$$

For numerical values in Fig. 9-6

$$P_o = \frac{(40 - 0)(300 - 0) \times 10^{-3}}{8} = 0.15 \text{ W or } 150 \text{ mW}$$

We next consider the input power from the collector bias source and the overall efficiency of the Class A amplifier. Input dc power from V_{CC} is calculated from the battery voltage and the average collector current:

$$P_i(\text{dc}) = V_{CC}I_C \quad (9\text{-}5)$$

For the circuit in Fig. 9-5 and characteristic curves of Fig. 9-6,

$$P_i = (20)(150 \times 10^{-3}) = 300 \text{ mW}$$

For a transistor-coupled power amplifier, the power dissipated by the transformer and R_E will usually be small compared with the power delivered to the ac load. If we ignore these power losses, the only lost power is that dissipated by the power transistor,

$$P_D = P_i - P_o \tag{9-6}$$

where P_D represents power dissipated as heat by the active device (here a transistor).

This equation seems simple enough, but it may obscure an important concept. The amount of power that is dissipated by the transistor is determined by the difference between the power supplied by the bias source and that taken by the load. If the actual load in the secondary of the transformer were disconnected, the transistor would still take input power but would be supplying zero output power. In this situation the transistor has to handle all of the input power (dissipate the most). Conversely, when the output power is greatest, the transistor dissipates the least and will operate at a lower temperature. Obviously, the safest dissipation rating for the transistor being used is the maximum input power. Since normal operation with the load connected is the condition for the least transistor dissipation, it is good practice to avoid disconnecting the load when the amplifier is in operation.

Power conversion efficiency is expressed as the ratio of output power and input power, expressed in per cent:

$$\text{Efficiency}, \eta = \frac{P_o}{P_i} \times 100\% \tag{9-7}$$

For the numerical values already calculated, the efficiency of this amplifier is $(150/300) \times 100\% = 50\%$. A practical amplifier would have power losses in the transformer, in emitter resistors, and in other components that may be connected in the output circuit. These losses would decrease the available power and thus decrease the efficiency of the amplifier. Further reductions in ac power would occur in a practical circuit because of the saturation and cutoff regions of the transistor characteristics. These regions exhibit nonlinear reproduction of input signals, which reduces the effective ac power at the frequency of the input signals. The value of 50 per cent efficiency is the maximum theoretical value for the transformer-coupled power amplifier. An actual circuit would have a power conversion efficiency smaller than 50 per cent; how much smaller would depend upon the amplitudes of signal swings along the ac load line and the presence of distortion normally expected from large excursions.

Drill Problem

D9-1 A transistor Class A amplifier supplies power to a transformer-coupled load. The load requires 1 W of power. The primary winding resistance is 1/10 as large as the ac load resistance in the collector circuit. (*a*) How much ac power must be developed by the transistor? (*b*) How much power is lost in the winding resistance?

9-5
Class B Push-Pull Operation

When two transistors of the same type are connected in an amplifier circuit as shown in Fig. 9-7a, they are said to be connected in *push-pull*. This alludes to the alternate operation of the transistors during successive half-cycles of the input signal. Each transistor supplies half of the ac power to the load. They are connected in the circuit so the input signal drives their collectors into

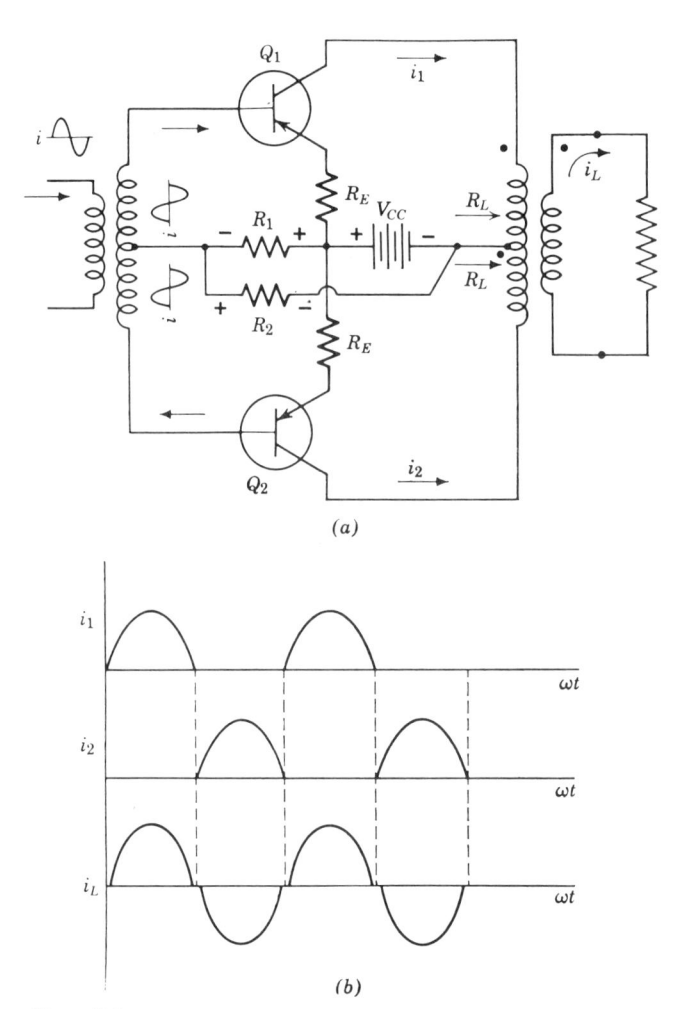

Figure 9-7
Transistor push-pull amplifier. (a) Schematic diagram. (b) Crossover distortion in load current without R_1 and R_2.

Sec. 9-5 Class B Push-Pull Operation 337

conduction during alternate half-cycles. Such operation is that already described as Class B. The load passes the two secondary currents out of phase.

The input transformer is usually the collector circuit of a Class A *driver stage*. The actual load is connected to the transistors through the center-tapped output transformer. A voltage-divider network, R_1 and R_2, which is shunting V_{CC} is used to eliminate *crossover distortion* that occurs because of the forward-biased emitter voltage at zero base current. The resistors R_E are temperature-compensating components for increased stability of the circuit.

For each input half-cycle to be equally amplified in the load, the input transformer and the circuit must be balanced. If the two transistors are equal in their operation, there will be no net average magnetic flux in the core of the output transformer. Thus, a smaller transformer may be used than for a comparable Class A amplifier; this is an advantage of the Class B push-pull operation of transistor power amplifiers.

The input signals cannot be applied to the power amplifier when their magnitudes are smaller than the potential difference across the emitter-base junction. During the small period of time when the input signals are going through zero, discontinuities in load current are shown in Fig. 9-7b as crossover distortion in the load current. The resistor R_1 is used to apply a potential in the base circuit that just equals the forward-biased junction potential. Thus, its effect is to eliminate the distortion.

Design Example. A push-pull Class B power amplifier is to be designed to deliver 10 W to a load of 8-Ω resistance. An output transformer is available that has 10-Ω primary resistance and 2-Ω secondary resistance. Its required turns ratio will be determined in the design.

The amplifier is to employ transistors that have the following ratings and specifications:

$$i_{C\,max} = 2\text{ A}$$
$$P_{C\,max} = 20\text{ W}$$
$$h_{fe} = 50 \text{ (assumed linear)}$$
$$BV_{CEO} = 65\text{ V}$$
$$\text{Emitter-base voltage (at } i_B = 0) = 0.6\text{ V}$$

We will use the circuit of Fig. 9-7 and assume the input signal is a signal-frequency sine wave. Each half of the output primary winding delivers the same power to the load and dissipates the same power internally. Similarly, each transistor supplies half of the total power delivered to the secondary winding and the load. Some of the available ac power will be dissipated in the resistance of the secondary winding.

338 Power Amplifiers Ch. 9

For the assumed linear h_{fe}, each transistor will deliver power to the transformer primary given by

$$P_o = \frac{I_{c\,max} V_{ce\,max}}{2} \quad \text{during alternate half cycles} \tag{1}$$

The power supplied will divide between the primary winding resistance and the reflected load:

$$P_o = \frac{I_{c\,max}^2}{2}\left(\frac{R_p}{2}\right) + \frac{I_{c\,max}^2}{2} R_L' \tag{2}$$

The power delivered to the secondary is given by the second term in this expression, and the power delivered to the load will be less by an amount of dissipation in the secondary resistance:

$$P_L = \frac{I_{c\,max}^2}{2} R_L' - \frac{I_{s\,max}^2}{2} R_s$$

$$= \frac{I_{s\,max}^2}{2}(R_L) \tag{3}$$

To meet the specifications of the design and deliver 10 W to R_L, the secondary must be supplied with the power

$$P_{s\,min} = 10 + \tfrac{2}{8}(10) = 12.5 \text{ W}$$

The maximum value of $V_{ce\,max}$ must be less than $\tfrac{1}{2} BV_{CEO}$, or less than 32.5 V. From Eqs. 1 and 2, the power supplied to the secondary is

$$P_{sec} = \frac{I_{c\,max} V_{ce\,max}}{2} - \frac{I_{c\,max}^2}{4} R_p \tag{4}$$

Equation 4 is graphed in Fig. 9-8 for the known values of R_p and $V_{ce\,max}$. Note that two values of $I_{c\,max}$ will yield the power needed in the secondary. The two values may be calculated from Eq. 4:

$$\frac{I_{c\,max}^2}{4} R_p - \frac{I_{c\,max} V_{ce\,max}}{2} + P_{sec} = 0$$

or

$$I_{c\,max} = \frac{\tfrac{1}{2} V_{ce\,max} \pm (\tfrac{1}{4} V_{ce\,max}^2 - R_p P_{sec})^{1/2}}{\tfrac{1}{2} R_p} \tag{5}$$

Two values of $I_{c\,max}$ result, each of which will yield the required P_{sec}. However, only one of them is likely to meet the transistor specification of $i_{c\,max} = 2$ A.

Inspection of Eq. 5 shows that some minimum collector voltage must be used, or the current expression will become a complex number, i.e.,

$$\frac{V_{ce\,max}^2}{4} \geq R_p P_{sec}$$

Class B Push-Pull Operation

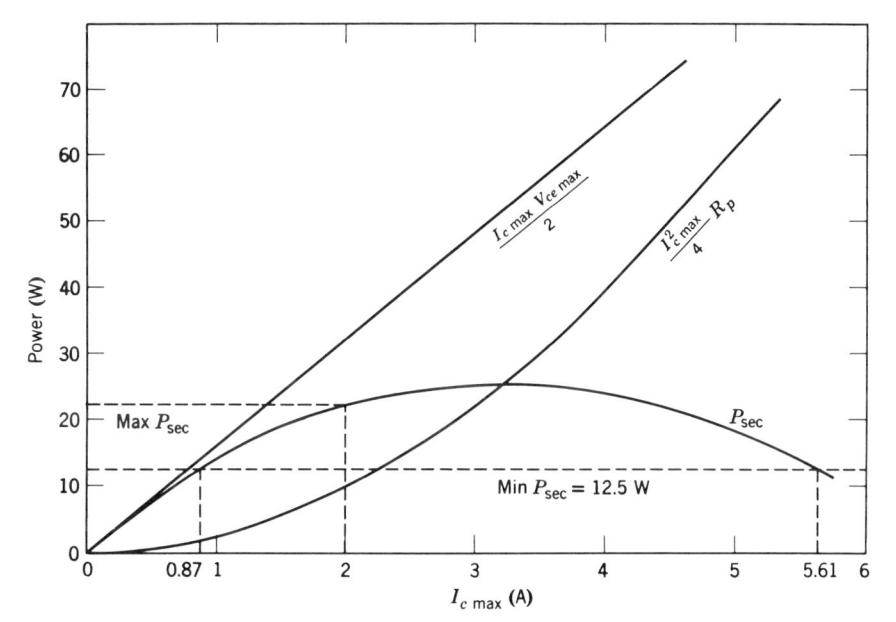

Figure 9-8
Plot of Eq. 4 in Design Example.

or the minimum voltage is

$$V_{ce\ max} \geq 2\sqrt{R_p P_{sec}} \qquad (6)$$

Substituting known values into Eq. 6 yields

$$\min V_{ce\ max} = 2\sqrt{10(12.5)} = 22.4\ V$$

which is less than the 32.5 V calculated as the maximum possible value. Any value of collector supply V_{CC} that is between these extremes may be used in this design.

The necessary turns ratio of the transformer can be calculated from the ratio of primary and secondary power requirements and from the ratio of load values in the primary and secondary windings:

$$\frac{P_{p\ max}}{P_{s\ max}} = \frac{I_{c\ max}^2 R_L'}{I_{s\ max}^2 (R_L + R_s)} = \left(\frac{n_s}{n_p}\right)^2 \left(\frac{R_L'}{R_L + R_s}\right) = 1$$

if the transformer is considered to have no losses except those in the resistive components. Thus,

$$\frac{n_p}{n_s} = \left(\frac{R_L'}{R_L + R_s}\right)^{1/2} \qquad (7)$$

Since

$$V_{ce\,max} = I_{c\,max}\left(R'_L + \frac{R_p}{2}\right) \quad (8)$$

the value of reflected load resistance can be found to calculate the necessary turns ratio, when the voltage and current values have been chosen in the design.

To complete the design, we calculate the two values of $I_{c\,max}$ in the figure; using $V_{c\,max} = 32.5$ V

$$I_{c\,max} = \frac{16.25 \pm \sqrt{(16.25)^2 - 10(12.5)}}{5}$$

$$= \frac{16.25 \pm 11.8}{5} = 5.61 \quad \text{or} \quad 0.87 \text{ A}$$

The value 0.87 A is well within the limiting value of 2 A set by the specifications.

Equation 8 and these limiting values of $I_{c\,max}$ may be used to calculate minimum and maximum values of R'_L:

$$R'_{L\,min} = \frac{32.5}{2} - 5 = 11.25 \, \Omega$$

$$R'_{L\,max} = \frac{32.5}{0.87} - 5 = 32.4 \, \Omega$$

Choosing a value of collector current at 1.2 A, the value of R'_L is

$$R'_L = \frac{32.5}{1.2} - 5 = 22.1 \, \Omega$$

From Eq. 4, the power supplied to the secondary will be

$$P_{sec} = \frac{(1.2)(32.5)}{2} - \frac{(1.2)^2(10)}{4} = 15.9 \text{ W}$$

Since only 12.5 W is required, this will provide more than the design requirements but may be used. The turns ratio is

$$\frac{n_p}{n_s} = \left(\frac{22.1}{8+2}\right)^{1/2} = 1.49$$

so that the power will be transformed to the load. A slightly smaller turns ratio could be used with the other calculated values, since the power available will be greater than the amount needed.

The transistor must be capable of dissipating 20 W of power. Since the load is taking 15.9 W and maximum collector current is 1.2 A, we may calculate the average power dissipated by the transistor. The average power over one cycle

Sec. 9-6   Push-Pull Power Amplifier Using Tubes

of input signal is

$$P_C = \frac{1}{2}\left[V_{CC}\frac{2i_{c\,max}}{\pi} - \frac{I_{c\,max}^2(R'_L + R_p/2)}{2}\right]$$

The first term must be less than 20 W:

$$\text{Power} = \left(\frac{1}{2}\right)(32.5)\frac{(2)(1.2)}{\pi}$$

$$= 12.4 \text{ W}$$

which is well within the specified maximum collector dissipation.

The R_1-R_2 network will be calculated to eliminate crossover distortion. The dc current in R_1 must be larger than the largest signal current in the base circuit. Since h_{fe} is given as 50, and $I_{c\,max}$ is 1.2 A, the base current will have a maximum value of

$$i_{b\,max} = \frac{1.2}{50} = 24 \text{ mA}$$

Then R_1 must be

$$R_1 \leq \frac{0.6 \text{ V}}{0.24 \text{ A}} = 2.5 \, \Omega \quad \text{to permit a dc current equal to } 10 \times i_{b\,max}$$

Choosing its value at 2.5 Ω, then R_2 will be

$$R_2 = \frac{(32.5 - 0.6)\text{V}}{0.24 \text{ A}} = 132.9 \, \Omega$$

Transformers designed for use as output devices for power amplifiers are usually rated for the load resistance and its corresponding *total* primary load resistance, i.e., the transformer used in the design example would be rated as 88.4 to 8 Ω. In terms of its turns ratio,

$$88.4 \, \Omega = \left(\frac{2n_p}{n_s}\right)^2 (8)\Omega$$

Drill Problem

D9-2   A certain output transformer is marked "400 Ω C.T. to 3.2 Ω." When used in a push-pull circuit, what will be the reflected load?

9-6
Push-Pull Power Amplifier Using Tubes

When two tubes of the same type are connected in an amplifier circuit as shown in Fig. 9-9b, they are said to be connected in push-pull. Their alternating

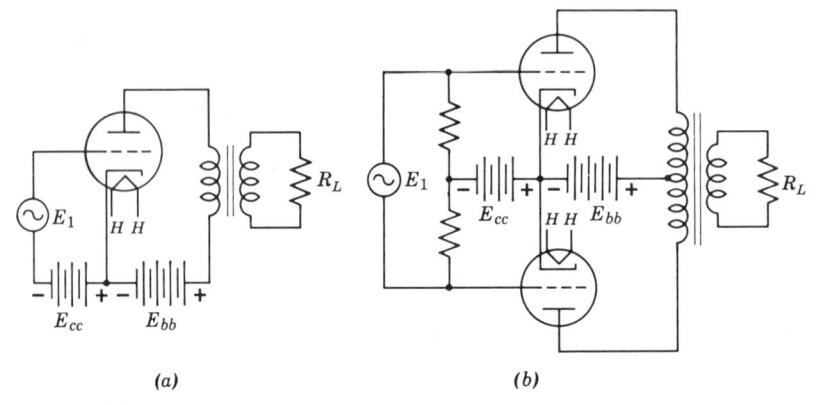

Figure 9-9
Power-amplifier circuits. (a) Single-ended operation. (b) Push-pull operation.

components of plate current flow in the same direction in the center-tapped primary winding of the transformer, *both either up or down.*

When the instantaneous plate current of one tube is at its maximum, the other tube has minimum plate current. When the plate current in the first tube is rising (increasing in value), the plate current in the second tube is falling (decreasing in value). The ac component of current of tube 1 is flowing into its plate while the ac component of current of tube 2 is flowing out of its plate. Thus *both tubes contribute current in the same direction in the primary winding.* Both tubes carry their own dc components of plate current, but these are equal in magnitude and flow in opposite directions through the output-transformer primary. The effect of this is the same as if there were no direct current in the transformer at all. This is a big advantage, as will be seen shortly.

Figure 9-10 shows the relation between the number of magnetic lines of flux in the core of a transformer and the current in its primary winding. With single-ended operation, as shown in Fig. 9-10a, there is a comparatively large dc component of current (I_{bo}), which magnetizes the core in the amount ϕ_o. With no signal voltage on the grid, the flux is constant at this value. When signal voltage is applied, the plate current alternatively rises above and falls below the I_{bo} value. The corresponding variations in flux above and below the ϕ_o value induce voltage in the secondary winding of the transformer and cause secondary current to flow in the load.

In push-pull action the absence of a net effective direct current in the primary winding makes it possible to use an ac component of current that is larger than twice the amount obtained with a single tube. This advantage, coupled with the automatic elimination of second harmonics (all even-numbered harmonics, in fact) of plate current, results in the realization of more than twice the power output obtainable from single-ended operation. Notice in Fig. 9-10b that the

Sec. 9-6   Push-Pull Power Amplifier Using Tubes   343

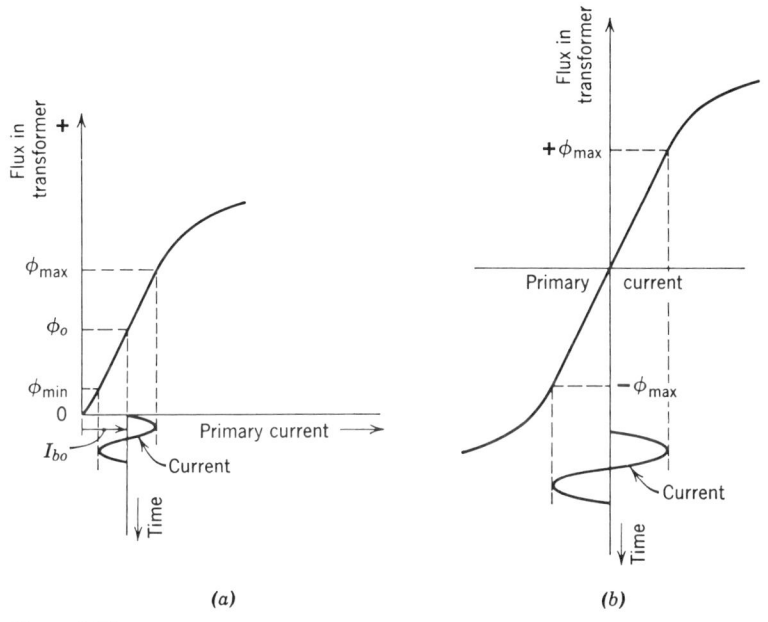

Figure 9-10
Flux-current relations in output transformers of power amplifiers. (a) Single-ended operation. (b) Push-pull operation. There is no net dc component of current in the transformer.

effective primary current, which is alternating current, is larger than twice the ac component of primary current in Fig. 9-10a. A smaller output transformer may be used than would be needed for even a single-ended amplifier employing the same type of tube because *there is no dc current in the windings to produce a large steady component of flux* in the transformer.

When the circuit is properly balanced, there is no signal-frequency component of current in the plate power supply. On the other hand, if there are any low-frequency components of current in the power supply, which are due to inadequate filtering, they will cancel out in the primary halves of the transformer and thus cause no objectionable effects in the output.

Cathode bias may be used; it is then unnecessary to use a bypass capacitor across the cathode resistor if the tubes are matched.

The automatic elimination of all even-order harmonics of plate current (2nd, 4th, 6th, and so forth) permits the operation of push-pull power amplifiers in Classes AB_1, AB_2, B_1, B_2 and C. In public-address systems, AB_1 and AB_2 are used extensively. Class B is used where larger power output is desired than can be obtained with AB_2, but in both of these types of operation the driving unit (the amplifier or other device that feeds the grids) must be capable of supplying the required grid-circuit input power without distortion. Class C operation is

344 Power Amplifiers Ch. 9

used when only one frequency, or a narrow band of frequencies, must be amplified. Class B and Class C amplifiers are used in the high-frequency circuits of radio and television transmitters.

The manner in which the plate currents of the tubes in push-pull AB_1 operation are produced and combined to form a large effective primary current in the output transformer is shown in Fig. 9-11. Since the grid of one of the tubes is being driven more negative while the other is being driven more positive, it has been found convenient by many authors to place the transfer-characteristic charts back to back at the common bias voltage ($E_c = E_{c1} = E_{c2}$).

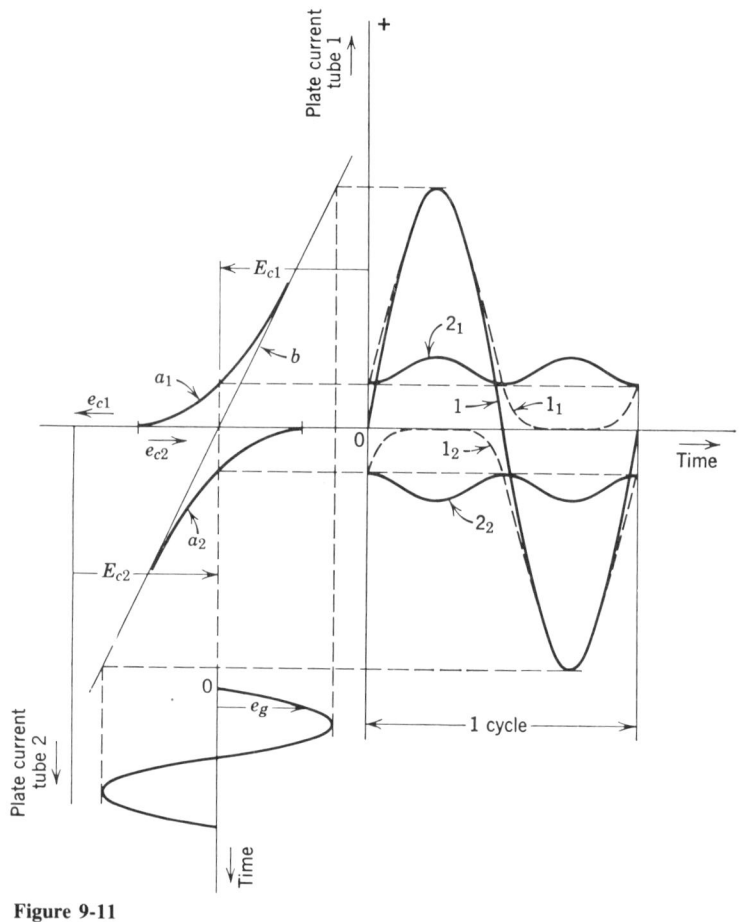

Figure 9-11
Graphical construction of plate-current waveforms push-pull amplifier, Class AB_1 operation. 1_1 is current in tube 1; 2_1 is second harmonic in 1_1; 1_2 is current in tube 2; 2_2 is second harmonic in 1_2; 1 is ac current in transformer primary.

This makes it convenient to draw the *composite dynamic transfer characteristic* (*b*), which is midway between the curves for the individual tubes, except when either tube's current is zero (in which case the composite characteristic coincides with the curve for the tube that is carrying current). Observe that the second-harmonic components, which flow simultaneously, are in phase opposition to each other and hence cancel in the transformer primary. That is also true of the other even harmonics.

There is reinforcement of the third-harmonic component effects, however; also of the fifth and other odd-numbered harmonics. Fortunately, the magnitude of even the third harmonic in each tube is appreciably less than that of the second, and the harmonic magnitudes higher than the third diminish rapidly. Nevertheless, the odd-harmonic distortion is a limiting factor in power output of push-pull stages. The symmetrical nature of the combined dynamic transfer characteristic is the underlying cause of the absence of even-numbered harmonics and the presence of odd-numbered harmonics.

A composite load line, drawn on a pair of plate-characteristic curve charts placed back to back, is used to analyze the performance of push-pull amplifiers. As an example we have chosen two type 6L6 tubes operating Class AB_1 at a plate-supply voltage of 350 V and with 22.5 V fixed bias. We did not choose 360 V because the graphical representation is somewhat clearer with 350 V.

Figure 9-12 shows two plate-characteristic charts placed back to back (or more precisely, bottom to bottom) so that the 350-V points on the plate-voltage axes coincide. This causes the other major divisions on the voltage scales to line up, and this is advantageous for studying the process. This would not occur if the operation were at 360 V on the plates, for then the light vertical lines on the graph sheet would not line up. Nevertheless, operation at 360 V is recommended; the analysis for that condition would require lining up the charts at the 360-V points.

The load line for the two tubes in push-pull operation, called the composite load line, is drawn so that its slope (tangent θ) is $+1/R_L$ and is equal to one-fourth of the effective load resistance. The effective load resistance, called the plate-to-plate resistance in the tube manual, is specified as 6600 Ω for Class AB_1 operation at highest possible output power using fixed bias. The load line is then drawn so that the acute angle θ with the voltage axis has a tangent value obtained by using

$$6600/4 = 1650 \ \Omega = R'_L$$

$$\tan \theta = \frac{1}{R'_L} = \frac{1}{1650} = \frac{\Delta i_b}{\Delta e_b} = \frac{\Delta i_b}{350}$$

$$\Delta i_b = \frac{350}{1650} = 0.212 \text{ A} = 212 \text{ mA}$$

The composite load line is therefore drawn through points $e_b = 350$, $i_b = 0$ and $e_b = 0$, $i_b = 212$ mA, extending over both charts.

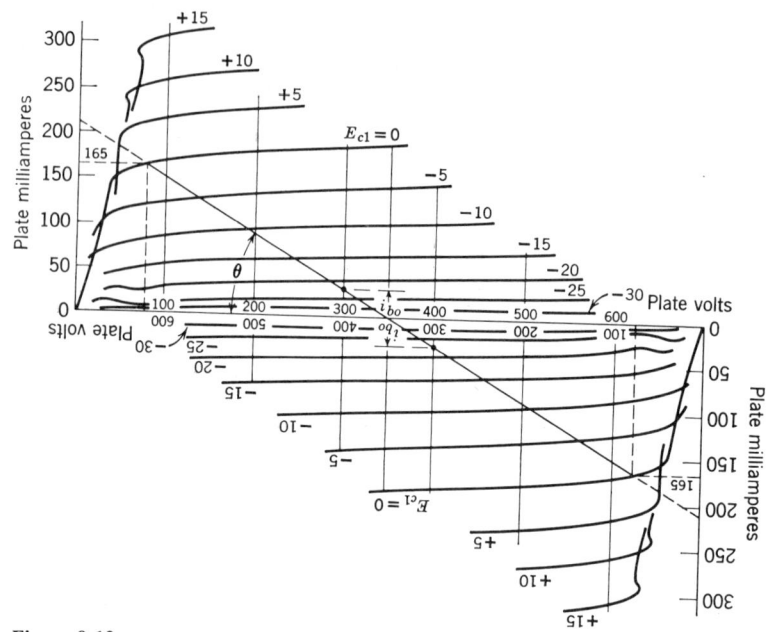

Figure 9-12
Push-pull power-amplifier analysis; two 6L6 tubes. $E_b = 350$ V; $E_c = -22.5$ V fixed bias; $E_{gm} = 22.5$ V.

When the grid-driving voltage is zero, the voltage on each plate is 350 V and each tube carries 30 mA (this is the reading at 22.5 V bias, and, although not exact, it is quite accurate). When a driving voltage of 22.5 V peak value is applied, one grid is driven to zero volts with respect to its cathode, and at the same time the other is driven to −45 V with respect to its cathode. The curves in the tube manual show that the plate current is practically zero when the grid voltage is driven below −35 V. This means that the current of each tube will be cut off during that part of the grid-voltage cycle in which the grid voltage is dropping from about −35 V to its lowest value (−45 V), and while it is coming up again to −35 V to start current flow.

The plate current in each tube rises to a peak value of 165 mA once each cycle. The plate voltage drops to about 77 V at the same time. With sine-wave grid excitation the ac components of plate current and plate voltage are nearly exact sine waves in this operation (total harmonic distortion is listed at only 2 per cent in the tube manual). We may therefore calculate the approximate power output on the basis of pure sine waves of current and voltage, and use their effective values.

$$P_o = E_p I_p = \left(\frac{350 - 77}{\sqrt{2}}\right)\left(\frac{0.165}{\sqrt{2}}\right) = 22.5 \text{ W}$$

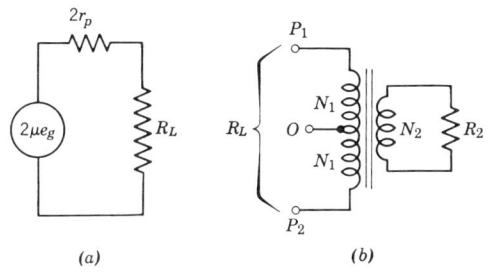

Figure 9-13
(a) Equivalent circuit for push-pull amplifier. (b) Output transformer and load, push-pull operation.

The push-pull circuit may be represented by the equivalent circuit of Fig. 9-13a. R_L is the load resistance after its reflection into the primary circuit of the output transformer. It is the plate-to-plate load resistance and is called R_{pp}. The transformer puts R_L into the circuit. The transformer turns ratio must be selected so that a given actual load (R_2) on its secondary will be reflected into the primary as the desired (R_{pp}) value. In Fig. 9-13,

$$R_L = \left(\frac{2N_1}{N_2}\right)^2 R_2$$

and this must be equal to R_{pp}, the plate-to-plate resistance of the push-pull tubes, for proper operation. Looking into only half the primary (between P_1 and O), a single tube "sees" $(N_1/N_2)^2 R_2$, which is only one-fourth of R_L (i.e., one-fourth of R_{pp}). Therefore the load line for push-pull operation (the composite load line) is drawn so that its slope is

$$-\frac{\Delta i_b}{\Delta e_b} = \frac{1}{\frac{1}{4}R_{pp}} = \frac{1}{R'_L}$$

In the discussion above, R_{pp} was 6600 Ω, so $R'_L = 6600/4 = 1650$ Ω was used to draw the composite load line.

9-7
Class B Push-Pull Amplification with Tubes

Operation of a push-pull amplifier in Class B is very similar to that of Class A and AB push-pull operation, except that the tubes are biased to cut off at the Q-point. The plate current in each tube has a large swing; as a result, the tube has more power output than in Class A. The efficiency is also improved. Figure 9-14 shows instantaneous grid-driving voltage and plate-current relations. Since the individual grids are driven alternately positive and grid current must flow, the driving amplifier must be capable of supplying the necessary power without

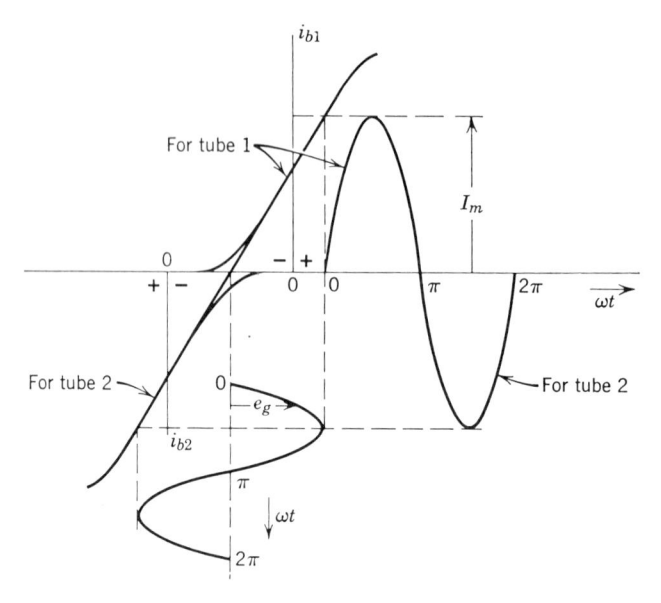

Figure 9-14
Class B push-pull action.

distorting the signal voltage on the grids. The power tubes are usually designed with dynamic characteristics that are linear well within the positive-grid region.

Theoretical analysis of Class B performance is possible if the *composite dynamic characteristic is assumed to be linear*; this will make the current waveform sinusoidal for sinusoidal input voltage. Assuming that the plate current of each tube is a half sine wave, the equivalent dc current *per tube* is

$$I_{dc} = \frac{I_m}{\pi} \qquad (9\text{-}8)$$

and the *total* dc plate current is twice this value. The effective ac load current is then

$$I_{rms} = \frac{I_m}{\sqrt{2}} \text{ A} \qquad (9\text{-}9)$$

The output power is

$$P_o = \left(\frac{I_m}{\sqrt{2}}\right)^2 R'_L = \frac{I_m^2 R'_L}{2} \text{ W} \qquad (9\text{-}10)$$

in which R'_L is the *load resistance on each tube*, which is one-fourth the plate-to-plate value. Using Eq. 9-8, the dc input power to both tubes is

$$P_{dc} = \frac{2}{\pi} I_m E_{bb} \qquad (9\text{-}11)$$

The ratio of output power to input power gives the plate-circuit efficiency:

$$\eta_p = \frac{I_m^2 R_L'/2}{2 I_m E_{bb}/\pi} = \frac{\pi}{4}\left(\frac{I_m R_L'}{E_{bb}}\right) \tag{9-12}$$

$I_m R_L'$ is the amplitude of the ac component of plate voltage and is equal to $E_{bb} - E_{b\,min}$ as always; therefore,

$$\eta_p = \frac{\pi}{4}\left(\frac{E_{bb} - E_{b\,min}}{E_{bb}}\right) \tag{9-13}$$

Theoretically, $E_{b\,min}$ should be made zero; then the maximum theoretical plate-circuit efficiency of a Class B push-pull amplifier is

$$\text{Maximum theoretical } \eta_p = \frac{\pi}{4} = 0.785 = 78.5\% \tag{9-14}$$

This assumes that plate voltage may vary from E_{bb} to zero along the load line. For vacuum tubes, this variation is unattainable; it may be approached in transistor circuits. The low saturation voltage of many transistors permits an efficiency very close to this theoretical maximum.

The plate dissipation is equal to the input power minus the output power. Since the input power ($E_{bb} I_{dc}$) is expressed in terms of the dc current, Eq. 9-8 may be used to express output power in terms of I_{dc}. The *total dc plate current* is $2I_m/\pi$, and I_m equals $\pi I_{dc}/2$, in which I_{dc} is the dc current *per tube*. The output power is

$$P_o = \left(\frac{\pi I_{dc}}{2}\right)^2 \frac{R_L'}{2} = \frac{\pi^2}{8} I_{dc}^2 R_L' \tag{9-15}$$

The plate dissipation is then

$$P_p = P_i - P_o = E_{bb} I_{dc} - \frac{\pi^2}{8} R_L' I_{dc}^2 \tag{9-16}$$

9-8
Complementary Symmetry

The transformers supplying, simultaneously, grid voltages of opposite polarities to tubes in push-pull and base currents of opposite direction of flow to transistors in push-pull perform *phase inversion* inherently. Note that although there is a single sine-wave current in the transformer primary of Fig. 9-7, there are two secondary currents. We may call the current in the upper half of the secondary a positive-polarity current flowing into the base of Q_1 and the current in the lower half a negative polarity current flowing into the base of Q_2.

By complementary symmetry is meant a principle of assembling a Class B transistor amplifier without requiring the center-tapped phase-inverting input

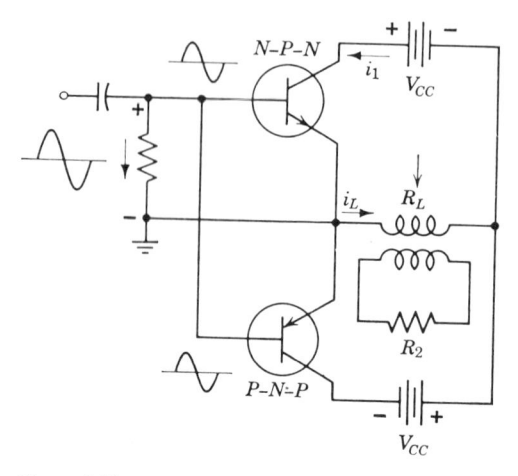

Figure 9-15
Transistor push-pull amplifier using complementary symmetry. Current arrows show direction of flow during positive half-cycle of input current.

transformers at the input and output of the stage. Fortunately, N-P-N and P-N-P transistors require opposite polarized bias voltages; the N-P-N common-emitter transistor must have a *positive* base-current drive, while the grounded-emitter P-N-P requires a *negative* base-current drive. If two such transistors having similar characteristics are connected back-to-back and biased at cutoff in a push-pull circuit, the N-P-N will amplify the current in the positive half-cycles, and the P-N-P in the negative, thus obviously eliminating phase inversion. This arrangement is shown in Fig. 9-15.

9-9
Transistor Heat Sinks

Individual power transistors are rated for maximum power dissipation from the junctions, particularly the base-collector junction. The temperature at which the junction may operate in a satisfactory manner also depends upon whether the transistor is constructed of germanium or silicon. Silicon units can be operated at higher temperatures than germanium units, for reasons discussed in Chapter 2. Typical maximum temperatures for these types of transistor are 100°C for germanium and 200°C for silicon.

The maximum power that may be handled by a particular transistor is related to its junction temperature. The power dissipation by the device will cause an increase in its operating temperature. If some means for controlling the temperature increase can be provided, the transistor can be operated closer to its maximum power rating. Such means often take the form of *heat sinks*, pictured in Fig. 9-16. These heat radiators are usually constructed of copper or

Sec. 9-9 Transistor Heat Sinks 351

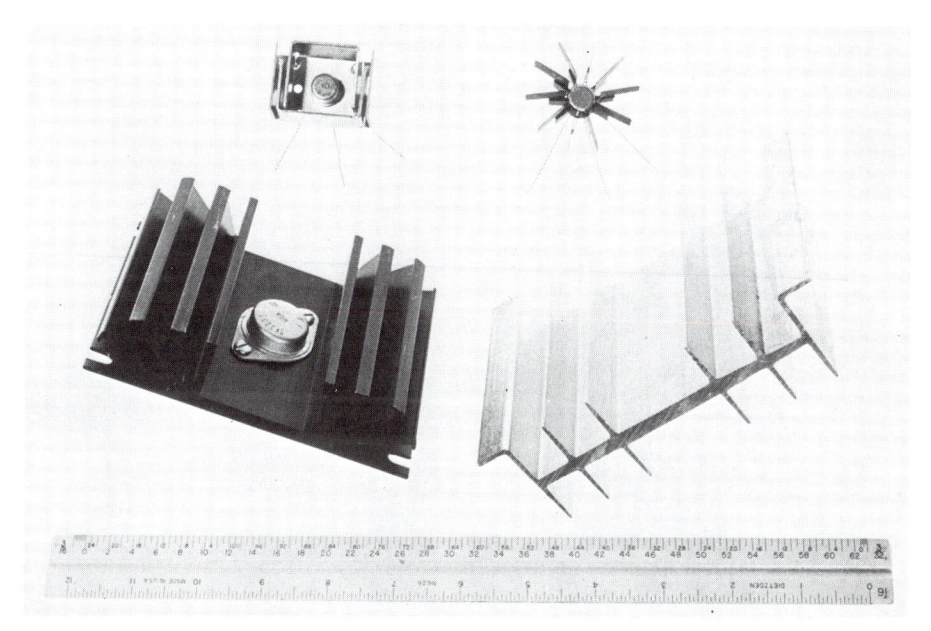

Figure 9-16
Typical heat sinks for transistors. Note their sizes relative to the size of the transistors attached to them.

aluminum. Their shape provides a large surface area to radiate heat from the transistor junctions to the surrounding air, which keeps the junction temperature from rising as much as it would without additional heat removal.

The average power dissipated is approximated as

$$P_C = V_{CE}I_C$$

A particular power transistor will be rated for some numerical value of average collector dissipation; for example, 10 W maximum. This amount of power can be safely dissipated for junction temperatures up to about 25°C for germanium and 75°C for silicon. Figure 9-17 shows the typical derating curves for power transistors that operate at junction temperatures above these values. For example, the curve for germanium is a graphical plot of the manufacturer's statement, "derate linearly to 100°C case temperature, 1 W/°C above 25°C."

The silicon transistor whose power dissipation curve is shown can be operated at its rated maximum power up to 50°C. The germanium unit, on the other hand, cannot be safely operated above two-thirds of its maximum ratings at this temperature. At higher temperatures, both types of transistor can handle power levels that are smaller than rated values. If power transistors are mounted on heat-dissipating surfaces, such as the described heat sinks, they may operate at lower temperatures and consequently at higher power levels.

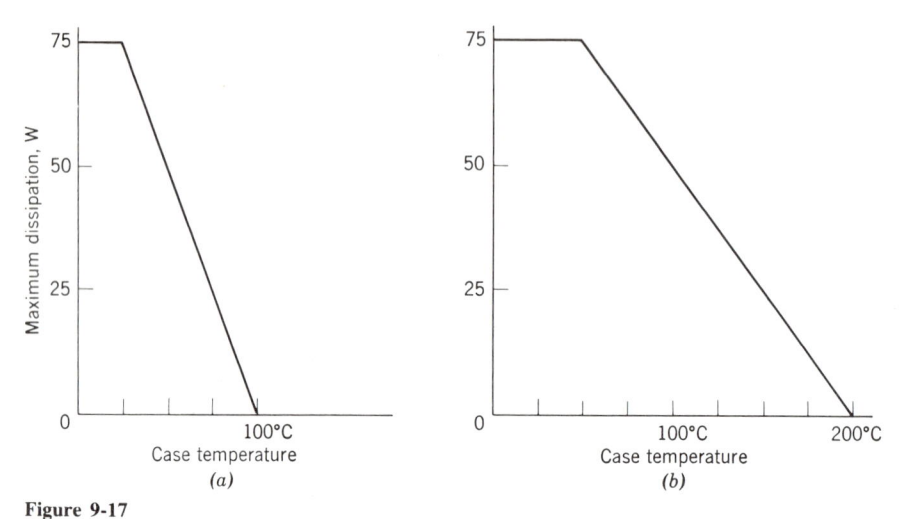

Figure 9-17
Typical power derating curves for transistors having 75 W maximum dissipation rating. (a) Germanium. (b) Silicon.

Questions

9-1 What is the basic difference between a voltage amplifier and a power amplifier?

9-2 Explain the action of an output transformer in making the actual load appear to be many times larger.

9-3 Discuss the use of an amplifier and output transformer to drive a loudspeaker, and explain how to determine the required turns ratio.

9-4 Explain how to locate by trial construction a load line that will give maximum power output from an amplifier without excessive distortion.

9-5 Describe second-harmonic distortion, and discuss how it arises in the operation of a power amplifier.

9-6 Discuss current flow in the primary winding of an output transformer used in a push-pull amplifier.

9-7 Compare the output power obtainable from a single-ended amplifier and from a push-pull arrangement.

9-8 What are the advantages of Class B push-pull operation over Class A? How do their efficiencies compare?

9-9 What causes transistors to increase their temperature during operation? Why should the load be connected when operating a push-pull circuit with a transformer?

Problems

9-1 Figure P9-1 shows common-base characteristics of an N-P-N power transistor. Choose a Q-point at $V_{CB} = 40$ V, $I_E = -25$ mA, and determine the optimum load resistance for maximum power output by constructing trial load lines, not allowing the collector current to go below 2 mA. Calculate the power output, using your optimum load.

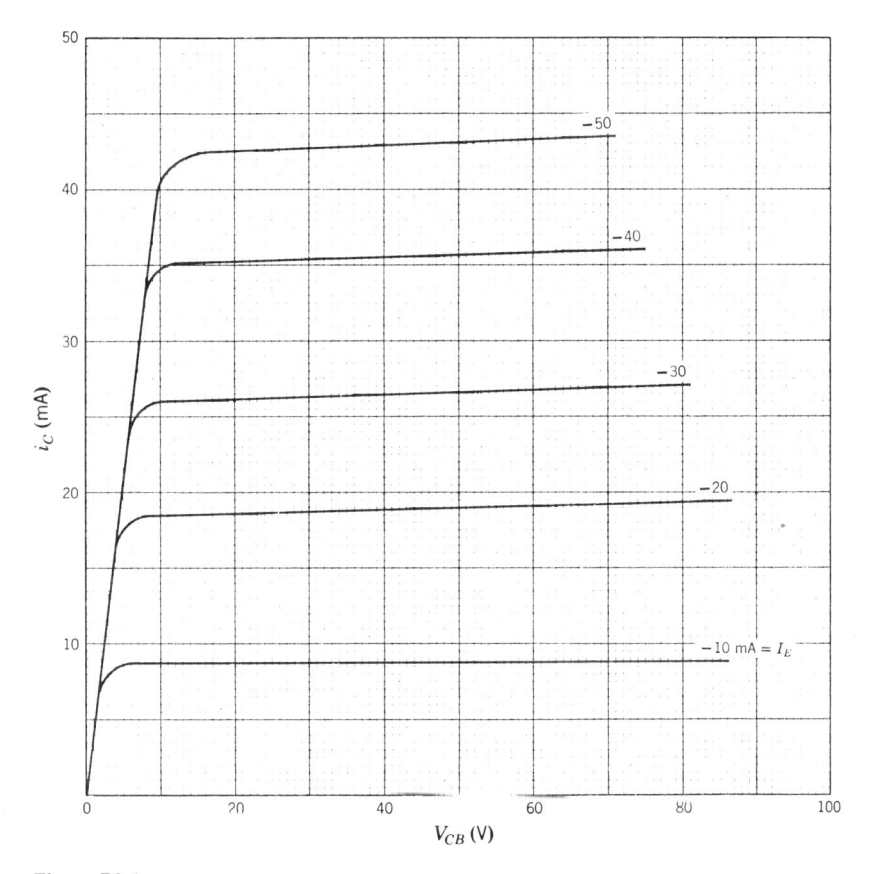

[Graph: i_C (mA) vs V_{CB} (V), with curves labeled -50, -40, -30, -20, -10 mA $= I_E$]

Figure P9-1

9-2 A 6L6 tube is used with a transformer-coupled load in a single-ended power amplifier with a fixed bias of -10 V and $E_{bo} = 200$ V. Find by trial the ac load resistance for maximum power output with not over 5 per cent second-harmonic distortion. The plate current should never go below 10 mA. How much would the plate dissipation be?

354   *Power Amplifiers*   Ch. 9

9-3 For the load resistance found in Problem 9-2, determine the transformer turns ratio needed to match a speaker voice coil having 8-Ω resistance. Calculate the maximum power and rms current that would be delivered to the voice coil, assuming no loss in the transformer. Suppose the resistance of the transformer primary is 100 Ω and the secondary 8 Ω. How much ac power is lost in the windings?

9-4 Two 6F6 pentodes are operated in Class B push-pull. E_{bb} = 350 V, E_{cc} = −38 V, e_g = 38 V peak. The plate-to-plate resistance is 6000 Ω. (*a*) Draw the composite load line. (*b*) Calculate the output power and plate-circuit efficiency.

9-5 A 6L6 tube in the second stage of an amplifier is to deliver 7.5 W to a 10-Ω load. An *R*-*C* coupled voltage amplifier using a 6J5 drives the 6L6 tube at frequencies between 160 and 5100 Hz. Cathode bias is to be used on both tubes, and the plate-supply voltage is 250 V. Draw a complete circuit diagram. Design the complete circuit for good efficiency and low distortion.

9-6 (*a*) In an undistorted Class A operation of a transistor, the collector dissipates a certain average power for no signal applied. When the input signal is applied, the collector dissipation decreases. For 35 per cent conversion efficiency, how large must the collector dissipation rating be for a load that takes 10 W? (*b*) The same transistor is used in Class B push-pull operation. The load requires the same 10 W of power. What is the collector dissipation when this power is supplied to the load? What is the collector dissipation when the signal is disconnected?

9-7 An input signal to a Class B push-pull, transformer-coupled amplifier can be expressed as $e_s = 20 \sin \omega t + 10 \sin 2\omega t + 5 \sin 3\omega t$. Assuming linear, undistorted amplification, what will be the frequency content of the output?

9-8 A power transistor having the following specifications is to be used in the design of a Class A amplifier driving a resistance load of 150 Ω in the collector circuit and operating as a common-emitter amplifier:

$$BV_{CEO} = 60 \text{ V}$$
$$P_{C \text{ max}} = 6 \text{ W}$$
$$h_{fe} = 50, \quad \text{assumed constant}$$

(*a*) Sketch its collector characteristic curves and locate the maximum power hyperbola. (*b*) Determine a *Q*-point that will permit 50 per cent efficiency in the collector circuit and hence develop maximum load power. (*c*) Under the conditions of (*b*), what is the load power? collector dissipation?

9-9 An audio power amplifier is to be designed to supply power to a 10-Ω load using a Class B push-pull transformer coupling. The transistors being used require a collector load of 400 Ω for maximum undistorted operation. What should be specified for the output transformer?

9-10 The circuit shown in Fig. P9-10 is used as a Class A driver for a Class B power amplifier. The diodes and 5-Ω resistors represent the input to the Class B stage. One diode will conduct when the collector current increases, the other when it decreases. It is designed to yield the maximum possible power in the Class B stage, with the smallest possible base current. Determine the required values of V_{CC}, turns ratio, and Q-point, for these ratings of the transistor:

$$BV_{CEO} = 40 \text{ V} \qquad i_{C\,max} = 70 \text{ mA} \qquad P_{C\,max} = 0.5 \text{ W}$$

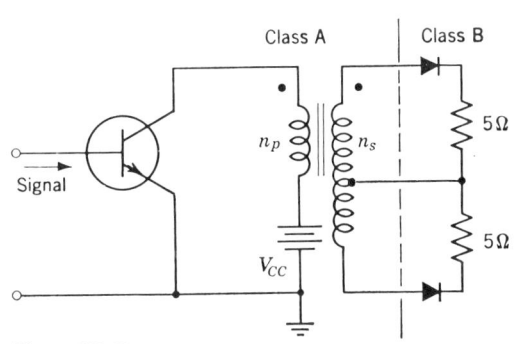

Figure P9-10

9-11 The coupling transformer shown in Fig. P9-11 is rated "400 Ω C.T. to 3.2 Ω." The transistors are rated at 120 mW maximum collector dissipation and 100 mA maximum collector current. (a) What is the value of the reflected load in the collector circuit? (b) What maximum collector

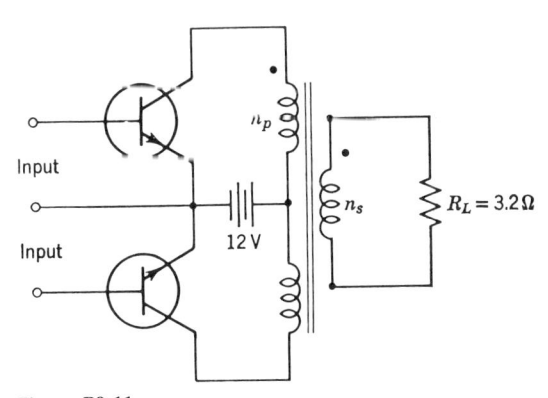

Figure P9-11

356   Power Amplifiers   Ch. 9

current can be used without exceeding any ratings? (c) Assuming linear operation of the transistors, calculate the turns ratio of the transformer for maximum efficiency, without exceeding any transistor ratings.

9-12 Show that the collector dissipation of a transistor operating in Class A common-emitter mode is minimum when power output is maximum.

9-13 A Class B push-pull circuit as shown in Fig. P9-13 uses 2N1485 transistors. Determine appropriate values for V_{CC}, n, R_1, and R_2, to provide maximum output power at minimum distortion. Assume operation at 25°C ambient temperature and the transistors mounted on efficient heat sinks.

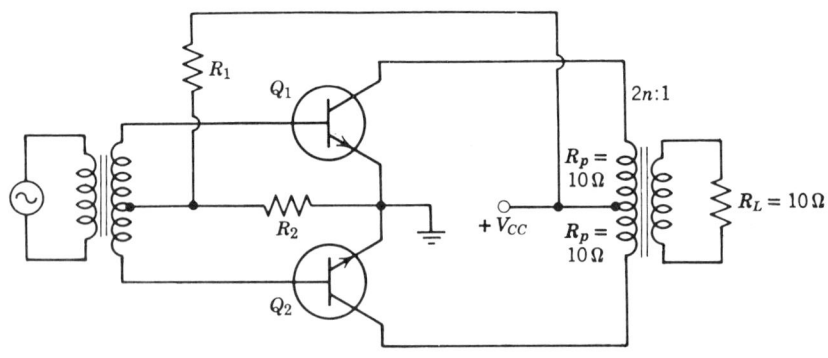

Figure P9-13

9-14 Assuming the values found in Problem 9-13 permit sinusoidal operation, calculate the collector efficiency and compare it with the theoretical maximum value for Class B.

9-15 From the results of Problem 9-13, determine the input power required for maximum output power. What is the power gain of the amplifier? (Assume the emitter-base junction may be approximated by a 0.7-V source in series with 100 Ω resistance.)

10
SPECIAL-PURPOSE AMPLIFIERS

In our studies of amplifiers so far, we have given attention to circuits that handle ac signals. We have considered the operation of R-C-coupled and transformer-coupled circuits, which can amplify ac signals over some bandwidth of frequencies. At low frequencies these circuits suffer from loss of gain due to the coupling elements; their bandwidth is limited by these effects and they do not respond at all to dc signals.

There are many applications of electronic circuits in which dc signals are important and some means of amplifying them is needed. In this chapter we shall examine circuits that have special properties designed in for particular applications. Some special-purpose circuits can amplify dc signals. Others are used for amplifying ac signals, utilizing properties that improve the performance of otherwise conventional amplifiers.

10-1
Differential Amplifier

The *differential amplifier* is used in a variety of applications. Its basic properties are related to general amplification in other circuits we have studied, but the way in which signals are connected provides special properties.

The basic circuit of a differential amplifier is given in Fig. 10-1. Two separate inputs and outputs are shown. Input signals are applied between ground and each base of the separate transistors, and output signals are taken between ground and the collector terminals of each transistor. Both emitters are

358 Special-Purpose Amplifiers Ch. 10

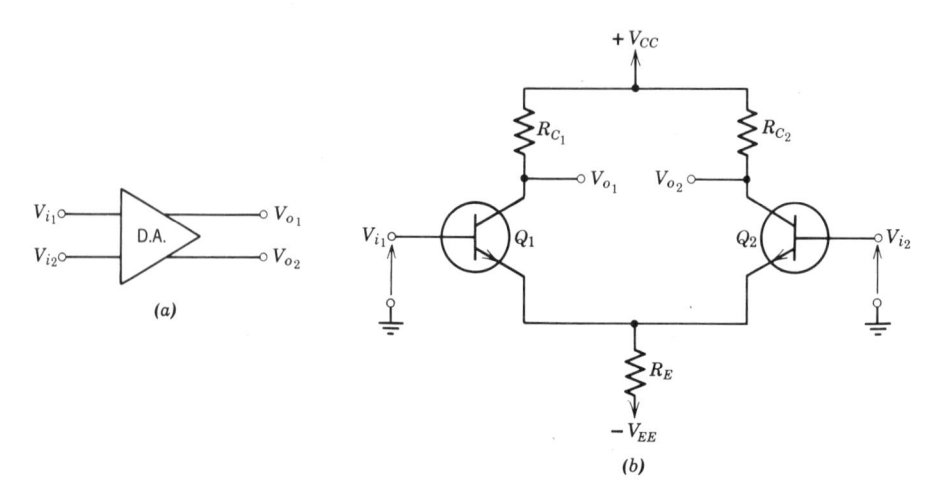

Figure 10-1
Basic differential amplifier. (a) Block symbol. (b) Simplified schematic diagram.

connected to a single resistor in the circuit, which will cause both output voltages to be affected by either input signal. It should be understood that both V_{CC} and $-V_{EE}$ are supply voltages that have their opposite polarities connected to the ground reference point.

When only one input terminal has a signal applied and the other terminal is grounded, the circuit is said to be operating in its *single-ended* mode. This mode of operation is illustrated in Fig. 10-2.

Let us analyze the circuit operation for the two cases in which a signal is applied first to one input terminal and then the other. These two cases are shown in Fig. 10-2a and b. Suppose $V_{i1} = 1$ Vdc, positive at the base of transistor Q_1, and the base of Q_2 is connected to ground. Collector current in Q_1 will increase, reducing the collector voltage by an amount based on the amplification in the circuit. Since collector current in Q_1 also passes through R_E, the voltage at the emitter will increase with respect to ground. The

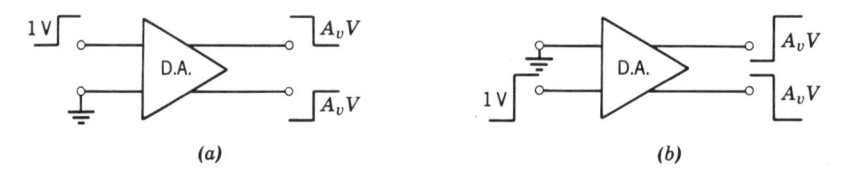

Figure 10-2
Differential amplifier operated in single-ended mode. (a) Input connected to one terminal; (b) input connected to other terminal. Phase relationships of input and output alternate in opposite connections.

magnitude of this voltage increase will be approximately equal to the 1-V increase at the base.

Since the input to Q_2 is grounded, it may seem that there will be no output at its collector. However, that conclusion is incorrect. Even though the base of Q_2 is grounded, its emitter is tied to R_E. At that point in the circuit the voltage has increased about 1 V with respect to ground. Thus, the emitter voltage of Q_2 will decrease its collector current, and its collector voltage will increase an amount based on the circuit amplification. If the amplifier is designed so that each transistor amplifies equally, the changes in collector voltage of both transistors will be equal in amplitude and opposite in phase. Output V_{o2} is in phase with the input signal, and V_{o1} is out of phase.

Now consider the case when the input signal is applied as V_{i2}, as shown in Fig. 10-2b. Similar action occurs in the circuit, but the phase relationships are just the opposite of the first case. Here again, though, the output voltages are amplified versions of the input signal.

It should be clear that a negative input signal will operate the circuit. The only change in its operation will be decreases where increases had occurred, and vice versa. Magnitudes and phase relationships will occur as before; relative polarities will be the opposite.

10-2
Differential Operation

It is possible to operate the differential amplifier with input signals applied to both input terminals, which will yield signals at the two output terminals. This mode of operation, referred to as *double-ended* or *differential* operation, is illustrated in Fig. 10-3. Input signals are shown as equal sine waves that are opposite in phase.

Each input signal will generate an output voltage that is related to the circuit amplification, as shown in Fig. 10-3a and b. By the principle of superposition, when both inputs are applied as in Fig. 10-3c, the output voltages will be twice the amplitude generated by one input acting alone. Both output signals will be twice as large.

It may appear that an amplification of four times the single-ended operation has resulted. However, that is not the case because both input signals are operating the circuit. Since these signals are equal in amplitude, the actual *differential input* is twice the amplitude of one signal, resulting in an overall amplification that is only *twice* that of the single-ended operation, not four times.

If the two input signals are in phase rather than of opposite phase, then the output voltages will be about equal in magnitude. The difference between outputs V_{o1} and V_{o2} would be practically zero, even though each output signal

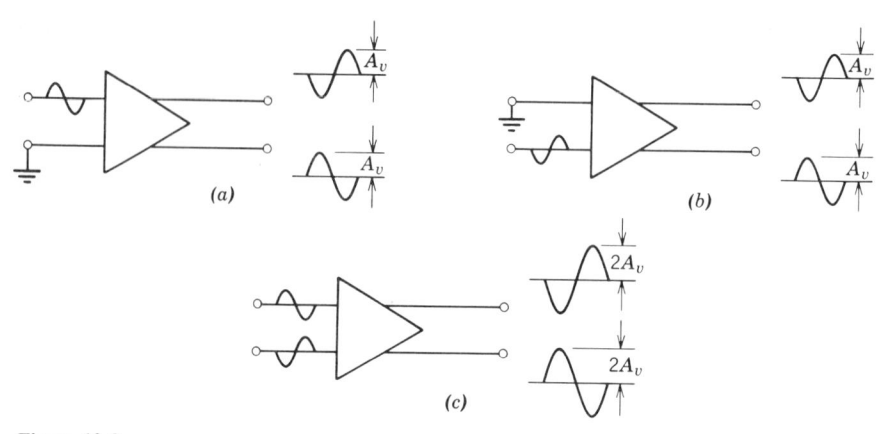

Figure 10-3
Operation of a differential amplifier (a) with one input, (b) with the other input, (c) with both input signals.

would exist. Then we may refer to the circuit as a *difference amplifier*. Only those signal components in V_{i1} and V_{i2} that are out of phase with each other will be amplified. Components that are in phase at the input terminals will produce no output signals. This resulting lack of output for in-phase input signals is often called *common-mode rejection*.

10-3
Common-Mode Rejection Ratio

An important property of a differential amplifier is its ability to cancel out unwanted voltage signals, which are referred to as "noise." Such signals may be superimposed on desirable signals, occurring because of stray electromagnetic fields, variations in power supply voltages, and other sources unrelated to the desired signals. Any such stray voltages that appear equally at the two input terminals will be rejected by the amplifier circuit. Ideally, the circuit will completely cancel common-mode noise signals.

A practical circuit cannot completely reject common-mode signals but will generate some small output voltage based on the unsymmetrical behavior of the two transistors making up the circuit. The amount of output signal that is produced by a practical circuit can be determined by connecting the same signal to both inputs and measuring the differential output voltage. Some value of output voltage may be measured; call it V_c. A comparison of this voltage to that produced by the same value of input signals, but which are connected out of phase, will be a measure of the circuit ability to reject noise voltages. Suppose the desired output voltage is labeled V_a. Then the ratio V_a/V_c is a

Sec. 10-4   Biasing for Maximum Signal Swing 361

measure of the ability to cancel out unwanted signals. Such a ratio is referred to as the *common-mode rejection ratio* (CMRR) of the amplifier. The larger the value of CMRR the better the circuit for amplifying signals that are subject to unwanted noise voltages. Some practical circuits may have a CMRR of 10^3 and higher.

10-4
Biasing for Maximum Signal Swing

In the operation of a differential amplifier, it may be desirable to permit a maximum variation of the output differential voltages. In this way, the amplifier may be used to advantage in obtaining as large a signal as possible without distortion due to cutoff or saturation. Similarly, the circuit should be designed so that it is biased in the center of its operating range.

In order to look at some practical limits on the design of a differential amplifier, let us analyze the circuit of Fig. 10-4. The two transistors are assumed to be silicon units, each having $h_{FE} = 50$ and well-matched characteristics. They should be matched to achieve good amplification and reasonable common-mode rejection.

For dc operation of the circuit, the source V_{EE} will forward bias the base-emitter diode of each transistor. For an assumed base current of 100 μA, the dc voltage across R_s will be 0.1 V. The base-emitter junction will be biased at about 0.7 V at this current, leaving 9.2 V across R_E. The current in R_E is then

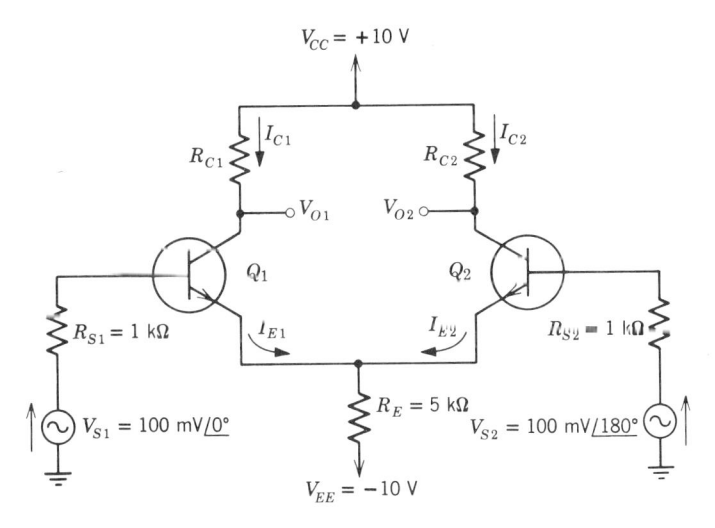

Figure 10-4
Basic differential amplifier circuit.

calculated as

$$I_E = \frac{9.2 \text{ V}}{5 \text{ k}\Omega} = 1.85 \text{ mA} \qquad (10\text{-}1)$$

This current consists of the separate emitter currents, half of it from each transistor. Then the emitter current in each will be 0.925 mA.

For these approximate voltage and current values, the base current may be determined to see if the approximations are valid:

$$I_{B1} = \frac{I_{E1}}{1 + h_{FE}} = \frac{0.925 \text{ mA}}{1 + 50} = 0.018 \text{ mA} = 18 \text{ }\mu\text{A} \qquad (10\text{-}2)$$

Since this value is smaller than assumed, the emitter current will be slightly larger.

Since collector current is the difference between emitter and base currents, $I_{C1} = 0.925 - 0.018 = 0.907$ mA (approximately). In order to achieve the maximum swing in output voltages, it is desirable to select a value for R_{C1} so that its dc voltage drop is about halfway between 0 V and V_{CC}, at 5 V. This value requires that $R_{C1} = 5 \text{ V}/0.907 \text{ mA}$, or about 5.6 k$\Omega$.

10-5
Input and Output Impedances

Now we shall turn our attention to the ac operation of this circuit. It should be noted that for matched transistors and balanced operation between them, the current in R_E will remain essentially constant. This is so because the input signals cause equal and opposite current changes in the two emitter currents, thereby maintaining a stable current in the emitter resistor. However, the collector currents will vary according to the current gain of the circuit, and will generate output voltages based on selected values for collector resistors R_{C1} and R_{C2}.

For assumed linear transistors, the ac and dc current gains will be equal: $h_{fe} = h_{FE} = 50$. Other typical ac parameters will be assumed for the transistors: $h_{ie} = 1$ kΩ, $h_{oe} = 1/50$ kΩ.

For the purpose of analysis, the partial ac circuit given in Fig. 10-5 will be used. For two signal inputs that are out of phase as shown, the two emitter currents will produce equal and opposite drops across R_E. Effectively, resistor R_E will not affect the value of input base current. Then the input impedance seen by signal V_{s1} is

$$Z_i = \frac{V_{s1}}{I_{b1}} = R_{s1} + h_{ie} \qquad (10\text{-}3)$$

For circuit components given in Fig. 10-4, the input impedance is 2 kΩ. For

Sec. 10-5   Input and Output Impedances   363

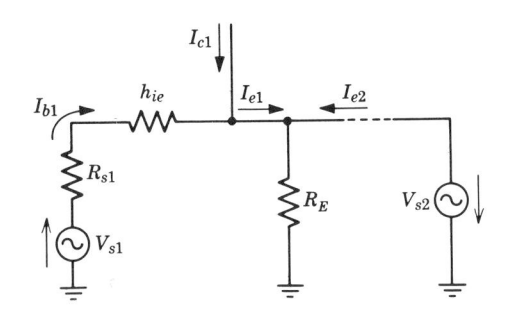

Figure 10-5
AC equivalent circuit for input signal.

$V_{s1} = 100$ mV, base current will be

$$I_{b1} = \frac{100 \text{ mV}}{2 \text{ k}\Omega} = 50 \text{ }\mu\text{A}$$

This level of base current will be too large for the bias value obtained in Eq. 10-2, which is 18 μA. Operation of the circuit will cause distortion due to clipping. A redesign of the circuit would include a reduction of the resistor R_E to permit a larger base bias current. An increase in voltage source V_{EE} would also permit larger variations in I_b, but other circuit designs can be achieved that will obtain this result as well as other beneficial characteristics. We shall look at some of these in subsequent sections.

An output impedance can be determined from analysis of the circuit in Fig. 10-6. The output voltage is developed across the parallel combination of R_C and $1/h_{oe}$, which is the output impedance:

$$Z_o = \frac{R_C(1/h_{oe})}{R_C + 1/h_{oe}} = \frac{R_C}{1 + h_{oe}R_C} \tag{10-4}$$

For components associated with Fig. 10-3, this impedance is about 5 kΩ.

Before looking at some applications of the differential amplifier in a special class of circuits called *operational amplifiers*, we shall discuss another circuit

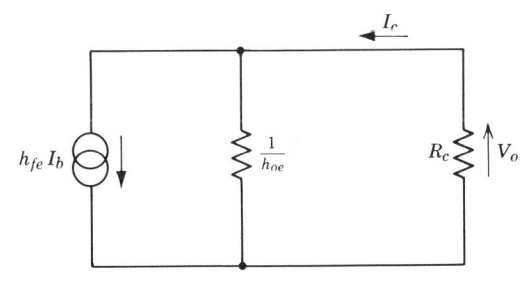

Figure 10-6
Equivalent circuit for output signal.

364 Special-Purpose Amplifiers Ch. 10

that is used extensively to adjust input and output impedances. Calculated values we have obtained so far are less than optimum to provide good characteristics in practical applications, where signal and load properties require different values. Usually it is desirable to have a high input impedance for an amplifier, to avoid loading its source, and a low output impedance. One circuit that has these properties is discussed in the next section. It is often found in modern integrated circuit packages.

Example 10-1. As an example of analysis of the ac behavior of a differential amplifier, we shall determine the ac differential voltage gain of the circuit in Fig. 10-4. Its h-parameter equivalent circuit is shown in Figure E10-1.

The differential voltage gain $A_v = (V_{o1} - V_{o2})/(V_{i1} - V_{o2})$. The output voltages are $V_{o1} = -I_{C1}R_C$ and $V_{o2} = -I_{C2}R_C$. For transistor Q_1,

$$I_{C1} = h_{oe}V_{o1} + h_{fe}I_{b1} = (-h_{oe}R_C)I_{C1} + h_{fe}I_{b1} = \frac{h_{fe}}{1 + h_{oe}R_C}I_{b1}$$

then the output voltage may be expressed

$$V_{o1} = \left(-\frac{h_{fe}R_C}{1 + h_{oe}R_C}\right)I_{b1}$$

but since

$$I_{b1} = \frac{V_{s1} - h_{re}V_{o1} - V_E}{R_s + h_{ie}}$$

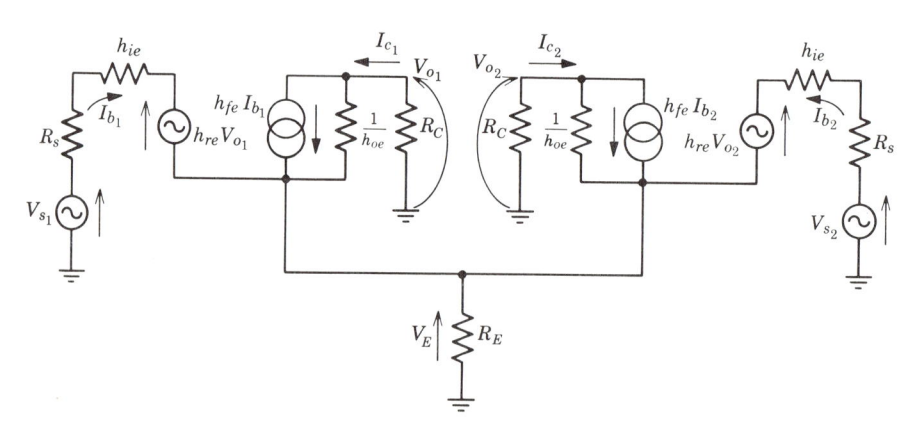

Figure E10-1
Equivalent circuit for Fig. 10-4.

$$V_{o1} = \left(\frac{-h_{fe}R_C}{1 + h_{oe}R_C}\right)\left(\frac{V_{s1} - h_{re}V_{o1} - V_E}{R_s + h_{ie}}\right)$$

$$= \left[\frac{-h_{fe}R_C}{(1 + h_{oe}R_C)(R_s + h_{ie}) - h_{re}h_{fe}R_C}\right](V_{s1} - V_E) \quad (1)$$

For identical transistors, source resistors, and collector loads, output voltage for Q_2 is

$$V_{o2} = \left[\frac{-h_{fe}R_C}{(1 + h_{oe}R_C)(R_s + h_{ie}) - h_{re}h_{fe}R_C}\right](V_{s2} - V_E) \quad (2)$$

and

$$(V_{o1} - V_{o2}) = [\text{bracketed factor in (1) and (2)}](V_{s1} - V_{s2})$$

Then

$$A_v = \frac{-h_{fe}R_C}{h_{ie} + (1 + h_{oe}R_C)R_s + DR_C} \quad (3)$$

Here it is evident that ac operation of the circuit is independent of the emitter resistor R_E, for identical components elsewhere in the amplifier. Resistor R_E is selected for placement of the dc bias and operating region rather than for its effect on voltage gain.

10-6
The Darlington Circuit

A circuit that has a high input impedance and a low output impedance is given in Fig. 10-7. This compound connection of two transistors is called a *Darlington circuit*. Some practical features of this circuit include the direct coupling between the transistors, using the emitter current of Q_1 as the base current of Q_2, and its small voltage gain but high current gain operation.

An equivalent circuit for ac operation, in Fig. 10-8, will be used to analyze the behavior of the Darlington circuit. It will be helpful to have a composite model for the two coupled transistors, so that previously determined relationships involving h-parameters can be used. We begin the analysis by writing equations for input voltage and output current in terms of input current and output voltage. Then h-parameters may be determined from these equations, which may then be used to analyze the circuit behavior.

Input voltage may be written as

$$V_i = h_{ie1}I_{b1} + h_{re1}V_{o1} + (V_o - V_{o1})$$
$$= h_{ie1}I_{b1} + (h_{re1} - 1)V_{o1} + V_o$$

366   Special-Purpose Amplifiers   Ch. 10

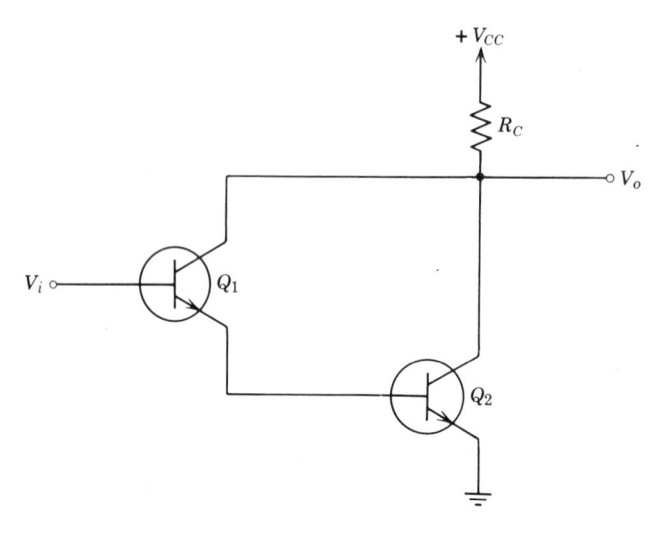

Figure 10-7
The Darlington circuit.

but since

$$V_{o1} = \frac{1}{h_{oe1}}[I_{b2} - (1 + h_{fe1})I_{b1}]$$

and

$$I_{b2} = -h_{oe2}V_o - h_{fe2}I_{b2} + I_{b1}$$

$$= \frac{-h_{oe2}V_o}{1 + h_{fe2}} + \frac{I_{b1}}{1 + h_{fe2}}$$

then

$$V_i = h_{ie1}I_{b1} + \frac{(h_{re1} - 1)}{h_{oe1}}\left[\frac{-h_{oe2}V_o}{1 + h_{fe2}} + \frac{I_{b1}}{1 + h_{fe2}} - (1 + h_{fe1})I_{b1}\right] + V_o$$

$$= \left[h_{ie1} - \frac{1 - h_{re1}}{h_{oe1}(1 + h_{fe2})} + \frac{(1 - h_{re1})(1 + h_{fe1})}{h_{oe1}}\right]I_{b1}$$

$$+ \left[1 + \frac{h_{oe2}(1 - h_{re1})}{h_{oe1}(1 + h_{fe2})}\right]V_o$$

$$= \left[h_{ie1} + \frac{1 - h_{re1}}{h_{oe1}}\left(1 + h_{fe1} - \frac{1}{1 + h_{fe2}}\right)\right]I_{b1}$$

$$+ \left[1 + \frac{h_{oe2}(1 - h_{re1})}{h_{oe1}(1 + h_{fe2})}\right]V_o \quad (10\text{-}5)$$

Sec. 10-6   The Darlington Circuit   367

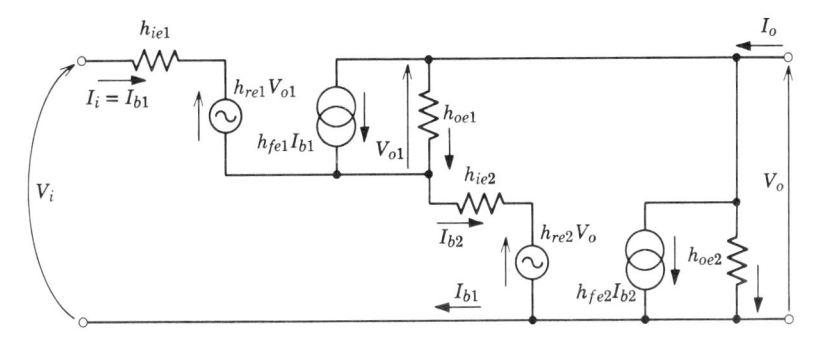

Figure 10-8
AC equivalent circuit for Darlington connection.

Output current may be expressed as

$$I_o = (1 + h_{fe2})I_{b2} + h_{oe2}V_o - I_{b1} \qquad (10\text{-}6)$$

The base current is

$$I_{b2} = (1 + h_{fe1})I_{b1} + h_{oe1}V_{o1}$$
$$= (1 + h_{fe1})I_{b1} + h_{oe1}[(1 - h_{re2})V_o - h_{ie2}I_{b2}]$$

Rearranging the terms,

$$I_{b2} = \left(\frac{1 + h_{fe1}}{1 + h_{ie2}h_{oe1}}\right)I_{b1} + \left(\frac{h_{oe1}(1 - h_{re2})}{1 + h_{ie2}h_{oe1}}\right)V_o$$

Substituting into Eq. 10-6,

$$I_o = \left(\frac{(1 + h_{fe1})(1 + h_{fe2})}{1 + h_{oe1}h_{ie2}} - 1\right)I_{b1} + \left(h_{oe2} + \frac{h_{oe1}(1 + h_{fe2})(1 - h_{re2})}{1 + h_{oe1}h_{ie2}}\right)V_o \qquad (10\text{-}7)$$

From general conditions for calculating h-parameters,

$$h'_{fe} = \left.\frac{I_o}{I_{b1}}\right|_{V_o=0} = \frac{(1 + h_{fe1})(1 + h_{fe2})}{1 + h_{oe1}h_{ie2}} - 1 \approx h_{fe1}h_{fe2} \qquad (10\text{-}8)$$

$$h'_{ie} = \left.\frac{V_i}{I_{b1}}\right|_{V_o=0} = h_{ie1} + \frac{1 - h_{re1}}{h_{oe2}}\left(1 + h_{fe1} - \frac{1}{1 + h_{fe2}}\right) \approx \frac{h_{fe1}}{h_{oe1}} \qquad (10\text{-}9)$$

$$h'_{re} = \left.\frac{V_i}{V_o}\right|_{I_{b1}=0} = 1 + \frac{h_{oe2}(1 - h_{re1})}{h_{oe1}(1 + h_{fe2})} \approx 1 \qquad (10\text{-}10)$$

$$h'_{oe} = \left.\frac{I_o}{V_o}\right|_{I_{b1}=0} = h_{oe2} + \frac{h_{oe1}(1 + h_{fe2})(1 - h_{re2})}{1 + h_{oe1}h_{ie2}} \approx h_{oe1}h_{fe2} \qquad (10\text{-}11)$$

368 Special-Purpose Amplifiers Ch. 10

A composite model is used for the circuit of Fig. 10-7, which is given in Fig. 10-9. Since this amplifier is connected in a common-emitter mode, we may use

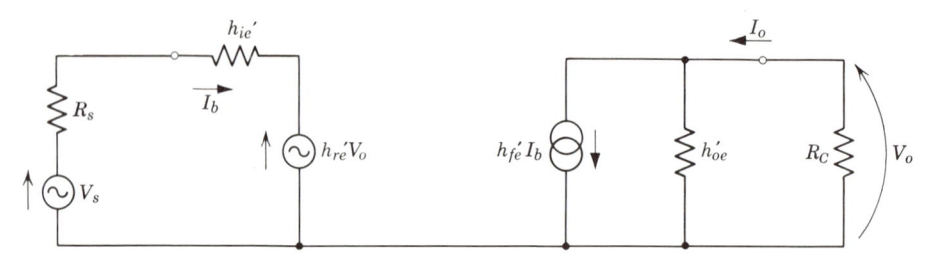

Figure 10-9
Composite model for Darlington circuit.

equations from Chapter 8 to analyze its behavior. Necessary calculations are worked out in the example that follows.

Example 10-2. For the purposes of these calculations, we will assume source and load resistors are each 10 kΩ, and that the two transistors are matched silicon units having these h-parameters: $h_{ie} = 5\,\text{k}\Omega$, $h_{fe} = 100$, $h_{re} = 700 \times 10^{-6}$, $h_{oe} = 20\,\mu$mhos.

Before our principal calculations may proceed using the equations in Table 8-2, values of the composite h-parameters and the factor D' are needed:

$$h'_{fe} = \frac{(1+100)(1+100)}{1+(20 \times 10^{-6})(5000)} - 1 = \frac{10{,}101}{1.1} - 1 = 9183$$

$$h'_{ie} = 5000 + \frac{(1 - 700 \times 10^{-6})}{20 \times 10^{-6}}\left(1 + 100 - \frac{1}{1+100}\right) = 5.1\,\text{M}\Omega$$

$$h'_{re} = 1 + \frac{(20 \times 10^{-6})(1 - 700 \times 10^{-6})}{(20 \times 10^{-6})(1+100)} = 1 + \frac{0.9993}{101} = 1.01$$

$$h'_{oe} = 20 \times 10^{-6} + \frac{20 \times 10^{-6}(1+100)(1 - 700 \times 10^{-6})}{1 + (20 \times 10^{-6})(5000)}$$

$$= 20 \times 10^{-6} + 1835 \times 10^{-6} = 1855 \times 10^{-6}\,\text{mho}$$

The factor D' may be determined:

$$D' = h'_{ie}h'_{oe} - h'_{re}h'_{fe} = 5.1 \times 10^{6} \times 1855 \times 10^{-6} - 1.01(9183)$$
$$= 9460 - 9275 = 185$$

Calculations for the quantities listed in Table 8-2 will use the primed parameters:

Current gain: $$A_i = \frac{h_{fe}}{h_{oe}R_C + 1} = \frac{9183}{(1855 \times 10^{-6})(10,000) + 1} = 480$$

Voltage gain: $$A_v = \frac{-h_{fe}R_C}{h_{ie} + DR_C} = \frac{-9183(10,000)}{5.1 \times 10^6 + 185(10,000)} = -13.2$$

Power gain: $$A_p = A_v A_i = (13.2)(480) = 5700$$

Input resistance: $$R_i = \frac{h_{ie} + DR_C}{h_{oe}R_C + 1} = \frac{5.1 \times 10^6 + (185)(10,000)}{(1855 \times 10^{-6})(10,000) + 1} = 350 \text{ k}\Omega$$

Output resistance: $$R_o = \frac{R_s + h_{ie}}{h_{oe}R_s + D} = \frac{10,000 + 5.1 \times 10^6}{(1855 \times 10^{-6} \times 10,000) + 185} = 25 \text{ k}\Omega$$

10-7 Operational Amplifiers

An *operational amplifier* is a dc amplifier having a very high gain that is often designed using differential amplifier stages to establish high input impedance and low output impedance. It is used in analog computers, in data-handling equipment, and in various forms of electronic instrumentation. Many useful units are available as commercial components in integrated-circuit form, which provide the properties desired in a small volume. This type of circuit component is often referred to as an "op amp."

The circuit of an operational amplifier usually provides two input terminals for use in a differential mode, and one output terminal. Its schematic symbol is given in Fig. 10-10. Note the similarity to the symbol for a differential amplifier. As shown in the drawing, one input terminal is referred to as the "inverting" input; the second input terminal as the "noninverting" input. These labels refer to the relative phase relationships between an input signal and the output voltage. An input signal applied to the "inverting" terminal is out of phase with the amplified output; one applied to the "noninverting" terminal is in phase.

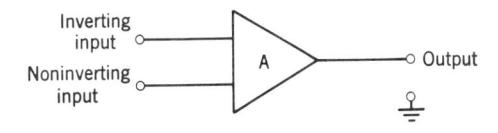

Figure 10-10
Schematic symbol for operational amplifier.

10-8
Applications of Operational Amplifiers

Operational amplifiers are intended for use with external circuit components that make the circuit behavior follow some preselected input/output relationship. The ideal op amp would have infinite gain, infinite input impedance, zero output impedance, and an infinite bandwidth extending from dc. Such a device would also have an infinite cost. Practical op amps approach these ideal characteristics, particularly in integrated-circuit form, and their cost is low compared with transistor prices of a few years ago. High quality and low cost have made possible the implementation of complex systems that would be impractical using discrete components. For example, the Zeltex ZEL-1 differential op amp has a dc gain of 5×10^5 (minimum) and CMRR of 20,000. Its input impedance is 1 MΩ for dc and 10 kΩ for ac; output impedance is 1 kΩ. It is designed to operate from voltage sources at ± 15 Vdc. The circuit is encased in solid plastic and packaged in a case that measures about 3 cm square and 1.5 cm high.

The fundamental theorem that permits the operational amplifier to be used successfully to perform mathematical operations is Kirchhoff's law for electric currents: "the sum of currents entering a node equals the sum of currents leaving the node." If the amplifier of Fig. 10-11 is assumed to have an infinite input impedance, then the currents at the input node (the summing junction) can exist only in the input resistors and the feedback resistor. Thus, the currents entering from the inputs must just equal the current leaving through the feedback path. This means that when an input voltage increases, the output voltage must decrease. This equality of currents will be maintained within the bandwidth of the amplifier, and permits the summing of ac as well as dc input signals. Since the summing junction is ideally maintained at zero potential, inputs of zero will result in an output of zero. When several input paths are provided at the summing junction, the amplifier will cause the feedback current

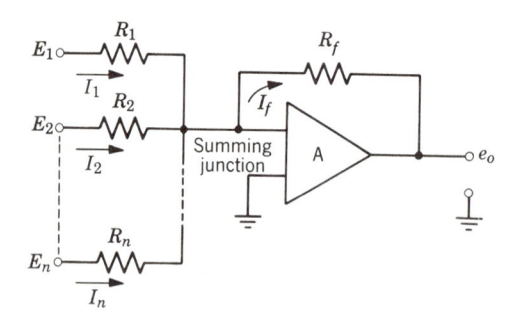

Figure 10-11
Op amp connected as a summing amplifier.

to equal the algebraic sum of all input currents, and the output voltage will be proportional to the sum of the input voltages. Written in terms of the voltages and resistances, the general equation for the circuit is

$$e_o = -R_f \left(\frac{E_1}{R_1} + \frac{E_2}{R_2} + \cdots + \frac{E_n}{R_n} \right)$$

This basic equation can also be used to analyze the behavior of the amplifier when the resistors are general impedance elements. For example, if the feedback resistor is replaced with a capacitor, the circuit will produce an output that is the integral of the inputs. If a current flows into a capacitor, the voltage across it will change at a rate that is proportional to the current

$$e_C = \frac{1}{C} \int i \, dt$$

Since the summing junction is at zero potential, the voltage across the feedback capacitor will be e_o, and the current summation may be written in terms of the voltages and impedance elements as

$$e_o = -\frac{1}{C} \int_0^t \left(\frac{E_1}{R_1} + \frac{E_2}{R_2} + \cdots + \frac{E_n}{R_n} \right) dt + E_o(t=0)$$

Similarly, the circuit may be used to subtract input signals. This only requires that the signals be of opposite phase and the ratio of feedback and input resistors be the same for the two signals. Of course, it can be seen that by a proper choice of resistance ratios, the circuit can multiply an input signal by any desired factor, which will be set by the resistance ratio. The process of division by a constant can be handled in the same way, since division is inverse multiplication.

The many applications of operational amplifiers are too broad for our discussion here. The reader is referred to the multitude of technical papers and textbooks on the use of these circuits in analog computers and data systems. This introduction should be sufficient to indicate the importance of these circuits in electronic applications.

Example 10-3. An operational amplifier is used to sum three voltages, using a circuit similar to that of Fig. 10-11. The amplifier has a gain of −1000 and may be considered to have infinite input impedance. Then the sum of currents entering the summing junction from the input sources equals the current in the feedback impedance.

Considering the summing junction to be at ground potential, we may write the current as

$$\frac{E_1}{R_1} + \frac{E_2}{R_2} + \frac{E_3}{R_3} = \frac{-e_o}{R_f} \qquad (1)$$

372 Special-Purpose Amplifiers

Suppose each of the resistors is chosen to be 1 MΩ. Then the output voltage will be the sum of the three input voltages, provided the sum does not exceed the saturation voltage of the operational amplifier. The saturation voltage will be determined by the dc supply used to power the operational amplifier. In some cases, it may be about 100 V, in others, 10 V or so. Higher saturation voltages are permissible in circuits that use vacuum tubes, whereas those using transistors typically have lower permissible values.

This circuit may also be used to scale certain input voltages, or apply "weighting factors," before the addition process. This is easily accomplished by adjusting the ratios of input and feedback resistors. Suppose it is desired that the input E_1 be reduced by a factor of 1/10 compared with the other two inputs before addition. Equation 1 may be rewritten,

$$\frac{R_f}{R_1} E_1 + \frac{R_f}{R_2} E_2 + \frac{R_f}{R_3} E_3 = -e_0 \qquad (2)$$

From R_f equal to 1 MΩ, R_1 would equal 10 MΩ, and R_2 and R_3 would be chosen equal to R_f, 1 MΩ. Then the sum of the voltages would include E_1 reduced by a factor of 1/10.

In effect, by choosing the ratio $R_f/R_1 = 1/10$, the input voltage E_1 was divided by 10. Similarly, multiplication of the input voltage may be obtained by choosing a ratio necessary for the required multiplication factor.

10-9
Amplifiers with Tuned Loads

When it is necessary to transmit signals over long distances, amplifiers with tuned loads are used. The circuits that are discussed in this chapter are adequate for audio amplifiers used in home entertainment and public address systems and for servomechanisms, electronic instruments, and a host of other confined applications. However, when the signals are transmitted from an earth satellite or manned space probe to a receiving station on the earth, or from city to city, the distances involved require the use of high-frequency carriers, radio frequencies, for effective use of the transmitting medium. The signals may include only a narrow band of frequencies. Tuned loads are used to amplify the narrow-band signals with the carrier and to separate them from other spurious frequencies.

Various systems of this kind operate throughout the frequency spectrum. The commercial broadcast industry, for example, operates within the range from 550 kHz to 1600 kHz to transmit voice and music signals. Each commercial transmitter is limited to a narrow-band spectrum of 5 kHz on either side of its assigned carrier frequency or a total bandwidth of 10 kHz that it may use to transmit its signals. A transmitter of this signal is required to operate within this restricted range, using tuned loads that resonate at the center of the bandwidth. Its operation must confine the signals to the region of 5 kHz above and below the carrier frequency. Similarly, receivers for the signals require frequency-selective circuits in order to distinguish between the many signals that may be present. Frequency-selective amplification can be achieved by using tuned loads, because their operation provides a form of filtering to eliminate unwanted signals.

The response of a single-tuned amplifier is compared in Fig. 10-12 with the ideal response needed to completely eliminate signals that are outside the desired passband. The circuit shown is one of the simplest narrow-band amplifiers, because it uses a single-tuned load. Double-tuning and stagger-tuning will be discussed after we have analyzed the behavior of the single-tuned amplifier.

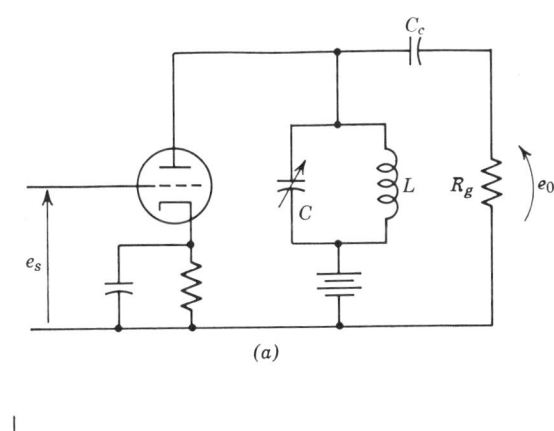

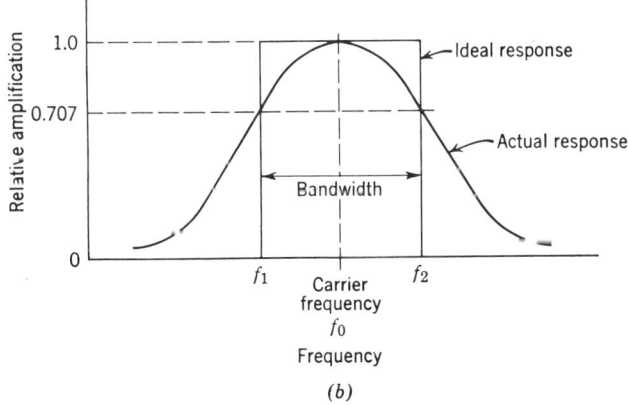

Figure 10-12
Single-tuned amplifier and its response. (a) Amplifier with tuned load. (b) Actual amplification compared with ideal response.

374 *Special-Purpose Amplifiers* Ch. 10

Some frequency of the input signal will permit the load of the circuit to be in resonance. At this frequency the amplification will be greatest, and it will be less than maximum at frequencies above and below this value. This is shown by the curve of actual response in the figure. The resonant frequency is tunable within limits, by varying the value of the capacitance in the plate load. The parallel connection L-C is often called a *tank circuit* because of its ability to store energy in the form of a circulating current. The principal disadvantage of this circuit is that the amplification decreases slowly with frequency about the carrier frequency, but the desired passband may be relatively narrow. Thus, it may not reject signals that are near the desired frequency range.

The voltage amplification of the circuit may be expressed as

$$A_v = -g_m Z_{\text{load}} = -g_m \frac{1}{j\omega C + 1/R + 1/j\omega L}$$

where R is the parallel combination of r_p and R_g. Let us examine only the expression for Z_{load}. This factor may be written

$$\frac{1}{Z_{\text{load}}} = Y_{\text{load}} = \frac{1}{R} + j\left(\omega C - \frac{1}{\omega L}\right)$$

This may be put into a more useful form by using the relation that $1/\omega_0 L = \omega_0 C$, and factoring $1/R$:

$$Y = \frac{1}{R}\left[1 + j\left(\frac{R\omega C \omega_0}{\omega_0} - \frac{R\omega_0}{\omega L \omega_0}\right)\right]$$

$$= \frac{1}{R}\left[1 + j\omega_0 RC\left(\frac{\omega}{\omega_0} - \frac{\omega_0}{\omega}\right)\right]$$

$$= \frac{1}{R}\left[1 - jQ'_o\left(\frac{f}{f_o} - \frac{f_o}{f}\right)\right] \qquad (10\text{-}12)$$

The parameter Q'_o is convenient because it includes the resonant frequency and all circuit components:

$$Q'_o = \frac{R}{\omega_0 L} = R\omega_0 C$$

We may also express any frequency f in terms of the resonant frequency

$$f = f_o + \alpha f_o$$

where α is a fractional part of f_o, so it follows that

$$\alpha = \frac{f - f_o}{f_o} = \frac{f}{f_o} - 1$$

Then Eq. 10-12 may be written

$$Y = \frac{1}{R}\left[1 + jQ'_o\left([1 + \alpha] - \frac{1}{1 + \alpha}\right)\right]$$

$$= \frac{1}{R}\left[1 + jQ'_o\alpha\left(\frac{2 + \alpha}{1 + \alpha}\right)\right] \quad (10\text{-}13)$$

The fraction $(2 + \alpha)/(1 + \alpha) = 2 - \alpha + \alpha^2 - \alpha^3 + \cdots$, or is ≈ 2 for $\alpha \ll 1$. Then the load admittance for $\alpha \ll 1$ can be written

$$Y_{load} = \frac{1}{R}[1 + jQ'_o 2\alpha] \quad (10\text{-}14)$$

and load impedance is

$$Z_{load} = \frac{1}{Y_{load}} = \frac{R}{1 + jQ'_o 2\alpha} \quad (10\text{-}15)$$

The voltage amplification can be expressed as

$$A_v = -g_m R \frac{1}{1 + jQ'_o 2\alpha} \quad (10\text{-}16)$$

The gain will be "down 3 dB" when $Q'_o 2\alpha = 1$, or at $\alpha = 1/2Q'_o$. Then the bandwidth of this amplifier is between the frequencies $f_o - (f_o/2Q'_o)$ and $f_o + (f_o/2Q'_o)$. These may be expressed as

$$f_o \pm f_o\left(\frac{1}{2Q'_o}\right) = f_o \pm \frac{1}{4\pi RC}$$

and the bandwidth, B, is $2\alpha f_o$, or

$$\frac{f_o}{2Q'_o} = \left(\frac{B/2}{f_o}\right)^{-1}$$

Then Eq. 10-16 becomes

$$A_v = -g_m R \left(1 + j2Q'_o \frac{B/2}{f_o}\right)^{-1}$$

$$= A_{v\,max}\left(1 + j2Q'_o \frac{B/2}{f_o}\right)^{-1} \quad (10\text{-}17)$$

Thus, the half-power frequencies are a measure of the circuit Q'_o: $Q'_o = f_o/B$.

Example 10-4. A pentode having $g_m = 5000\ \mu$mho is to be used in a single-tuned amplifier. The signal to be amplified is centered at 1 MHz and has a half-power bandwidth of 10 kHz. Determine required values of R, L, and C, and the amplification at resonance.

Solution: For the specified center frequency, $\omega_o^2 = 1/LC$, and the bandwidth in terms of R and C is $B = 1/2\pi RC$,

$$\omega_o^2 = (2\pi)^2(10^6)^2 = \frac{1}{LC}$$

$$2\pi B = (2\pi)(10^4) = \frac{1}{RC}$$

Once the value of C is chosen, the other two components are fixed by the specified frequencies. The value of C will depend on parasitic wiring* and interelectrode capacitances as well as on the actual component used for C in the circuit. Suppose its value is taken as 200 pF for calculation; then the other components are

$$R = \frac{1}{2\pi(10^4)(200)(10^{-12})} = 80 \text{ k}\Omega$$

$$L = \frac{1}{(2\pi)^2(200)} = 0.126 \text{ mH}$$

The amplification at resonance is $A_v = -g_m R = -(5)(10^{-3})(80)(10^3) = -400$.

10-10
Double and Stagger Tuning

Figure 10-13 shows a double-tuned amplifier, which is probably the most widely used. The two tuned circuits are coupled by the mutual inductance

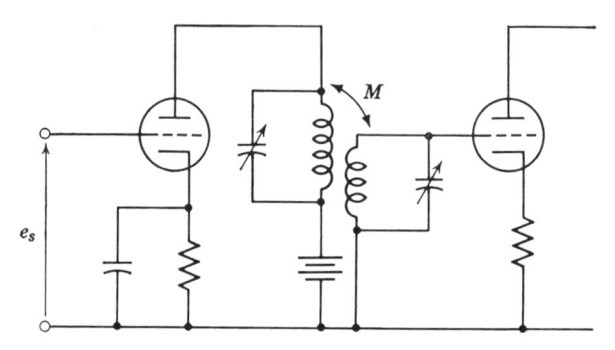

Figure 10-13
Double-tuned amplifier.

---

* Stray capacitance is present between connecting wires in the amplifier circuit.

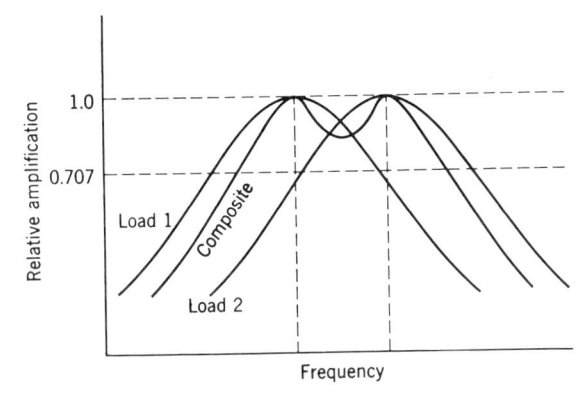

Figure 10-14
Relative amplification response in a stagger-tuned amplifier.

between the coils. Each tank circuit is tuned to the same center frequency, and if each has the same amplification and filtering characteristics in the bandwidth of incoming signals, the overall performance is superior to that of the single-tuned circuit. The double-tuned circuit provides (1) selective filtering and amplification, and (2) impedance matching for maximum amplification of signals.

When several tuned stages are cascaded for greater amplification, the bandwidth is reduced by the successive amplifications. It may be necessary or desirable to expand the bandwidth at some point in the cascaded amplification. This can be accomplished by *stagger tuning*, i.e., using a different center frequency for two successive tuned loads. Then the overall performance generates a composite amplification as shown in Fig. 10-14. Each stage of amplification may generate a sharply defined bandwidth, but the composite more closely resembles the ideal response of a frequency-selective amplifier.

Questions

10-1 In a differential amplifier circuit, why does one input signal generate an output at both output terminals?

10-2 When operating a differential amplifier in its single-ended mode, why is it necessary to connect one input terminal to ground?

10-3 What is common-mode rejection?

10-4 Why is a high CMRR desirable in a differential amplifier?

10-5 What is the significance of a high input impedance in an amplifier?

10-6 Discuss the possible use of a Darlington circuit in a differential amplifier.

10-7 Identify some possible use for a differential amplifier.

378   Special-Purpose Amplifiers   Ch. 10

Problems

10-1   Sketch the output waveforms at both output terminals of the differential amplifier in Fig. P10-1.

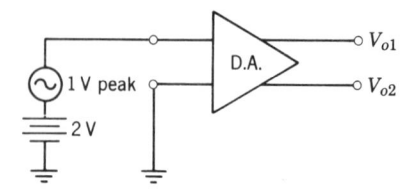

Figure P10-1

10-2   Sketch the output waveforms for the circuit of Problem 10-1, for the input signals reversed.

10-3   Draw the ac equivalent circuit for the amplifier in Fig. 10-4, using h-parameters for the transistors.

10-4   For the circuit of Fig. 10-4, assume the collector resistors and transistors are identical, and having these values: $R_C = 5 \text{ k}\Omega$, $h_{ie} = 1 \text{ k}\Omega$, $h_{fe} = 100$, $h_{re} = 10^{-5}$, $h_{oe} = 100 \,\mu$mhos. Calculate the ac differential voltage gain.

10-5   The Darlington circuit described in Example 10-2 is used for a voltage source having $20 \text{ k}\Omega$ resistance and feeds a collector load of $100 \text{ k}\Omega$. Determine for the circuit those quantities listed in Table 8-2.

10-6   A Darlington circuit is used in a common-collector configuration, feeding a 1-kΩ emitter load. Compute the quantities listed in Table 8-2, assuming the transistors are identical to those described in Example 10-2, and the source resistance is $10 \text{ k}\Omega$.

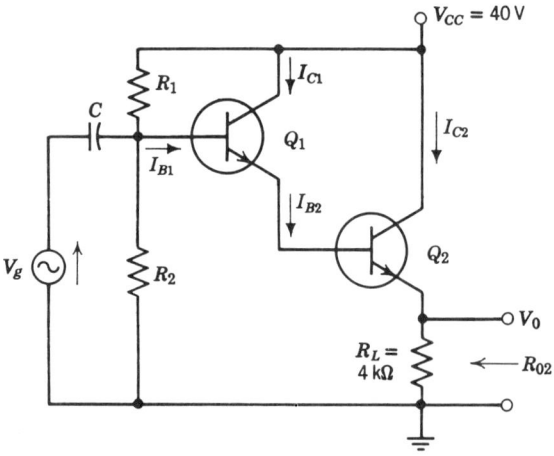

Figure P10-7

10-7 For the direct-coupled emitter-follower amplifier in Fig. P10-7, determine the quiescent operating points for both transistors.

10-8 An operational amplifier is being used as a simple voltage adder. The three input voltages are 10 Vdc, 5 Vdc, and 5 sin ωt, and the feedback resistor $R_f = 1\,M\Omega$. What must be the values of input resistors?

10-9 Sketch the waveform of output voltage from the amplifier in Problem 10-8. Compare it with the output waveform if all input resistors are chosen to be equal to $0.5\,R_f$.

10-10 An operational amplifier is to be used to change the algebraic sign of an input voltage but must not alter its amplitude. Show the circuit that is used, and specify the ratio of input and feedback impedances.

10-11 The three input voltages of Problem 10-8 are to be added together and appear at the output with the same phase as the input. Show a circuit that may be used to sum the voltages and maintain them in the same phase at the output.

10-12 A single tuned-circuit amplifier has its maximum amplification at 1 MHz and a half-power bandwidth of 10 kHz. A pentode having $g_m = 5000\,\mu$mho and $r_p = 1\,M\Omega$ is used to drive the tank circuit. The output voltage is taken across a resistor $R_g = 100{,}000\,\Omega$, as in Fig. 10-12. Determine the required values of L and C and the amplification at the half-power frequencies.

10-13 If the value of C in Problem 10-12 is variable by ± 50 per cent, how will its variation affect circuit operation? When it is adjusted to its extreme values, what will be the circuit bandwidth? Through what range of frequencies will the variable C change the resonant frequency?

11
FEEDBACK AMPLIFIERS AND OSCILLATORS

Some of our previous discussions of amplifiers developed the concept of *feedback*, an example being the reduction in voltage gain at low frequencies because of poor bypassing of cathode bias resistors. In transistor circuits, feedback was introduced into emitter bias networks to stabilize the circuit operation against temperature changes in its parameters. In those cases where the signal that was fed back counteracted the input signal, the circuit was said to have *negative* or *degenerative feedback*.

It is also possible to introduce feedback signals that will reinforce input signals. This leads to *positive* or *regenerative feedback* and can develop oscillations in the circuit, at a frequency determined by components of the circuit. A circuit that is designed to perform in this manner is called an *oscillator*.

This chapter introduces the general concept of feedback applied to an amplifier for stability of its operation, to decrease noise and distortion, and to broaden its bandwidth. We shall also discuss some basic features of electronic oscillators.

11-1
Basic Considerations of Feedback

The process of applying to the input of an amplifier a signal that is proportional to either current or voltage in the output circuit of the amplifier is called *feedback*. Feedback affects the gain, input and output impedances, and

381

frequency response of the amplifier, and under certain circumstances it will cause the amplifier to sustain an oscillation that is independent of any applied input signal.

Figure 11-1 shows the effects of feedback in its simplest form. A_v represents the voltage amplification of the amplifier without feedback, and the block designated as β is the feedback network.* The input signal, E_i, is to be amplified to form an output signal, E'_o. However, a feedback signal, E_{fb}, is introduced at the amplifier in addition to E_i. The feedback signal is a fraction of the output voltage E'_o:

$$E_{fb} = \beta E'_o \tag{11-1}$$

and the signal at the output can be expressed as the sum of E_i and E_{fb}. The output signal becomes

$$E'_o = (E_i + E_{fb})A_v$$
$$= (E_i + \beta E'_o)A_v \tag{11-2}$$

Rearrangement of this equation shows that the gain with feedback is related to the gain without feedback in this way:

$$\text{Feedback gain, } A'_v = \frac{E'_o}{E_i} = \frac{A_v}{1 - \beta A_v} \tag{11-3}$$

The term βA_v is called the *feedback factor*. We can see that its value compared to unity is important in determining the feedback gain.

The amount of feedback is often expressed in decibels and is usually given by

$$\text{Decibel of feedback} = 20 \log_{10} \frac{1}{|1 - \beta A_v|}$$
$$= -20 \log_{10} |1 - \beta A_v|$$

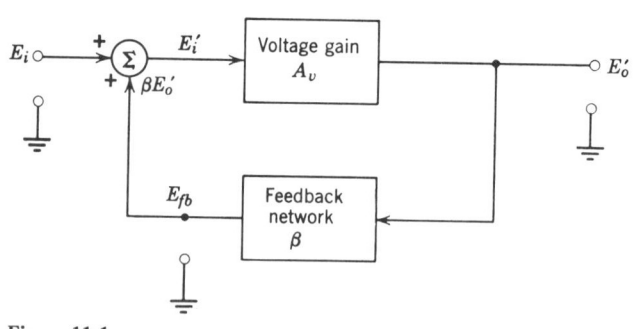

Figure 11-1
Basic block diagram of a feedback amplifier.

* This symbol for feedback circuits does not refer to the beta of transistors in the circuits.

Observe that the ratio of the gain with feedback to the gain without it is

$$\frac{A'_v}{A_v} = \frac{1}{1 - \beta A_v}$$

The *decibel of feedback* is given by 20 $\log_{10}$ of this voltage ratio.

Example 11-1 An amplifier has a gain without feedback expressed as 20 dB. What value of feedback factor is required to reduce the gain (with feedback) to 10 dB?

Solution: The amount of feedback is obviously 10 dB (20 dB − 10 dB = 10 dB). Then

$$10 \text{ dB} = -20 \log_{10} |1 - \beta A_v|$$

$$-\tfrac{1}{2} = \log_{10} |1 - \beta A_v| = \log_{10} 0.317$$

This requires that the feedback factor be equal to 1.000 − 0.317 = 0.683.

Since the gain without feedback is 20 dB, its numerical value is $A_v = 10$. This means that the value of β is negative, because the gain was reduced.

The feedback is termed *negative* or *degenerative* when it reduces the gain and *positive* or *regenerative* when it increases the gain. These statements can be symbolized as

$$|1 - \beta A_v| > 1 \quad \text{for negative feedback}$$
$$|1 - \beta A_v| < 1 \quad \text{for positive feedback}$$

However, it is seen that a special case arises when $|1 - \beta A_v| = 0$, or $\beta A_v = 1$. The feedback gain under this condition becomes infinitely large, from Eq. 11-3. This means, in a practical sense, that the amplifier provides its own input signal from its operation and breaks into self-sustaining oscillation. This condition is purposely introduced in a class of circuits called *oscillators*, the subject of later sections of this chapter. It is important to note here that an amplifier circuit employing feedback is subject to oscillation, and its design must take this possibility into account. A circuit design should be checked closely for the presence of feedback loops that might initiate oscillation where it is not wanted.

11-2
Negative (Inverse) Feedback in Amplifiers

The circuit designer may choose to have the feedback voltage either in phase or out of phase with the input signal. If they are in phase, then β in Eq. 11-3 is a *positive* number and *positive* or *regenerative feedback* results. When the feedback voltage is out of phase with the input signal, the value of β will be negative and *negative feedback* results.

Let us demonstrate that when feedback is added to an amplifier, its gain with frequency changes, i.e., its frequency response is affected. Consider an amplifier that has a voltage gain of 10 in the midband range without feedback, so that $E_i = 1/\underline{0°}$ when $E_o = 10/\underline{180°}$

$$E_i = 1 + j0 \qquad E_o = -10 + j0$$

If $\beta = -0.1$, from Eq. 11-3,

$$A'_v = \frac{10}{1-(-0.1)(10)} = 5$$

This means a 50 per cent drop in midband gain caused by adding feedback $\beta = -0.1$.

We now investigate the effect of feedback at the half-power frequencies. The output voltage *without feedback* is down to 7.07 V and, since there is a $+45°$ phase shift, E_o is represented at low frequencies as

$$E_o = 7.07/\underline{225°} = -5 - j5 \text{ V}$$

The feedback voltage is $\beta E_o = 0.5 + j0.5$ V, and the *new input voltage* is $E_i + \beta E_o$.

$$E'_i = E_i + \beta E_o$$
$$= 1 + j0 + 0.5 + j0.5 = 1.5 + j0.5 \text{ V}$$

The gain with feedback is E_o/E'_i:

$$A'_v = \frac{-5-j5}{1.5+j0.5} = \frac{7.07/\underline{225°}}{1.58/\underline{18.4°}} = 4.47/\underline{206.6°}$$

This shows that the gain at the low-frequency *half-power point* has fallen only to 89.4 per cent of the midband gain of 5, whereas without feedback the gain fell 29.3 per cent, from 10 to 7.07. The phase shift is an angle of $206.6° - 180° = 26.6°$ instead of $45°$ without feedback. The same amount of shift occurs at the upper half-power frequency. This means that the bandwidth is extended at a sacrifice in gain, when negative feedback is used.

With high-mu triodes and pentodes, the sacrifice in midband gain due to feedback is not a problem, because even with feedback the gain is very satisfactory. That the gain holds up so well at low and high frequencies means that the new midband range is much wider than the old one.

The decrease in phase-shift effect means that *pulse distortion* in amplifiers is appreciably diminished by feedback. Whether there is amplitude or phase distortion in an amplifier without feedback, we can visualize how it is

Sec. 11-3 **Effect of Voltage Feedback on Input and Output Impedances** 385

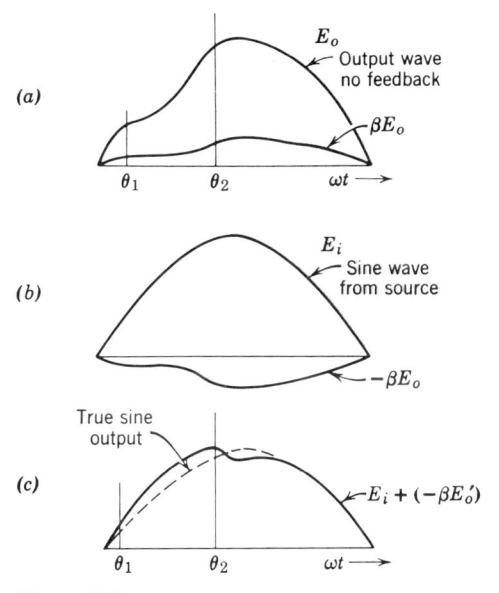

Figure 11-2
Voltages of negative-feedback amplifier. $E_i + (-\beta E_o')$ is the input voltage with feedback.

diminished by considering the following. Assume that an amplifier distorts a sine wave and delivers an output wave E_o as in Fig. 11-2a. The wave shape indicates that the gain of the amplifier is below normal from θ_1 to θ_2 on the ωt axis. The wave βE_o, which is a fraction of E_o, is picked off at the feedback tap and fed back in inverse phase to the grid, where it combines with the pure sine wave from the source (Fig. 11-2b) and modifies it to the shape in Fig. 11-2c of $E_i + (-\beta E_o)$. When the amplifier performs its "faulty-gain" procedures on this wave, its below-normal effects from θ_1 to θ_2 have larger than normal values to work on, and so these are reduced to normal and the result is a true sine-wave output.

11-3
Effect of Voltage Feedback on Input and Output Impedances

In Fig. 11-3, the *input impedance* of the amplifier *with feedback* is

$$Z_{if} = \frac{E_i}{I}$$

But $E_i = E_{gk} + E_{ka}$, so that

$$Z_{if} = \frac{E_{gk} + E_{ka}}{I}$$

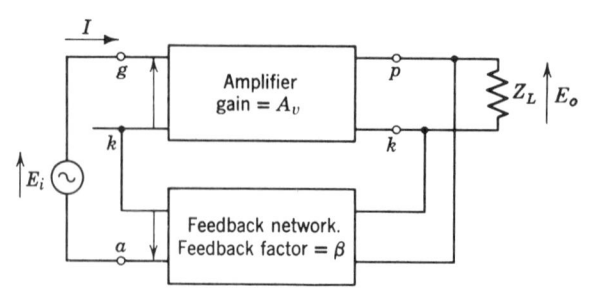

Figure 11-3
General block diagram for voltage-feedback amplifier.

Note that $E_{ka} = -E_{gk} \times A'_v \times \beta$, so that *the input impedance of an amplifier with negative feedback* is

$$Z_{if} = \frac{E_{gk}(1 - \beta A'_v)}{I} \tag{11-4}$$

Now the input impedance of the amplifier *without feedback* is

$$Z_i = \frac{E_{gk}}{I} \tag{11-5}$$

so that the relation between the two impedances is

$$Z_{if} = Z_i(1 - \beta A'_v) \tag{11-6}$$

Since β is always negative for inverse, or negative, feedback, it is evident that *adding inverse-voltage feedback increases the input impedance.*

As an example, suppose an amplifier has an input impedance without feedback of 10,000 Ω and a voltage gain of 40. Adding inverse feedback for which $\beta = -0.15$ increases the input impedance to 70,000 Ω. Negative-voltage feedback is responsible for the high input impedance of the cathode follower.

Negative-current feedback also increases the input impedance of an amplifier. Since voltage proportional to current is fed back, it does not matter where or what the source is.

To obtain an expression for Z_{of}, *output impedance with feedback*, it is necessary to recall that Thévenin's theorem enables us to represent an amplifier stage as shown in Fig. 11-4. In Fig. 11-4a, instead of looking back from the output terminals and seeing a series impedance—formerly called Z' but now Z_o, the output impedance of the amplifier itself without feedback—we now *apply a voltage E'_o* to the output terminals. The feedback branch inside the amplifier picks off the β fraction of the terminal voltage E'_o and sends it back to the input in inverse phase as $-\beta E'_o$. The gain of the amplifier (used here like μ of the tube) then makes the generated voltage in the equivalent circuit become $-\beta A_v E'_o$. This is all *without an input signal* at the front end of the stage.

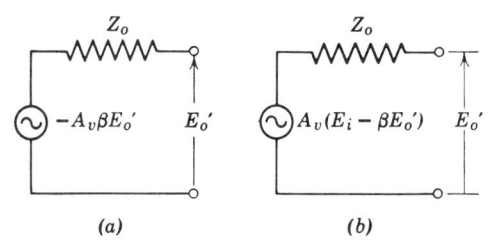

Figure 11-4
Equivalent-output circuits of voltage amplifier with inverse feedback. (a) No input signal, but E'_o is applied at output terminals. (b) Input signal E_i at front end; this produces E'_o at output.

When an input signal E_i is applied at the front end, it too is amplified, and the equivalent circuit becomes that in Fig. 11-4b. E'_o shows up of its own accord; i.e., we need not apply an external voltage at the output, of course. But let us use this value of E'_o and apply it to the output terminals after first removing the input signal and its source.

The current in the series circuit of Fig. 11-4a is then

$$I'_L = \frac{(E'_o - A_v \beta E'_o)}{Z_o}$$

from which we can get $E'_o / I'_L = Z'_o$, the output impedance with feedback. The *output impedance of an amplifier with negative voltage feedback is*

$$Z'_o = \frac{E'_o}{I'_L} = \frac{Z_o}{1 - \beta A_v} \qquad (11\text{-}7)$$

This means that the impedance looking back into an amplifier from the output terminals is reduced (remember that β is negative) by adding voltage feedback. Changes in external-load impedance, and therefore in current, will not cause as much change in load terminal voltage when inverse (negative) feedback is used, and the larger the product βA_v, the greater the reduction in Z'_o and the more stable the output voltage.

As an example, suppose an amplifier has an output impedance of 35,000 Ω and a voltage gain of 40 without feedback. Adding inverse feedback for which $\beta = -0.15$ decreases the output impedance to 5000 Ω. We will note later that current feedback increases the output impedance.

Drill Problems

D11-1 A voltage amplifier has a gain of -100 without feedback. What value of β will increase the input impedance from 100 kΩ to 500 kΩ?

D11-2 What value of β in the amplifier of D11-1 will decrease the output impedance from 50 kΩ to 10 kΩ?

D11-3 (*a*) How much change occurs in the output impedance in D11-1? (*b*) How much does the input impedance change in D11-2?

11-4
Feedback in Transistor Amplifiers

The effects of inverse feedback on the impedances of an amplifier are much more important in transistor circuits than in tube circuits. The inherently low input impedance of transistor amplifiers, particularly the common-base and common-emitter circuits, is a handicap in amplifier design. Inverse-voltage feedback therefore helps solve that problem. Similarly, the reduction in output impedance of common-emitter and common-base transistor amplifiers through the use of inverse feedback is conducive to better matching possibilities in *R*-*C* coupled circuits.

Transistor circuits using negative feedback are shown in Figs. 11-5 and 11-6. R_f is the feedback resistor in both circuits. Current feedback, generally termed *series* feedback, is achieved in Fig. 11-5 in the following manner. First we assume static conditions; this means there is no input signal ($E_i = 0$). The dc current direction in R_f is then *upward*, making the emitter terminal negative with respect to the positive terminals of the power sources. The collector current flows *out* of the collector terminal.

A positive input signal, E_i, will be in opposition to this polarity and will reduce the emitter and collector current. This causes the ac component of the collector current (i.e., the output current in R_L) to flow *into* the collector terminal and *downward* through R_f. Therefore the net input signal from emitter to base has been reduced and the feedback is negative. Because it is proportional to the ac current in R_f, it is termed *current* or *series* feedback.

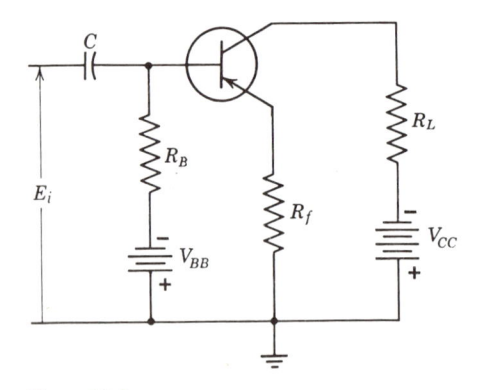

Figure 11-5
Transistor amplifier stage using negative current feedback.

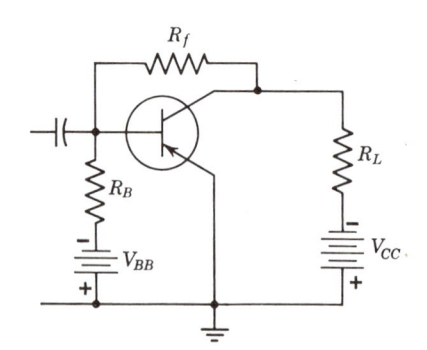

Figure 11-6
Transistor amplifier using negative voltage feedback.

To explain negative-voltage (*shunt*) feedback in a transistor amplifier stage, we refer to Fig. 11-6. When a positive input signal is applied, an *amplified current* (ac) appears in the output resistor R_L due to the current amplification. This current flows into the collector terminal, as explained for the circuit of Fig. 11-5, with the result that the potential of the upper end of R_L, and of the collector terminal, is given a negative increment. This drop in potential is fed to the base of the transistor and thus opposes the positive-input signal.

Example 11-2. We will analyze the circuit of Fig. 11-5 to determine the effects of feedback on gain and input impedance. The equivalent circuit for Fig. 11-5 is shown in Fig. 11-7.

Solution: From Chapter 8 the voltage gain of the common-emitter amplifier without the feedback across R_f is

$$A_{ve} = \frac{-h_{fe}R_L}{h_{ie}(1 + h_{oe}R_L) - h_{re}h_{fe}R_L}$$

and its input impedance is

$$Z_{in} = h_{ie} - \frac{h_{re}h_{fe}R_L}{1 + h_{oe}R_L}$$

We will calculate the voltage gain of the circuit by first writing the equations that define its operation, based on the equivalent circuit:

$$E_i = h_{ie}I_b + h_{re}E_c + E_{fb} \qquad (1)$$

$$E_{fb} = (I_b + I_c)R_f \qquad (2)$$

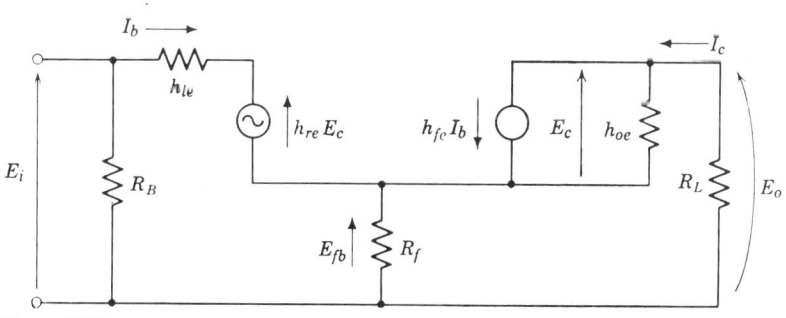

$$E_c = \frac{1}{h_{oe}}(I_c - h_{fe}I_b) \qquad (3)$$

$$E_o = E_{fb} + E_c \qquad (4)$$

$$E_o = -I_c R_L \qquad (5)$$

Figure 11-7

390 Feedback Amplifiers and Oscillators Ch. 11

By eliminating E_{fb}, E_c, and E_o, we form a determinant from these equations:

$$E_i = I_b h_{ie} + h_{re}\left[\frac{I_c - h_{fe}I_b}{h_{oe}}\right] + (I_b + I_c)R_f$$

$$0 = (I_b + I_c)R_f + \left[\frac{I_c - h_{fe}I_b}{h_{oe}}\right] + I_c R_L$$

$$\begin{vmatrix} I_b & I_c & \\ h_{ie} - \dfrac{h_{re}h_{fe}}{h_{oe}} + R_f & \dfrac{h_{re}}{h_{oe}} + R_f & E_i \\ -\dfrac{h_{fe}}{h_{oe}} + R_f & \dfrac{1}{h_{oe}} + R_f + R_L & 0 \end{vmatrix}$$

The output voltage is

$$E_o = -I_c R_L = -\frac{\begin{vmatrix} h_{ie} - \dfrac{h_{re}h_{fe}}{h_{oe}} + R_f & E_i \\ -\dfrac{h_{fe}}{h_{oe}} + R_f & 0 \end{vmatrix}}{\begin{vmatrix} h_{ie} - \dfrac{h_{re}h_{fe}}{h_{oe}} + R_f & \dfrac{h_{re}}{h_{oe}} + R_f \\ -\dfrac{h_{fe}}{h_{oe}} + R_f & \dfrac{1}{h_{oe}} + R_f + R_L \end{vmatrix}} \times R_L$$

and the feedback gain is, after considerable manipulation,

$$A'_v = \frac{E_o}{E_i}$$

$$= -\frac{(h_{fe}R_L)\left(\dfrac{1}{h_{oe}}\right)\left(1 - \dfrac{R_f h_{oe}}{h_{fe}}\right)}{\dfrac{1}{h_{oe}}[h_{ie}(1 + h_{oe}R_L) - h_{re}h_{fe}R_L + R_f(1 + h_{oe}R_L + h_{oe}h_{ie} - h_{re}h_{fe} - h_{re} + h_{fe})]}$$

(6)

For $R_f = 0$, this reduces to the expression for A_{ve}. We should put this in the form of Eq. 11-3:

$$A'_{ve} = \frac{A_{ve}}{1 - \beta A_{ve}} = \frac{N/D}{1 - \beta(N/D)}$$

where N is the numerator and D the denominator of the expression for A_{ve}, without feedback. If we note the similarities of expressions for A_{ve} and A'_{ve}, we may write A'_{ve} in a "shorthand" notation as

$$A'_{ve} = \frac{NB}{D + C}$$

or

$$A'_{ve} = \frac{N/D}{1 + \left(\dfrac{C - BD + D}{NB}\right)\left(\dfrac{N}{D}\right)} \tag{7}$$

Then

$$\beta = -\left(\frac{C - BD + D}{NB}\right)$$

Substituting the circuit parameters from Eq. 6 yields for A'_{ve}, after considerable manipulation,

$$A'_{ve} = -\frac{\dfrac{h_{fe}R_L}{h_{ie}(1 + h_{oe}R_L) - h_{re}h_{fe}R_L}}{1 - \left[+\dfrac{R_f}{R_L}\left(1 + \dfrac{1 + h_{oe}R_L}{h_{fe}}\right)\left(\dfrac{(h_{fe} + \Delta_h)}{1 - h_{oe}R_f}\right)\right]\dfrac{h_{fe}R_L}{h_{ie}(1 + h_{oe}R_L) - h_{re}h_{fe}R_L}}$$

where $\Delta_h = (h_{oe}h_{ie} - h_{re}h_{fe})$. Thus,

$$\beta = +\frac{R_f}{R_L}\left(\frac{h_{fe} + \Delta_h}{1 - h_{oe}R_f}\right)\left(1 + \frac{1 + h_{oe}R_L}{h_{fe}}\right)$$

In a practical circuit, $h_{oe}R_L \ll 1$, and

$$\beta \approx +\frac{R_f}{R_L}(h_{fe} + \Delta_h) \tag{8}$$

The input impedance is $Z'_{in} = E_i/I_b$, which may be calculated from the determinant above Eq. 6:

$$Z'_{in} = \frac{E_i}{I_b} = \frac{\Delta_h}{\dfrac{1}{h_{oe}}(1 + h_{oe}R_L + h_{oe}R_f)}$$

$$= \frac{\left(h_{ie} - \dfrac{h_{re}h_{fe}}{h_{oe}} + R_f\right)(1 + h_{oe}R_f + h_{oe}R_L) + \left(\dfrac{h_{fe}}{h_{oe}} - R_f\right)(h_{re} + h_{oe}R_f)}{1 + h_{oe}R_L + h_{oe}R_f}$$

$$= \frac{[h_{ie}(1 + h_{oe}R_L) - h_{re}h_{fe}R_L] + R_f(1 + h_{oe}R_L - h_{re}h_{fe}) + R_f(h_{oe}h_{ie} - h_{re} + h_{fe})}{[1 + h_{oe}R_L] + h_{oe}R_f}$$

The bracketed terms are the input impedance *without feedback*. We note that Z'_{in} can be larger than Z_{in} if the numerator increases more than the denominator. For the values of h-parameters expected for a practical transistor and circuit element values expected, e.g., $R_L \gg R_f$, $1/h_{oe} \gg R_f$, the input impedance with feedback may be approximated:

$$Z'_{in} \approx h_{ie} - \frac{h_{re}h_{fe}R_L}{1 + h_{oe}R_L} + \frac{R_f(1 + h_{fe})}{1 + h_{oe}R_L} \tag{9}$$

392 Feedback Amplifiers and Oscillators   Ch. 11

A 2N334 transistor may have these typical values of common-emitter parameters: $h_{ie} = 2 \times 10^3 \ \Omega$, $h_{oe} = 16 \ \mu\text{mho}$, $h_{fe} = 30$, $h_{re} = 10^{-3}$. We will calculate the effect of R_f on voltage gain when the feedback resistor is increased from 0 to 100 Ω, and its effect on input impedance. For these calculations we will assume the load is $R_L = 5 \ \text{k}\Omega$.

The voltage gain without feedback is

$$A_{ve} = \frac{-h_{fe}R_L}{h_{ie} - R_L\Delta_h}$$

where Δ_h is defined as $(h_{ie}h_{oe} - h_{fe}h_{re})$

$$A_{ve} = \frac{-30 \times 5 \times 10^3}{2 \times 10^3 - 5 \times 10^3(2 \times 10^3 \times 16 \times 10^{-6} - 30 \times 10^{-3})}$$

$$= \frac{-150 \times 10^3}{2 \times 10^3 - 10} = -\frac{150{,}000}{1990}$$

$$= -75.4$$

From Eq. 8, when $R_f = 100 \ \Omega$

$$\beta \approx \frac{R_f}{R_L}(h_{fe} + \Delta_h)$$

$$\approx \frac{100}{5 \times 10^3}(30 + 2 \times 10^{-3})$$

$$\approx 0.6$$

The voltage gain will be reduced because of the feedback

$$A'_{ve} = \frac{A_{ve}}{1 - \beta A_{ve}}$$

$$= \frac{-75.4}{1 - 0.6(-75.4)} = -\frac{75.4}{46.24}$$

$$= -1.63$$

The emitter resistor has a significant effect on the voltage gain of the common-emitter amplifier. Even if R_f were increased to only 2 Ω, the feedback would affect the gain:

$$\beta \approx \frac{2 \times 30}{5 \times 10^3} = 0.012$$

and the gain would reduce to

$$A'_{ve} = \frac{-75.4}{1 + 0.012(75.4)} = \frac{-75.4}{1.905}$$

$$= -39.5$$

Sec. 11-5   Criterion for Sustained Oscillations

A resistor is often used in the emitter circuit to stabilize circuit operation, using the resulting negative feedback. However, the value of the emitter resistor will be exceedingly small compared with other circuit components because of its great effect on amplifier gain.

The input impedance without feedback ($R_f = 0$) may be calculated:

$$Z_{in} = h_{ie} - \frac{h_{re}h_{fe}R_L}{1 + h_{oe}R_L}$$

$$= 2 \times 10^3 - \frac{30(10^{-3})(5 \times 10^3)}{1 + 16 \times 10^{-6}(5 \times 10^3)}$$

$$= 2 \times 10^3 - \frac{150}{1 + 80 \times 10^{-3}}$$

$$Z_{in} \approx 2000 - 150 = 1850 \, \Omega$$

Because of the feedback in the emitter circuit when $R_f = 100 \, \Omega$, the input impedance will increase, according to Eq. 9,

$$Z'_{in} \approx Z_{in} + \frac{R_f(1 + h_{fe})}{1 + h_{oe}R_L}$$

$$= 1850 + \frac{100(1 + 30)}{1 + 16 \times 10^{-6}(5 \times 10^3)}$$

$$Z'_{in} \approx 1850 + 3100 = 4950 \, \Omega$$

Drill Problems

D11-4   A transistor having these parameter values is used in a common-emitter amplifier circuit that has a voltage gain of 100 without feedback: $h_{ie} = 500 \, \Omega$, $h_{re} = 3 \times 10^{-4}$, $h_{fe} = 200$, $h_{oe} = 15 \, \mu\text{mho}$. The load is $R_L = 5 \, \text{k}\Omega$ and emitter feedback resistor $R_f = 100 \, \Omega$. Calculate the voltage gain with feedback and the value of β.

D11-5   For the amplifier of D11-4, calculate the change in input impedance caused by R_f.

11-5
Criterion for Sustained Oscillations

Figure 11-8 is a block diagram of a feedback oscillator with gain $A_v = E_2/E_1$ and a feedback voltage $E_f = \beta E_2$. Since E_1 is the required input voltage to produce the output voltage E_2, the following is a requirement for maintaining operation:

$$\beta E_2 = E_f = E_1 \qquad (11\text{-}8)$$

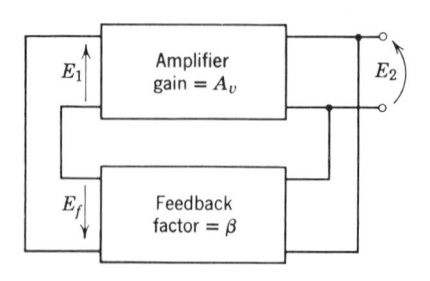

Figure 11-8
Block diagram of a feedback oscillator.

This means that the feedback network must produce this required magnitude and its phase must be the same as the phase of E_1. This can be achieved in a practical circuit, so we can write

$$A_v = \frac{E_2}{\beta E_2} = \frac{1}{\beta}$$

$$\beta A_v = 1 \tag{11-9}$$

a requirement for sustained oscillation.

Since β can never be greater than unity, A_v must not be less than unity in a feedback oscillator.

Example 11-3. Calculate the necessary conditions and frequency of oscillation in the tuned-collector oscillator, shown in Fig. 11-9.

Solution: Set up circuit equations:

$$E_i = I_b h_{ie} + h_{re} E_c \tag{1}$$

$$I_c = I_b h_{fe} + h_{oe} E_c \tag{2}$$

For the transformer,

$$E_c = (j\omega L_1 + R_1)I_1 - j\omega M I_2 \tag{3}$$

$$E_2 = -j\omega M I_1 + (j\omega L_2 + R_2)I_2 \tag{4}$$

The currents I_1 and I_2 are

$$I_1 = -I_c - j\omega C E_c \tag{5}$$

$$I_2 = -\frac{E_2}{Z_i} = -\frac{E_2}{h_{ie}} \tag{6}$$

Substituting (2), (5), and (6) into (3) and (4) yields

$$E_c = -(R_1 + j\omega L_1)(I_b h_{fe} + h_{oe} E_c + j\omega C E_c) + j\omega M \left(\frac{E_2}{h_{ie}}\right) \tag{7}$$

$$E_2 = j\omega M (I_b h_{fe} + h_{oe} E_c + j\omega C E_c) - (R_2 + j\omega L_2)\left(\frac{E_2}{h_{ie}}\right) \tag{8}$$

Sec. 11-5   Criterion for Sustained Oscillations

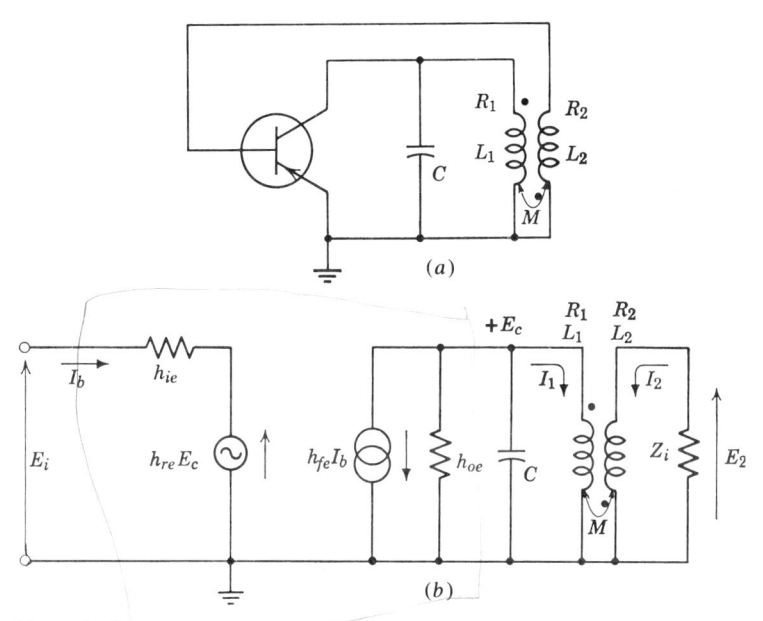

Figure 11-9
Tuned-collector oscillator. (a) Circuit schematic. (b) Equivalent circuit.

and with (1)

$$E_i = I_b h_{ie} + h_{re} E_c \tag{1}$$

a determinant may be written for the coefficients of E_2, E_c, and I_b.

	E_2	E_c	I_b	
Eq. 1:	0	h_{re}	h_{ie}	E_i
Eq. 7:	$\dfrac{j\omega M}{h_{ie}}$	$-1-(R_1+j\omega L_1) \times (h_{oe}+j\omega C)$	$-h_{fe}(R_1+j\omega L_1)$	0
Eq. 8:	$-1-\dfrac{R_2+j\omega L_2}{h_{ie}}$	$j\omega M(h_{oe}+j\omega C)$	$j\omega M h_{fe}$	0

We then calculate the value of the determinant D; after a lengthy manipulation, it is

$$D = -h_{ie} - R_2 + R_1 \Delta\left(1 - \frac{R_2}{h_{ie}}\right)$$
$$+ \omega^2\left[\frac{\Delta}{h_{ie}}(L_1 L_2 - M^2) + C(h_{ie}L_1 + R_2 L_1 + R_1 L_2)\right]$$
$$+ j\omega\left[\omega^2(L_1 L_2 - M^2)C + L_2\left(R_1 \frac{\Delta}{h_{ie}} - 1\right)\right]$$

where $\Delta = h_{ie}h_{oe} - h_{fe}h_{re}$.

396 Feedback Amplifiers and Oscillators Ch. 11

Similarly, we calculate E_2:

$$E_2 = -\frac{j\omega M h_{fe} E_i}{D}$$

The oscillator criterion (section 11-5) requires that $E_2/E_1 = 1$, so we may set the real parts equal and the imaginary parts equal. The real parts determine the frequency of oscillation, and the imaginary parts set the conditions necessary for sustained oscillation. Thus, from the real part,

$$0 = -h_{ie} - R_2 + R_1\Delta\left(1 - \frac{R_2}{h_{ie}}\right)$$
$$+ \omega^2\left[\frac{\Delta}{h_{ie}}(L_1L_2 - M^2) + C(h_{ie}L_1 + R_1L_2 + R_2L_1)\right]$$

The frequency of oscillation is

$$f = \frac{1}{2\pi}\left[\frac{h_{ie} + R_2 + R_1\Delta\left(\frac{R_2}{h_{ie}} - 1\right)}{\frac{\Delta}{h_{ie}}(L_1L_2 - M_2) + C(h_{ie}L_1 + R_1L_2 + R_2L_1)}\right]^{1/2}$$

In a practical circuit, the resistances R_1 and R_2 will be small compared to the reactances of L_1 and L_2, that is, the Q will be high. Further, if the two windings are closely coupled ($k \approx 1.0$), since $M = k\sqrt{L_1L_2}$, the frequency may be approximated by

$$f \approx \frac{1}{2\pi}\left(\frac{1}{L_1C}\right)^{1/2} \tag{9}$$

Thus, the oscillation frequency will be determined by the tuned circuit in the collector and primary of the transformer.

The condition for oscillation derives from the imaginary parts:

$$-j\omega M h_{fe} = j\omega\left[\omega^2(L_1L_2 - M^2)C + L_2\left(R_1\frac{\Delta}{h_{ie}} - 1\right)\right]$$

Substituting from Eq. 9 yields

$$-Mh_{fe} = \frac{1}{L_1C}(L_1L_2 - M^2)C + L_2\left(R_1\frac{\Delta}{h_{ie}} - 1\right)$$

Assuming R_1 small compared to the reactance of L_1, as before, this reduces to

$$h_{fe} = \frac{M}{L_1} \tag{10}$$

If the circuit is to oscillate at a frequency of 1 MHz, using an R-F transformer having $L_1 = 1$ mH, the required value of C may be calculated by

Eq. 9, assuming high Q and close coupling in the transformer:

$$f = \frac{1}{2\pi}\left(\frac{1}{L_1 C}\right)^{1/2}$$

$$\sqrt{C} = \frac{1}{2\pi f}\left(\frac{1}{L_1}\right)^{1/2} = \frac{1}{2\pi \times 10^6}\left(\frac{1}{10^{-3}}\right)^{1/2}$$

then

$$C = \frac{1}{4\pi^2 \times 10^{12}}\left(\frac{1}{10^{-3}}\right) \approx 25 \text{ pF}$$

Drill Problems

D11-6 A tuned-collector oscillator is designed to have a variable frequency from 200 kHz to 1 MHz, using an R-F transformer having $L_1 = 1$ mH. Assuming high Q and close coupling, determine the necessary variable range of C.

D11-7 In the oscillator of D11-6, the coefficient of coupling between primary and secondary windings of the transformer is $k = 0.95$ and $L_2 = 10$ mH. (*a*) If the transistor has $h_{fe} = 10$ or more, will the circuit generate oscillations? (*b*) As h_{fe} is increased, say by using a different transistor in the circuit, what must be done at the transformer to continue oscillations?

11-6
Resistance-Capacitance Oscillators

These oscillators are used as *signal generators* from the lowest audio range up to around 10 MHz. Before studying the circuit, we should consider the basic *phase-shift oscillator* in which sustained oscillations are produced without the aid of inductive-capacitive resonance.

We have found that in a series path containing capacitance and resistance a shift in phase takes place, and the voltage between the terminal common to the two units and either one of the ends has a phase that is different from the phase of the voltage across the ends. In the circuit of Fig. 11-10, it has been found that with three R-C networks a total phase shift of 180 degrees can be achieved without losing the signal. This means that the output voltage, which is actually a small fraction of the ac voltage across R_L, is applied to the grid of the tube in proper phase to maintain oscillation.

A practical circuit for a phase-shift oscillator is that of Fig. 11-11. The tube is a twin triode with a large amplification factor, such as a 6SC7 or a 6SL7. The variable R_k adjusts the gain of the first stage so that oscillation may be started

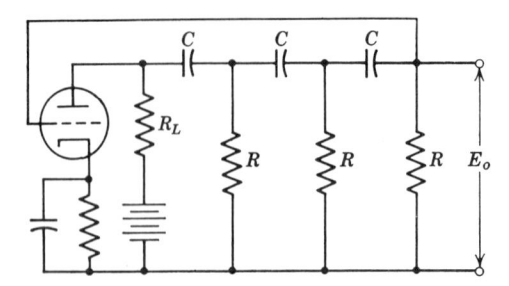

Figure 11-10
Basic circuit of a phase-shift oscillator.

or controlled. The second stage is a cathode follower which isolates the phase-shift network from the first stage.

The *resistance-capacitance* oscillator is a two-stage R-C coupled amplifier with special features for feeding voltage back to the first grid through a phase-sensitive network. In Fig. 11-12, R_f and C_f are the phase-sensitive feedback components, and they form a *coupling network* with R_1 and C_1. The net phase shift through the amplifier is zero; therefore it is evident that for oscillation to be maintained the *phase shift through the coupling network must be zero*. It can be shown that this occurs at the frequency given by

$$f = \frac{1}{2\pi\sqrt{R_f C_f R_1 C_1}} \tag{11-10}$$

It is practical to make $R_f = R_1$ and $C_f = C_1$. Calling them simply R and C, we have

$$f = \frac{1}{2\pi RC} \tag{11-11}$$

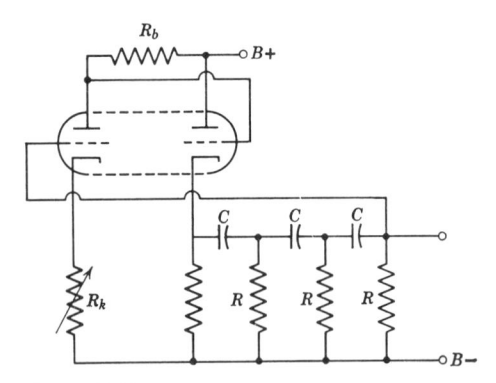

Figure 11-11
Practical circuit for a phase-shift oscillator.

Sec. 11-6   Resistance-Capacitance Oscillators   399

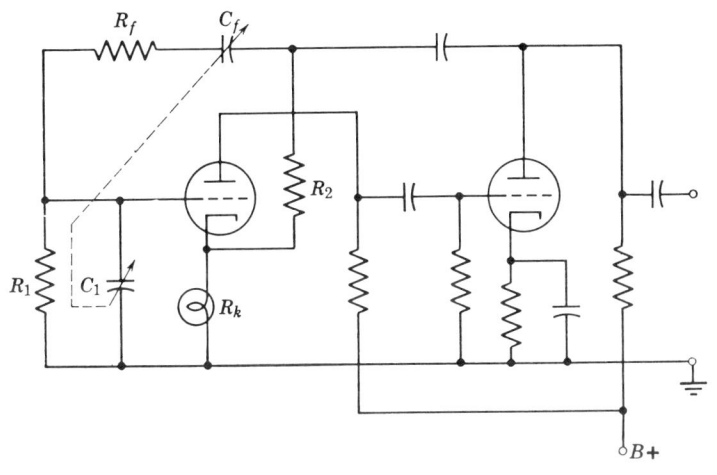

Figure 11-12
Circuit of resistance-capacitance oscillator.

For these conditions, $\beta = 1/3$ for the feedback and this means the amplifier must have a gain of at least 3.*

In addition to this positive feedback, a *negative* feedback voltage is developed across a cathode resistor R_k that has a special feature that is very valuable here. The required ohms value of R_k *increases rapidly* as its current increases, and therefore the negative bias increases *much more rapidly than the current*. An ordinary 3-W 115-V light bulb with a tungsten filament may be used as R_k. Its resistance depends on temperature, whether developed by $I_{dc}^2 R_k$ or $I_{eff}^2 R_k$.

A change in amplitude, and therefore in effective value, of signal current through R_k will change the negative feedback and the gain of the amplifier in a direction to oppose the change in signal level. The signal current in R_k has nearly the same effective value as the dc plate current. The result is extremely stable output voltage, good frequency stability, and excellent output-voltage waveform.

The resistance-capacitance oscillator is capable of *continuous variation of frequency* by means of a single tuning arrangement consisting of C_f and C_1 as variable air capacitors mounted on a single shaft carrying a frequency dial. Multiplication of the frequency values of scale divisions of the dial by 10, 100, 1000, etc., is made possible by switching in different pairs of R_f and R_1, which are fixed resistors.

* The proof is given in the Appendix.

Drill Problem

D11-8 The circuit of Fig. 11-12 is used to generate audio frequencies from 20 Hz to 40 kHz in 3 ranges: 20 to 400 Hz, 200 to 4000 Hz, and 2 to 40 kHz. The frequency is varied by adjusting C_1 and C_f, and the range is chosen by switching in different values for R_1 and R_f. If C_1 and C_f can vary from 100 to 2000 pF, what are the required values of R_1 and R_f in each of the three ranges, assuming $R_1 = R_f$?

Questions

11-1 What is meant by *feedback* in an amplifier?

11-2 Distinguish between negative and positive feedback.

11-3 Why is negative feedback called degenerative?

11-4 Why is positive feedback called regenerative?

11-5 What does β represent? When is it a negative number and when is it a positive number?

11-6 What is the significance of the product of β and the gain without feedback?

11-7 What is the effect of inverse feedback on the gain of an amplifier? on the frequency-response curve?

11-8 How does inverse feedback affect phase shift through an amplifier stage?

11-9 How does inverse feedback affect the input impedance and the output impedance of an amplifier? Are these effects advantageous in transistor amplifiers? Explain.

11-10 How is feedback provided in a resistance capacitance oscillator? What about phase shift?

11-11 What makes possible a *continuous frequency band* of operation of a resistance-capacitance oscillator? How is the frequency band made so extensive, i.e., from low-audio to the megacycle range?

Problems

11-1 The gain of an amplifier with *negative*-voltage feedback is 100. Assume the input impedance without feedback to be 50 kΩ. Calculate data for plotting a curve of input impedance with feedback vs. β from $\beta = -0.005$ to $\beta = -0.15$.

11-2 On the same curve sheet used for Problem 11-1, plot the output impedance of the amplifier vs. β over the same range, using the output impedance *without* feedback as $Z_o = 500$ kΩ.

11-3 A resistance-capacitance oscillator uses two variable capacitors mounted on the same shaft for tuning. At one setting of the decade multiplier the frequency range is from 200 to 2000 Hz. (*a*) What values of the product RC in Eq. 11-11 are required for the two extremes of the frequency range? (*b*) Assume that 0.0001 μF is the minimum value possible for C. Calculate the value of R. (*c*) Since R is fixed, C must be changed to obtain a frequency at the other end of the range. Calculate the value of C.

11-4 In the circuit of Fig. 11-5, V_{CC} = 20 V, R_L = 10 kΩ, and the circuit has a voltage gain of -100 when R_f = 0. Parameters h_{ie}, 500 Ω; h_{re}, 3 $\times$ 10^{-4}; h_{fe}, 200; h_{oe}, 15 μmho. Calculate the change in voltage gain when R_f = 10 Ω. What is the value of feedback β?

11-5 In the circuit of Problem 11-4, what is the change in input resistance caused by R_f?

11-6 Examine qualitatively the effects of R_f on voltage gain and input resistance in the circuit of Problem 11-4.

12
BASIC MODULATION AND DETECTION IN COMMUNICATIONS

An alternating current, or voltage, that has an unchanging waveform cannot convey any information other than that which is related to its frequency and magnitude. Some physical property of the wave—e.g., its amplitude, frequency, or phase—must be varied if it is to convey information (intelligence). The variation must be caused by, and therefore correspond to, the intelligence that is to be transmitted. When such variations are produced in the waveform the wave is said to be *modulated*.

Let us visualize a current or a voltage that has simple sine-wave form (Fig. 12-1a). It may be a *carrier wave* whose shape can be changed to conform to the tones of the voice or to some other kind of intelligence. When this is done, the waveform may change to something like that in Fig. 12-1b. It is then called a *modulated carrier wave*.

12-1
Amplitude Modulation (A-M)

As the name implies, amplitude modulation is the process of changing the instantaneous values of the voltage or current from those of the steady sine-wave contour to other values with the result shown in Fig. 12-1b. The amplitude of each cycle of the wave might thus be changed and *amplitude modulation* would exist.

The frequency of the carrier wave is much higher than the frequency of the voice wave or other modulating wave. The *contour* of the modulated carrier

(a)

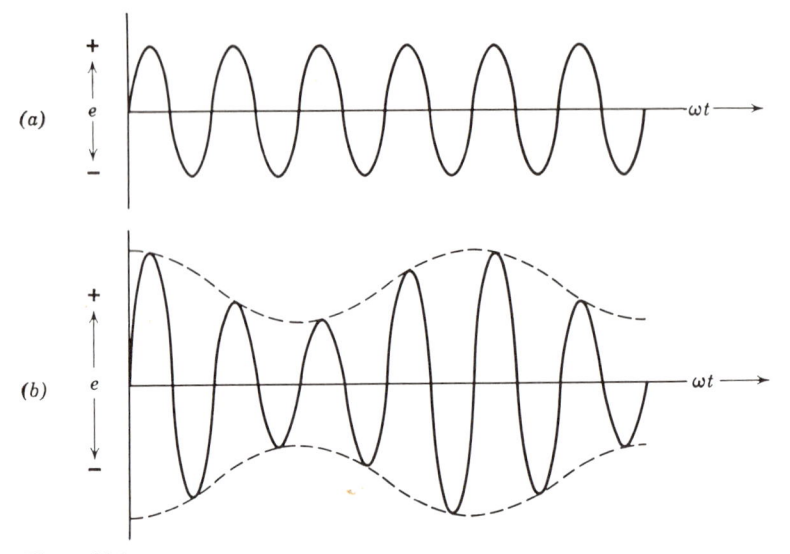

(b)

Figure 12-1
Voltage waves radiated by the antenna of an amplitude-modulated radio transmitter. (*a*) Carrier wave. (*b*) Modulated carrier wave.

(shown as a dashed line along the envelope) varies at *modulation frequency*. For example, if modulation is done by a 256-Hz sine wave (the frequency of middle C on the musical scale), the *variations in the envelope* occur at 256 Hz.

12-2
Frequency Modulation (F-M)

The electromagnetic wave radiated by commercial television antennas and by some radio antennas is frequency modulated. The instantaneous frequency of the antenna current and, simultaneously, of the antenna voltage is varied in accordance with the intelligence transmitted. Only the frequency of the carrier is changed; the amplitude remains practically constant, as shown in Fig. 12-2.

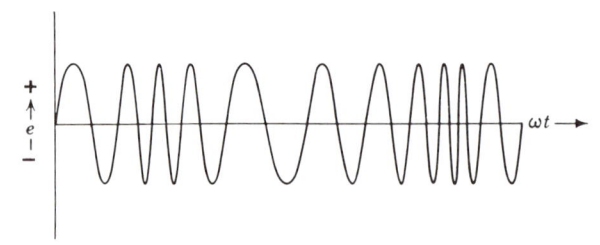

Figure 12-2
Voltage wave radiated by the antenna of a frequency-modulated radio transmitter.

12-3
Pulse-Code Modulation

The modulating signal may be in the form of pulses (sudden bursts of voltage) that are coded to represent the elements of information to be transmitted. The information may be the temperature inside an orbiting space vehicle, for example. A prearranged *code* would be employed to identify variations in the pulses that represent temperature values which can be *telemetered* by radio to an earth station. The *telemetry* transmission system converts temperature readings into electrical signals and arranges them into a *pulse code*. In this form they alter the high-frequency carrier wave, by varying either the amplitude or the frequency.

12-4
Demodulation (Detection)

After the modulated carrier wave is received and converted into an amplified signal, the intelligence must be recovered from it. The process of information recovery is called *demodulation* or *detection*.

The transmission of information by modulation and subsequent detection is usually achieved by *frequency translation*. The modulated carrier, being a *complex wave*, has components with frequencies different from the frequency of the original, unmodulated carrier. Some components have frequencies higher than carrier frequency and some have lower frequencies. These components are related to the lower frequency components in the original modulating signal that carries the intelligence. For example, an A-M carrier at 100-kHz frequency modulated by a 1-kHz signal will have frequency components at 99 and 101 kHz in addition to the 100-kHz carrier. The 1-kHz signal has been *translated in frequency* so that it appears in the form of two much higher frequencies, one above carrier frequency and one below. In the detection process the original 1-kHz signal is recovered in its original form from the modulated components.

12-5
Classification of Radio Waves

Electrical energy travels in space in the form of electromagnetic waves. *Radio waves* are those whose frequencies are within the 30 kHz to 30,000 MHz range. They travel at the velocity of light: 3×10^8 m/s (186,000 miles per second). The distance, along the line of travel, between two successive crests of a wave is called the wavelength, usually denoted by λ, the Greek letter *lambda*. Physicists usually use the letter c to represent the velocity of light. We

406 Basic Modulation and Detection in Communications Ch. 12

shall use v to represent the velocity of a radio wave in free space or in air. The relation between wavelength, velocity, and frequency (f) is

$$v = f\lambda \qquad (12\text{-}1)$$

Frequencies (cycles per second) are now expressed in hertz, but the frequencies of radio waves are in the kilohertz (kHz) or megahertz (MHz) ranges. Wavelength is expressed in meters or, when the frequency is extremely high, in centimeters (see Table 12-1).

Table 12-1
Frequency Ranges and Classification

Classification	Frequency Range	Wavelength Range	Typical Uses
Low frequency (LF)	30–300 kHz	10,000–1000 m	Long-distance point-to-point, marine, and navigation aid systems
Medium frequency (MF)	300–3000 kHz	1000–100 m	Commercial broadcast, marine communication, navigation, harbor telephone
High frequency (HF)	3–30 MHz	100–10 m	Moderate and long distance communications of various sorts
Very high frequency (VHF)	30–300 MHz	10–1 m	Television, F-M radio broadcast, radar, aircraft navigation
Ultra-high frequency* (UHF)	300–3000 MHz	100–10 cm	Television, radar, telemetry, telephone relay systems
Super-high frequency (SHF)	3000–30,000 MHz	10–1 cm	Radio relay, radar, space communications

* Waves with frequencies higher than 2000 MHz are usually called *microwaves*.

Radio waves are classified into groups (or bands) according to their frequencies. Classification and typical applications are listed in Table 12-1.

Radio waves in the various frequency ranges are important in the study of modulation and detection, because they are commonly used as carriers. From the table it is seen that typical uses include commercial A-M and F-M broadcasting and television.

Drill Problem

D12-1 Calculate (*a*) the wavelength of a 60-Hz power line signal, (*b*) the frequency of a 40-m wave.

12-6
Separation of Radio Carriers

Radio waves are sent out into space from radiating systems called *antennas*. An antenna is designed with specific dimensions that are related to the wavelength of the carrier wave that is radiated, although an alternating current radiates a certain amount of energy in the form of electromagnetic waves from an open-wire circuit of any configuration. The amount of radiated energy is much greater, however, if all dimensions of the circuit approach the wavelength dimension. In the same way, a radio wave is captured by an antenna having dimensions that approach the wavelength of passing waves.

The electromagnetic energy of a radio wave, in passing over a conducting antenna, induces a voltage in the antenna that varies in the same manner as the passing wave and produces a current at the frequency of the wave. The induced voltage and associated current carry energy that is absorbed from the wave. Since every radio wave passing over an antenna produces its own voltage in the antenna, the receiving equipment must be capable of separating the signal of a desired frequency from all the others that are present. The separation of one signal from the others is made on the basis that each transmitting source uses a different carrier frequency. In commercial A-M broadcast systems, each station is allotted a *bandwidth* of 10 kHz about its carrier frequency, and carrier frequencies are separated by 10 kHz. For example, one broadcast station may be assigned a carrier frequency of 700 kHz and other stations assigned 710 kHz, 720 kHz, etc. Television stations in the United States are assigned carrier frequencies that are 6 MHz apart in frequency because of the larger bandwidth needed in the much higher frequency range to transmit information.

Resonant circuits can be made to discriminate between radio waves that have different frequencies. Resonant circuits respond very strongly in favor of a particular frequency. The ability of a system to separate waves of different frequencies is referred to as *selectivity*. A high selectivity means that frequency separation is achieved for signals that are close together in frequency. The process of adjusting a resonant circuit to select first one frequency and then another is called *tuning*. For example, a radio broadcast receiver can be tuned to pick up any station in the A-M range, from 550 to 1600 kHz, by varying a capacitance in a resonant circuit of the receiver.

12-7
Amplification of Modulated Carriers

The voltage induced in a receiving antenna may be very small, either because the transmitted signal is weak or because the signal at the receiving antenna is small. Much more satisfactory operation of a receiving detector is possible from relatively large signals. Modulated carriers may be amplified by circuits in the receiver before detection, or the information signal may be amplified after detection. The first case is referred to as *radio-frequency* (R-F) amplification, performed by amplifiers specifically designed to handle the high frequencies. Amplification of the detected information is handled by *audio amplifiers*.

Sometimes it is necessary to amplify signals within the receiver that are at some *intermediate frequency* between the carrier and the information frequencies. These intermediate frequencies (I-F) are generated in the frequency translation process, combined with the information frequencies conveyed by the carrier. The reasons for amplification of modulated carriers will be understood better as we progress through the study of modulation and detection processes.

12-8
Fundamental Modulation Process

A sinusoidal* voltage may be written as

$$e = E_c \cos \phi = E_c \cos(\omega t + \phi_o) \qquad (12\text{-}2)$$

and represented on the complex plane as a phasor rotating in time at a constant angular velocity. Its instantaneous amplitude is a projection on the real axis of the complex plane, as shown in Fig. 12-3.

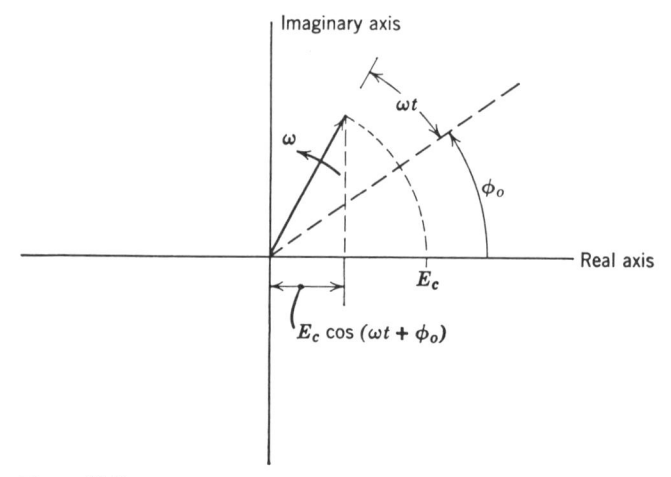

Figure 12-3
Rotating phasor on complex plane representing a time-varying sinusoid.

---

* Because a cosine wave has the same shape as a sine wave, and because it is more convenient to work with cosine terms in this application, we are using cosines. The only difference between sine and cosine functions of time is the choice of the instant at $t = 0$.

The concept that a sinusoid can be considered as a rotating phasor is sometimes lost in repeated study of steady-state, constant-frequency networks. We will examine the notation of Eq. 12-2 to show that the argument of the cosine function, $\omega t + \phi_o$, can be used to modulate the voltage waveform. We will also see that the amplitude, E_c, may be altered to achieve modulation.

The instantaneous angle of the phasor in Fig. 12-3, referred to the real axis, is changing in time because of the constant angular velocity, ω. Then we may write

$$\omega = \frac{d\phi}{dt}$$

$$\phi = \int \omega \, dt + \text{a constant}$$

The constant may be zero, or an angle ϕ_o at $t = 0$. In general, then,

$$\phi = \int \omega \, dt + \phi_o \tag{12-3}$$

This may be used in the form of Eq. 12-2 to express the voltage as

$$e = E_c \cos\left(\int \omega \, dt + \phi_o\right) \tag{12-4}$$

Equation 12-4 indicates that two factors may be used to modulate the sinusoidal voltage waveform: (1) the amplitude, E_c, and (2) the instantaneous angle, $\int \omega \, dt + \phi_o$. Variations in these two factors result in the two basic types of modulation. Variation of amplitude, E_c, with time is amplitude modulation (A-M). Variation of the angle with time results in two types of angle modulation: (1) time variation of $\int \omega \, dt$ yields frequency modulation (F-M), and (2) time variation of ϕ_o yields *phase modulation*. In our discussions of this chapter, we shall be interested only in A-M and F-M. Phase-modulation systems are used in special applications, but A-M and F-M are by far the most prevalent.

12-9
Analysis of Amplitude Modulation (A-M)

We shall first consider the process of *altering the amplitude* of a sinusoidal waveform to generate A-M signals for transmission. We shall rewrite Eq. 12-2 to show a sinusoidal variation of its amplitude:

$$e = (E_c + E_s \cos \omega_a t) \cos(\omega_o t + \phi_o) \tag{12-5}$$

in which the angular velocity symbol ω_o represents the constant frequency of the carrier wave, and ω_a is 2π times the frequency of the modulating signal.

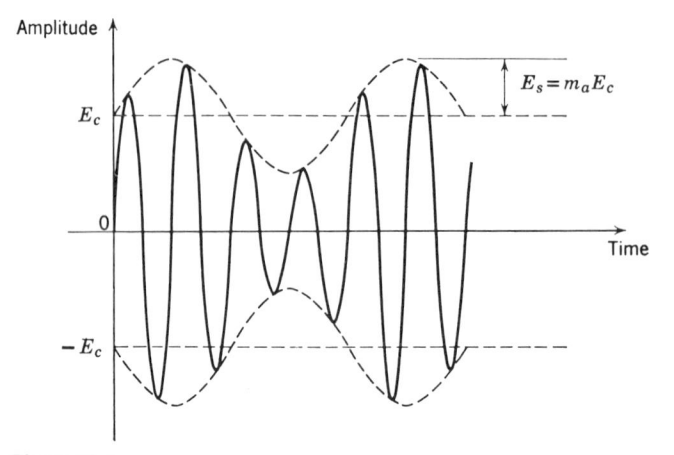

Figure 12-4
Sketch of A-M wave, showing envelope developed by modulating signal.

The amplitude expression is usually written as

$$A = E_c + E_s \cos \omega_a t$$
$$= E_c \left[1 + \left(\frac{E_s}{E_c}\right) \cos \omega_a t \right]$$
$$= E_c (1 + m_a \cos \omega_a t)$$

The factor m_a is the ratio of the modulating signal and carrier amplitudes. It is called the *degree of modulation* or *modulation factor*. Why it has this name may be better explained by reference to Fig. 12-4, where Eq. 12-5 is graphed as a function of time. It can be seen that the amplitude of the modulating signal is important in determining the resultant waveform of the modulated carrier. The alterations in carrier amplitude are directly related to the changes in modulating signal. The modulating signal produces an *envelope* on both positive and negative portions of the carrier. As stated earlier, it is this envelope that represents the information being conveyed by the modulated carrier.

The factor m_a is often used to designate the degree of modulation as "modulation percentage." This number results when the value of m_a is multiplied by 100%. It is usually desirable to maintain $m_a < 1.0$, i.e., <100%. If the modulation is allowed to become greater than this, distortion of the signal may result so that detector circuits cannot retrieve the original information.

The A-M voltage wave may be written as

$$e = E_c (1 + m_a \cos \omega_a t) \cos \omega_o t \qquad (12\text{-}6)$$

dropping the constant angle, ϕ_o, as having no significance in the variations. This equation may be shown to contain *three* frequency components, even though it was derived from the combination of only *two*, a modulating signal and a carrier frequency. A familiar trigonometric identity permits the A-M signal to be written as*

$$e = E_c \cos \omega_o t + \frac{m_a}{2} E_c \cos (\omega_o + \omega_a)t + \frac{m_a}{2} E_c \cos (\omega_o - \omega_a)t \quad (12\text{-}7)$$

This shows that the A-M signal is a composite of three separate component phasors, each at a different frequency. The first term of Eq. 12-7 is the carrier-frequency component. Added to it are two other frequency components, one at a frequency equal to the *sum* of the carrier and modulating frequencies and the other at a frequency equal to their *difference*. Thus, *three* frequency components make up the total A-M wave.

As long as the carrier frequency is greater than the modulating frequency, Eq. 12-7 represents the A-M signal. Should the modulating signal frequency become equal to or greater than the carrier frequency, then this expression is not valid. The final term would generate a "negative frequency." In typical applications of A-M the carrier frequency is many times greater than the modulating signal frequency. For example, a radio broadcast station is limited to a bandwidth of 10 kHz about its carrier. This means that its transmitted signal, represented by Eq. 12-7, must not contain frequencies outside the limits (carrier frequency + 5 kHz) and (carrier frequency − 5 kHz). For the station that uses 1000 kHz as a carrier, the ratio of frequencies of carrier and modulating signal is 1000 to 5 (200 to 1) as a limit. The form of the graph in Fig. 12-4 indicates that the A-M signal will contain 200 cycles of carrier *for each cycle* of 5 kHz modulating signal. The graph actually shows a 4 to 1 ratio.

Drill Problems

D12-2 Make graphical sketches for these two expressions, plotting at least one cycle of the lowest frequency, on a common time scale. Which one describes A-M? Why doesn't the other describe A-M? (*a*) $e = \cos t + \cos 10t$. (*b*) $e = \cos 10t + \cos t \cos 10t$.

D12-3 An A-M wave of frequency 1 kHz is modulated at 50 per cent by a signal frequency of 100 Hz. Sketch the waveform over one cycle of the modulating signal.

D12-4 An A-M broadcast station is assigned a carrier of 1 MHz and is limited to a bandwidth of 10 kHz. What range of information frequencies may it transmit?

---

* $\cos A \cos B = \frac{1}{2}[\cos (A + B) + \cos (A - B)]$.

12-10
Generation of Sidebands in A-M

A constant-frequency carrier of a broadcast transmitter is modulated by signals that originate as vocal or musical sounds. Seldom is the modulating signal a single-frequency sinusoid. It is usually a combination of several components having different frequencies and acting simultaneously.

The principle of superposition may be used to show that the A-M signal is composed of the *carrier, a sum component,* and a *difference component* for *each individual frequency* that is a part of the modulating signal. The sum component is called the *upper sideband* component, and the difference component is called the *lower sideband* component of the A-M wave. They are called sideband components because they exist in *bands of frequencies* above and below the carrier on the frequency scale. All of the information being transmitted is contained in the sideband components, none in the carrier. The carrier (or central) component of the complex wave acts only as the vehicle that "carries" the information-bearing sideband components. Because signals representing voice and music contain many frequencies and their harmonics, a band of frequencies on each side of the carrier frequency exists in the A-M wave.

An advantage of amplitude modulation is that low-frequency signals may be translated to higher frequencies for ease of transmission, and then retranslated at a receiver to the lower (original) frequency range for reclaiming the information. Because the carrier may be produced at any desired frequency, the information may be carried at a convenient frequency at any location in the frequency spectrum. Figure 12-5 shows the placement of the sidebands about a

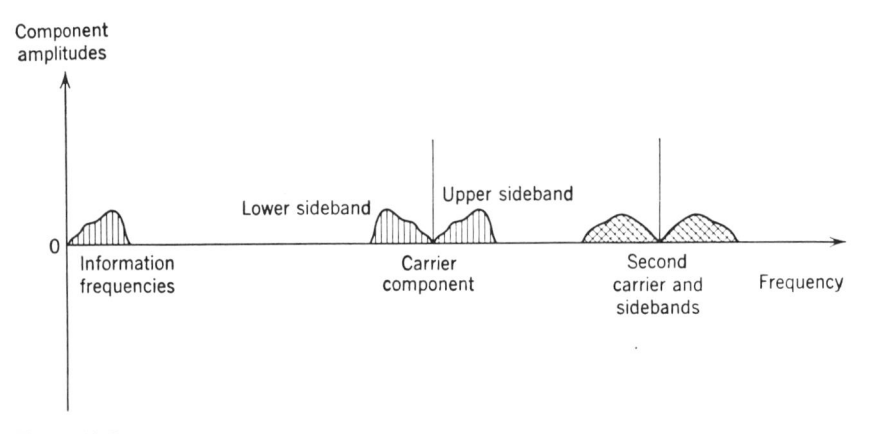

Figure 12-5
Frequency translation in A-M system.

carrier in A-M transmission. The shaded sidebands about a second carrier frequency indicate that several transmitting stations may broadcast simultaneously without interference, provided the sidebands do not overlap. A detector must have a bandwidth large enough to include the sideband frequencies it is tuned to receive. If another carrier has sideband components that overlap those of the desired carrier, the detector will accept both signals and produce a distorted output.

In a *superheterodyne* radio receiver, the carrier and its sidebands are translated from the carrier frequency to some intermediate frequency (I-F), amplified at the I-F, and then translated again to the information frequency band by a detector. This employment of an intermediate frequency makes the superheterodyne receiver more selective than any other kind. That is, its circuit is more efficient than that of any other receiver in rejecting unwanted signals that have sidebands very near to the one desired.

12-11
Bandwidth Requirements in A-M

Each frequency component of a modulating signal produces a pair of sideband frequencies that are spaced symmetrically about the carrier frequency. The new frequencies describe a bandwidth about the carrier that is twice the frequency of the modulating signal. It is apparent that the bandwidth of the A-M wave will be twice the highest frequency contained in the modulating signal. The bandwidth that is required in a given application to convey information by A-M depends on the type of intelligence involved. The bandwidth will be greater for rapidly varying signals (containing high-frequency components) than for signals made up of low-frequency components. Bandwidth requirements for voice transmission are less stringent than for television signals, because of the different ranges of frequencies involved.

From a listener's point of view, the quality of a voice transmission can be measured in terms of two characteristics, intensity and intelligibility. These two parameters, although they are virtually independent of each other over a broad frequency range, together determine the quality of reproduced sound. Most of the *energy* in a voice signal, and hence the intensity, is contained in the lower frequencies, while the higher frequency harmonics and overtones contribute intelligibility.

The curves of Fig. 12-6 illustrate the roles of frequency and intensity in the intelligibility of voice signals. If all frequencies of a voice signal below 1000 Hz are eliminated, about 85 per cent of the intelligence would be understood, even though the energy content would be only about 17 per cent of the original voice energy. Similarly, if the frequencies above 1000 Hz were eliminated, 83 per cent of the energy would remain but only about 45 per cent of the signal would

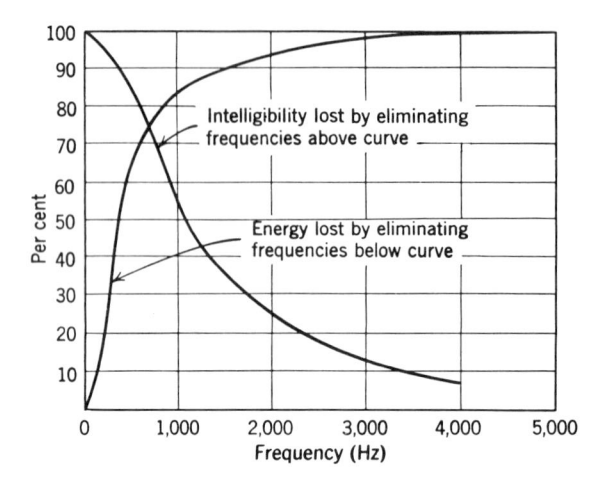

Figure 12-6
Curves show that energy in voice signals is concentrated at low frequencies, and intelligibility is concentrated in high frequencies.

be understood. This means that voice transmission must include both low- and high-frequency components for acceptable operation.

Because bandwidths are limited in many A-M systems, some compromise is usually necessary. In a telephone relay system, in which voice signals are the primary information, a smaller bandwidth may be used than for A-M radio broadcast in which signals are both voice and music. The compromise in A-M radio is shown by the 5-kHz bandwidth allowed for each station (10 kHz about the carrier, because the total bandwidth is twice the highest frequency of information). This bandwidth is insufficient for high-fidelity reproduction of music signals, which may have frequency components above 15 kHz, but it is acceptable for most listeners. In a telephone transmission system designed to carry voice signals only, a smaller bandwidth may be used because the normal speaking voice generates significant frequencies up to about 3500 Hz. Typically, the range of frequencies that is transmitted is from 200 to 3200 Hz. Table 12-2 lists several signals and the bandwidth that is used for A-M transmission.

12-12
Power Contained in A-M Sidebands

When a modulating signal is a single-frequency sinusoid, the power in the A-M wave will consist of three parts, one for the carrier voltage and two for the sideband voltages. Equation 12-7, rearranged here, is useful in a study of the

Table 12-2
Typical Signals and Amplitude-Modulation Frequencies

Type of Signal	Transmitted Frequency Range (Hz)
Telegraph	
Morse code at 100 words per minute	0–170
Voice	
Typical broadcast program	100–5000
Long-distance telephone	200–3200
Intelligible but poor quality	500–2000
Television	
Standard 525-line picture, interlaces, 30 frames per second rate	60–4,500,000
Pulses	
1 μs duration	0–1,000,000

distribution of power in the A-M wave:

$$e = E_c \left[\cos \omega_o t + \frac{m_a}{2} \cos (\omega_o + \omega_a)t + \frac{m_a}{2} \cos (\omega_o - \omega_a)t \right] \quad (12\text{-}7)$$

Since power associated with a sinusoidal voltage is proportional to the square of the rms voltage, total power in the A-M wave is

$$\text{Power} = kE_c^2 \left(1 + \frac{m_a^2}{4} + \frac{m_a^2}{4} \right) = K \left(1 + \frac{m_a^2}{2} \right) \quad (12\text{-}8)$$

The power contained in the sidebands is $m_a^2/2$ times the carrier power. For a modulation percentage of 100 per cent ($m_a = 1$), sideband power is only half the carrier power. For lower modulation percentages, the portion of total power contained in the sidebands is even less than this fraction. The ratio of sideband power to total power can be expressed as follows:

$$\frac{\text{Sideband power}}{\text{Total power}} = \frac{m_a^2/2}{1 + m_a^2/2} \quad (12\text{-}9)$$

When $m_a = 1.0$, the ratio of sideband power to total power is 1/3, i.e., the sidebands contain only one-third of the total power. At 10 per cent modulation, the ratio is 1/201; the sideband power is less than half of 1 per cent of total power. The fraction of the total power that represents information carried by the sidebands decreases rapidly as the modulation percentage is reduced. For this reason, a high percentage of modulation is generally desired.

12-13
Methods of Generating A-M

We have discussed the possibility of slowly varying the envelope of a sinusoidal carrier voltage in accordance with a modulating signal. We may now ask: How is A-M physically produced? To answer this question, we note that the mathematical expression for A-M (Eq. 12-5) is a *product* function, in which two quantities are multiplied together. The simplest solution would be to devise a method for multiplying the two quantities and obtain an output that is their product. A circuit that performs this function is called a *product modulator*.

Before looking at some circuits that may be used to generate A-M, we need to discuss some characteristics that are commonly used as a basis for their operation. Modulating devices and circuits may be classified in one of two categories: (1) their terminal (voltage–current) characteristics are *nonlinear*, or (2) the device contains a *switch* independent of the signals that changes operation from one linear characteristic to another, *piecewise linear* characteristics.

We know that linear networks are characterized by two basic properties: the response to a sum of inputs is the sum of responses to each input applied separately, and multiplication of an input by a constant also multiplies the response by the same constant. The first property is a statement of the superposition principle. The second property reinforces the concept of linearity, implying that the terminal conditions are independent of any unique property of the network. As an example, consider the steady-state behavior of a simple R-L-C network in response to a sinusoidal input voltage. The network merely modifies the amplitude and phase of the voltage. *No new frequencies are generated in the linear network.* Thus, a linear network cannot be used to modulate a carrier signal.

A nonlinear network, on the other hand, distorts input signals and generates frequency components that are not present in the input. Similarly, a *piecewise linear* network or device, such as the diode rectifier studied in Chapter 3, can generate harmonic frequencies based on the input signal frequency. When it is operated so that its region of operation extends over two separate linear portions of its characteristic, harmonic frequencies are generated. Thus, its overall operation would be nonlinear.

As an example of A-M generation by a nonlinear device, we will use a semiconductor diode whose characteristic is given in Fig. 12-7. Assume that this characteristic curve can be expressed by

$$i = a_1 e + a_2 e^2 + a_3 e^3 + \cdots \tag{12-10}$$

The first term is a linear relationship between current and voltage, and the other terms show the presence of curvature in the graph of i vs. e. The first two terms represent the only significant portions of the current, because the higher

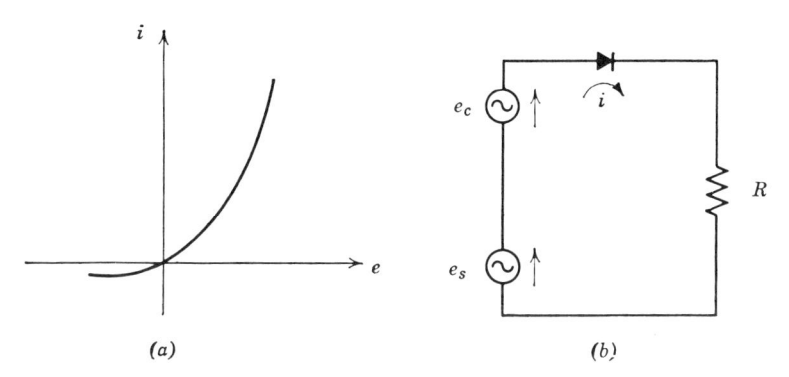

Figure 12-7
Diode characteristic and circuit used to generate A-M from nonlinear operation. (*a*) Diode characteristic. (*b*) Circuit.

powered terms represent components that have frequencies too far from those of the sideband components. They are filtered out of the system and thus do not affect the shape of the modulated output wave.

The voltage contains a carrier component and a signal-frequency component. Its equation is

$$e = E_c \cos \omega_o t + E_s \cos \omega_a t$$

where $E_s < E_c$. Substituting this into Eq. 12-10, we get the expression for the current

$$i = a_1 E_c \cos \omega_o t + a_1 E_s \cos \omega_a t + a_2 E_c^2 \cos^2 \omega_o t$$
$$+ 2a_2 E_c E_s \cos \omega_a t \cos \omega_o t + a_2 E_s^2 \cos^2 \omega_a t + \cdots \quad (12\text{-}11)$$

The components represented by the second, third, and fifth terms are eliminated by filters in the modulating unit. The remaining components, represented by the first and fourth terms, give the following equation for the current:

$$i = a_1 E_c \left(1 + \frac{2a_2}{a_1} E_s \cos \omega_a t\right) \cos \omega_o t \quad (12\text{-}12)$$

This has the form of Eq. 12-6 and describes an A-M current wave. The current in an antenna will radiate an electromagnetic wave that has components in the sidebands and a carrier component.

12-14
Modulated Class C Amplifier

Commercial broadcast transmitters commonly use plate-modulated Class C amplifiers to generate A-M signals for transmission. The basic method is shown

418 Basic Modulation and Detection in Communications Ch. 12

in Fig. 12-8a, and the circuit in Fig. 12-8b. The modulation process involves the extreme nonlinearity of Class C operation of the amplifier, because of the high negative grid bias. The tuned circuit at the plate provides the necessary filtering to allow only the carrier and sidebands in the output.

Conduction in the tube of a Class C amplifier occurs only at the peaks of the carrier input signal. Pulses of current at the carrier frequency are fed to the tank circuit. Thus, the tube acts as a switch to turn the carrier signal on and off. Since the amplitudes of the current pulses are nearly linearly related to the plate supply voltage, varying the effective value of E_{bb} by means of the modulating signal generates current pulses that are proportional to the modulating signal amplitude.

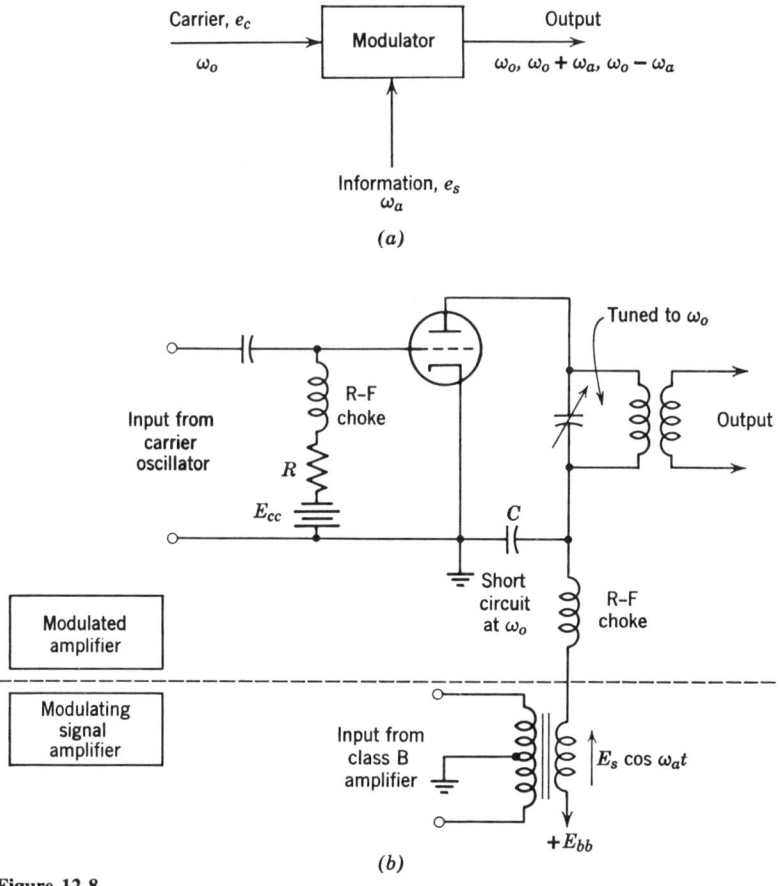

Figure 12-8
Basic concept and circuit to generate A-M. (a) Basic method for A-M. (b) Plate modulated Class C amplifier.

Sec. 12-14   Modulated Class C Amplifier   419

If the plate supply voltage is varied slowly compared to a cycle of carrier frequency (the resonant frequency of the tuned load), then the oscillating current in the tank circuit will have an envelope corresponding to the waveform of the modulating signal. Thus, modulation of the carrier will occur at the rate and amplitude of the modulating signal, generating the required waveform for A-M.

The effective plate supply voltage will be a combination of E_{bb} and modulating signal

$$E_b = E_{bb} + E_s \cos \omega_a t$$

or

$$E_b = E_{bb}(1 + m_a \cos \omega_a t)$$

which is the usual form representing the amplitude of the A-M wave.

Solid-state components are used in modern systems. They are found in low-power, portable transmitter-receiver (*transceiver*) equipment, and in some high-power systems. The circuitry of a typical Class C transistor A-M modulator is shown in Fig. 12-9.

Unmodulated carrier signal is applied in the base circuit and amplified in the collector circuit. Note that the collector supply voltage is applied through the

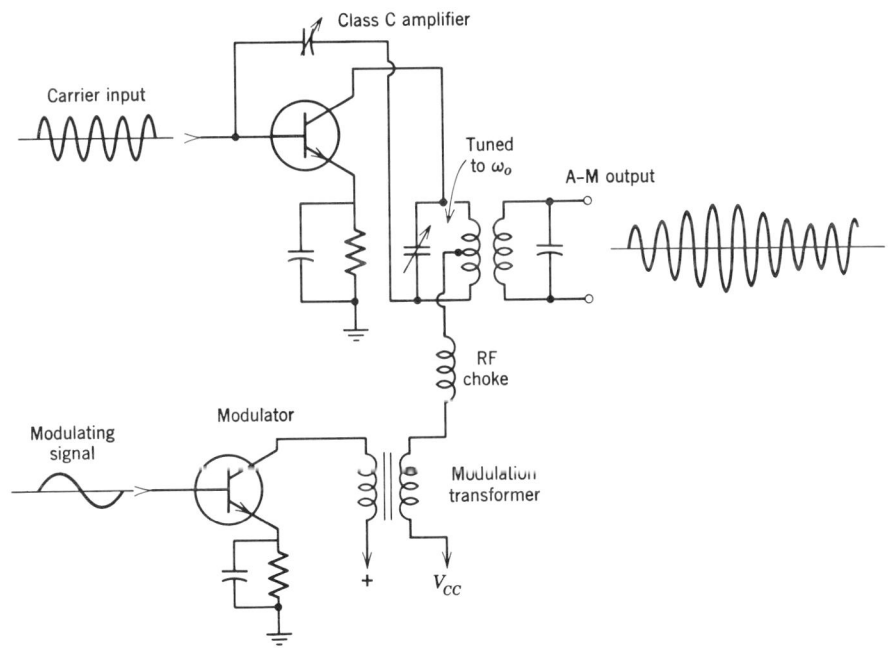

Figure 12-9
Transistor A-M modulator.

420   Basic Modulation and Detection in Communications   Ch. 12

modulation transformer. Changes occurring in the modulation transformer will vary the collector voltage, causing the amplitude of the amplified carrier to be modulated. Since its amplitude is modulated, the resulting waveform is an A-M signal.

12-15
Balanced Modulator and Single-Sideband (SSB) Operation

We have seen that the information being carried in A-M operation is contained entirely in either one of the two sidebands. The carrier itself conveys no information, and the information in one of the sidebands is rejected. Only one sideband needs to be transmitted to convey all of the information contained in the modulating signal. Further, transmission of the carrier and a second sideband is a waste of power, particularly carrier power. It is also a waste of frequency space, because transmission of both sidebands requires double the bandwidth needed for a single sideband. This has resulted in the use of single-sideband (SSB) transmission for radiotelephony in fixed and mobile radio services at frequencies below 25 MHz, by regulations issued by the Federal Communications Commission (FCC). In this type of transmission the carrier and one sideband are suppressed, and the remaining sideband is transmitted.

A modulator that generates the two sidebands but suppresses the carrier is the *balanced modulator*. A simplified circuit diagram is given in Fig. 12-10.

In the bridge modulator, the diodes act as switches to turn the output signal on and off, operating at the carrier rate. With identical diodes (balanced operation) the bridge is balanced when the modulating signal is zero, and there will be no output. When the carrier signal has the polarity shown, the diodes are essentially short-circuits, and the output is again zero. But when the carrier

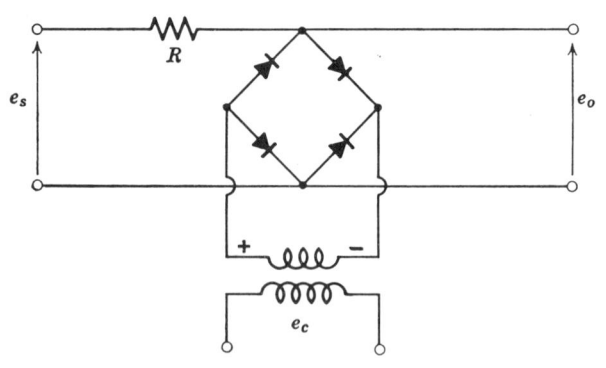

Figure 12-10
Balanced modulator for suppressed carrier operation.

signal changes polarity, the diodes are essentially open circuits, and the output will equal the modulating signal. Thus, the output is alternately zero and equal to the modulating signal, and is switched at the carrier frequency rate. The mathematical analysis of this type of modulator is complex and will not be attempted here. The important result is that the carrier frequency will not be a component of the output; only the two sidebands remain. The output signal is sometimes referred to as a *double-sideband, suppressed carrier* (DSB/SC) signal.

The unwanted sideband may be removed by a bandpass filter that passes only the desired sideband. The balanced modulator and sideband filter are the heart of SSB transmitters commonly used. The modulator is operated at low power levels in SSB operation, then followed by linear power amplifiers to produce the level necessary for transmission. This is in contrast to conventional A-M, where modulation is frequently performed at high power levels in Class C amplifiers.

Example 12-1. We will calculate the peak power required in single-sideband operation compared to that in conventional A-M, where each carries equal intelligence at 100 per cent modulation.

In ordinary A-M operation, the peak A-M voltage at 100 per cent modulation will be

$$\text{Peak voltage} = 2E_c$$

and the peak power will be proportional to the voltage squared, or

$$\text{Peak voltage} = k(4E_c^2)$$

The peak voltage in one sideband is $E_c/2$, and the peak power at 100 per cent modulation is $k(E_c^2/4)$. Then the peak power required in conventional A-M is 16 times as great as the peak power required in SSB operation, or

$$\frac{k4E_c^2}{kE_c^2/4} = 16$$

12-16
Frequency Conversion and Mixing

The block diagram of a superheterodyne A-M receiver is illustrated in Fig. 12-11. The block labeled *mixer* represents a circuit that is used to translate the carrier and sidebands to a new frequency, the I-F of the receiver. The local oscillator and R-F amplifier are simultaneously tuned so that the output of the mixer occurs at the I-F, regardless of the incoming carrier frequency. Such receivers usually have the I-F at about 455 kHz.

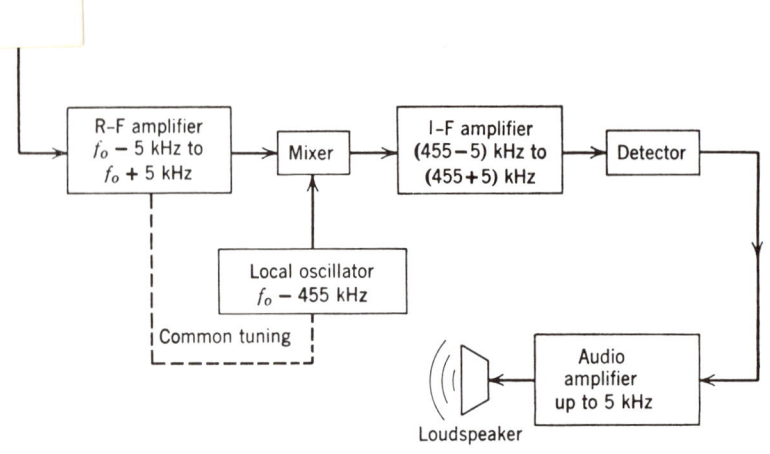

Figure 12-11
Block diagram of A-M broadcast receiver.

The basic operation of a frequency converter may be understood in terms of the circuit in Fig. 12-12. The vacuum tube is usually referred to as a *pentagrid converter*, and the circuit operates as a *frequency mixer*. The cathode, first grid, and second grid act as a triode to alter the plate current at a rate corresponding to the local oscillator frequency, ω_l. The third grid acts as the normal control grid of a conventional triode, altering the plate current at a rate corresponding to the R-F carrier frequency and sidebands. The tuned load is designed to resonate at the I-F and is broad enough to include the sideband frequencies. The mixer acts as a frequency converter, shifting the incoming signal down to the intermediate frequency of the receiver.

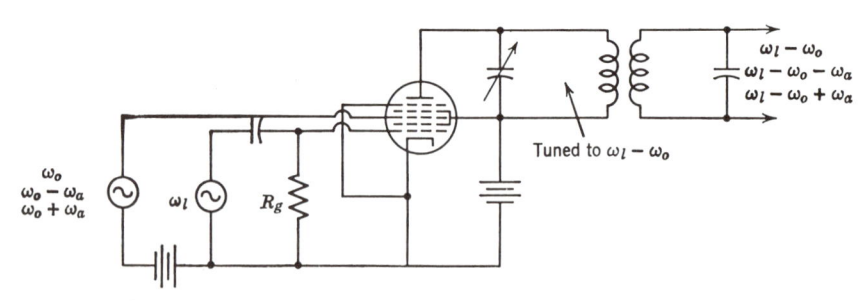

Figure 12-12
Pentagrid mixer as a frequency converter.

Several stages of amplification at the I-F level may be used to increase the magnitudes of information signals. They are then detected to recover the original information. (Some inexpensive portable receivers use only one I-F stage.)

12-17
Single-Sideband Receiver

Because of the economy of power and frequency space offered by SSB, it may be wondered why it has not been used for standard radio broadcasting. Part of the reason is the complexity required in a receiver. Mass production of inexpensive home receivers has not been possible.

A block diagram of a typical SSB receiver is given in Fig. 12-13. The complexity arises because the carrier must be reinserted at the receiver in order that the detector circuits may operate properly to retrieve the transmitted information signals. The receiver contains a local oscillator that is used to reinsert the carrier. Deviations of its output cause distortion in the sideband signals and in the information being conveyed. The precision frequency control

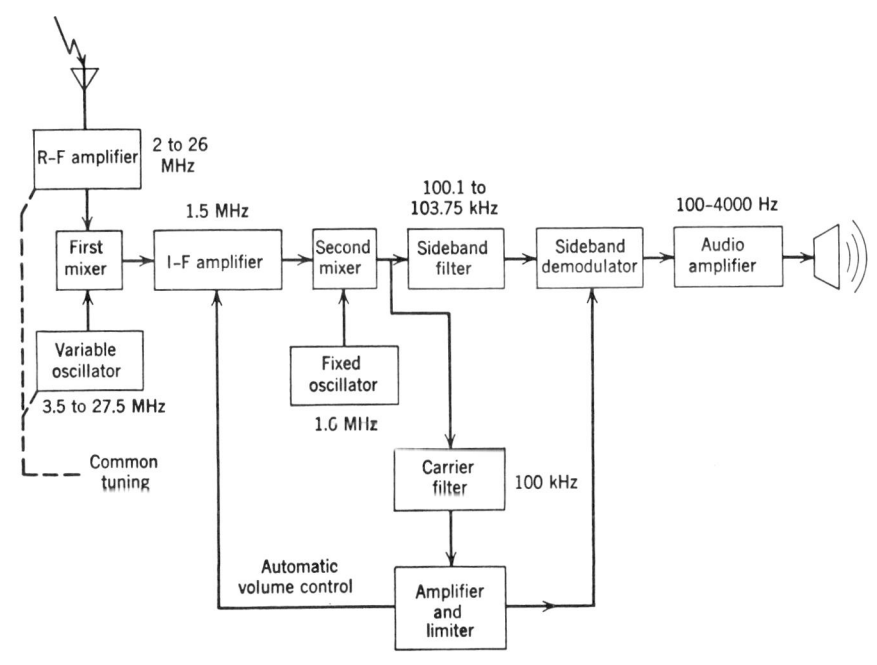

Figure 12-13
Block diagram of SSB receiver. (Adapted from M. Schwartz, *Information Transmission, Modulation, and Noise*, McGraw-Hill.)

that is required in the receiver, as well as in the transmitter, is the basic deterrent to the widespread use of SSB in radio broadcasting.

A basic difference in the SSB receiver compared with the conventional A-M receiver is the method of detection. The SSB system uses frequency mixing and translation of the information frequencies by shifting them down to the audio range. A conventional A-M system uses *envelope detection*, a process that will be discussed in subsequent sections of the chapter. Generally, the A-M wave has an envelope that varies in time with the information signal, but an SSB wave bears no relationship in time with the information signal.

12-18
The Envelope Detector for A-M Signals

The process of retrieving information from a modulated carrier is called *demodulation* or *detection*. A detector operates basically as the inverse of a modulator; frequency translation brings the information signals down in frequency to their original locations in the frequency space. It requires nonlinear circuit elements for proper operation. Two general types of nonlinear detectors are used, the so-called *square-law detector* and *piecewise linear detector*. First, we shall discuss the piecewise linear type.

In conventional A-M detection the information is recovered by applying the A-M wave to a half-wave rectifier, as in Fig. 12-14. The output is then filtered to recover the envelope. Such a detector-filter combination is called an *envelope detector*, for obvious reasons.

The output of an envelope detector can be a close replica of the original information signal. Depending on the time constants of the R-C network that acts as a filter to remove the carrier component, the amount of distortion can be kept small. Figure 12-15 illustrates the effect of the RC time constant on the distortion that may be introduced by the filter. If the RC product is too small, the time constant will be too short, and the output waveform will have a ragged edge that essentially follows the envelope. Some frequency components will be

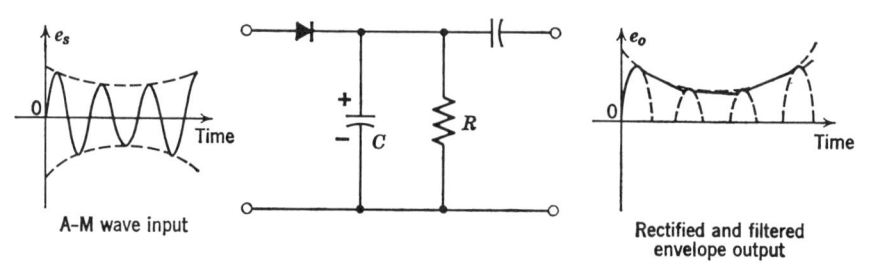

Figure 12-14
Linear diode envelope detector for A-M.

Sec. 12-18   *The Envelope Detector for A-M Signals*   425

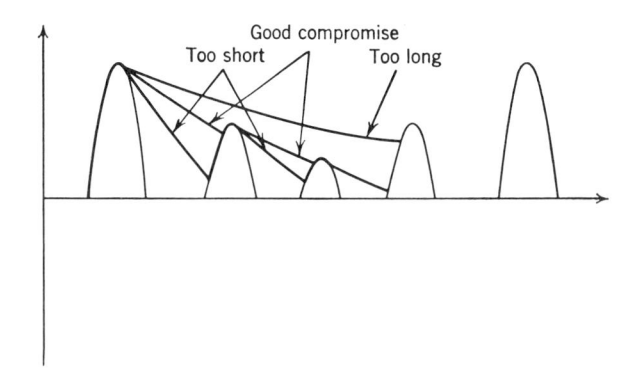

Figure 12-15
Effect of time constant on the output waveform of envelope detector.

present that are not portions of the original information signal. Conversely, if the time constant is too long, the filter will not be able to follow rapidly varying envelopes. Some compromise will be designed into the filter to permit good reproduction of the envelope, without introducing extraneous frequencies or "clipping" some wanted frequencies. A short time constant introduces higher frequency components than are contained in the envelope, and a long time constant eliminates some of the high-frequency components of the original signal. Either, of course, is a form of distortion and should be avoided where possible.

In the design of the filter, the value of C is chosen to have a small reactance compared to the value of R at the carrier frequency. This will generate a slowly varying voltage across R at the modulating frequency. Compared to the variation of an individual cycle of carrier signal, the output voltage remains constant except for variations at the modulating signal frequency. This is especially true if the carrier frequency is many times greater than the modulating frequency, which is usually true in conventional A-M radio broadcasting. The value of R should be large compared to the forward resistance of the diode, so that the charging time constant permits the output voltage to follow the envelope variations. However, its value is also important in the discharging time constant.

Example 12-2. It is desirable to know how the values of R and C in the envelope detector are related to the highest modulation frequency and modulation percentage. Both of these factors affect the discharge rate of C so that the output voltage may follow the envelope variations. The problem is most serious at the highest frequency and modulation percentage, because the discharge rate of C must be less than the slope of the envelope at this instant.

426 Basic Modulation and Detection in Communications Ch. 12

Solution: For the A-M wave, the rate of change of the envelope is

$$\frac{de}{dt} = \frac{d}{dt}[E_c(1 + m_a \cos \omega_a t)]$$

$$= -E_c m_a \omega_a \sin \omega_a t \tag{1}$$

The capacitor C will discharge from a voltage E according to the relation

$$e_C = E\epsilon^{-t/RC}$$

and the rate of change of capacitor voltage will be

$$\frac{de_C}{dt} = -\frac{E}{RC}\epsilon^{-t/RC}$$

$$= -\frac{e_C}{RC} \tag{2}$$

When the capacitor voltage is at the peak of one of the rectified carrier pulses, the voltage rate of change is

$$\frac{de_C}{dt} = -\frac{E_c(1 + m_a \cos \omega_a t)}{RC} \tag{3}$$

In order that the voltage across C may be able to follow the envelope variations, the envelope should decrease slower than the capacitor voltage. A critical condition occurs when the two voltages have the same rate of decrease

$$\frac{de}{dt} = \frac{de_C}{dt}$$

$$E_c m_a \omega_a \sin \omega_a t = \frac{E_c(1 + m_a \cos \omega_a t)}{RC}$$

$$RC = \frac{1}{\omega_a}\left(\frac{1 + m_a \cos \omega_a t}{m_a \sin \omega_a t}\right) \tag{4}$$

Because the time constant should be equal to or less than this value, this equation describes a maximum value of RC. In order to eliminate the variable t, methods of calculus may be used to determine the maximum value of the expression on the right with respect to t. This process yields

$$\cos \omega_a t = -m_a$$

and (5)

$$\sin \omega_a t = \sqrt{1 - m_a^2}$$

Substituting these in Eq. 4 yields the relationship of RC, m_a, and ω_a to be satisfied in the design of the filter circuit

$$RC \le \frac{1}{\omega_a}\frac{\sqrt{1-m_a^2}}{m_a} = \frac{\sqrt{1-m_a^2}}{2\pi f_a m_a} \tag{6}$$

Even though the RC time constant should approach zero for 100 per cent modulation, a value not greater than $1/2\pi f_a m_a$ has been found satisfactory in practice.

As an example of the required limiting time constant in A-M detection, consider the broadcast signal of 5 kHz and 50 per cent modulation. For this case the RC product should be

$$RC \le \frac{\sqrt{1-0.5^2}}{2\pi(5)(0.5)10^3} = \frac{0.866}{5\pi} \times 10^{-3}\ \text{s}$$

$$\le 0.055\ \text{ms}$$

Using the approximation resulting from practice, the value is

$$RC \le \frac{1}{2\pi(5)(0.5)10^3} = 0.064\ \text{ms}$$

For a 1000-Ω resistor R, the value of C should be less than 0.064 μF. Larger values of resistance require smaller values of capacitance, of course.

12-19
Analysis of Frequency Modulation (F-M)

In section 12-8 we saw that the mathematical expression for a sinusoidal voltage can be written as

$$e = E_c \cos\left(\int \omega\, dt + \phi_o\right) \tag{12-4}$$

repeated here for convenience. It was stated that time variation of the angular frequency results in frequency modulation. Since an analysis of this equation is too complex for the purposes of this chapter, we will look at some of the similarities and differences of A-M and F-M signals without rigorous mathematical analyses.

To compare the waveforms of A-M and F-M signals that are modulated by a single-frequency sine wave, refer to Fig. 12-16. The A-M wave varies its amplitude in time according to the changing amplitude of the modulating signal, and *the F-M wave varies its frequency* according to the changing amplitude of the modulating signal. The *amplitude* of the F-M "carrier" remains unchanged, however; carrier frequency occurs in the sketch at the instants when the modulating signal amplitude is zero.

Frequency modulation is a nonlinear process, and we should expect that new

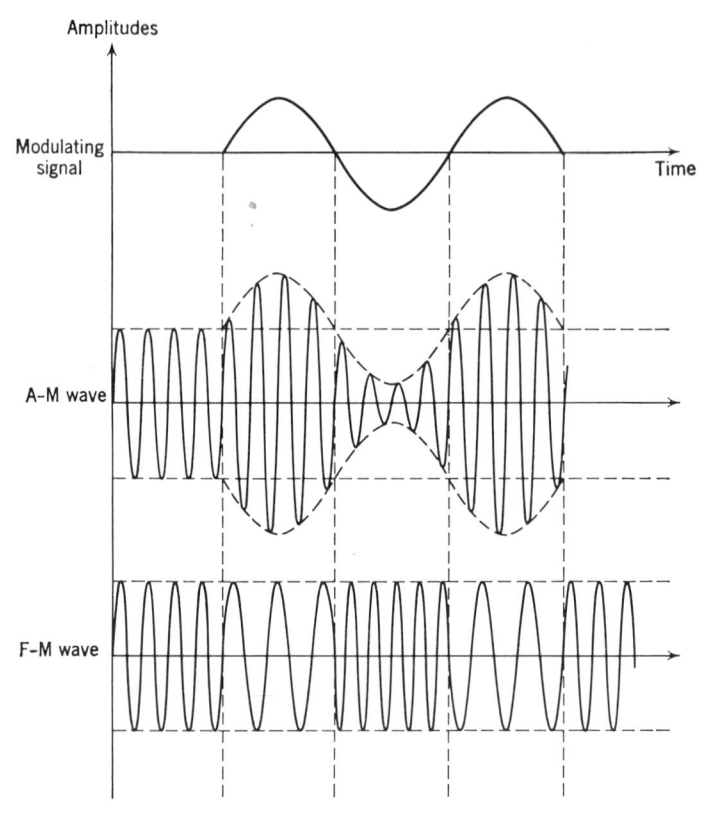

Figure 12-16
Comparison of A-M and F-M waveforms when modulating signal is a sine wave.

frequencies will be generated in the process. Further, since the frequency oscillates about a nominal "carrier" frequency, the new frequencies will define a bandwidth for the F-M signal.

12-20
Bandwidth of F-M Signals

Suppose the modulating signal, a single-frequency cosine wave, is made to vary the angular frequency according to

$$\omega = \omega_o [1 + k_f \cos \omega_f t]$$

where k_f is the amount of modulation of ω_o, the *center frequency*, and ω_f is the modulating frequency. Substituting into Eq. 12-4 and simplifying yields this result:

$$e = E_c \cos\left(\omega_o t + \frac{\omega_o k_f}{\omega_f} \sin \omega_f t\right) \quad (12\text{-}15)$$

The coefficient of the sine term represents the deviation of the F-M signal from its center frequency, and in terms of frequency it can be written as

$$\frac{\omega_o k_f}{2\pi} = \Delta f, \quad \text{the maximum } \textit{frequency deviation}$$

A *modulation factor*, or *deviation ratio*, is also defined for the F-M signal:

$$\frac{\omega_o k_f}{\omega_f} = \frac{\Delta f}{f_f} = m_f, \quad \text{the deviation ratio} \quad (12\text{-}16)$$

The deviation ratio will have a *different value for every modulating frequency*. Because of this nonlinear behavior of F-M signals, the principle of superposition cannot be used when the modulating signal contains more than a single frequency. Of course, voice and music signals contain many frequency components, and an F-M signal that transmits them is very complex. We shall see that the F-M signal is much more complicated than that of the A-M signal, even for a signal of only one frequency.

The general expression for the F-M signal can be written in terms of the deviation ratio:

$$e = E_c \cos(\omega_o t + m_f \sin \omega_f t) \quad (12\text{-}17)$$

The familiar identity* for the cosine of the sum of two angles may be used to express the F-M signal as

$$e = E_c [\cos \omega_o t \cos(m_f \sin \omega_f t) - \sin \omega_o t \sin(m_f \sin \omega_f t)] \quad (12\text{-}18)$$

This "simplification" results in unusual forms of trigonometric functions. In fact, it leads to what mathematicians call "Bessel functions," the use of which is beyond the scope of this book. For our purposes, we will write the F-M signal in its final form and work an example to show how to determine the bandwidth of a typical signal.

The F-M signal is expressed in terms of Bessel functions (the J factors) in this form

$$\begin{aligned} e = E_c [&J_o(m_f) \cos \omega_o t \quad J_1(m_f)\{\cos(\omega_o - \omega_f)t - \cos(\omega_o + \omega_f)t\} \\ &+ J_2(m_f)\{\cos(\omega_o - 2\omega_f)t + \cos(\omega_o + 2\omega_f)t\} \\ &- J_3(m_f)\{\cos(\omega_o - 3\omega_f)t - \cos(\omega_o + 3\omega_f)t\} \\ &+ J_4(m_f) \cdots] \end{aligned} \quad (12\text{-}19)$$

---

* Cos$(A + B)$ = cos A cos B − sin A sin B.

430   Basic Modulation and Detection in Communications   Ch. 12

This expression continues as an *infinite series*, but, in general, higher order terms contribute small components. The important result is the generation of *sideband frequencies at all of the modulating signal harmonics*. This means that the F-M bandwidth must be considerably greater than that for an A-M signal carrying the same information. This fact is a fundamental difference between the operations of A-M and F-M systems.

In Eq. 12-19, the coefficients $J_n(m_f)$ may have widely different values, depending on the deviation ratio, m_f. The coefficients are listed in the Appendix. These numbers are important in determining the relative amplitudes of the sideband frequencies and the required bandwidth. It can be seen that for small values of m_f, the amplitudes decrease rapidly. Therefore, if the center frequency does not change very much, the bandwidth will be small in order to include significant frequencies.

Example 12-3. It will be instructive to determine the bandwidth required for a 5-kHz modulating signal, for two different deviation ratios. Suppose the modulating signal has such amplitudes that it causes a frequency deviation of first 50 and then 10 kHz, about a center frequency of 100 mHz in a F-M transmission system. This corresponds to a 5-kHz tone that decreases in amplitude.

Equation 12-19 is used for each of the two conditions separately, because the bandwidth is different for the two cases, even though the modulating frequency is the same in both. In order to compare the two cases directly, the F-M coefficients $(J_n(m_f))$ will be graphed on a frequency scale in Fig. 12-17.

In the case of 50 kHz frequency deviation, the value of m_f is

$$m_f = \frac{50 \text{ kHz}}{5 \text{ kHz}} = 10$$

The coefficients may be read from the Appendix for $m_f = 10$: -0.2459, 0.0443, etc. The minus sign for a coefficient simply means that the frequency component is shifted by 180 degrees from those with plus sign. These coefficients are plotted in the figure.

Similarly, the second case has $m_f = 10/5 = 2$. The coefficients can be read from the Appendix and plotted as shown in the figure. It will be noted that, even though the sideband components are spaced 5 kHz apart, the bandwidth required for the larger m_f (the larger frequency deviation) is much greater than for the smaller m_f. In fact, more than the 100-kHz bandwidth shown should be used, because the coefficients have not decreased significantly to permit other higher sideband components to be eliminated. To remove these higher harmonics would generate distortion in the transmission. Conversely, the bandwidth required for $m_f = 2$ could be limited to about 50 kHz without significant

Sec. 12-20   Bandwidth of F-M Signals   431

Figure showing spectrum (a): 100 kHz is insufficient bandwidth, with spectral lines at 100 MHz (0.0443), 0.2546, -0.2459, and 5 kHz spacing.

(a)

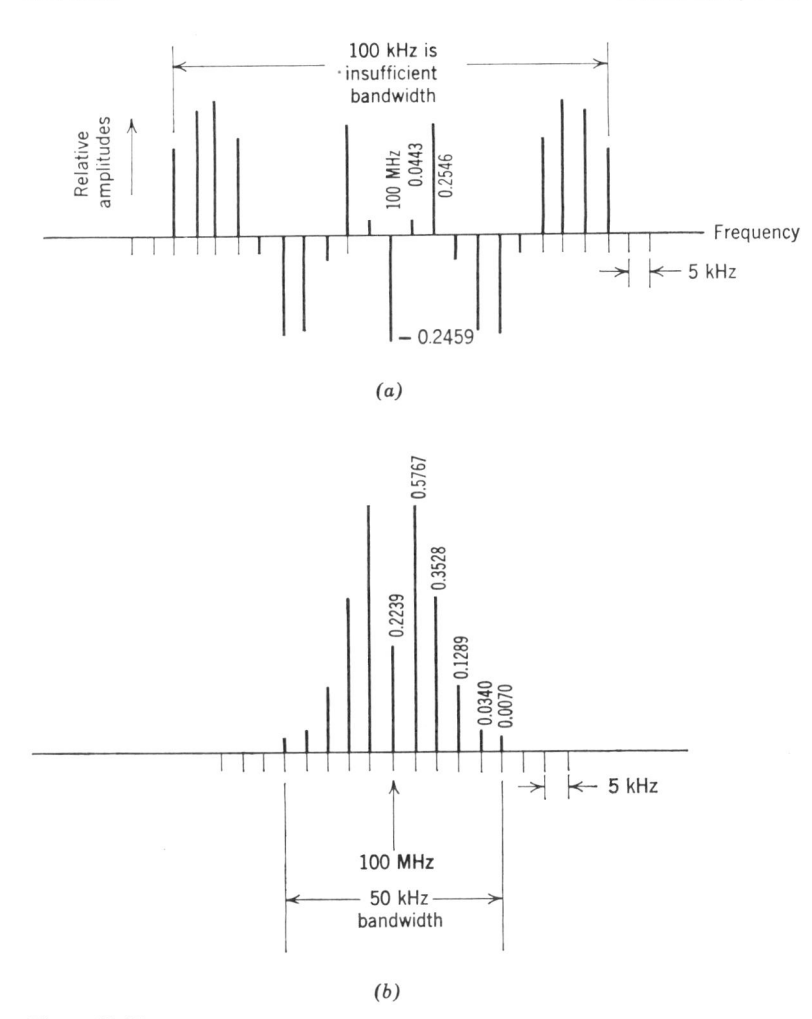

Figure showing spectrum (b): 50 kHz bandwidth centered at 100 MHz, with amplitudes 0.2239, 0.5767, 0.3528, 0.1289, 0.0340, 0.0070, and 5 kHz spacing.

(b)

Figure 12-17
Comparison of bandwidth requirements in F-M for deviation ratios of (a) $m_f = 10$; (b) $m_f = 2$.

distortion. The transmission system should then be able to pass frequencies from $(100 - 0.025)$ MHz to $(100 + 0.025)$ MHz, in the second case.

Drill Problem

D12-5 An F-M broadcast station is assigned a channel from 92.1 to 92.3 MHz. (a) What is its center frequency? (b) What is its permissible bandwidth for transmission? (c) What is the maximum permissible deviation ratio for a modulating frequency of 10 kHz? (*d*) For this f_I, how many sideband components could exist on each side of the center frequency and still remain in the bandwidth? (*e*) Plot the frequency spectrum, similar to Fig. 12-17, for this application, using the maximum deviation ratio calculated above.

12-21
Basic F-M Radio Receiver

A block diagram of a typical F-M radio receiver for the normal broadcast band from 88 to 108 MHz is given in Fig. 12-18. Except for the blocks labeled *limiter* and *discriminator*, the receiver is similar to the conventional A-M receiver. The intermediate frequency is commonly 10.7 MHz in such a receiver. The audio amplifier has a much wider bandwidth than that of the A-M system, because F-M transmission can carry high-fidelity music signals within the assigned bandwidth, whereas the A-M signal is limited to signals up to 5 kHz.

The limiter is essentially a saturated amplifier that clips signals whose amplitudes exceed the level permitted by the operation of the limiter. Since amplitude variations carry no information in the F-M system but may exist because of noise pulses and atmospheric disturbances, the limiter may remove them without removing any information. Another purpose is to eliminate them before the I-F signal is applied to the discriminator, where they could cause distortion of the information signals.

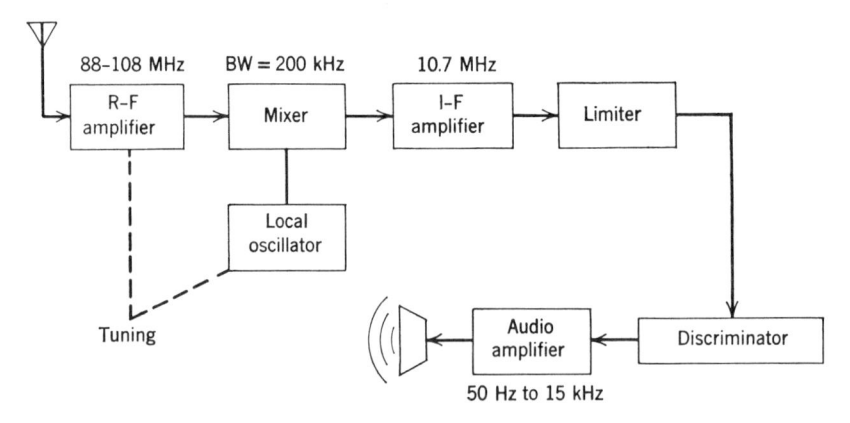

Figure 12-18
Typical F-M receiver.

12-22
The F-M Discriminator

Demodulation, or detection, of F-M signals is the process of converting them to their original location in the frequency spectrum. It may be performed by a circuit whose output voltage (or current) amplitude varies linearly with the frequency deviation of the F-M signal. The name *frequency discriminator* is commonly used to describe the circuit that *converts frequency to amplitude.* The desired characteristic of an ideal discriminator is shown in Fig. 12-19. The output amplitude varies linearly, plus and minus values, for frequency variations above and below the center frequency.

A tuned circuit may be used to approximate the operation of the ideal discriminator. If the tank is detuned so that the center frequency occurs below the resonant frequency, as shown in Fig. 12-20, the sloping part of the amplitude-frequency curve is essentially linear over a small region that may

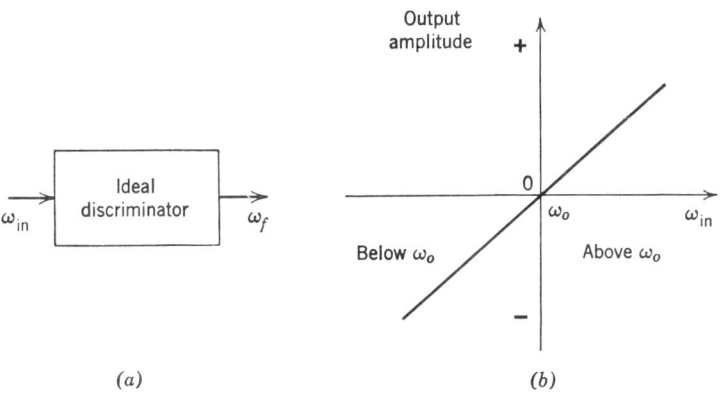

Figure 12-19
Desired response of a frequency discriminator. (*a*) Ideal discriminator. (*b*) Output characteristics.

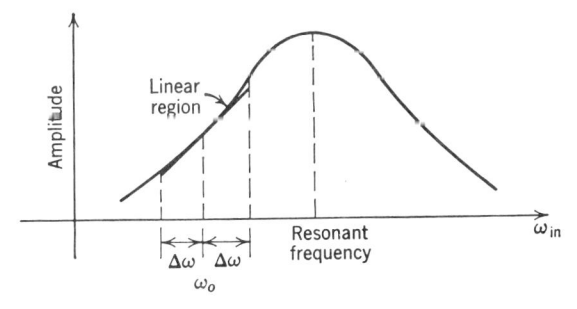

Figure 12-20
Detuned resonant circuit characteristic for use as a frequency demodulator.

434 Basic Modulation and Detection in Communications Ch. 12

include the center frequency and the maximum frequency deviation (±75 kHz for commercial F-M broadcasting). As the modulated F-M carrier is applied to the tuned circuit, the output amplitude will vary according to the frequency variations. This circuit is not very practical, though, because amplitude variations are superimposed on the large amplitude representing the center frequency, which carries no information. Of course, two circuits might be used in such a way that this amplitude is balanced out of the final output, similar to the operation of the balanced modulator.

A more practical arrangement is the circuit of Fig. 12-21. This circuit is known as the *Foster-Seeley balanced discriminator.* It uses a double-tuned input with transformer coupling to the diode and capacitor combinations that act as peak detectors. The output resistors operate with the associated capacitors to generate an output voltage that follows the envelope of voltage variations occurring on the two capacitors. The upper capacitor charges to follow the peak values of $E_1 + E_2$, and the lower capacitor follows $E_1 - E_2$. The output signal is then

$$e_o = |E_1 + E_2| - |E_1 - E_2| \tag{12-20}$$

so that it is a low-frequency signal having an amplitude proportional to the input frequency variation. When the voltage amplitudes in Eq. 12-20 are equal (for the center frequency), the output is zero.

Both the primary and secondary of the input transformer are tuned to the unmodulated, center frequency. The coupling capacitor C_c is chosen to have negligible reactance at the carrier frequency, and the voltage E_2 will lead the voltage E_1 by about 90 degrees. As the F-M signal varies, however, the phase difference varies. This occurs because the phase characteristic of a tuned circuit varies in the vicinity of its resonant frequency as the frequency varies. For small deviations about the resonant frequency, the phase-frequency curve is almost linear. The phase change that occurs for modulated signals is shown

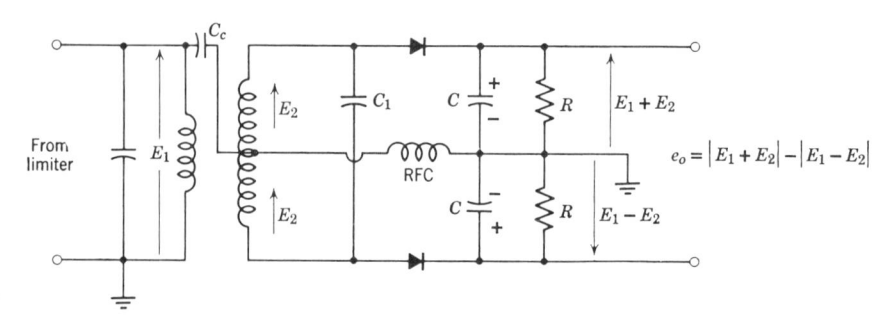

Figure 12-21
Foster-Seeley phase discriminator.

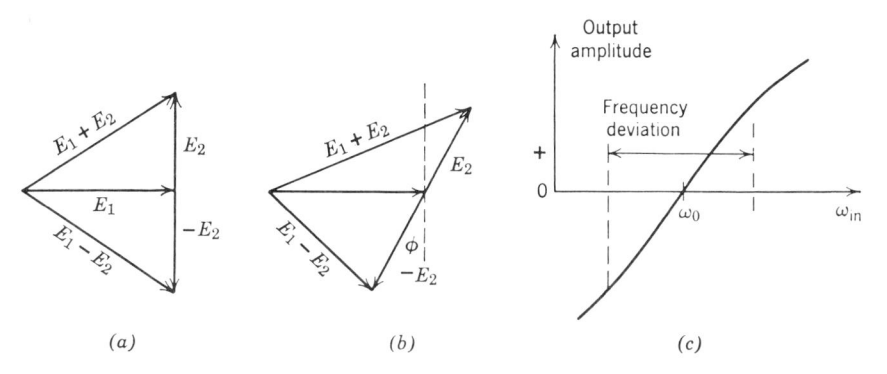

Figure 12-22
Effect of frequency deviation on output of frequency discriminator. (a) Phase diagram, no modulation. (b) Phase diagram, below ω_0. (c) Discriminator characteristic.

in Fig. 12-22, and its effect on the output voltage is evident. The output voltage will be nearly proportional to the frequency deviation.

The discriminator circuit is sensitive to amplitude changes of the input F-M signal, because of the way voltage E_1 enters its operation to generate an output. For this reason, the amplifier immediately ahead of a discriminator is operated as a limiter. It clips positive and negative amplitudes of the signal and produces square-topped pulses of uniform amplitudes but of varying frequency. Since the frequency variation carries the modulating information, the limiter has no effect on the information content but eliminates amplitude variations caused by atmospheric noise.

The circuit of Fig. 12-21 is suitable for discussing the operation of the discriminator, but more practical arrangements are possible. The fact that the

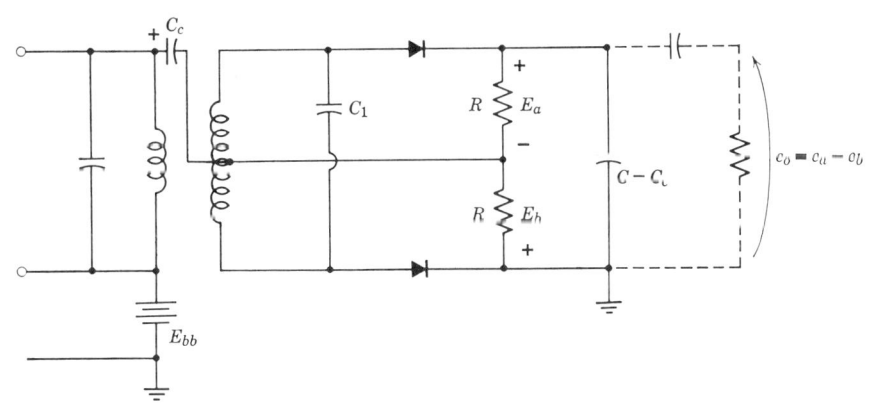

Figure 12-23
Modified phase discriminator.

output voltage is "floating" about the circuit ground connection may lead to problems in the receiver. The circuit is rearranged in Fig. 12-23 to eliminate the floating ground. It will also be noted that fewer circuit components are needed. One of the output capacitors and the RFC have been removed, because they are unnecessary in the modified circuit.

It is also customary to obtain the output voltage without its dc component, so that it will represent the original modulating signal. The R-C network shown by dashed lines may be used in practice. The capacitor C blocks the dc component and passes the slowly varying envelope corresponding to the original signal information.

12-23
Pulse-Code Modulation (PCM)

Many information-transmitting systems cannot operate from continuous time-varying signals because either they are not available or it is not feasible. For example, a radar system obtains target information only when the antenna scans it once each revolution. In the measurement of slowly varying quantities, such as the temperature of a chemical process, it is only necessary to sample the variable periodically rather than continuously. Continuous sampling would not yield much more information than readings taken at widely spaced points in time, particularly if the variable changes very slowly.

The famous *sampling theorem*[*] of modern communication theory states that any two independent pieces of information about a single period of a periodically recurring variable will completely characterize the variable. For a single-frequency sinusoid, this means that two independent samples of its amplitude will completely describe its information content. This is shown in Fig. 12-24a, where the sine wave is approximated by two samples, one for the positive half-cycle and another for the negative half-cycle. It is seen that only two samples during one period yield a very rough approximation to the original sine wave, but according to the theorem, they are sufficient to characterize the wave. In Fig. 12-24b, several samples show that a much closer relationship exists between the samples and the sine wave.

In a pulse-code modulation system, waveforms are sampled at intervals and the amplitude samples are arranged in a *binary code* for transmission. As an example of a simple code based on measured amplitude samples, consider the sketch of Fig. 12-25. The signal amplitude range is divided into eight levels, including zero level.

In order to have a unique pulse code for each amplitude level, a *binary* system consisting of a series of pulses of uniform height can be devised. Each separate signal amplitude level is then characterized by a coded sequence of

---

[*] H. S. Black, *Modulation Theory*, D. Van Nostrand, New Jersey, 1953.

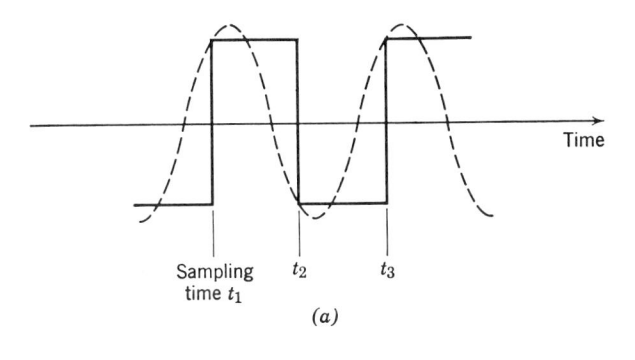

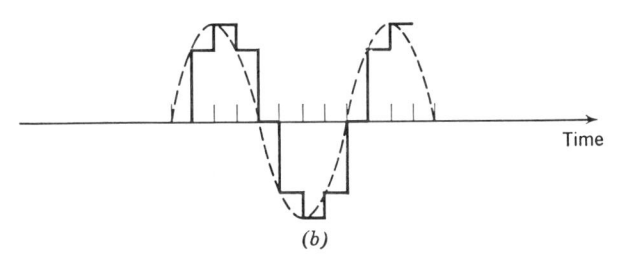

Figure 12-24
Sample rate effect on reconstructed waveform. (a) Two samples per cycle approximation to a sine wave. (b) Eight-sample approximation to a sine wave.

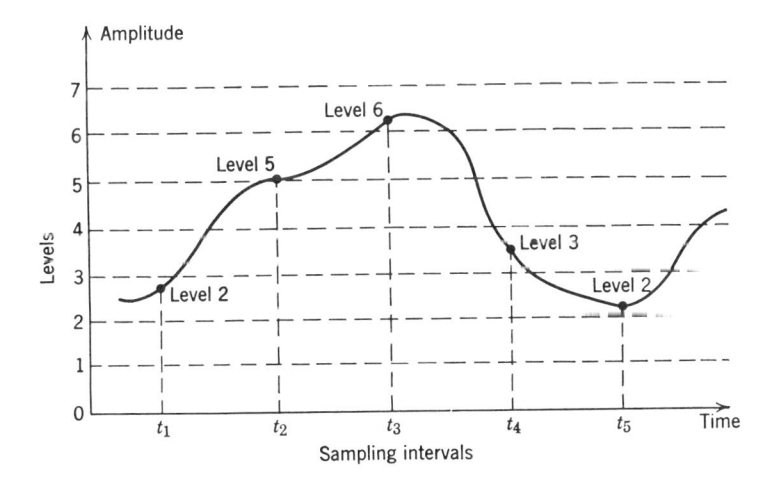

Figure 12-25
Amplitude levels preselected to form the eight-level pulse code.

pulses. In the binary system, the amplitudes are coded to represent the powers of 2 according to the arrangement of Table 12-3. As an example, the level 3 equals $2^1 + 2^0 = 2 + 1 = 3$; level 4 is $2^2 = 4$, and has code 100.

Table 12-3
Eight-Level Binary Code and Pulse Sequence

Level	Powers of 2 $2^2 + 2^1 + 2^0$	Pulse Sequence for Pulse Code	Levels
0	0 0 0	————————	0
1	0 0 1		
2	0 1 0	——————⬜—	1
3	0 1 1		
4	1 0 0	———⬜————	2
5	1 0 1		
6	1 1 0	———⬜—⬜——	3
7	1 1 1		etc.

It should be obvious that a code based on many more levels would more closely resemble the original waveform, if it is desired to reconstruct the waveform. A continuous waveform has an infinite number of discrete amplitudes. In order to represent it by a pulse code, it is necessary to limit the number of levels used. Sampling the amplitudes according to preselected levels is called "quantizing" the waveform. The difference between the quantized level and the actual amplitude of the signal waveform leads to distortion at a receiver, called "quantizing noise." It may be minimized by using as many levels as possible, since this reduces the difference between the actual waveform and its reconstruction based on quantized samples. This was pointed out in connection with the sketch of Fig. 12-24. This increases the number of code pulses, however, because the number of levels equals 2^n, where n is the number of code pulses. In the example above, the eight-level code required 3 pulses: $2^3 = 8$.

As the number of code pulses increases, so does the bandwidth needed to transmit them. Since a waveform needs to be sampled at least twice during each cycle, the bandwidth will be twice that for the original signal. A compensating advantage of PCM, however, is its ability to sense the presence or absence of a pulse in time, even though the pulse may be severely distorted. Regenerative repeaters that are capable of reshaping distorted pulses on long-distance channels can restore the code to its original state for effective communication over great distances. The same distortion applied to the original continuous signal would mask its character so completely that communication would be impossible. Even though bandwidth requirements are greater for

PCM than for conventional A-M, the ability to recognize and regenerate a pulse code more than compensates for this disadvantage in many applications.

For a more complete discussion of PCM and its many and varied applications, the reader is referred to the vast literature that is available. It is included in this chapter to show only an introduction to the use of sampling and coding as another method for transmitting information from one point to another.

Questions

12-1   Compare the two fundamental methods of modulating a carrier, insofar as they alter the waveform of the carrier over a period of time.

12-2   What is "frequency translation" as it occurs in the process of modulation?

12-3   How is "frequency translation" manifested in the detection process?

12-4   What fundamental characteristic of a radio wave is independent of its frequency or wavelength?

12-5   How are the frequency and wavelength of a radio wave related?

12-6   Why do you think television transmission uses the VHF and UHF frequency bands rather than the MF band?

12-7   What is the significance of "selectivity" in a radio receiver?

12-8   Explain the process of "tuning" a radio receiver to separate a desired signal from all that are present at its antenna.

12-9   What is the need for I-F in a radio receiver?

12-10 Sketch from memory your concept of the waveshape of a general A-M wave that is modulated by a sine wave.

12-11 Why is it important that the carrier be much higher in frequency than the highest-frequency component of its modulated *envelope*?

12-12 What happens to the waveform of the A-M wave if modulation exceeds 100 per cent?

12-13 Why is the bandwidth of A-M transmission required to be *twice* the highest frequency of modulating signals?

12-14 What are upper and lower *sidebands* of the A-M carrier? How are their frequencies related to the modulating signal frequency?

12-15 Why must voice transmission include both low- and high-frequency components?

12-16 Explain why the carrier power in the A-M wave is so much greater than the power of sidebands.

12-17 In a plate-modulated Class C amplifier, how does the tuned circuit at the plate provide *filtering* so that the output is only the carrier and sidebands?

440 Basic Modulation and Detection in Communications   Ch. 12

12-18 Although SSB is an efficient method of conveying information, what are some of the problems that arise in the operation of an SSB receiver?

12-19 In a superheterodyne radio receiver, what is the purpose of the mixer?

12-20 What should be the bandwidth of mixer and I-F stages in a superheterodyne receiver that is designed for commercial A-M reception?

12-21 Explain the basic operation of an envelope detector.

12-22 Compare the time-varying alterations in A-M and F-M waves that are modulated by the same signal.

12-23 Why must the bandwidth of an F-M system be greater than that of an A-M system that is modulated by the same information signal?

12-24 Some of the coefficients of F-M frequency components (Appendix) are negative. What does this signify?

12-25 What is the purpose of the "limiter" in an F-M receiver that uses a frequency discriminator as a detector?

12-26 Explain the basic operation of a frequency discriminator as a detector for F-M.

12-27 What is "quantizing noise" in a pulse code?

12-28 Show that four amplitude levels of a signal, including zero, require two pulse positions to describe the signal levels, whereas 16 levels require only four pulse positions.

12-29 What is the primary advantage of PCM over continuous signal transmission, particularly for long-distance communication?

Problems

12-1 The amplifier in Fig. P12-1 generates a load voltage that is related to the grid signal by $e_L = a_1 e_g + a_2 e_g^2$. (a) Determine all frequencies present

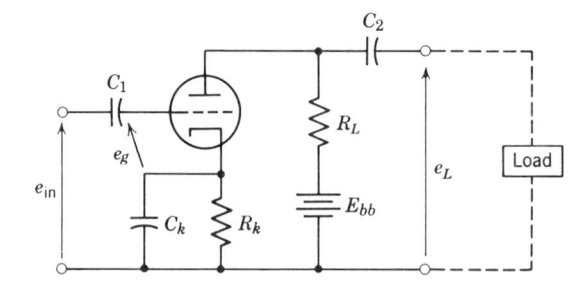

Figure P12-1

in the load voltage when $e_g = \cos \omega_a t$. (b) What changes or additions to the circuit would permit it to be used as a second-harmonic generator for the input frequency? (c) Assume $a_1 = 10$ and $a_2 = 1$. Calculate the ratio of second-harmonic amplitude to fundamental amplitude in the output. (This is "second harmonic distortion.")

12-2 In the circuit of Problem 12-1, the input signal is: (a) $2 \cos 100t$. Calculate the frequency components and their amplitudes in the load. (b) $2 \cos 100t + \cos 500t$. Calculate the frequency components and amplitudes. What new frequencies are present that were not in (a)? How did they arise?

12-3 A nonlinear amplifier has an output expressed by

$$e_o = 20[1 + 0.5 \cos (2\pi)(1000t)][\cos (2\pi)(10^5 t)]$$

(a) Determine all frequency components of the wave. (b) What is the degree of modulation? (c) Calculate the relative power levels for all frequency components.

12-4 An A-M wave of $f_o = 1$ kHz is modulated at 50 per cent by $f_a = 500$ Hz. Plot a time graph of three frequency components over one cycle of the lowest frequency and add them point by point, to obtain the A-M wave. Consider ways of measuring m_a, e.g., with an oscilloscope.

12-5 Calculate the wavelength of a 100 Hz wave; of a 1 MHz wave.

12-6 Calculate the frequency range of waves having wavelengths from 1 to 5 m.

12-7 What are the wavelengths of frequencies known as "microwaves"?

12-8 An A-M signal consists of a carrier voltage $10 \sin \omega_o t$ and sidebands generated by the modulating signal $5 \sin \omega_a t$. (a) What is the modulation percentage? (b) Sketch the envelope over a modulation cycle. Label with numerical values of significant points. (c) What average power will be delivered to a 100-Ω load? What peak power?

12-9 Assume that the carrier in Problem 12-8 is suppressed after modulation. How much may the average power of the sidebands be increased and not exceed the peak power level of Problem 12-8?

12-10 An A-M broadcast station is assigned a carrier of 1 MHz and is limited to a 10-kHz bandwidth. (a) What range of information signal frequencies may it transmit? (b) If its total radiated power is 1 kW, how much of this represents signal information?

12-11 The frequency spectrum of an A-M wave may be plotted by vertical bars representing amplitudes placed at the frequency component on a frequency scale. A certain A-M wave has an amplitude of 3 at 10 kHz, 1 at both 8 and 12 kHz, and 0.5 at 6 and 14 kHz. Sketch the frequency spectrum and write the time equation for this waveform.

442 Basic Modulation and Detection in Communications Ch. 12

12-12 A 5-kHz signal is used to modulate a 100-kHz carrier, and the upper sideband is transmitted in a SSB system. At a receiver the incoming signal is mixed with a local oscillator operating at 95 kHz before detection. Assuming the mixing and detecting operations regenerate the lower sideband, discuss the output signal from an audio amplifier following the detector. (What happens if the local oscillator varies from 95 to 100 kHz in a random manner?)

12-13 Show that the time constant RC defined for an envelope detector in Example 12-2 will be satisfactory for all frequencies lower than the maximum signal frequency and for all modulation percentages, if the filter is designed for maximum frequency and 100 per cent modulation as the "worst-case" condition.

12-14 An A-M transmitter is tested as shown in Fig. P12-14. The receiver includes a linear detector having a bandwith of 1 kHz. The R-F amplifier in the receiver is swept successively through the range from 100 kHz to 1 MHz. With no audio input to the transmitter, the wattmeter reads 1 kW average power. The peak-reading VTVM indicates 10 V at 800 kHz. (*a*) For an audio signal of 10 V peak at 2 kHz, the wattmeter indicates 1.5 kW average power. What are the voltage peaks indicated by the VTVM? (*b*) At what frequencies will the voltmeter indicate an output?

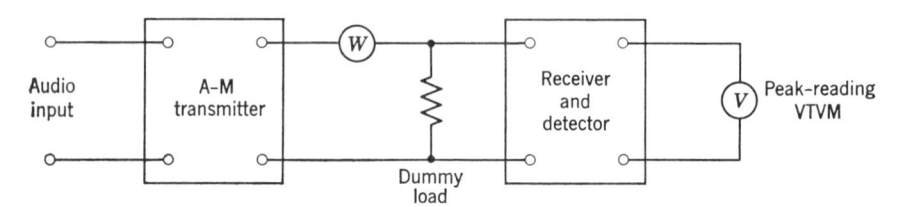

Figure P12-14

12-15 An F-M system uses 1 MHz as a center frequency. The modulating signal at 1 kHz has an amplitude such as to generate a maximum frequency deviation of 2 kHz. (*a*) Find the required bandwidth to pass the F-M signal. (*b*) Repeat (*a*) for a 2 kHz signal of the same amplitude. (*c*) Find the bandwidth necessary to pass the signal for a modulating signal having twice the amplitude at 2 kHz.

12-16 A modulating signal of amplitude 1 V at 1 kHz is used to modulate a 1-MHz carrier in both A-M and F-M systems. The 1-V amplitude produces a 2-kHz frequency deviation in the F-M system. (*a*) Compare the bandwidths required in the R-F amplifiers and the audio amplifier in both A-M and F-M receivers. (*b*) Repeat (*a*) for signal having 3-V amplitude.

12-17 Two F-M signals are tuned into a receiver alternately. One has an audio modulating signal at 2 kHz and requires a bandwidth of 20 kHz. The other requires a bandwidth of 100 kHz for a 10 kHz modulating signal. The carriers of both have the same intensity. Compare the amplitudes of the audio signals.

12-18 In a general F-M transmission system, how many sideband frequency components are generated? In the frequency spectrum for the F-M wave, what determines the spacing between adjacent spectral amplitudes? What determines their amplitudes? Under what conditions could the carrier amplitude be zero?

12-19 An F-M signal has a maximum frequency deviation of 50 kHz and is modulated by a 10 kHz signal. If a receiver has a bandwidth of 50 kHz in its R-F amplifiers, is this adequate for good reproduction of the 10 kHz audio signal?

12-20 A "slope detector" utilizes a detuned tank circuit to obtain a modulating signal from an incoming F-M signal, as given in the block diagram of Fig. P12-20. Assuming a sine-wave modulating signal, sketch the waveforms at points A, B, and C.

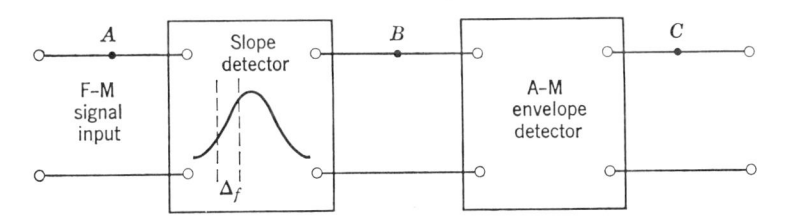

Figure P12-20

12-21 Show that a general waveform can be approximated by taking samples of its amplitude at equal time increments. Suppose that ten quantizing levels are used to generate a pulse code for the waveform samples. How many pulses are needed in the code, assuming it to be the simple binary code discussed in Section 12-24?

13
ELECTRONIC POWER CONVERSION

We have studied many situations in which power must be converted from ac to dc. Transistors and vacuum tubes need sources of dc power for both signal circuits and biasing networks. These needs are commonly met by half-wave and full-wave rectifiers of various circuit designs. This method of converting power may handle several amperes, using either semiconductor or vacuum diodes.

In circuits where the current demand exceeds a few amperes, it may not be practical to use these rectifier elements, except possibly semiconductor stack rectifiers. However, it is possible to extend the power-handling capability to several amperes with other circuit elements. Some practical applications of electronics may not require filtered dc power, but instead need a means of controlling the amount of power that is converted. Circuit elements that can fulfill this need include the *semiconductor controlled rectifier* (SCR).

In this chapter we will study the SCR as a power converter, its operating characteristics, and some typical applications in practical circuits. Extensions of its characteristics in the construction of other, similar devices will be discussed, to show the growing dependence upon semiconductor rectifiers in electronic power conversion.

Brief descriptions of the *solar cell* and *laser* are included as an introduction to these important devices. It will be seen that these elements are basically power converters.

13-1
The Semiconductor Controlled Rectifier (SCR)

The SCR is one member of the more general class of semiconductor devices referred to as the *P-N-P-N* or four-layer family. We have studied two-layer diodes and three-layer triodes operating as rectifiers and transistors for controlling and converting power up to a few watts. The SCR can convert power up to thousands of watts. However, this capability is not its most useful property.

The family of *P-N-P-N* devices, including the SCR, *function as switches.* Their power-handling capability extends rectification of ac power to much higher ranges than is possible using *P-N* devices, but their most important property is their ability to latch in either an "on" or "off" state, similar to the action of a simple switch. They also require very little power to operate, often achieving 99 per cent conversion efficiency.

The SCR evolved as a practical device from the pioneering work of Bell Telephone Laboratories in 1954 and 1955, under the direction of J. L. Moll.* This group built the first working *silicon P-N-P-N* device, which has served as the basis for all subsequent developments.

Controlled rectifiers are used in a broad span of applications ranging from circuit protection to power switching and control. Examples are speed control of motors in hand tools, light dimming in theaters, light flashers, ignition systems for automobiles, and temperature controllers. Before looking in detail at some typical applications, we shall develop the underlying physical principles of operation.

13-2
Construction of the SCR

Basically, the SCR consists of four alternating *P* and *N* layers of semiconductor, three junctions, and three external terminals for making connections in an operating circuit. The layer construction is shown in Fig. 13-1.

The terminals labeled *anode* and *cathode* perform the same function as their counterparts in a diode rectifier. The load in which power is to be controlled is connected in series with the anode-cathode terminations and the power source. Current will pass in the SCR only from anode to cathode as in the diode, but it may exist only when a bias current is applied to the *gate* terminal. Once conduction begins, the gate loses control of the current. Anode current will stop only when its source is disconnected.

The SCR initially blocks both negative and positive currents at its anode. However, by applying a suitable gate current as a bias, the device can be

---

* J. L. Moll *et al.*, "*p-n-p-n* Transistor Switches," *Proc. I.R.E.*, **44**, 1174–82 (1956).

Sec. 13-3   *Theory of Operation; Two-Transistor Analogy* 447

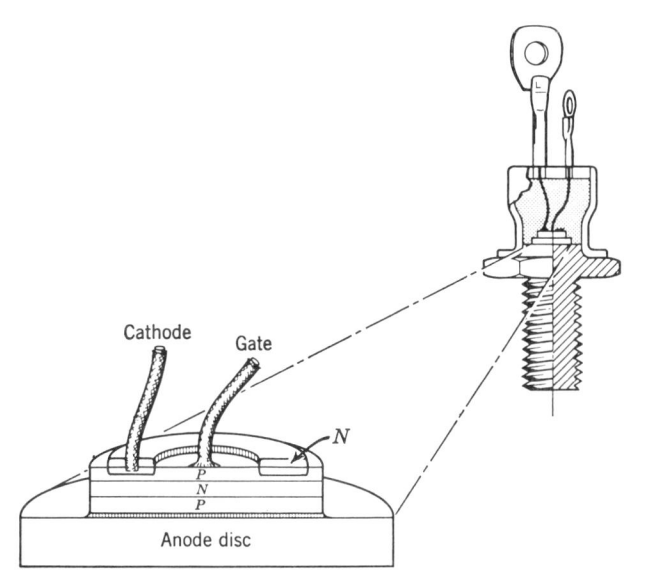

Figure 13-1
Internal construction of SCR.

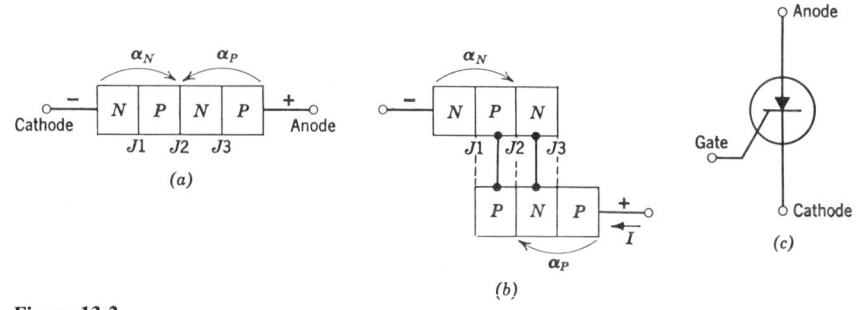

Figure 13-2
Two-transistor analogy for four-layer SCR and its schematic circuit symbol. (*a*) Four-layer SCR. (*b*) Two-transistor analogy. (*c*) Circuit symbol for SCR.

"broken down" into conduction at any point on a positive voltage waveform at the anode.

13-3
Theory of Operation; Two-Transistor Analogy

The four layers of semiconductor material making up the SCR may be regarded as two transistors as shown in Fig. 13-2. Three junctions are formed by the four layers, but their effects on one another are better understood from the two-transistor analogy.

The SCR has forward and reverse characteristics that are widely different, determined by the internal operation at the various junctions and their effects on one another. *Forward bias* for the composite device is defined as shown in Fig. 13-2. Junctions $J1$ and $J3$ are forward biased and $J2$ is reverse biased; i.e., both the N-P-N and P-N-P are biased in the conventional manner for common-emitter operation. When operated in this way, each transistor has associated with it a current gain α (alpha). Let us define the current gain of the N-P-N transistor as α_N, and that of the P-N-P unit as α_P. The current gain is the fraction of electron current (for the N-P-N) or hole current (for the P-N-P) injected at the emitter that reaches the collector.

At the junction that is common to both transistors ($J2$), current will be made up of three components—the electron current from the N-end region, the hole current from the P-end region, and the leakage current. The total current in an external circuit must be equal to the current at $J2$:

$$I_{J2} = I = I\alpha_N + I\alpha_P + I_{CO} \qquad (13\text{-}1)$$

Written in another form,

$$I - I\alpha_N - I\alpha_P = I_{CO}$$

or

$$I(1 - \alpha_N - \alpha_P) = I_{CO}$$

so that the total current may be written in terms of I_{CO}:

$$I = \frac{I_{CO}}{1 - (\alpha_N + \alpha_P)} \qquad (13\text{-}2)$$

From this expression, it is seen that the circuit current can become very large as the sum of the current gains approaches unity. Circuit current will be limited only by any external resistance, when $\alpha_N + \alpha_P = 1$. Thus, increasing the current gains will "turn on" the SCR. When the current gains are very small, the current in $J2$ will be approximately the leakage current I_{CO}. This condition corresponds to the "off" state, or blocking state, of the SCR.

When the emitter current of a transistor is small, its current gain α is small. One way to increase the current gain is to increase the emitter current. Figure 13-3a shows the variation of current gain with emitter current. Another method of increasing the current gain is to increase the collector-emitter voltage. This is shown in Fig. 13-3b.

If the voltage between the collector and emitter is increased sufficiently, the junction $J2$ will break down and permit a large reverse current to pass through it. This effect is called *avalanche breakdown*. When $J2$ avalanches, the current in $J2$ increases and in turn increases both α_N and α_P. The SCR thus goes into its "on" state.

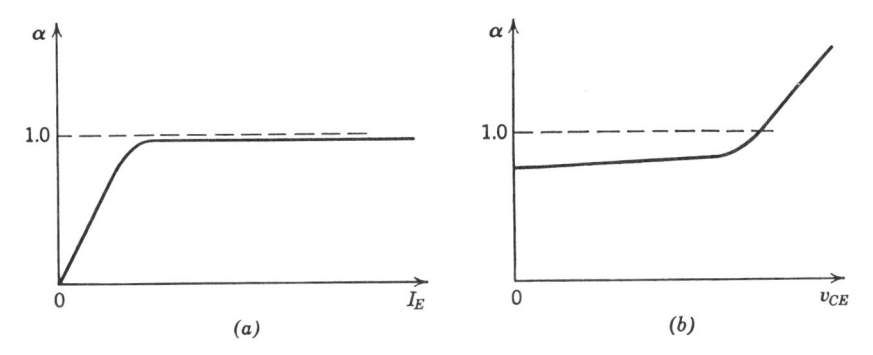

Figure 13-3
Variations of α of a silicon-controlled rectifier with current and voltage. (a) Current gain vs. emitter current. (b) Current gain vs. collector-emitter voltage.

The anode-cathode voltage that causes these events in the SCR is called the *breakover voltage*, BV_F. After the device is turned on by the breakover voltage, it will remain in its "on" state as long as the current through $J2$ (external circuit current) is large enough to make $\alpha_N + \alpha_P$ equal to unity. The minimum current that will keep the SCR "on" is called the *holding current*, I_H. This is the current in $J2$, i.e., the *external* current.

When the polarity of the external bias is reversed, in order to cause $J1$ and $J3$ to be reverse-biased and $J2$ to be forward-biased, the SCR behaves much as a diode in reverse bias. The voltage–current curves for the SCR are shown in Fig. 13-4 for forward and reverse bias.

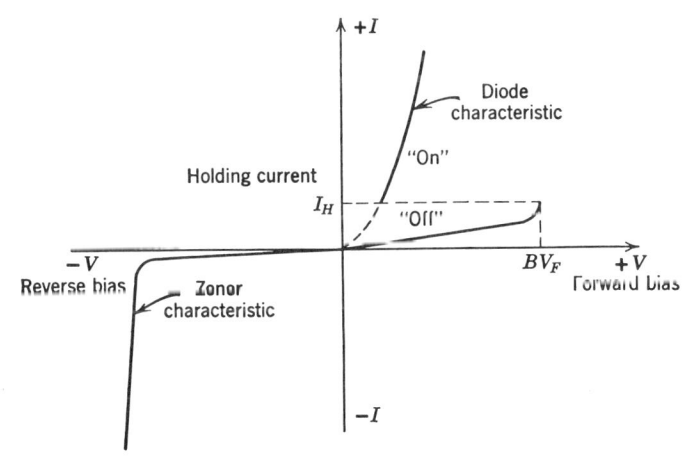

Figure 13-4
Forward and reverse bias characteristics of SCR.

13-4
Effect of the Gate

The discussion so far has concerned the behavior of the SCR as a two-terminal device, as forward and reverse bias are applied to it. In a practical application of the SCR, however, the gate terminal is used to control the anode current over some range of anode voltage amplitudes. We shall now see how the gate current can be used to control the action of the SCR.

The two-transistor analogy of the SCR is used again to describe the action of the gate in controlling anode current. In Fig. 13-5, the gate terminal is shown connected to the P-region of the N-P-N transistor.

If a positive voltage is applied to the gate-cathode junction, $J1$, it is forward biased and current will pass through it. Then many electrons appear at $J2$, and because of its reverse bias they will pass into the N-region between $J2$ and $J3$. When these electrons reach $J3$, which is forward biased, they recombine with holes in the anode P-region. This will trigger the generation of an electron–hole pair at the anode terminal to restore equilibrium, and an electron leaves the anode to contribute to circuit current. Some of the holes that diffuse across the $J3$ junction during conduction will not recombine with electrons there. They drift on into the N-region between $J2$ and $J3$. As a hole reaches $J2$ it is swept into the P-region between $J2$ and $J1$. It then drifts until it reaches $J1$, where it combines with an electron from the emitter (cathode).

Thus, a current from the gate terminal into the junction $J1$ increases the current in the entire device. This occurs because the current gains, α_N and α_P, increase. As the gate current continues to increase, the breakover voltage, BV_F,

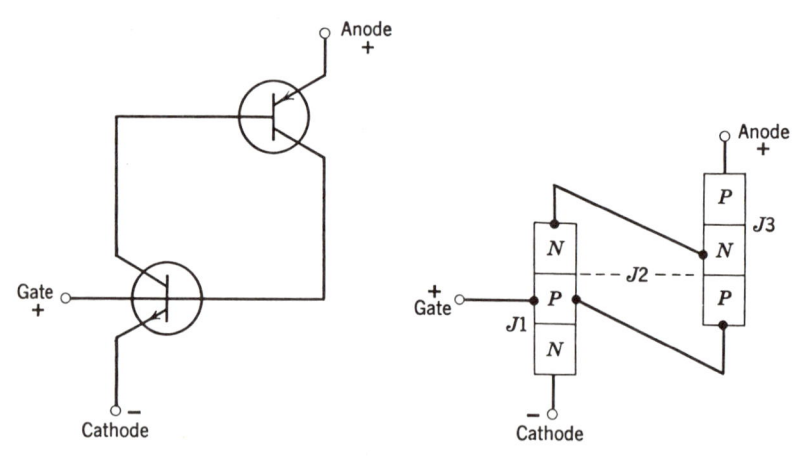

Figure 13-5
Gate connection to two-transistor analogy.

Sec. 13-4   *Effect of the Gate*   451

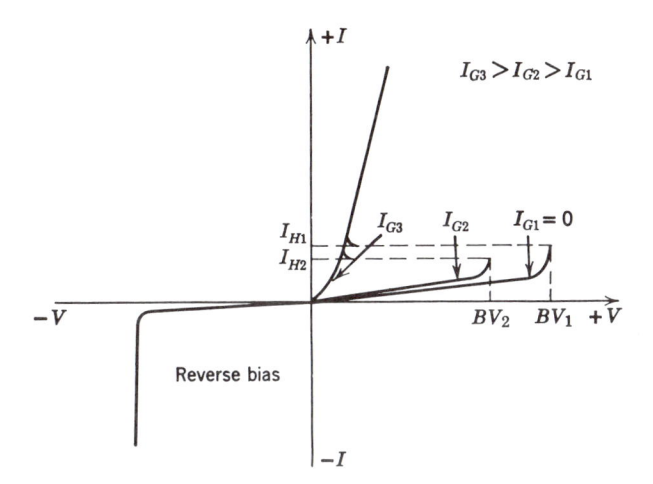

Figure 13-6
Effect of gate current on breakover voltage and SCR behavior in forward bias.

is decreased significantly. This is shown in Fig. 13-6. The gate current can become large enough to cause the SCR to act as a P-N junction diode.

We see that the SCR remains in its "off" state when it is forward biased by a voltage less than the breakover voltage. Upon application of sufficient gate current, the breakover voltage will decrease and permit conduction in its "on" state. Once the device is turned on, the gate no longer has any control over the anode current. To return it to its "off" state, anode current must be reduced below the value of holding current. This may be accomplished either by reducing the anode current or by removing the forward anode voltage. If the anode voltage is a sine wave of voltage, it will reverse its polarity each cycle and automatically return the SCR to its "off" state every cycle. It is this characteristic that is important in power control applications. It will be illustrated in connection with application of the SCR in following sections.

Since the gate current increases α_N, which in turn increases α_P, the gate current gain of the SCR is equivalent to that of a grounded-emitter N-P-N transistor connected to a P-N-P transistor, as illustrated in Fig. 13-7. As gate current flows into the base of the transistor, it is amplified and appears at the collector as $I_G\beta_N$, where $\beta = \alpha/(1 - \alpha)$. This amount of collector current enters the base of the P-N-P transistor and appears as a collector current of $I_G\beta_N\beta_P$. If α_P, the current gain of the P-N-P transistor, is large, a regenerative action occurs and the SCR anode current increases. A small gate current can then trigger the SCR into its "on" state, thus controlling a current that may be thousands of times larger than the gate current. To turn on the SCR, it is necessary that the gate current be large enough to trigger the regenerative action.

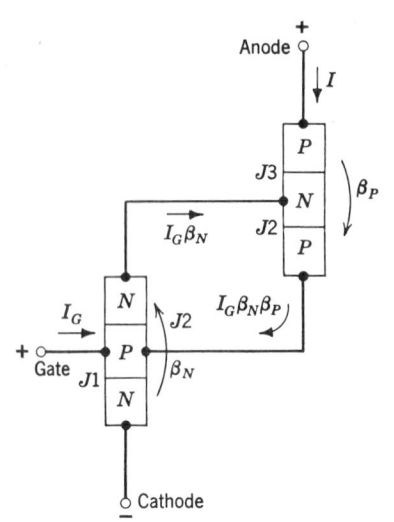

Figure 13-7
Current regeneration in forward-biased SCR.

13-5
Ratings and Characteristics of the SCR

The following ratings and characteristics of the SCR refer to their designations in Fig. 13-8a. Manufacturers' data sheets include numerical values for these parameters.

RATINGS

V_{FM}	Peak forward blocking voltage	The maximum instantaneous voltage that may be applied to the anode and will not switch on the SCR.
V_{FR}	Continuous forward blocking voltage	The maximum continuous voltage that may be applied to the anode and not switch on the SCR.
V_{RM}	Peak reverse blocking voltage	The maximum instantaneous reverse voltage that will not break over the SCR in a reverse mode.
V_{RR}	Continuous reverse blocking voltage	The maximum reverse voltage that may be applied continuously and not break over in a reverse mode.

PFV	Peak forward voltage	The maximum instantaneous forward voltage that may be used to switch on the SCR. If other specified maximum ratings are not exceeded and applied voltage is lower than PFV, the SCR will not be damaged.
I_F	Maximum continuous anode current	The maximum current that may pass in the forward direction without exceeding rated junction temperatures.
I_O	Maximum average forward current	Maximum current averaged over a full cycle that may be applied continuously under single-phase, 60-Hz half-sine-wave operation with a resistive load. Average current and peak reverse voltage ratings apply simultaneously.

CHARACTERISTICS

V_F	Static forward voltage	The forward voltage drop when the SCR is conducting in the forward direction.
V_{GT}	Gate trigger voltage	The gate voltage required to produce the gate trigger current.
I_{GT}	Gate trigger current	The minimum gate current required to switch the SCR from its "off" state to "on" state.
I_H	Holding current	Minimum principal current required to maintain "on" state following conduction of steady-state forward current.
I_{RR}	Static anode reverse current	Principal current for negative anode-to-cathode voltage.

As an example of ratings and characteristics furnished by a manufacturer, these numerical values are published for a 2N685 type of SCR:

Min. fwd. breakover voltage, BV_F	200 V
Peak reverse voltage, V_{RM}	300 V
Continuous reverse voltage, V_{RR}	200 V

Maximum Allowable Ratings

Continuous anode current, I_F	35 A rms, all conduction angles
Maximum average anode current, I_o	Depends on conduction angle (see Fig. 13-8b)
Peak one-cycle surge current	150 A
Peak forward gate voltage	10 V
Peak reverse gate voltage	5 V

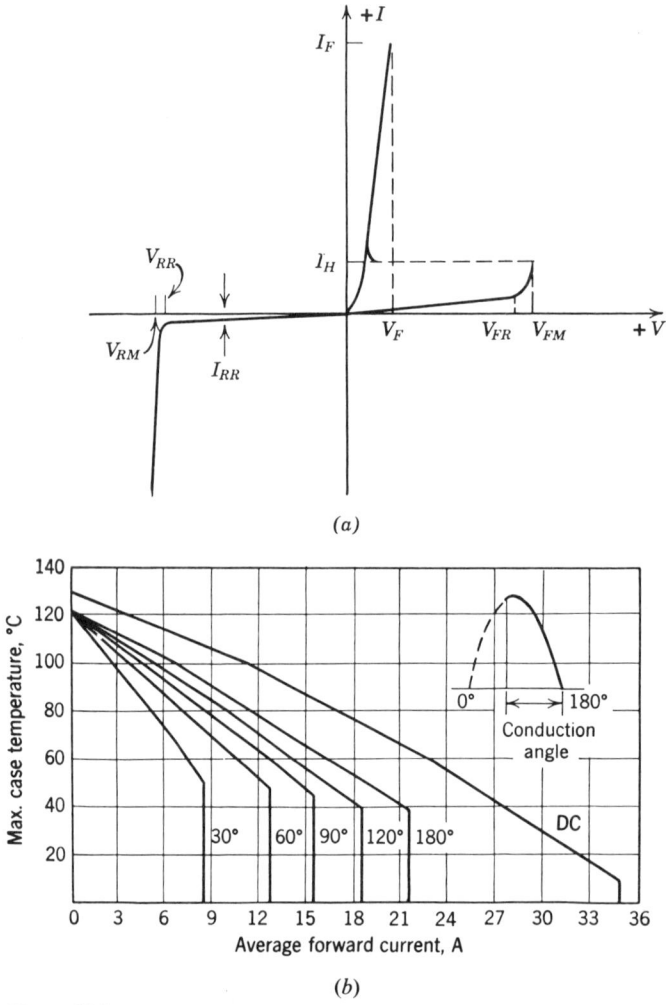

Figure 13-8
Characteristics of the silicon-controlled rectifier. (a) Identification of symbols. (b) Maximum case temperature for sinusoidal current waveforms, at various conduction angles, SCR type 2N685.

Characteristics

Gate trigger current, I_{GT}	15 mA typ., 40 mA max.
Gate trigger voltage, V_{GT}	1.5 V, typ.; 3.0 Vdc, max.
Holding current, I_H	10 mA dc, typ.; 100 mA dc, max.

Example 13-1. The type 2N685 silicon controlled rectifier is used in a single phase resistive-load circuit. If the maximum case temperature is maintained *at 80°C or less*, what average forward current can it safely carry?

Solution: From the curves of Fig. 13-8b, we see that different values of average current are permissible for different conduction angles. If the SCR is triggered as soon as its anode goes positive, current will flow during the entire 180 degrees of a sinusoidal voltage. At this conduction angle the average current can be almost 12 A. For a conduction angle of 120°, the gate does not energize the anode until after the first 60 degrees of the sinusoidal voltage, and average current can be about 10 A. At 30° conduction angle, the permissible average current is only 5 A.

The curves in Fig. 13-8b have definite endpoints (maximum *average current*) for each conduction angle. They occur because they represent the same rms values for each of the waveforms. For example, the 2N685 is rated at 35 A *rms*, at all conduction angles. For a conduction angle of 180° the average of the *half sine wave* is 0.318 times the maximum amplitude, and the rms value is 0.500 times the maximum amplitude. The ratio of rms to average values is 1.57, so that the average current is limited to 35/1.57 = 22.3 A for 180° conduction angle. If the *average current* does not exceed 22.3 A, the *rms value* will not exceed the rated 35 A.

13-6
Gate Trigger Characteristics

The gate trigger characteristics of the SCR are usually presented in the graphical form in Fig. 13-9. These characteristics apply to the type 2N685. The SCR will trigger at a definite point on its gate characteristic. The shaded areas of the curve contains all of the possible trigger points, based on rated values of I_{GT} and V_{GT}. In order to achieve reliable triggering, the trigger circuit must simultaneously supply I_{GT} and V_{GT} values that lie *outside* the shaded area.

For the simple on-off switching of the SCR in a dc circuit, a continuous gate signal of amplitude required to furnish V_{GT} and I_{GT} can be used. However, when dc is used on the anode of the SCR, once the gate has triggered the device into conduction, the gate will have no more effect on the operation. Conduction will continue until the anode current falls below the holding current I_H. In order

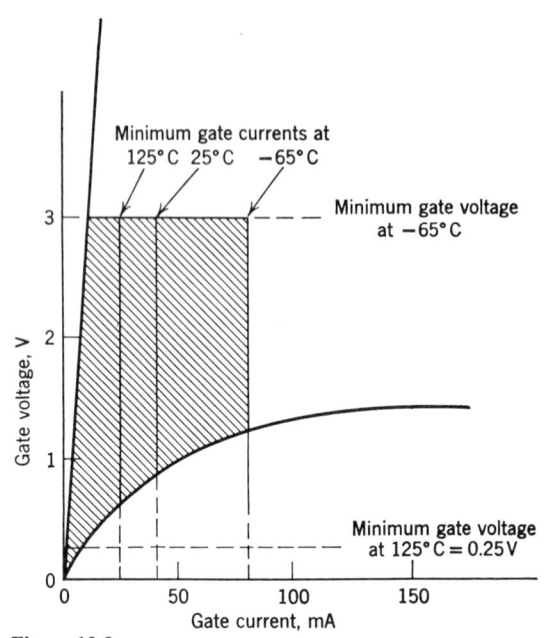

Figure 13-9
Critical gate firing voltage and current for 2N685. Points for a firing condition (V_{GT}, I_{GT}) must fall *outside* the shaded area.

to stop anode current, the dc anode supply may be disconnected. This can be done in an operating circuit by some mechanical switch (push-button, reed switch, etc.), actuated either manually or by transducers in response to heat, light sources, pressure, etc. If an ac source is used on the anode, a dc gate signal may be used, and conduction will cease during negative half-cycles of the anode voltage. The gate can control conduction each time the anode voltage becomes positive.

A simple arrangement of a gate trigger circuit is shown in Fig. 13-10a, where $e_s = E_{max} \sin \theta$. Gate current is supplied directly from the anode-voltage supply through a suitable current-limiting resistor, R. The resistor should be chosen for proper operation within the ratings of the SCR. For a worst-case design, the switch SW closes at the peak value of supply voltage, and the value of R will be

$$R \geq \frac{E_{max}}{I_{G\ max}} \quad (13\text{-}3)$$

The SCR will trigger into conduction when suitable values of V_{GT} and I_{GT} occur simultaneously or when the instantaneous anode voltage is

$$e = V_{GT} + I_{GT}R + V_F \quad (13\text{-}4)$$

where V_F is the forward drop of the diode D.

Sec. 13-6 Gate Trigger Characteristics 457

Figure 13-10b is another form of the anode triggering circuit. Resistor R is used as a control device for phase control of the SCR. When $R = 0$, the circuit is the same as in Fig. 13-10a, and the SCR will trigger according to Eq. 13-4, in which $R = R_{min}$ of the control circuit. The value of R_{min} is selected from Eq. 13-3.

As the value of R is increased, in the control circuit of Fig. 13-10b, the SCR will trigger at larger firing angles (θ in electrical degrees) until the anode voltage equals E_{max}. Since the SCR cannot wait for I_{GT} to occur at some time *after* E_{max}

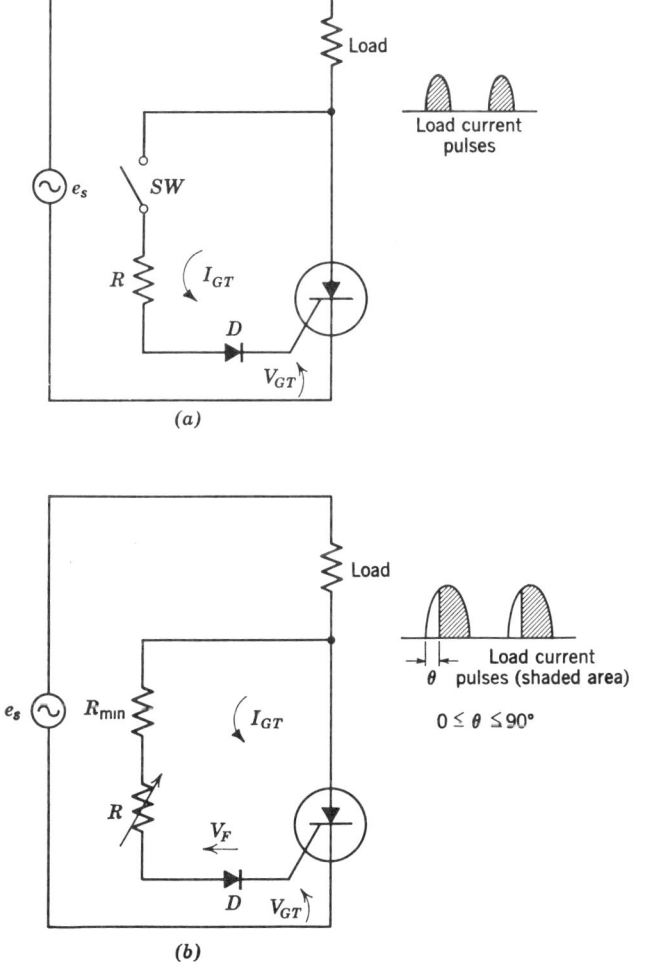

Figure 13-10
Anode resistor trigger circuits. (a) Simple anode trigger. (b) Phase-control trigger.

458   Electronic Power Conversion   Ch. 13

is reached but will trigger the first time I_{GT} occurs, conduction can be controlled only for $0 \leq \theta \leq 90°$. Therefore, the circuit of Fig. 13-10b provides continuous control of conduction from the full 180 degrees of half-wave operation to 90 degrees of half-wave.

Drill Problems

D13-1   In the circuit of Fig. 13-10a, assume the diode forward voltage drop is 1 V, and the gate triggers at $V_{GT} = 1$ V, $I_{GT} = 10$ mA. For $R = 10$ kΩ, at what anode voltage will the SCR trigger? What must be the maximum gate current rating, $I_{G\,max}$?

D13-2   In the circuit of Fig. 13-10b, $I_{G\,max} = 10$ mA, $V_F = 1$ V, $V_{GT} = 1$ V, $I_{GT} = 1$ mA. When $R + R_{min} = 0$, what anode voltage will trigger the SCR? What will be the firing angle of the SCR?

The range of control provided by this simple circuit can be extended by the circuit in Fig. 13-11a. The resistor–capacitor combination in the gate circuit permits control of conduction through the entire 180 degrees of half-wave operation. The action of the R-C circuit delays the necessary V_{GT} for triggering to occur. During positive half-cycles of supply voltage, the capacitor will charge to the trigger voltage in a length of time determined by the R-C time constant and the frequency of the supply voltage. On the negative half-cycle the capacitor charges to the negative maximum of supply voltage through diode D_2. This resets it for the next charging cycle.

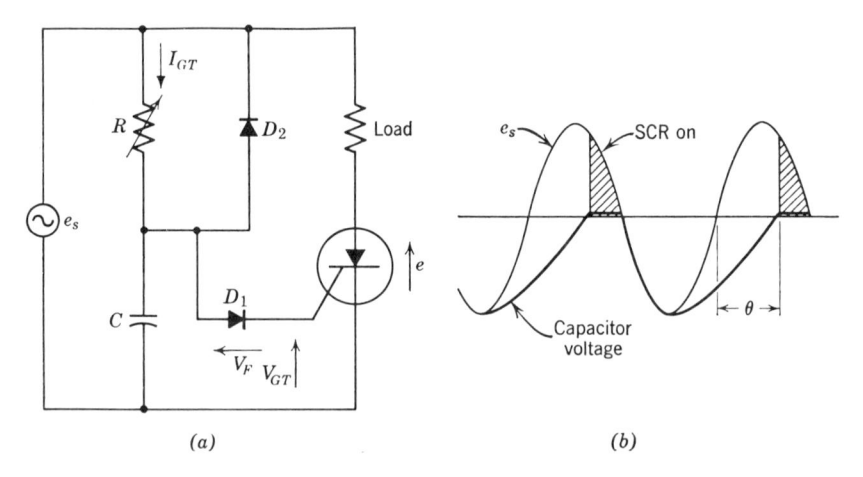

Figure 13-11
Full half-cycle phase control of SCR. (a) R-C phase control. (b) Voltage waveforms. Firing is delayed θ degrees after anode goes positive.

Because of the spread of gate characteristics, as in Fig. 13-9, an exact solution for values of firing angle θ is not practical. The value of resistor R is selected to allow trigger current I_{GT} when the capacitor voltage is large enough to trigger the SCR. When the capacitor voltage is equal to $V_{GT} + V_F$, resistor R must allow trigger current, I_{GT}, to flow, and the anode voltage is

$$e \geq I_{GT}R + V_{GT} + V_F$$

requiring a maximum value for resistor R given by

$$R \leq \frac{e - V_F - V_{GT}}{I_{GT}} \tag{13-5}$$

Figure 13-11b shows a typical waveform of capacitor voltage and the resulting firing angle of the SCR. The gate circuit will trigger the SCR into conduction whenever the gate voltage reaches its rated value.

By adjusting the value of the resistor R, the firing angle θ can be shifted to control the conduction angle anywhere in the range from 0° to 180°, giving full half-cycle control over the SCR and its load.

13-7
Full-Wave Phase Control of the SCR

The single-phase, half-wave circuits of Section 13-6 are often inadequate for supplying necessary power to the load of an SCR. Two devices may be used for full-wave control of load power in the circuit of Fig. 13-12a. In this circuit, two separate gate control circuits are required. If they provide equal conduction angles for the two SCRs, the load voltage will be symmetrical as shown in Fig. 13-12b.

An alternate form of full-wave control is shown in Fig. 13-12c. Only one gate control circuit is required to operate the circuit. This is sometimes an advantage, because the gates and cathodes have a common connection, and either may be grounded in a practical application. The diodes prevent negative voltage across the SCR during alternate half-cycles of the input voltage.

13-8
RMS and Average Load Voltage Calculations

The SCR is rated in terms of average current, but most ac loads require certain rms, or effective, currents and voltages to meet power requirements. Because the SCR is a switch that can proportion power according to its conduction angle, the load voltage will have average, rms, and peak values that depend on the conduction angle. The power in the load will be determined by the rms values of current and voltage. For a resistance load, the load power will be $I_{rms} \times V_{rms}$.

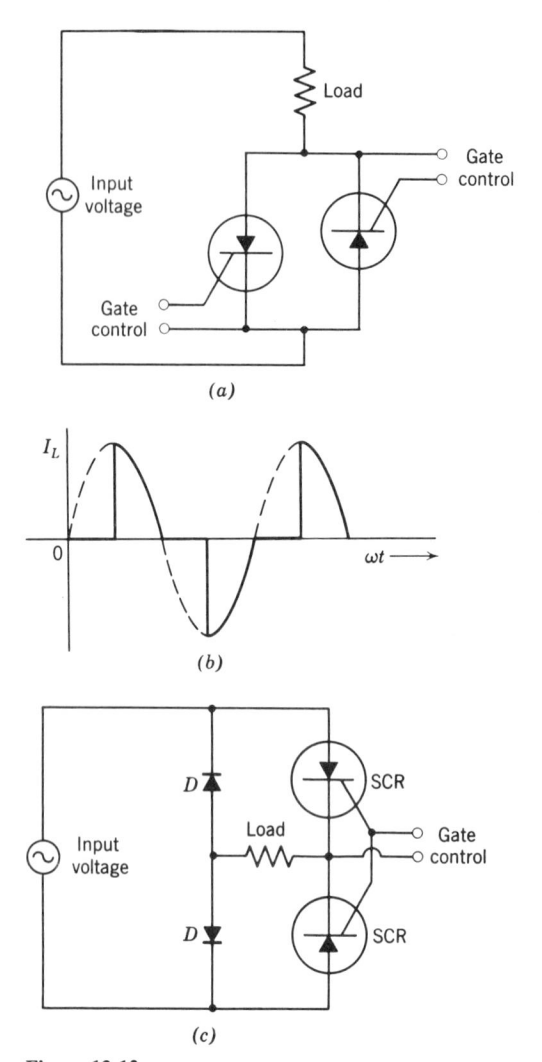

Figure 13-12
Full-wave control of SCR load power. (a) Full-wave control of SCRs. (b) Symmetrical load voltage waveform, at 90° conduction angle. (c) Alternate full-wave control circuit.

Figure 13-13 illustrates how the average, rms, and peak values of load voltage vary with conduction angle for both half-wave and full-wave phase control.

Example 13-2. It is desired to supply 1 kW to a resistive load rated at 110 V from a supply of 220 V sinusoidal waveform. Since the power supplied is proportional to the square of the applied voltage, connecting it to the 220-V

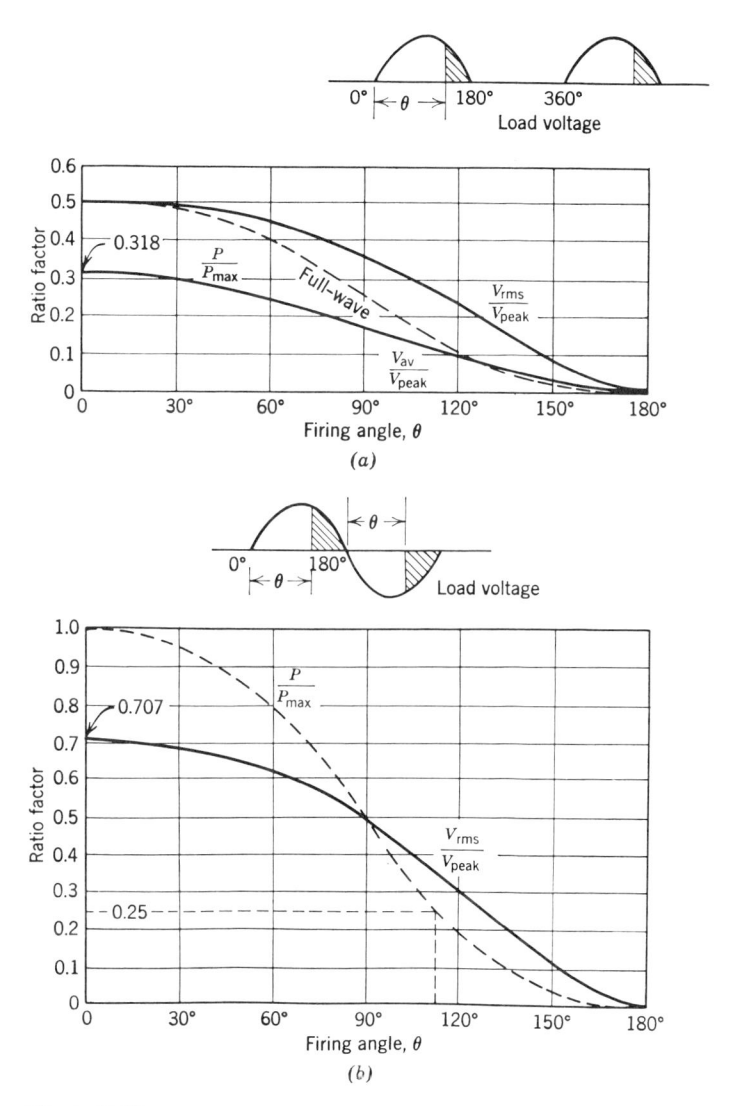

Figure 13-13
Variations of rms and average voltages with changing firing angle. (a) Half-wave phase control. (b) Full-wave symmetrical phase control.

supply would result in 4 kW of power. In order to furnish only the required 1 kW, either half-wave or full-wave phase control may be used with an SCR.

Starting with the half-wave operation, we see from Fig. 13-13a that one-fourth power (0.25 on the ordinate) occurs for a firing angle of 90°. The peak voltage is $1.414 \times 220 = 312$ V. Average voltage is $0.159 \times 312 = 50$ V and the rms voltage is $0.353 \times 312 = 110$ V. The ratios of rms to peak values and average to peak values are read from the graph at a conduction angle of 90°.

The load resistance has a value $110^2/1000 = 12.1\ \Omega$. Therefore the average load current is $50\ \text{V}/12.1\ \Omega = 4.13$ A. This value would be indicated by a dc ammeter in series with the load. The SCR must therefore be rated to handle at least 4.13 A at 90° conduction angle. The load must also be capable of operating at the high values of peak voltage and peak current.

Now we will see how these values may be changed in a full-wave phase control connection of the SCR. In a symmetrical full-wave circuit, one-fourth power occurs at a firing angle of about 115°. This means that each half-cycle conducts for only $(180° - 115°) = 65°$ in order to supply 1 kW to the load. The peak voltage is $312 \times \sin(115°) = 312 \times 0.82 = 256$ V, and the rms voltage is again $0.353 \times 312 = 50$ V. The average voltage will be zero for the symmetrical waveform. To determine the rating of each SCR, consider each as operating as a half-wave source at a firing angle of 115°. From Fig. 13-13a, the average voltage is $0.095 \times 312 = 29.6$ V. Then the average current in each SCR is $29.6/12.1 = 2.44$ A. Each SCR must be rated to carry this average current at a firing angle of 115°.

It is important to note that the first and last 30 degrees or so of each half-cycle contribute a small part of the total power (about 5 per cent only). Consequently, a triggering range of firing angles from 30° to 150° will yield a range of power control from about 2.5 to 97.5 per cent of total power. This range of firing angles provides the most effective control of power in the load.

13-9
Bilateral SCR, the TRIAC

An important shortcoming of the SCR is its inability to conduct current through a load in both directions. In its "on" state it is limited to current in one direction. Of course, two SCRs can be connected in a full-wave bridge or center-tapped transformer mode, but such operation not only requires two SCRs but also makes the design of gate trigger circuits a further problem.

The fabrication techniques used to build four-layer P-N-P-N devices have led to the development of a novel, bilateral SCR, the TRIAC (triode, ac operated). It is sometimes referred to as a bilateral triode switch. Its fabrication and internal behavior are beyond the scope of this book, but we will discuss briefly its features that are similar to the SCR in controlling load power.

The internal structure of this device permits it to trigger into conduction from a single gate terminal, from either positive or negative gate signals. Also, it is capable of conducting current in either direction. Trigger and phase-control circuits permit the power in a load to be varied from pure sinusoidal input

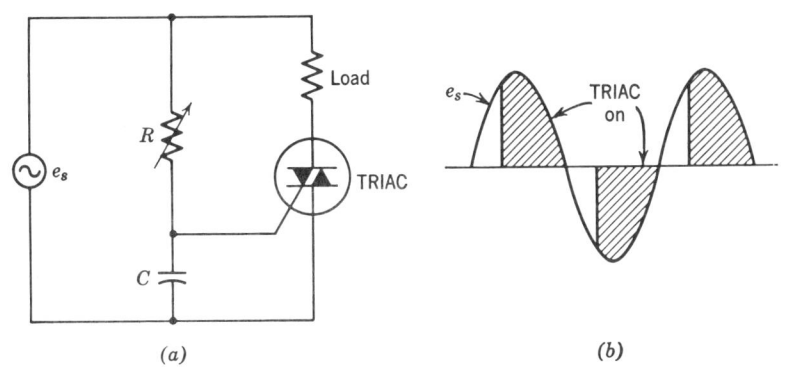

Figure 13-14
Bilateral circuit using TRIAC. (a) Circuit. (b) Voltage waveforms.

voltage to nonconduction in both half-cycles. Figure 13-14 is a typical circuit arrangement and load voltage waveform for the bilateral triode switch controlling power in a resistive load.

The curves developed in section 13-8 for full-wave SCR control apply to the operation of the TRIAC. The difference is that the TRIAC will have average current and voltage values that are twice their counterparts in the SCR circuit, because they occur twice in each cycle in the same device, whereas two SCRs shared this behavior.

13-10
Photovoltaic Action; The Solar Cell

Photovoltaic action occurs in semiconductors constructed to absorb incident light energy. An internal potential difference is generated across a *P-N* junction in the material making up the device. The potential difference can generate a current in an external load connected to the photovoltaic device. The most common photovoltaic energy converter is the *solar cell*. Solar cells have been used extensively as electrical power supplies in artificial earth satellites and space probes. They have been used to convert the sun's energy into useful electrical energy for powering and controlling the equipment on board.

Two similar semiconductor materials are used to form a photovoltaic solar cell. The materials differ mainly in the nature of impurity atoms present in them. One of them contains impurity atoms that make it an *N*-type semiconductor and the other contains *P*-type atoms. The cell consists of a thin slice of single-crystal *P*-type silicon up to $2 \, cm^2$ into which a layer (about $0.5 \, \mu$) of *N*-type material is diffused. The *N*-type material is usually phosphorus, although other materials have been used in experimental designs.

464 *Electronic Power Conversion* Ch. 13

Electrical connections are made to the P-region and N-region of the cell. An external load may then be connected to the cell. The arrangement shown in Fig. 13-15 is sometimes used. Incident light strikes the upper face of the cell and generates the internal potential difference that will force current through an external load. The load is connected between the bottom surface (labeled "positive contact") and the negative contact along one edge of the upper surface. The contact on the dark side of the cell can have a large area, covering the entire surface as a continuous layer of solder. However, the negative contact should have a small area so that it does not block the incident light. It should permit maximum exposure to incident light but at the same time make good electrical contact with the surface. To meet these conflicting requirements in a cell having a large surface, a grid electrode arrangement is sometimes used as shown in the figure. These grids are spaced so that a low-resistance connection can be made to the illuminated surface but are narrow enough to permit exposure to light. Very small cells make contact along one edge only.

Where the two doped semiconductor materials join is a *P-N* junction. It is a thin, neutral region separating the two materials. The junction region contains both types of impurities and is essentially electrically neutral. An inherent property of such a *P-N* junction is an internal electric potential that is developed across it.

Incident light photons generate electron–hole pairs in the cell, as shown in Fig. 13-16. The impinging photon can pass into the cell and react with an atom in either the N-region or P-region of the cell. An electron may absorb the energy of the photon, increasing its own energy sufficiently to release it from its

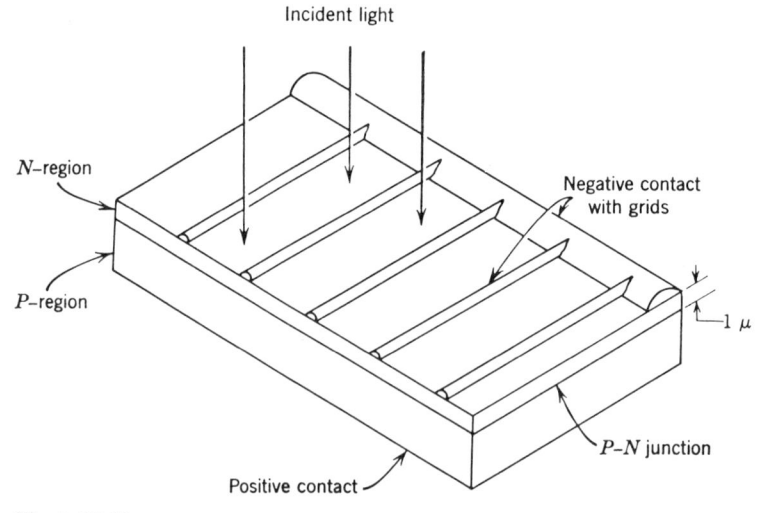

Figure 13-15
Negative contact grid structure on exposed surface of a cell.

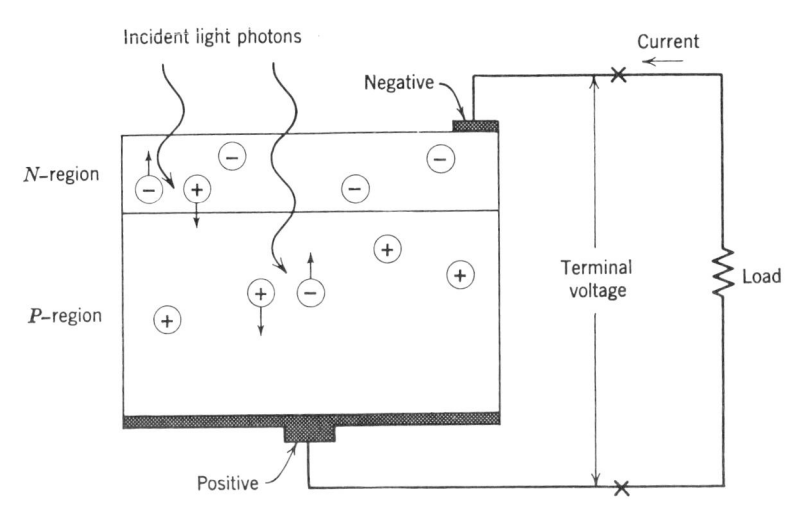

Sec. 13-10   *Photovoltaic Action; The Solar Cell*   465

Figure 13-16
Incident light generates electron–hole pairs, causing current in the load.

parent atom. The electron may then be conducted about in the cell by diffusion and drift, leaving a hole in its atom. The electron and hole thus generated are in excess of the number needed for thermal equilibrium in the cell. In reestablishing equilibrium in the cell, they may recombine in the region of their generation or they may wander randomly about until they are forced across the junction by the electric field there. Those that cross the junction contribute to current in the cell and its external load.

The electrical behavior of a typical commercial cell is given in Fig. 13-17. The load current and terminal voltage were measured for varying values of load resistance from short-circuit to open-circuit conditions. The open-circuit voltage is less than 0.350 V, and short-circuit current is only a few milliamperes for the range of illuminations up to 800 fc. The voltage–current relations are marked for load lines representing 50, 200, and 1000 Ω.

For a given illumination there will be some load resistance that absorbs maximum power. Constant-power hyperbolas may be drawn on the figure to determine the maximum power for the selected illumination. The constant-power hyperbola that is tangent to the illumination curve will indicate the maximum power condition. Then, the load line that passes through the tangent point will represent the load resistance that will absorb the amount of power. The maximum power at 400 fc is approximately 290 μW.

In applications to furnish power from the sun to operate electrical equipment in space, a single cell such as this is certainly inadequate. Its voltage, current, and power are minuscule. However, many similar cells may be connected in series and parallel in large arrays, either on the surface of the space vehicle or

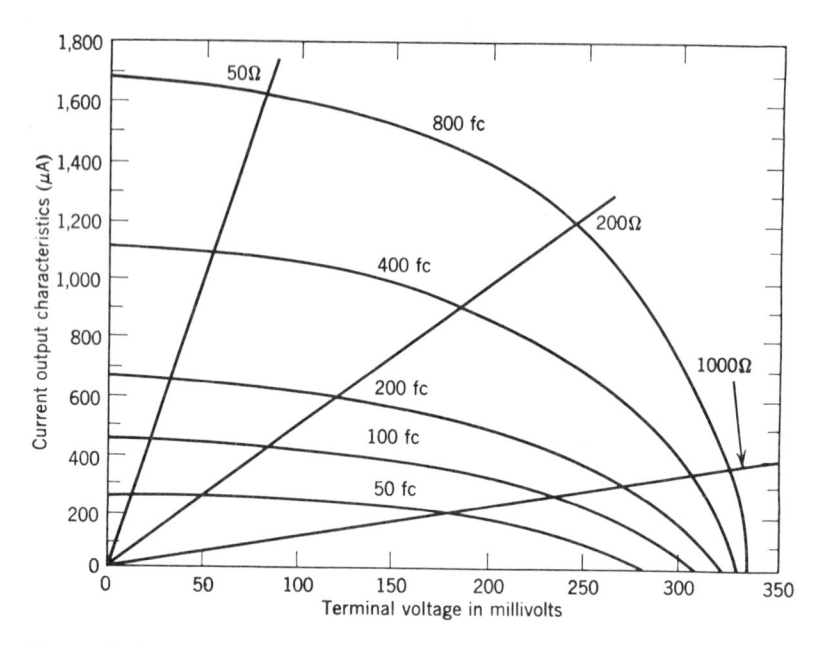

Figure 13-17
Voltage–current characteristics of a solar cell for various external loads and illumination levels.

on movable "paddles" that extend from it. Movable paddles covered with solar cells can be positioned by the vehicle so that maximum exposure to the sun is maintained regardless of the position and attitude of the vehicle.

13-11
Light Amplification by Laser

One of the most exciting developments in photo technology has been the *laser*. From the pioneering work of T. H. Maiman in 1960,* lasers have fired the imagination of engineers and scientists around the world. They have been used to cut holes in diamonds, weld metals almost instantly, perform delicate surgery of the human body, and carry communications channels from earth to space vehicles. Their ability to concentrate power in a narrow beam of light is an important property in many applications.

The word *laser* is an abbreviation of "*L*ight *A*mplification by *S*timulated *E*mission of *R*adiation." The laser controls a process of generation of coherent light by stimulating emission of radiation from the atoms in a solid or gaseous

---

*T. H. Maiman, "Optical Maser Action in Ruby," *Brit. Commun. Electron.*, **7**, 674 (1960).

material. Lasers are sometimes referred to as "optical masers," where the word maser refers to the same action, but for microwaves rather than lightwaves.

Transfer of energy by radiation in an atom can be described by three different types of radiation:

1. *Absorption*—occurs as an atom reacts with a photon of incident light. The photon must have sufficient energy to increase the energy of an electron and move it to a higher orbit in the atom. During the process the electron increases its energy by an amount equal to the absorbed energy of the incident photon. Any excess energy of the photon is lost in the process, either as fluorescence or as heat. The electron may acquire only discrete amounts of additional energy, even though the impinging photon may have more than is needed to raise the energy of an electron. The important result is the excited state of the atom when an electron has absorbed additional energy. As in all physical systems, the atom will tend to return to its original lower energy state.
2. *Spontaneous emission*—occurs when the atom is in an excited, unstable state. An electron may spontaneously return to its original energy level, or it may lose only part of its increased energy. In the latter case, it will take up an orbit position somewhere between that of its excited state and its original position. It is then in a *metastable state*. Whenever an electron loses some energy, a photon containing the amount of energy lost may be emitted.
3. *Stimulated emission*—occurs in excited atoms that have electrons in metastable states. An electron may exist in such a "semistable" state for some period of time. An incident photon at the same energy level as the metastable electron may stimulate the electron to return to its original stable state. If the photon is not absorbed, but only passes through the atom, then *two photons* at the same energy may be emitted from the process. An "amplification" has occurred in the atom because of the *stimulated emission of radiation*. This action is the basis of laser operation.

13-12
Coherency of Laser Radiation

In the process of stimulated radiation from an atom, the two photons that derive from a single incident photon will have the same energy. They will also have the same frequency of oscillation and the same phase relationship in time. Such waves are said to be *coherent*. If two waves have different frequencies or phase, or both, they are incoherent. An incoherent wave is difficult to modulate with information. It may be used only in on-off operation. A coherent wave, on the other hand, is easily modulated with information, either as amplitude, frequency, or phase variations as discussed in Chapter 12.

468 Electronic Power Conversion

The output of a laser is coherent radiation at frequencies in the visible light region. This means that a laser has a bandwidth of about 10^{14} Hz. This bandwidth is greater than is required for all radio and television transmitting stations in the United States. A single laser could carry simultaneously all of the information being transmitted by these stations.

13-13
Types of Lasers

Since the pioneering development of the optical maser by Maiman, other types of lasers have been produced. The first laser was a ruby rod doped with chromium ions as its active element. Recent developments use a gas as the active element, and semiconductors have been used in others. Basic differences in the types of lasers occur in their output power levels.

The energy level diagram for chromium in Fig. 13-18 will be used to discuss the internal operation of a ruby rod laser. An unexcited electron will be in the stable energy state, labeled level 1. During absorption of an incident photon the electron may absorb the energy of the photon and take an energy level somewhere in the broad level 3, provided the photon had an energy great enough. Next, the electron may emit part of its energy and make a transition to level 2, an intermediate, metastable state of the chromium atom. The final step in the process is a transition from level 2 to level 1. This final step can occur as either spontaneous or stimulated emission. In either case it will emit light energy equivalent to the difference in energy of the two levels.

Figure 13-19 represents a typical pattern of photon motion inside the ruby rod. The first atom undergoes spontaneous emission, yielding a photon. It

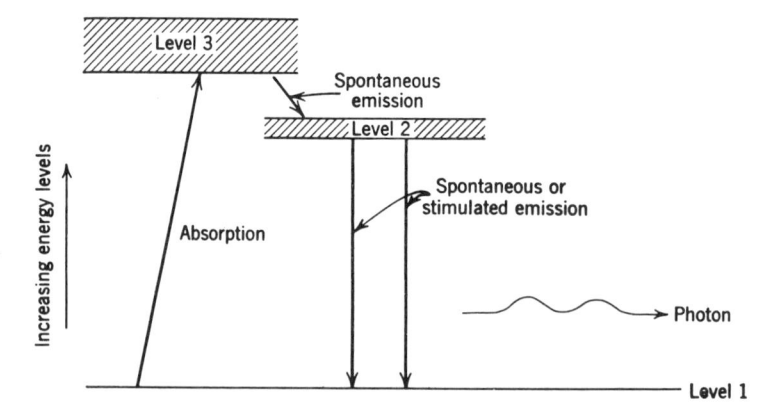

Figure 13-18
Energy levels in chromium.

Sec. 13-13	Types of Lasers 469

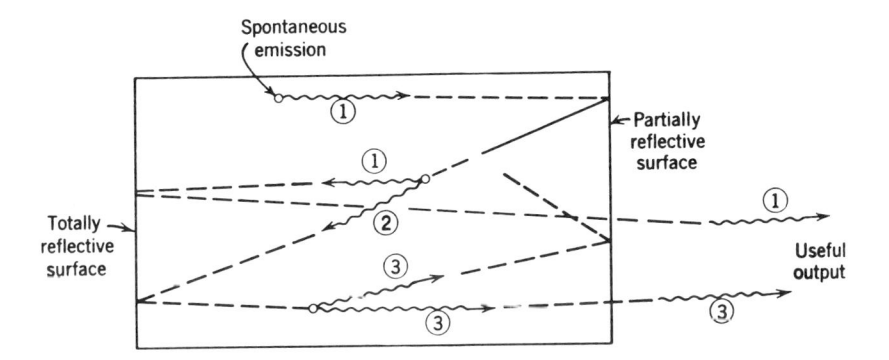

Figure 13-19
Typical photon amplification in ruby rod.

strikes the partially reflective surface at the end of the rod and is reflected, stimulating a second atom to emit a photon as it travels back through the rod. Two photons are now present to continue the process. The intensity of the photon beam moving through the rod has been amplified. Further reflections and stimulations increase the number of photons moving back and forth in the rod. Some of them will pass through the partially reflective end of the rod and become the useful output of the laser.

Figure 13-20 is a sketch of a ruby laser system. The power supply drives a flash lamp, generating photons that enter the ruby rod to initiate laser action. The light from the flash lamp excites some of the chromium atoms. A few will emit spontaneous radiation to start the laser. Their emission triggers stimulated emission from other atoms, continuing the process. Further flashes are synchronized with the emission from the laser to continue the operation.

Similar systems are capable of stimulating coherent radiation in gaseous matter. An explanation of their internal behavior is far beyond the scope of this

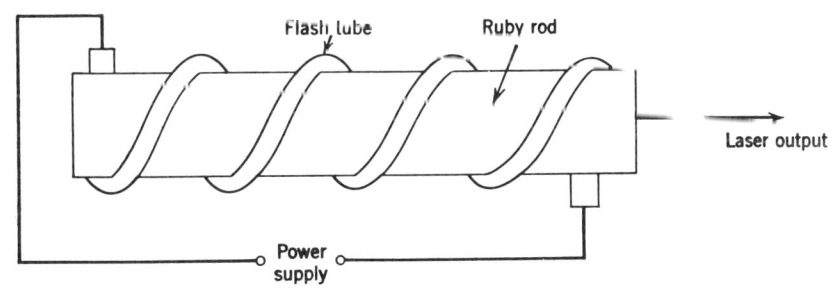

Figure 13-20
Sketch of ruby rod laser system.

discussion. The same is true of semiconductor types of lasers. We will limit our study to a brief look at the comparable parameters of the types of lasers, listed in Table 13-1.

Table 13-1
Operating Characteristics of Lasers

Laser Type	Operating Mode	Practical Efficiency (Percent)	Power Output
Solid (ruby)	Pulsed	1	50 mW
Gas-discharge	Continuous	1	10 W
Semiconductor	Either pulsed or continuous	10	500 mW

Ionized-argon lasers have been used as an "optical torch" in several experiments. They provide a continuous beam of light as a high-energy density over a small area. They have been used to mechanically trim microcircuit resistors and to make incisions in laboratory animals. Probably their best known application is in connection with the Gemini 7 space flight. Argon lasers were installed at geographically separated ground locations to serve as aiming beacons or visual targets for the astronauts. Even though the power output may be small, the laser is by far the brightest of controllable, man-made light sources, because of its coherent output.

Questions

13-1 What is the basic function of the SCR?

13-2 How is the SCR more efficient than a thyratron in controlling power in a load?

13-3 How does the gate function as the controlling element of the SCR?

13-4 What prevents the SCR from conducting internally from cathode to anode?

13-5 Sketch the schematic symbol of the SCR.

13-6 What is the significance of a "holding current" in SCR operation?

13-7 In dc operation of the SCR, the gate loses control of conduction once it starts. Why doesn't this happen in ac operation of the SCR?

13-8 What advantages does the TRIAC have over the SCR in controlling load power?

13-9 What do the letters of the word *laser* indicate?

13-10 How is light "amplified" by a laser?

13-11 What characteristics of a laser are of primary importance in potential applications?

Problems

13-1 A resistance load that requires 1 kW of power at 100 V is being supplied by a single SCR from a sine-wave source of 200 V rms. If half-wave phase control is being used for the SCR, what can be its maximum conduction angle without exceeding the 1 kW rating of the load?

13-2 Repeat Problem 13-1 for full-wave phase control using two SCRs.

13-3 Sketch the waveforms of voltage across the SCR and its resistive load, for half-wave phase control of a 100-V sine-wave source, for firing angles of 30°, 60°, 90°, 180°. (Assume the forward drop of the SCR is negligible.)

13-4 A TRIAC is used to proportion power to a resistive load of 1 kW rated at 100 V. The power source is a 200-V sine wave. Determine the required maximum average voltage and current ratings of the TRIAC.

14
INTEGRATED CIRCUITS

The advent of the Space Age, with its missiles, rockets, and satellites, has required that electrical and electronic components *and whole systems* be built to occupy only a small fraction of the space required by those constructed with miniature tubes and transistors of conventional physical sizes. The needs of the aerospace industry have accelerated improvements in electronic packaging. This industry uses some of the most complex electronic systems ever developed, and reliability, size, and weight are of the utmost importance. Every extra pound of load placed in a rocket, for example, adds many extra pounds to the weight of the rocket. Such things as additional structural material and more fuel needed for lift-off greatly increase the costs of the system.

Instead of an oscillator, an amplifier, or a whole radio transmitter-receiver system occupying several cubic inches of space, each must require only a small fraction of that amount. Furthermore, these systems must operate at extremes of temperature, acceleration, and radiant-energy fields and be highly reliable. The failure of a single component worth only a few cents may mean that millions of dollars and months of effort have been wasted.

Microminiaturization in electronics has produced circuits and systems that combine discrete miniature components, such as resistors and capacitors, with semiconductor "chips" on which transistors and diodes are fabricated. These combinations are referred to as "hybrid" or "multichip" circuits. The techniques used to minimize space requirements in this type of circuit construction led to the development of *whole circuits*, including passive resistor and capacitor elements and active semiconductor diodes and transistors, on a common material called a *substrate*. These circuits are called *integrated circuits*, because they include the required active and passive components as a *monolithic* structure. The fabrication of elements on a common substrate depends on *thin-film* technologies using materials such as tantalum, Nichrome, and various oxides in sophisticated manufacturing processes.

This chapter presents a description of several types of microminiaturization techniques used to fabricate integrated circuits and examples of the structure of components and systems commercially available. Integrated circuits are mass produced as components of computer systems and have been announced as the working elements in consumer television sets, portable radios small enough to fit in an earring, and hearing aids that weigh only a few grams, including the battery used as a power supply.

14-1
Fabrication of Integrated Circuits

The fundamental requirement in the fabrication of an integrated circuit is that circuit components be processed simultaneously from common materials. Capacitors and resistors operating as linear networks can be effectively fabricated by thin-film technologies using Nichrome, tantalum, and tin oxide. Manufacturing methods developed for cadmium sulfide are capable of producing both field-effect transistors and passive components on a common substrate.

The technology presently used to build integrated circuits that include transistors and resistive and capacitive circuit elements, however, is based on methods developed for silicon planar semiconductor components. Its popularity for fabricating integrated circuits is based on its ability to provide high-quality active devices.

Because the basic process for integrated circuit construction is similar to that used to make transistors, these components may be formed to have characteristics that are almost identical to those of discrete units. The major difference between discrete and integrated transistors is the capacitance associated with the separation of the transistor and the substrate on which it is formed.

The basic result of the silicon planar process for fabricating integrated circuits is shown in Fig. 14-1. The starting material is a polished wafer cut from a single-crystal N-type or P-type silicon, which is the starting *substrate* for the process that leads to a *monolithic integrated circuit*. The circuit elements are built up as layers of doped semiconductor and silicon dioxide, SiO_2 (quartz).

The process illustrated here represents about 40 major manufacturing steps. Each is completed under precise control of parameters that affect the "yield," the number of good units that result. If each of the 40 steps is 97 per cent correct, on the average, the complete yield will be about 30 per cent.

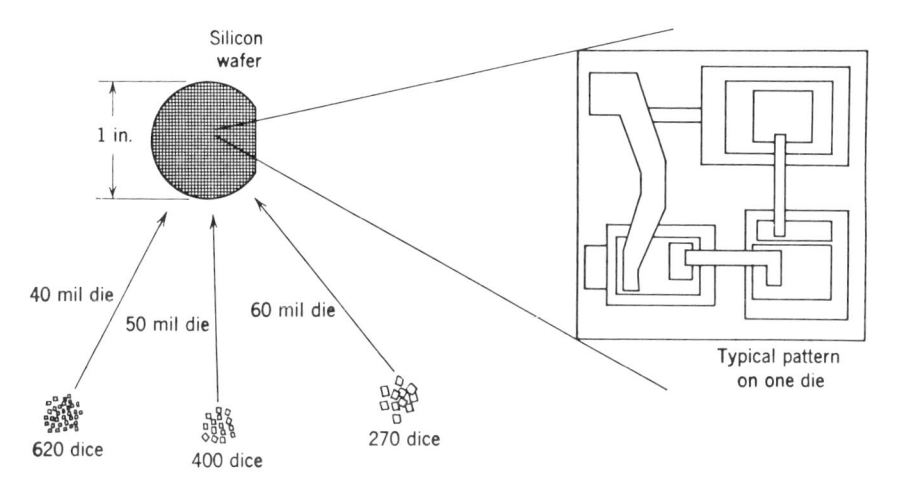

Figure 14-1
Several identical circuits may be fabricated on a single wafer. The number of dice depends on the size of an individual die. Circuits requiring few components may use a smaller die size.

A normal production of small integrated circuits (ICs) might include as many as 100 to 700 circuits side by side on the starting wafer. Each die, representing a single circuit, contains 10 to 100 active and passive circuit components. The ability to fabricate simultaneously all the components of many circuits permits close matching of finished ICs.

The major steps in the fabrication of ICs, starting with the 1-in. diameter slice of silicon, are as follows:

1. Surface preparation.
2. Epitaxial growth of starting layer.
3. Diffusion of subsequent semiconductor layers.
4. Metalization and connections.

14-2
Use of Photomasks

In order to control the fabrication of the IC structure, two separate and distinct features are important. One of them, the depth or vertical dimensions from the surface of a slice, is determined by a sequence of operations, e.g., epitaxial and diffusion processes. The second feature determines the lateral dimensions and shapes of the various layers that are built up during these processes. Photographic and chemical processes are used to fit the patterns required for the various components that are built into a circuit. Photomasks are the key to the making of high-performance ICs.

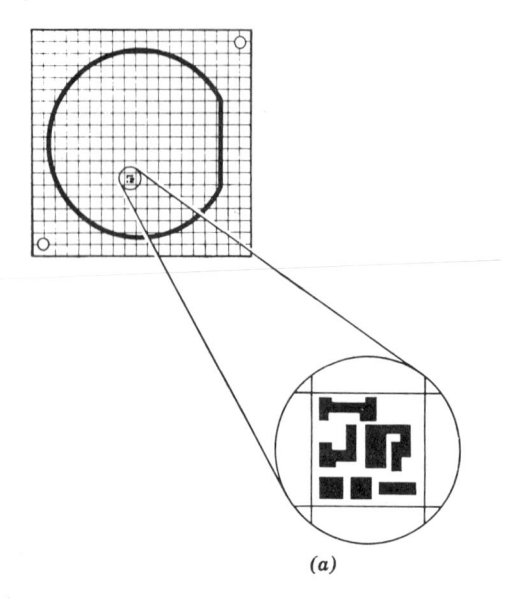

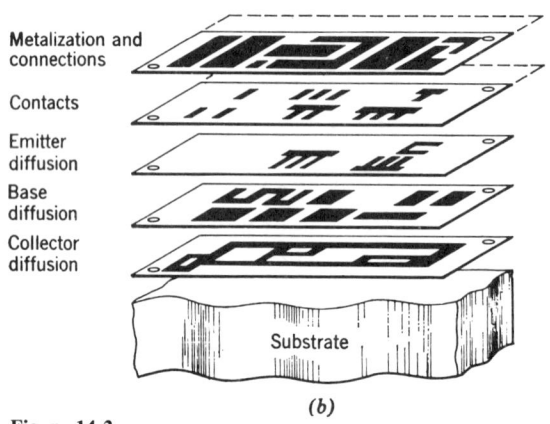

Figure 14-2
Photomask patterns used to fabricate ICs. (*a*) Typical mask pattern. (*b*) Set of photomasks.

A manufacturer may go from the production of one type of circuit to another by using a different set of photomasks. The same diffusion and assembly processes may be used to fabricate a different circuit. A set of photomasks is the only tooling required for a new device or circuit configuration. In essence, the photomasks restrict the processing steps to specific locations on the wafer.

In a series of repeated photographic, etching, diffusion, and other processing steps, the components of an individual IC may be formed and repeated simultaneously side by side for the 100 to 700 dice on a single wafer. To do this, the processes are performed in minute selected areas over the entire wafer, while the remainder of the surface is not affected. Patterns on the photomask that is used in a particular step of the manufacture limit the process in that step to specific shapes and locations. The use of photolithography to produce the required patterns is sketched in Fig. 14-2.

Lateral sizes are set and controlled by the patterns of the photomasks. Photomasks may be made to produce 0.0004-in. line widths and a location accuracy of 0.0002 in. over a 1-in. area. The masks that are used in production start as a layout of components that will make up the final circuit. When the circuit design is complete, the pattern of the circuit is accurately drawn and photographed to make a large-scale negative of it, on which the photographer may make minor corrections to line widths and stray spots. A reduction in size and creation of a matrix of identical patterns completes the photomask.

A set of accurately registered photomasks can produce good quality ICs. A manufacturer can produce a whole family of circuits just by using a different set of photomasks for each circuit, using the same diffusion, metalization, packaging, and other processes. Changes in a specific circuit may be made by redesigning one or more of the masks of a given set. Accurate sets of photomasks are vital to making complex, yet reliable, integrated circuits.

14-3
The Epitaxial Layer

An integrated circuit comprising 10 to 100 components and their interconnections might be fabricated on a piece of silicon that is only 40 to 60 mil^2 (0.040–0.060 in.). Typically, the silicon slice would be about 5 or 6 mil thick, or slightly thicker than this page. Many of the advantages of ICs arise from the microscopic size of the silicon chips and the circuit components, but this size also creates problems in manufacture.

Figure 14-3 shows two enlarged views of the sizes and locations of various areas on a 6-mil chip. The areas labeled "N" are N-type semiconductor on which the circuit components are fabricated. The small area at the left corresponds to the cross section needed for a typical transistor in the IC. The magnified view directly above this section shows the depth dimensions of the transistor. The emitter and base regions have depths of only 2 to 3 μ. Other circuit components have about the same depth relationships, seldom requiring more than 1 mil (25 μ) of thickness.

Since the circuit components only require about 1 mil of thickness, it might seem desirable to use silicon slices that are not much thicker than this. However, as the silicon wafers are made thinner, they break more easily during manufacturing operations. Losses during the fabrication processes due to wafer breakage are substantial, even for 5- to 6-mil slices, but this thickness has been found to be practical for economical use of silicon crystals. The bulk of a

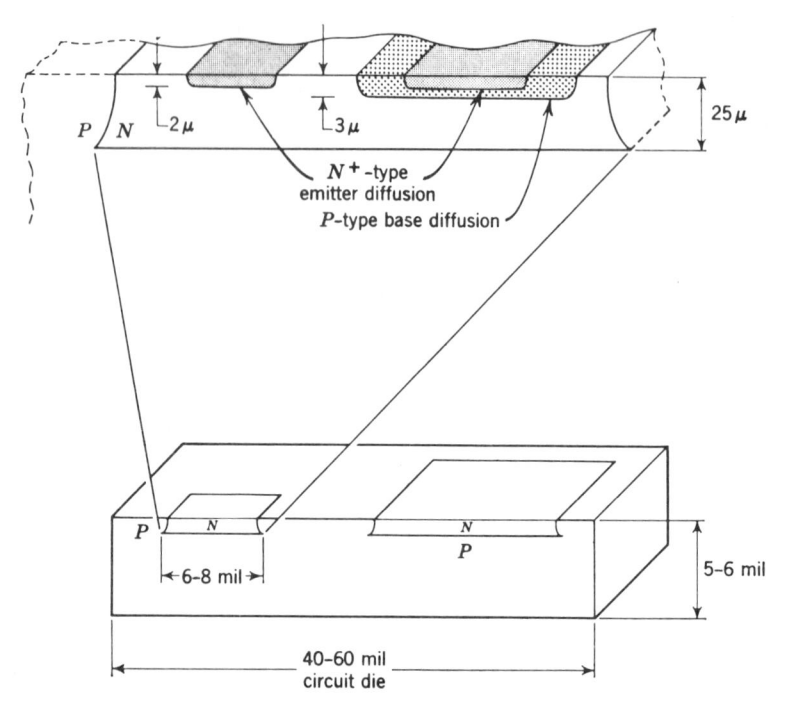

Figure 14-3
Epitaxial layers used to fabricate circuit components.

slice is not used for circuit components but is mechanically important as a "handle" during the manufacture of ICs. It is also electrically important because of its influence on the circuit properties. The bulk material is called the substrate.

Epitaxy is the process of growing one material on the surface of another. As used in manufacturing ICs, epitaxy refers to the growth of an additional semiconductor on a wafer by deposition from a vapor phase. In this process, a thin layer of additional single-crystal silicon is deposited on the polished surface of the starting wafer. While it is being deposited, the desired resistivity is obtained by controlling the amount of N-type impurity present in the silicon vapor phase. A photomask is used to restrict the epitaxial layer to the desired locations and areas on the wafer, as shown in Fig. 14-3.

14-4
Diffusion of Additional Layers

Diffusion of additional layers of materials is the key process in forming the desired circuit components. *Diffusion* is the substitution of N- or P-type atoms for silicon atoms within the crystal structure of a material. It results from heating the wafer to a temperature of 900° to 1300°C in the presence of controlled amounts of impurity. The temperature required depends on the particular dopant being used and is chosen to maintain close control over the diffusion process. The penetration of dopant increases by a factor of about 5 for every 100°C temperature rise above ambient. Typical diffusion time at 1200°C is 1 to 6 h.

The diffusion processes form P-N junctions in the epitaxial layers at locations permitted by placement of photomasks, as shown in Fig. 14-3. Precise locations on the surface may be held to tolerances as tight as ±15 millionths of an inch. The additional layers of materials on the IC form the various diodes, transistors, resistors, and capacitors of the circuit.

The materials used as dopants modify the silicon material by providing regions of N-type and P-type characteristics. Phosphorus (P) and arsenic (As) are widely used N-type dopants; boron (B) is usually used as a P-type dopant. The depth of penetration of a dopant depends on the temperature, length of time the slice is exposed, and properties of the doping material. Some dopants have diffusion rates that are orders of magnitude faster than others; for example, boron diffuses in silicon more than 10 times as fast as arsenic.

As a minimum, IC fabrication requires diffusion layers for an isolation from the substrate, a base for transistors, and an emitter diffusion. More complex circuit components may be made with additional diffusion processes. For example, P-N-P transistors may be formed in addition to N-P-N types, controlled layers may form Zener diodes with specific breakdown characteristics, and different values of resistor and capacitor elements may be provided.

The regions of the silicon wafer and layers into which diffusion of additional materials occurs are controlled by a combination of oxidation, photomasking, and chemical etching. Before each diffusion the slice is prepared by exposing it to oxygen at temperatures in excess of 900°C. A layer of silicon dioxide (quartz) can easily be grown in this environment and is usually allowed to form a thickness ranging from 2000 Å to 10,000 Å. This oxide layer is not easily penetrated by the common dopants and can be used as a mask to prevent diffusion where it is not desired in further processing. Parts of the oxide coating may be selectively removed by photomasking and etching processes, to form "windows" that set the size and location of additional circuit elements. On the photomask, the areas to be etched are opaque and those to be left with an oxide coating are transparent. The patterns of the mask are transferred to the silicon slice using a thin, uniform, photosensitive film that is coated on the surface. Exposure to light through the transparent photomask exposes the photographic film in these areas. The unexposed areas are removed by a developing step. A hydrofluoric acid etch removes the oxide, leaving the silicon surface exposed for further diffusion steps. The result is an accurate pattern of etched areas through the oxide coating to permit diffusion in the selected areas.

14-5
Metalization to Form Electrical Connections

A metal coating is vacuum deposited onto the diffusion coatings to interconnect the circuit elements and form the desired circuit. Metalization provides contacts for external connections to the circuit, as well. The most common metallic film used is aluminum, because it adheres well to silicon dioxide and is easy to handle. Other metals used alone, or in combination, include gold, silver, chromium, and nickel. When resistors and capacitors are required, titanium, tantalum, and tin may be used.

Figure 14-4 illustrates the formation of circuit components by the various processes in the fabrication of ICs, concluding with the metalization connections as the outermost layer. To form a resistor, the N-type emitter diffusion is omitted and two ohmic contacts are made to the P-type region that is formed with the base diffusion. In forming a capacitor, the oxide itself may be used as the dielectric, and metalization provides the electrodes. A typical circuit and its interconnections are shown in the figure.

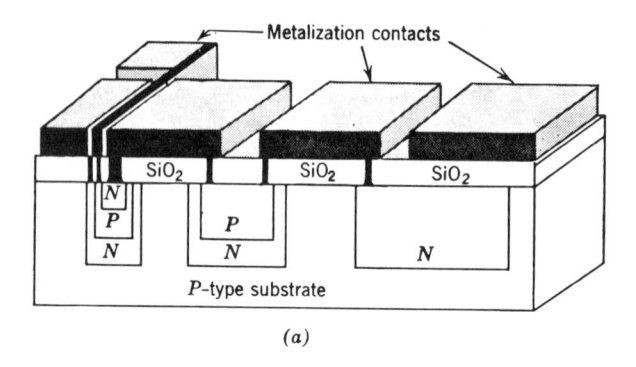

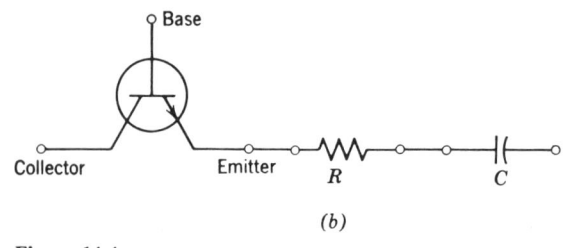

Figure 14-4
Completed IC comprising N-P-N transistor, resistor, and capacitor. (*a*) Layer construction of circuit components. (*b*) Circuit constructed in (*a*).

14-6
Cost Factors

Integrated resistors and capacitors are significantly different from their discrete counterparts. Discrete resistors and capacitors are usually made in standard physical sizes, corresponding to power and voltage ratings, and different values are obtained by varying the characteristics of the materials making up the components. A discrete resistor may be formed to have a given value by changing the resistivity of the material. The capacitance of a capacitor can be changed by using a different dielectric or by changing the surface area of its plates. However, in integrated circuits, the resistivity of the material cannot be varied to obtain different values of resistance, because it is optimized to obtain the best performance from the base diffusion of transistor structures. An integrated resistor, then, depends primarily on its geometry to set its resistance value. Its value is determined by the "square resistance," resistivity of a volume of material and the ratio of its length and width. As a result, resistors having large ohmic values are long and narrow, and small ohmic values are formed by short, squat dimensions. Similarly, the capacitance of capacitors directly depends on their surface area, because the thickness of the dielectric is a constant in well-controlled production.

Most of the cost of making monolithic ICs is associated with the processing steps in their manufacture. Because these costs are the same for any wafer, smaller circuits and more circuit dice from a single wafer lower the cost of an individual circuit. Therefore, circuit area is an important consideration in the design of an initial layout for manufacturing ICs. The *relative areas* required for integrated components are approximately 1 for a transistor or diode, 2 for a 1000-Ω resistor, and 3 for a 10-pF capacitor. Thus, for the economical production of ICs, each individual circuit should minimize passive components and use active components whenever possible. This is exactly the opposite of the economic rule for circuits using discrete components.

In a conventional transistor circuit, the transistors are relatively expensive compared with resistors and ceramic disc capacitors. In IC construction, the least expensive components are transistors. The resistors and capacitors of the IC are much more expensive.

Resistors must be considered differently in integrated and discrete forms. One undesirable characteristic of semiconductor resistors is their temperature dependence. A relatively large variation in ohmic value with temperature makes it difficult to achieve close tolerances on absolute values. However, resistance ratio values can be closely controlled by the geometry of photomasks. In circuits where resistance ratios, rather than absolute values, are important, integrated resistors have a number of advantages over discrete units. Because they receive almost identical processing and are closely matched in characteristics, temperature effects can be minimized over a wide

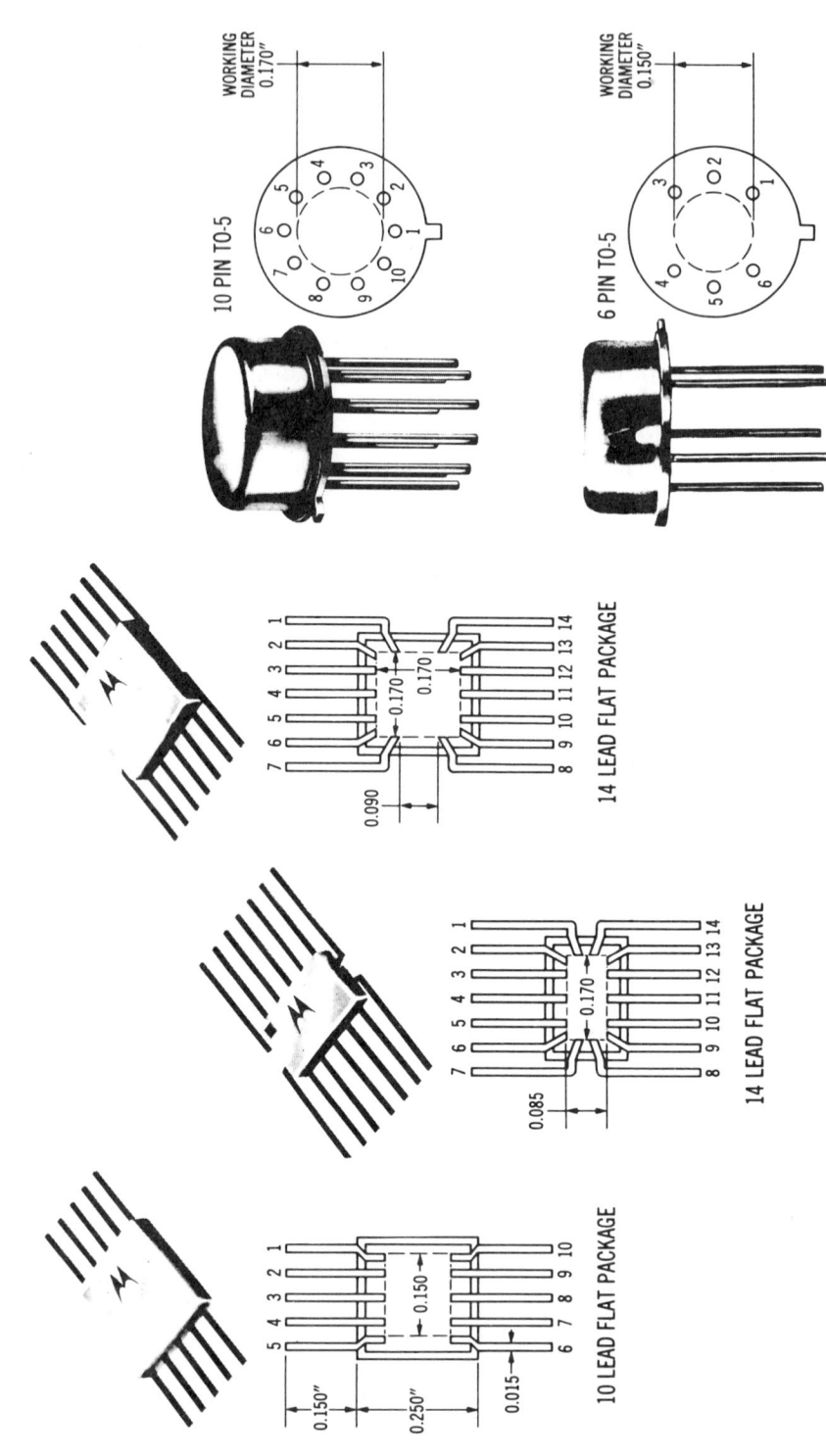

Figure 14-5
Flat packaging and TO-5 can configuration of finished integrated circuits. (Courtesy Motorola, Inc., Semiconductor Products Div., Phoenix, Arizona.)

482

operating range. As a result of their proximity, temperature differences that occur between components will be small.

Cost factors may be minimized in IC design by maximizing the use of transistors and other active components, by using resistance ratios instead of absolute values, and by taking advantage of matched component characteristics.

14-7
IC Packaging

The size and shape of the package in which the IC is housed is an important consideration in its use by circuit and system designers. A flat package with external leads extending from the ends of the package, as in Fig. 14-5, permits the placement of several units in confined spaces.

The circular package uses a TO-5 transistor configuration, with up to ten external pins provided for connecting the IC in a conventional socket. Both types are available with a variety of pin numbers and with different dimensions. In general, the flat package is more expensive than the round can package, because of higher manufacturing costs.

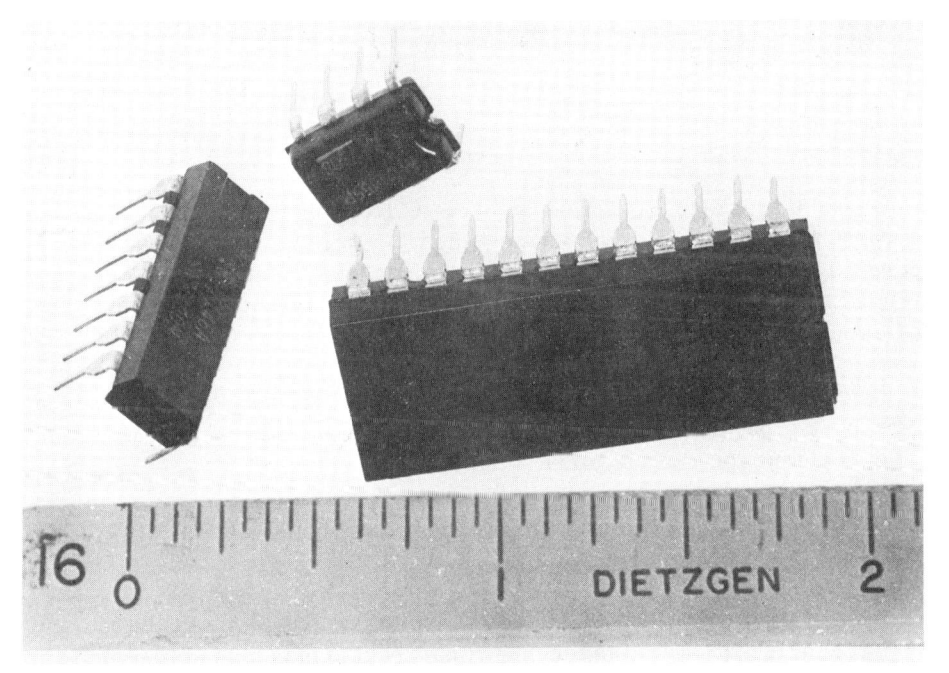

Figure 14-6
Typical encapsulated IC packages.

Figure 14-7
Internal connections from IC to external leads in a flat package IC. (Courtesy Motorola, Inc., Semiconductor Products Div., Phoenix, Arizona.)

Typical IC packages are shown in Fig. 14-6. Here the IC is housed in a plastic encapsulation, with several external leads arranged in rows along opposite edges. These leads are either inserted into sockets or soldered into place on printed circuit boards. The number of external leads is determined by the type of circuit contained in the IC package and the necessary external electrical connections.

Figure 14-7 illustrates the interior connections that are brought out to the pins of a flat package. The wires connecting the IC to external pins are usually gold or aluminum, and are about 1 mil in diameter.

14-8
Component Ratings and Unit Values

Because of their microscopic dimensions, integrated components have not matured to rival discrete components with respect to maximum ratings and unit values. Table 14-1 shows the ranges of values and ratings of components that are formed in the monolithic fabrication of ICs.

Table 14-1
Integrated Component Ratings and Values

Component	Characteristic	Ratings	Typical Values
Resistor	Ohms per square Maximum power Range of values	0.1 W	2.5 or 100–300 15 Ω–30 kΩ
Capacitor	Maximum capacitance Breakdown voltage Q (at 10 MHz) Voltage coefficient Tolerance	20 Vdc ± 20%	0.2 pF/mil^2 1–10 $C = KV^{-1/2}$
Transistor	BV_{CBO} BV_{CEO} BV_{EBO} h_{FE}/I_C	35 V 7 V 7 V	 40/1 mA 60/10 mA

14-9
Thin-Film and Multichip Construction of ICs

Thin-film circuits consist of a passive substrate, such as glass or ceramic, on which resistors and capacitors are deposited as thin patterned films of conducting and nonconducting layers. Thin-film resistors may have values up to 100 kΩ or so, and thin-film capacitors, using a tantalum oxide dielectric, may have a capacitance of 2.5 pF/per square mil of surface area. When active components (transistors and diodes) are added in discrete form to thin-film passive components, the resulting fabrication is called a *multichip circuit*. Figure 14-8 shows a typical arrangement of separate components and two ICs mounted on a TO-5 base.

Such a combination of monolithic and thin-film elements permits the fabrication of more complex circuits than the monolithic approach alone. There is a limit to the maximum size of an individual monolithic chip because of cost factors of economic yield, so that circuits having many components may be built on two or more chips. Further, circuits that need complementary transistors, or transistors having widely different characteristics, can be designed on separate chips as a practical solution. Multichip circuit construction can greatly improve circuit performance and permit combinations of components that cannot be fabricated by the monolithic process.

Figure 14-8
Internal structure of a multichip circuit, using two ICs and other discrete components on individual chips. (Courtesy Motorola, Inc., Semiconductor Products Div., Phoenix, Arizona.)

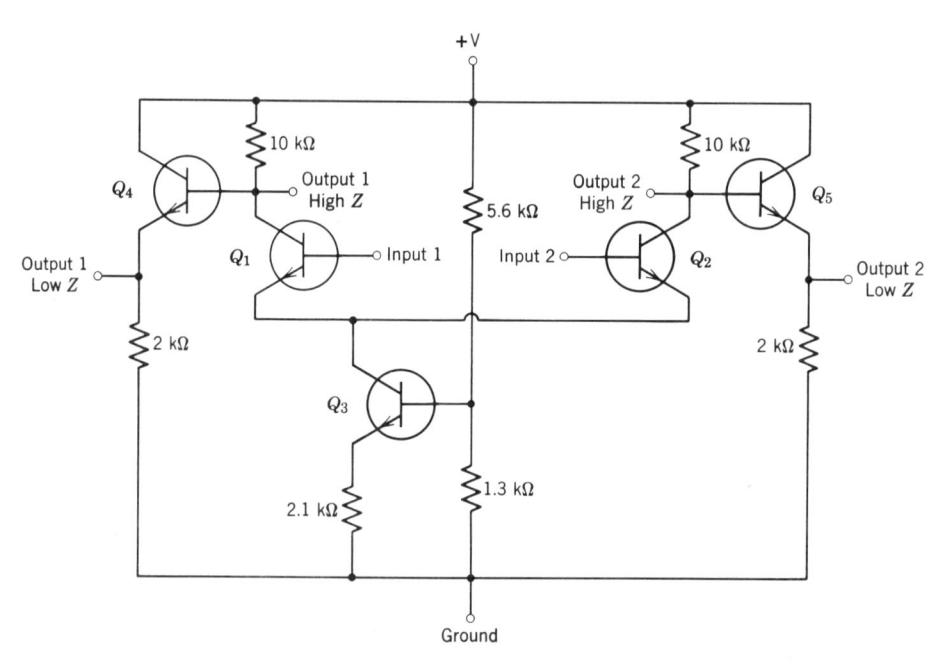

Figure 14-9
Schematic diagram of IC differential amplifier.

486

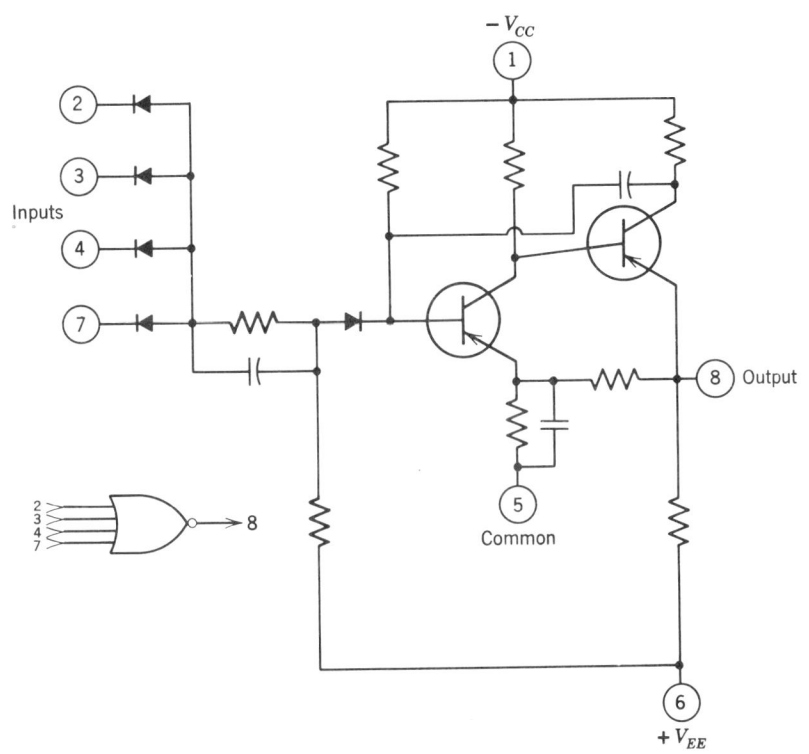

Figure 14-10
Circuit for IC 4-input NOR gate.

14-10
Circuits Constructed as ICs

Today there are hundreds of different circuits available in IC form. They include the linear differential amplifier whose schematic is given in Fig. 14-9. This package is designed in TO-99 case with eight-pin lead configuration. The unit can be operated from a 12-V source of dc power, and delivers 5 V output, peak to peak. Its input impedance (differential mode) is 5 kΩ and has a rated CMRR of 70 dB.

Other IC units include waveshaping and logic circuits, such as the 4-input NOR gate shown in Fig. 14-10. Note that this circuit has integrated resistors and capacitors, as well as diodes and transistors.

Questions

14-1 Why is the term "monolithic" applied to the fabrication of integrated circuits on a common substrate?

14-2 How does the monolithic structure differ from multichip construction?

14-3 Describe the steps used to fabricate a monolithic integrated circuit.

14-4 What is the purpose of photomasks in the monolithic fabrication process?

14-5 Describe the epitaxial process for growing layers of semiconductors.

14-6 What is the purpose of forming oxide layers in certain regions of an integrated circuit?

14-7 Why is it desirable to use transistors and diodes whenever possible in an integrated structure, rather than passive resistors and capacitors?

14-8 Compare some of the advantages and disadvantages of discrete and integrated resistors.

APPENDIX

A-1
Calculation of Effective (RMS) Value of a Sinusoidal Waveform

The effective value of an ac current is said to be 1 A when the average heating effect in a resistor is the same as that produced by 1 Adc in the same resistor. The heating effect caused by an ac current consists of two parts, one for each of the alternating positive and negative half-cycles. The maximum amplitude does not tell how the current varies between zero and maximum, nor does it indicate the average heating effect in a resistor. What is needed is the mathematical statement relating the effective value of the current and a direct current that causes the same average heating:

$$I^2 R = I_{dc}^2 R \qquad (A1\text{-}1)$$

where I = effective value of ac current, in amperes; I_{dc} = equivalent direct current, in amperes; R = resistance, in ohms.

The average heat generated in the resistor during one cycle may be expressed

$$\text{Average heat} = \frac{1}{T} \int_0^T R i^2 \, dt \qquad (A1\text{-}2)$$

where T = period of one cycle, in seconds; and i = instantaneous value of current, in amperes. This average heat is the same as in Eq. A1-1:

$$I_{dc}^2 R = I^2 R = \frac{1}{T} \int_0^T R i^2 \, dt \qquad (A1\text{-}3)$$

Appendix

Solving for I gives

$$I = \left(\frac{1}{T}\int_0^T i^2\,dt\right)^{1/2} \quad \text{(A1-4)}$$

This equation is applicable to a periodic current of any waveform. It indicates that the effective value is the square *root* of the *mean* of the *squares*; for this reason, it is often called the *rms* (root-mean-square) value of the current.

Sinusoidal waveforms of current and voltage are used extensively in ac circuit analysis. It is convenient to know the rms value of a sinusoid in terms of its amplitude. For a current expressed as $i = I_m \sin(\omega t + \phi)$,

$$I = \left(\frac{1}{T}\int_0^T i^2\,d(\omega t)\right)^{1/2} = \left(\frac{1}{T}\int_0^T I_m^2 \sin^2(\omega t + \phi)\,d(\omega t)\right)^{1/2}$$

$$= \left(\frac{I_m^2}{2T}\int_0^T [1 - \cos^2(\omega t + \phi)]\,d(\omega t)\right)^{1/2}$$

$$= \left(\frac{I_m^2}{2}\right)^{1/2}$$

$$I = \frac{I_m}{\sqrt{2}} = 0.707\,I_m \quad \text{(A1-5)}$$

A cosine wave has the same effective value as the sine wave; $\cos \omega t = \sin(\omega t + 90°)$.

A-2
Determinants

The determinant method of solving simultaneous algebraic equations has the advantage that it permits writing the equations in a "shorthand" notation. The variables of the equations may be obtained easily by following a few simple rules for using determinants.

Consider these simultaneous equations:

$$\begin{aligned} a_1 x + a_2 y &= C \\ b_1 x + b_2 y &= D \end{aligned} \quad \text{(A2-1)}$$

To eliminate x we first multiply each equation by the coefficient of x in the other equation:

$$\begin{aligned} a_1 b_1 x + a_2 b_1 y &= b_1 C \\ a_1 b_1 x + a_1 b_2 y &= a_1 D \end{aligned}$$

Then, subtracting the first equation from the second, we obtain

$$a_1 b_2 y - a_2 b_1 y = a_1 D - b_1 C$$

or

$$(a_1 b_2 - a_2 b_1)y = a_1 D - b_1 C \quad \text{(A2-2)}$$

Sec. A-2   Determinants   491

To eliminate y, we follow a similar procedure. First,
$$a_1b_2x + a_2b_2y = b_2C$$
$$a_2b_1x + a_2b_2y = a_2D$$

Then, subtracting the second equation from the first:
$$(a_1b_2 - a_2b_1)x = b_2C - a_2D \tag{A2-3}$$

The coefficients of x and y in (A2-2) and (A2-3) are the same quantity, $a_1b_2 - a_2b_1$. This quantity may be written in "shorthand" notation as

$$a_1b_2 - a_2b_1 = \begin{vmatrix} a_1 & a_2 \\ b_1 & b_2 \end{vmatrix} \tag{A2-4}$$

This symbol, which is called the *determinant* of the equations in (A2-1), consists of the coefficients of the variables x and y.

We may write (A2-2) and (A2-3) in determinant form:

$$\begin{vmatrix} a_1 & a_2 \\ b_1 & b_2 \end{vmatrix} y = \begin{vmatrix} a_1 & C \\ b_1 & D \end{vmatrix}$$

$$\begin{vmatrix} a_1 & a_2 \\ b_1 & b_2 \end{vmatrix} x = \begin{vmatrix} C & a_2 \\ D & b_2 \end{vmatrix}$$

The variables x and y may be written as the ratio of two determinants:

$$x = \frac{\begin{vmatrix} C & a_2 \\ D & b_2 \end{vmatrix}}{\begin{vmatrix} a_1 & a_2 \\ b_1 & b_2 \end{vmatrix}} \quad \text{and} \quad y = \frac{\begin{vmatrix} a_1 & C \\ b_1 & D \end{vmatrix}}{\begin{vmatrix} a_1 & a_2 \\ b_1 & b_2 \end{vmatrix}} \tag{A2-5}$$

Note that in the numerators the constants from the original equations are substituted for the coefficients of the variable being solved. The denominator is the determinant of the original equations, made up of the coefficients of the variables x and y.

Determinants of three equations take the form

$$\begin{vmatrix} a_1 & a_2 & a_3 \\ b_1 & b_2 & b_3 \\ c_1 & c_2 & c_3 \end{vmatrix} = \Delta = a_1b_2c_3 + a_2b_3c_1 + a_3b_1c_2 - a_1b_3c_2 - a_2b_1c_3 - a_3b_2c_1$$

which includes the coefficients of three variables x, y, and z. These variables may also be solved for as the ratio of two determinants

$$x = \frac{\begin{vmatrix} C & a_2 & a_3 \\ D & b_2 & b_3 \\ E & c_2 & c_3 \end{vmatrix}}{\Delta} \quad y = \frac{\begin{vmatrix} a_1 & C & a_3 \\ b_1 & D & b_3 \\ c_1 & E & c_3 \end{vmatrix}}{\Delta} \quad z = \frac{\begin{vmatrix} a_1 & a_2 & C \\ b_1 & b_2 & D \\ c_1 & c_2 & E \end{vmatrix}}{\Delta} \tag{A2-6}$$

A-3
Derivation of Charging Current in a Capacitor

Consider an uncharged capacitance C to which a constant dc voltage E is suddenly applied. There will be some resistance R in series, even if it is only in the connection wires.

By Kirchhoff's voltage law,

$$E = iR + \frac{1}{C}\int i\,dt$$

in which i is the current at any instant. Differentiating with respect to t,

$$0 = R\frac{di}{dt} + \frac{i}{C}$$

separating the variables,

$$\frac{di}{i} = -\frac{1}{RC}dt \tag{A3-1}$$

$$\ln i = -\frac{1}{RC}t + k_1$$

By the *definition of logarithm*,

$$i = \epsilon^{[-(1/RC)t + k_1]}$$

$$i = \epsilon^{-t/RC}\epsilon^{k_1} = k_2\epsilon^{-t/RC}$$

At the instant E is applied, $t = 0$ and $i = E/R$,

$$\frac{E}{R} = k_2$$

Therefore, at any time t after $t = 0$,

$$i = \frac{E}{R}\epsilon^{-t/RC}$$

A-4
Discharge Current in a Capacitor

When a capacitor charged to a voltage E is suddenly (at $t = 0$) connected to a resistance R, discharge current immediately starts to flow. *At any instant* the current is i, the charge on the capacitor is q, and the Kirchhoff voltage equation is

$$e_C + e_R = 0$$

Denoting the capacitance by C, this equation becomes

Sec. A-5   Voltage Applied to an Inductor   493

$$\frac{q}{C} + iR = 0$$

Differentiating and transposing,

$$R\frac{di}{dt} = -\frac{1}{C}\frac{dq}{dt} = -\frac{1}{C}i$$

Separating the variables,

$$\frac{di}{i} = -\frac{1}{RC}dt$$

This is Eq. A3-1, and the solution

$$i = \frac{E}{R}\epsilon^{-t/RC}$$

is obtained exactly as before.

A-5
Voltage Applied to an Inductor

When a coil having inductance L and resistance R has a constant voltage E suddenly applied, the Kirchhoff voltage equation, in which i is the instantaneous value of current, is

$$E = iR + L\frac{di}{dt}$$

Transposing and separating the variables,

$$L\frac{di}{dt} = E - iR$$

$$\frac{di}{E - iR} = \frac{dt}{L}$$

Multiplying both sides by $-R$,

$$\frac{-R\,di}{E - iR} = -\frac{R}{L}dt$$

Integrating,

$$\ln(E - iR) = \frac{-Rt}{L} + k_1$$

From the definition of logarithm,

$$E - iR = \epsilon^{[-(R/L)t + k_1]}$$
$$E - iR = k_2\epsilon^{-(R/L)t}$$

At $t = 0$, $i = 0$ because current in an inductance cannot change in zero time, so that k_2 is found:

$$k_2 = E$$

Substituting,

$$E - iR = E\epsilon^{-(R/L)t}$$

from which

$$i = \frac{E}{R}(1 - \epsilon^{-(R/L)t})$$

A-6
Current Decay in an Inductor

When an inductance L carrying current i is suddenly connected* across a resistance R, the instantaneous current i and the initial voltage ($E = L\,di/dt$) are related by Kirchhoff's voltage law:

$$iR + L\frac{di}{dt} = 0$$

Transposing and separating the variables,

$$\frac{di}{i} = -\frac{R}{L}\,dt$$

$$\ln i = -\frac{R}{L}t + k_1$$

$$i = \epsilon^{-(R/L)t}\epsilon^{k_1}$$

$$i = k_2\epsilon^{-(R/L)t}$$

At $t = 0$, $i = E/R$, so that

$$\frac{E}{R} = k_2$$

and

$$i = \frac{E}{R}\epsilon^{-(R/L)t}$$

* These conditions are electrically identical with *short circuiting*, with a zero-resistance switch, a coil of inductance L, and resistance R while it is carrying current. Draw the circuit, showing the coil in series with another resistance R_2 and a battery. E is the *EMF of self-induction* of the coil ($L\,di/dt$) at the instant the switch is closed.

A-7
Proof of Maximum Power Transfer Theorem

Figure A-1 represents a generator with internal impedance $R_G + j0$ and its load $R_L + j0$. The generated voltage is E_G (ac or dc). The load current is

$$I_L = E_G/(R_G + R_L)$$

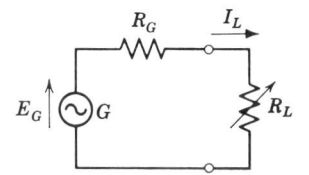

Figure A-1

The power delivered to the load is

$$P_L = E_G^2 R_L/(R_G + R_L)^2 \qquad (A7\text{-}1)$$

Since R_L can be varied until P_L is a maximum, the rate of change of P_L with respect to R_L may be expressed mathematically and set equal to zero, thus satisfying conditions for maximum P_L.

$$\frac{dP_L}{dR_L} = \frac{E_G^2(R_G + R_L)^2 - 2E_G^2 R_L(R_G + R_L)}{(R_G + R_L)^4} = 0$$

Therefore,

$$E_G^2(R_G + R_L)^2 = 2E_G^2 R_L(R_G + R_L)$$
$$R_G + R_L = 2R_L$$
$$R_G = R_L$$

Section 1-21 treats the case where reactance is present. The equation for power in the receiver circuit (1-60) reduces to Eq. A7-1 when X_G and X_R are equal but opposite in sign. R_R there and R_L here represent the same quantity, i.e., load resistance.

A-8
Temperature Conversion Factors

The various temperature scales used in the text are based on the freezing point and boiling point of water. On the Celsius scale (often called Centigrade), the freezing point of water is taken as a starting point, 0°C. The boiling point is then set 100 degrees higher, 100°C. The Fahrenheit scale (the common temperature scale in use in the United States) uses 32°F for the freezing point and 212°F for the boiling point of water. This represents a total of 212 − 32 = 180 degrees between the two extremes, while the Celsius scale uses only 100 degrees for the same temperature range. Then, for each °C there are 1.8°F. In addition, when converting from °C to °F, there must be added another 32 degrees because of the different starting points on the separate scales.

$$°F = 1.8°C + 32°$$
$$= \tfrac{9}{5}°C + 32°$$

Similarly, temperature given in °F may be converted to °C

$$°C = \tfrac{5}{9}(°F - 32°)$$

The Kelvin scale of temperatures uses 0° for the *absolute zero of temperature*. Each degree Kelvin (°K) represents the same temperature change as each degree on the Celsius scale (°C). However, absolute zero occurs at about −273°C (0°K), making 0°C = 273°K. To convert temperature given in °C to °K, simply add 273° to the temperature in °C. For example, 27°C = (27 + 273) = 300°K.

A-9
Bessel Function Coefficients

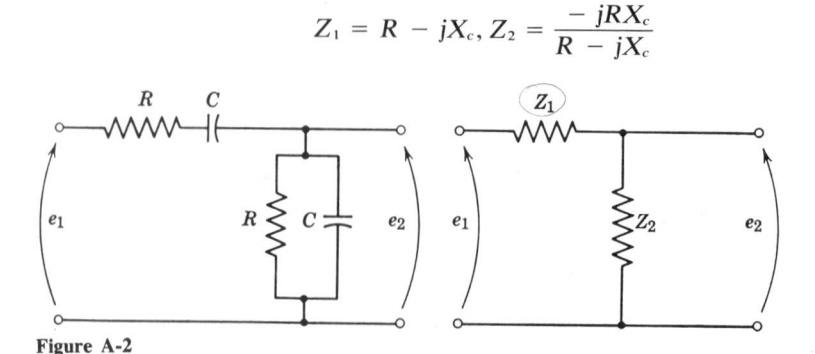

Table A9-1 lists Bessel function coefficients for selected values of F-M coefficients.

A-10
Required Gain of Phase-Shift Oscillator

The following demonstrates proof that a two-stage phase-shift oscillator must have a gain of at least 3 to sustain oscillations, when the feedback circuit shown in Fig. A-2 is used.

$$Z_1 = R - jX_c, \quad Z_2 = \frac{-jRX_c}{R - jX_c}$$

Figure A-2

Table A9-1
Bessel Function Coefficients for Selected Values of F-M Coefficients

m_f	$J_0(m_f)$	$J_1(m_f)$	$J_2(m_f)$	$J_3(m_f)$	$J_4(m_f)$	$J_5(m_f)$	$J_6(m_f)$	$J_7(m_f)$	$J_8(m_f)$	$J_9(m_f)$	$J_{10}(m_f)$
0	1.0000	0	0	0	0	0	0	0	0	0	0
1	0.7652	0.4401	0.1149	0.0196	0.0025	0.0003	0	0	0	0	0
2	0.2239	0.5767	0.3528	0.1289	0.0340	0.0070	0.0012	0.0002	0	0	0
3	−0.2601	0.3391	0.4861	0.3091	0.1320	0.0430	0.0114	0.0026	0.0005	0	0
4	−0.3971	−0.0660	0.3641	0.4302	0.2811	0.1321	0.0491	0.0152	0.0040	0.0009	0.0002
5	−0.1776	−0.3276	0.0466	0.3648	0.3912	0.2611	0.1310	0.0534	0.0184	0.0055	0.0015
6	0.1506	−0.2767	−0.2429	0.1148	0.3576	0.3621	0.2458	0.1296	0.0565	0.0212	0.0070
7	0.3001	−0.3047	−0.3014	−0.1676	0.1578	0.3479	0.3392	0.2336	0.1280	0.0589	0.0235
8	0.1717	0.2346	−0.1130	−0.2911	−0.1054	0.1858	0.3376	0.3206	0.2235	0.1263	0.0608
9	−0.0903	0.2453	0.1448	−0.1809	−0.2655	−0.0550	0.2043	0.3275	0.3051	0.2149	0.1247
10	−0.2459	0.0435	0.2546	−0.0584	−0.2196	−0.2341	−0.0145	0.2167	0.3179	0.2919	0.2075
11	−0.1712	−0.1768	0.1390	0.2273	−0.0150	−0.2383	−0.2016	0.0184	0.2250	0.3089	0.2804
12	0.0477	−0.2234	−0.0849	0.1951	0.1825	−0.0735	−0.2437	−0.1703	0.0451	0.2304	0.3005
13	0.2069	−0.0703	−0.2177	0.0023	0.2193	0.1316	−0.1180	−0.2406	−0.1410	0.0670	0.2338
14	0.1711	0.1334	−0.1520	−0.1768	0.0762	0.2204	0.0812	−0.1508	−0.2320	−0.1143	0.0850

where
$$X_c = \frac{1}{\omega C}$$

From voltage division,

$$\frac{e_2}{e_1} = \frac{Z_2}{Z_1 + Z_2} = \frac{-jX_c}{R - jX_c} \div \left(R - jX_c - \frac{jRX_c}{R - jX_c}\right)$$

$$\frac{e_2}{e_1} = \left(\frac{-jRX_c}{R - jX_c}\right)\left(\frac{R - jX_c}{(R - jX_c)^2 - jRX_c}\right) = \frac{-jRX_c}{R^2 - X_c^2 - 3jRX_c}$$

$$\frac{e_2}{e_1} = \frac{1}{3 + j(R^2 - X_c^2)/RX_c} = \frac{1}{3 + j(R/X_c - X_c/R)}$$

When $R = X_c$,

$$\frac{e_2}{e_1} = \frac{1}{3} \quad \text{or} \quad e_2 = \frac{e_1}{3}$$

The two-stage amplifier must therefore have a voltage gain of 3 so that the ac voltage between the plate and cathode of the second stage will be adequate to provide enough input voltage to the first stage to sustain oscillations. The operating frequency is such that $X_c = R$.

A-11
General-Purpose Curves and Data

P-N-P Alloy-Junction Germanium Transistor
Type 2N404
(Courtesy Texas Instruments Incorporated)

Maximum Ratings (25°C):
    Collector-base voltage   −25 V
    Collector-emitter voltage   −24 V
    Emitter-base voltage   −12 V
    Collector current   −100 mA
    Total device dissipation (see note 1)   150 mW

Note 1: Derate linearly to 85°C free-air temperature, 2.5 mW/°C.

Characteristics (25°C):
    Collector-base breakdown voltage, BV_{CBO} −25 V
    Common-emitter *h*-parameters
        (taken at $V_{CE} = -6$ V, $I_C = -1$ mA): h_{fe} 135
                                                      h_{ie} 4 kΩ
                                                      h_{oe} 50 μmho
                                                      h_{re} 7×10^{-4}

The common-emitter characteristic curves are shown in Fig. A-3.

Sec. A-11 General-Purpose Curves and Data 499

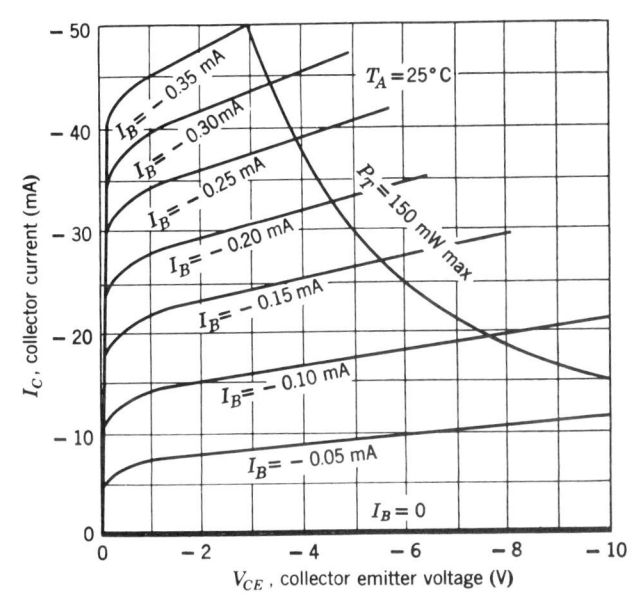

Figure A-3
Common-emitter curves for 2N404 transistor.

N-P-N Grown Junction Silicon Transistor
Type 2N334
(Courtesy Texas Instruments Incorporated)

Maximum Ratings:
    Collector-base voltage   45 V
    Emitter-base voltage   1 V
    Collector current   25 mA
    Emitter current   −25 mA
    Device dissipation (25°C)   150 mW
                        (100°C)   100 mW
                        (150°C)   50 mW

Characteristics—Common Base:
    Collector breakdown voltage BV_{CBO} 45 V
    h_{ib}, input impedance   30–80 Ω
    h_{ob}, output admittance   0–1.2 μmho
    h_{rb}, feedback voltage ratio   0–1000 × 10^{-6}
    h_{fb}, current transfer ratio   −0.948 to −0.989

The characteristic curves (operation at 25°C) are shown in Fig. A-4.

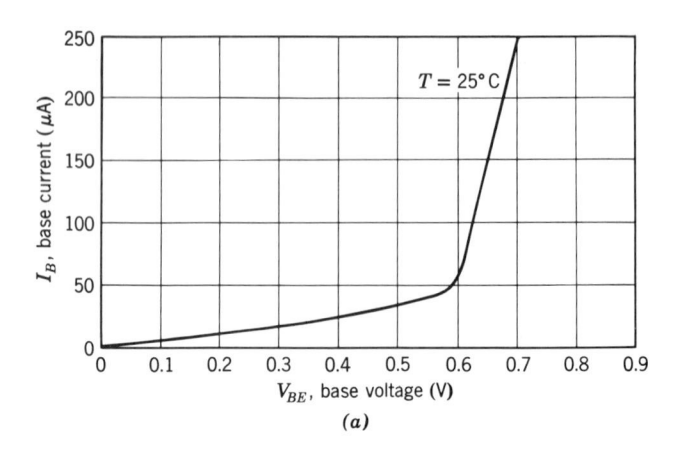

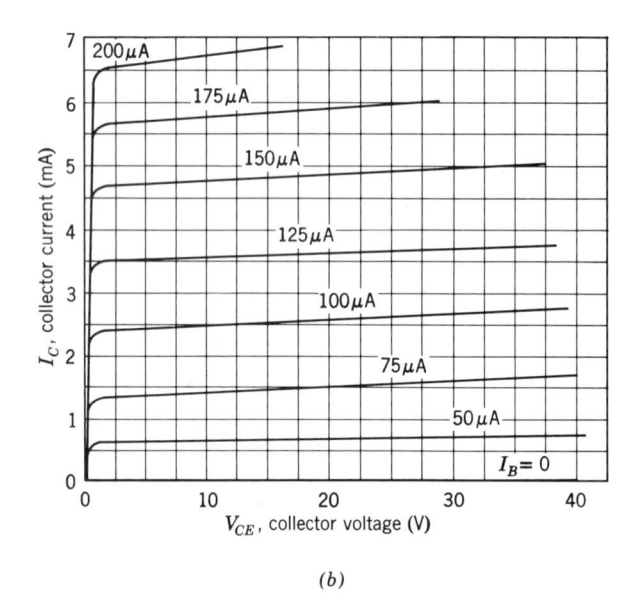

Figure A-4
Voltage and current relationships for Texas Instruments 2N334 transistor. (*a*) Input characteristic. (*b*) Output characteristics.

Complementary Alloy-Junction Germanium Transistors
N-P-N Types 2N1302, 1304, 1306, 1308
P-N-P Types 2N1303, 1305, 1307, 1309
(Courtesy Texas Instruments Incorporated)

Maximum Ratings (25°C):	N-P-N Types	P-N-P Types
Collector-base voltage	25 V	30 V
Emitter-base voltage	25 V	25 V
Collector current	300 mA	300 mA
Total device dissipation (see note 1)	150 mW	150 mW

Note 1: Derate linearly to 85°C free-air temperature: 2.5 mW/°C.

Characteristics:
(N-P-N Types)
Common-base h-parameters
(taken at $V_{CB} = 5$ V, $I_E = -1$ mA):

	2N1302	2N1304	2N1306	2N1308
h_{ib}	28	28	28	28 Ω
h_{rb}	5×10^{-4}	5×10^{-4}	5×10^{-4}	5×10^{-4}
h_{ob}	0.34	0.34	0.34	0.34 μmho
Common-emitter forward current transfer ratio, h_{fe}.	105	120	135	170

Characteristics:
(P-N-P Types)
Common-base h-parameters
(taken at $V_{CB} = -5$ V, $I_E = 1$ mA):

	2N1303	2N1305	2N1307	2N1309
h_{ib}	29	29	29	29 Ω
h_{rb}	7×10^{-4}	7×10^{-4}	7×10^{-4}	7×10^{-4}
h_{ob}	0.40	0.40	0.40	0.40 μmho
Common-emitter forward current transfer ratio, h_{fe}	115	130	150	190

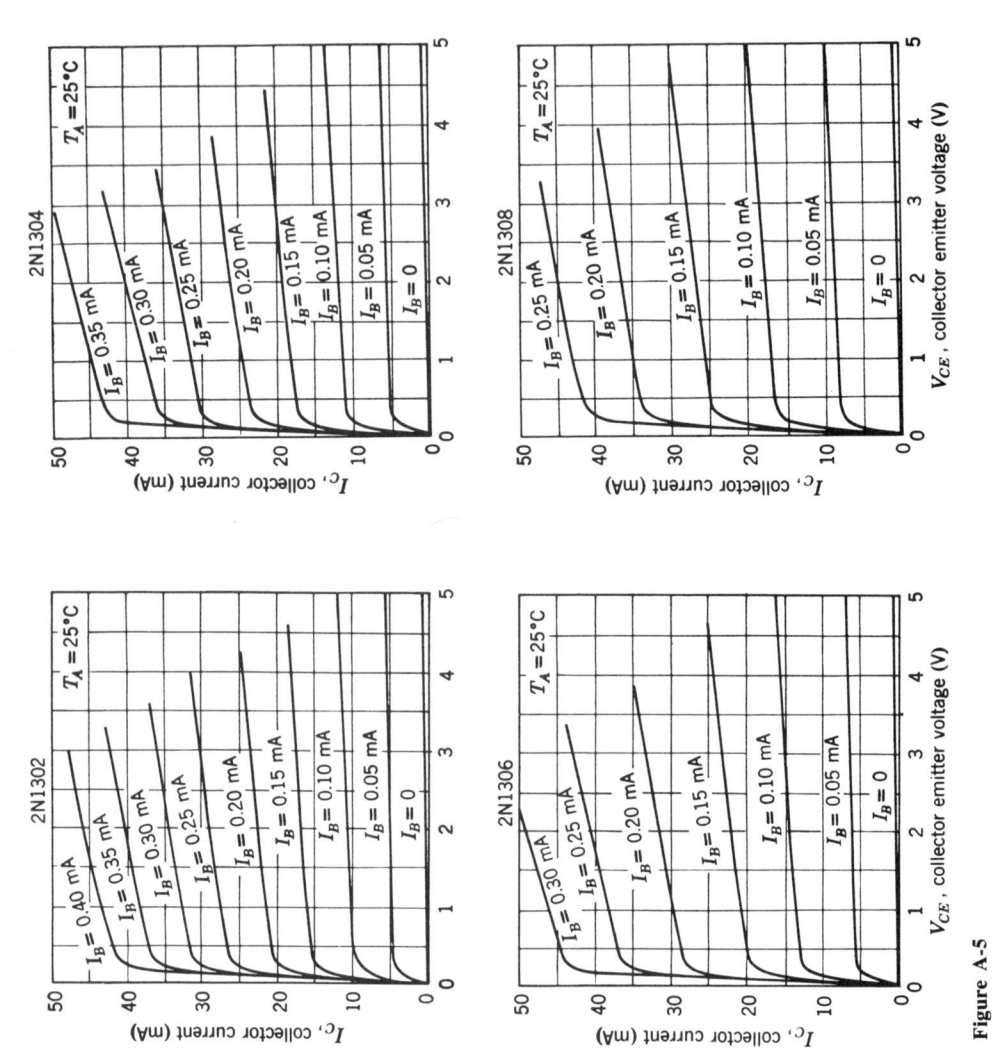

Figure A-5
Common-emitter characteristics N-P-N types 2N1302, 1304, 1306, 1308.

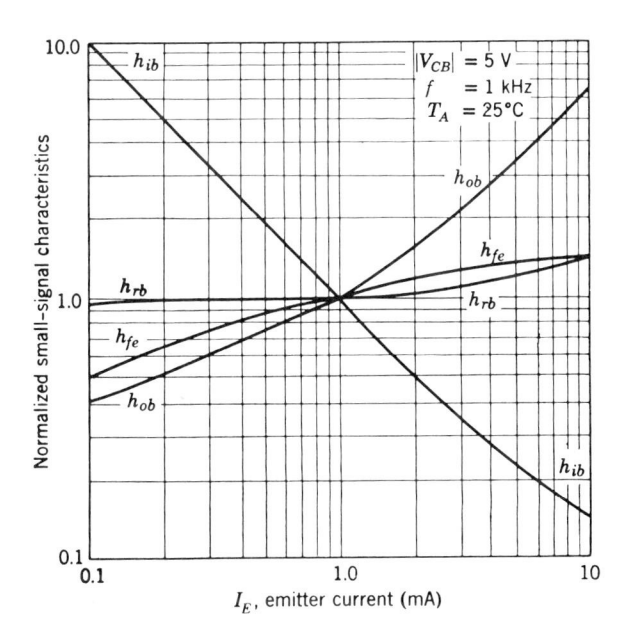

Figure A-5 (*contd.*)
Normalized *h*-parameters, types 2N1302-1309.

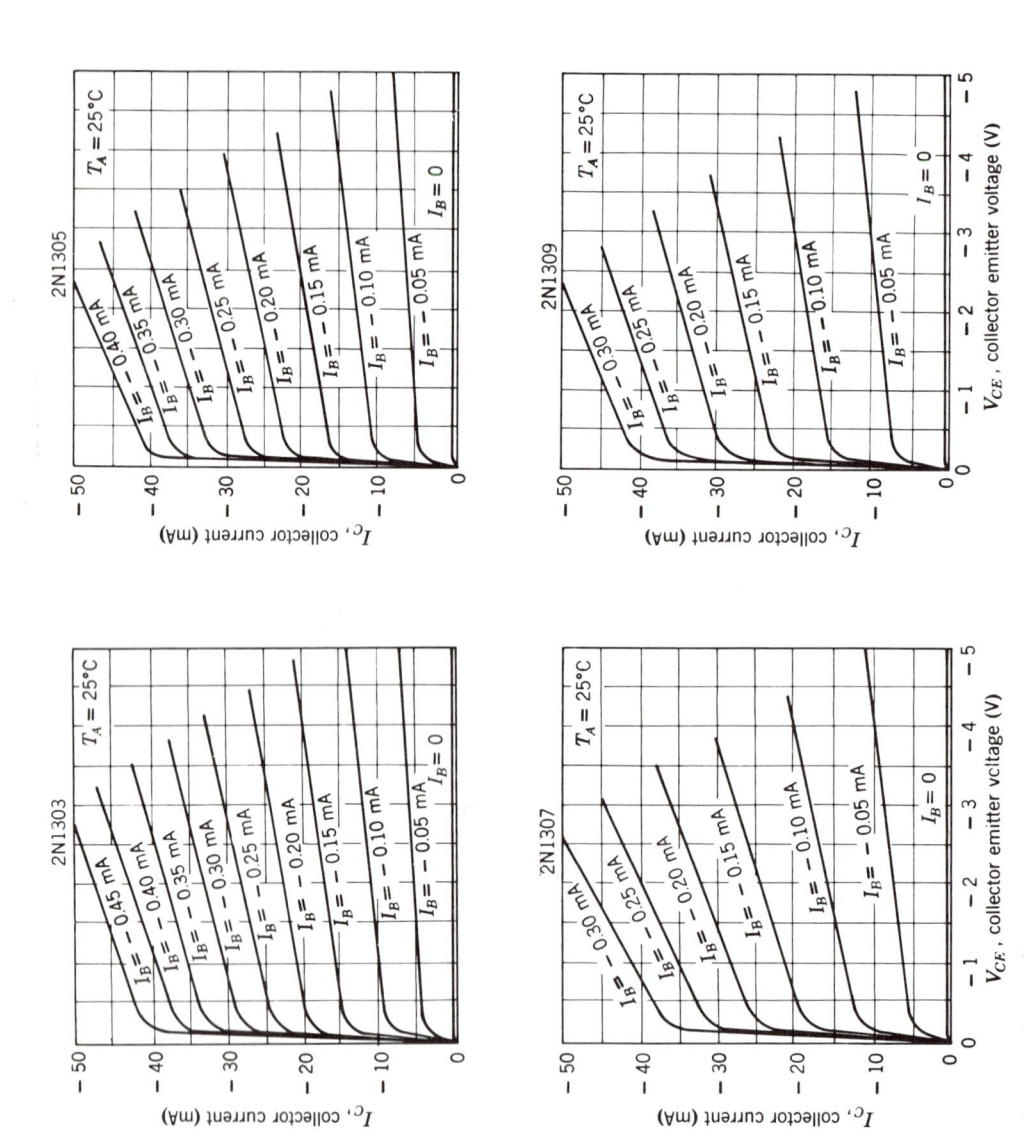

Figure A-5 (contd.)
Common-emitter characteristics, P-N-P Types 2N1303, 1305, 1307, 1309.

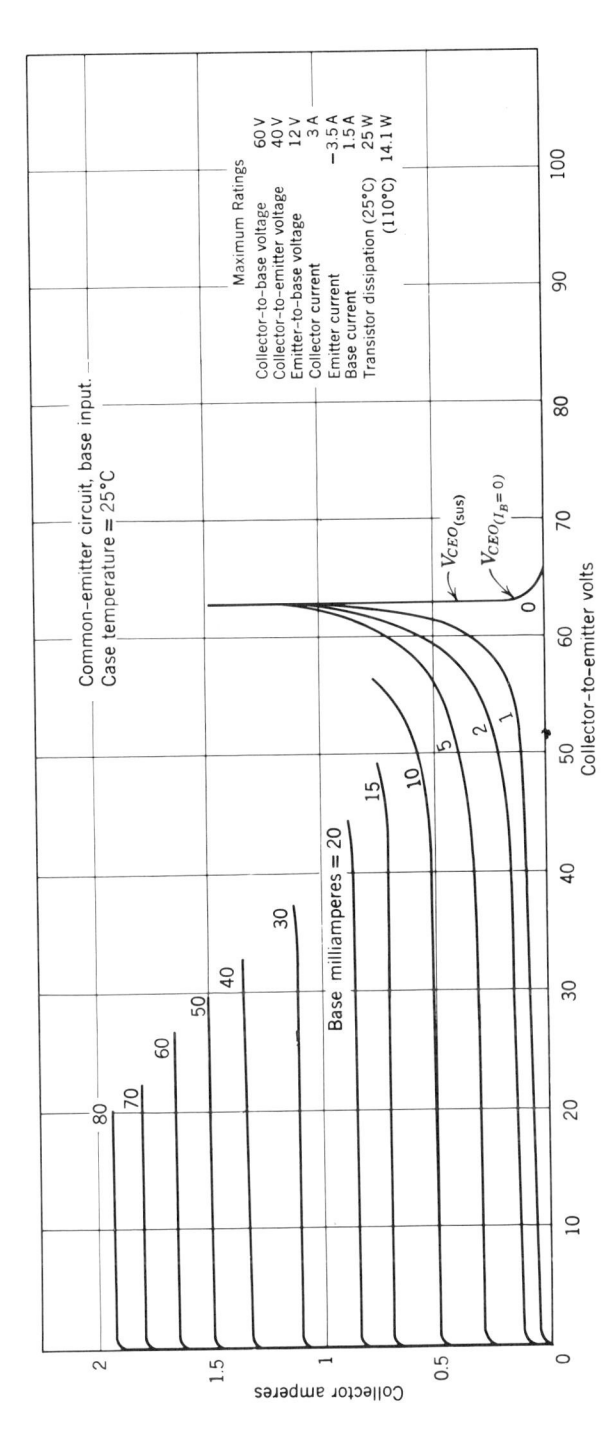

Figure A-6
Collector characteristics of silicon *N-P-N* diffused-junction transistor, type 2N1485 (courtesy Radio Corporation of America).

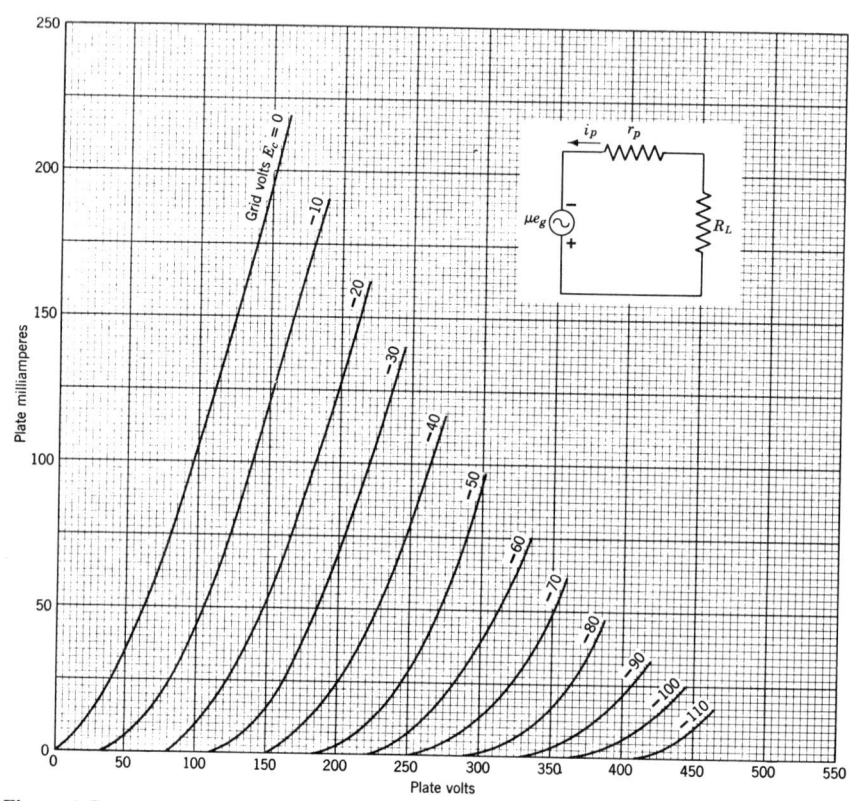

Figure A-7
Average plate characteristics, 2A3 power triode. $E_{bo} = 250$ V, $I_{bo} = 60$ mA, $E_c = -43.5$ V, $r_p = 800\,\Omega$, $\mu = 4.2$, $g_m = 5250\,\mu$ mho.

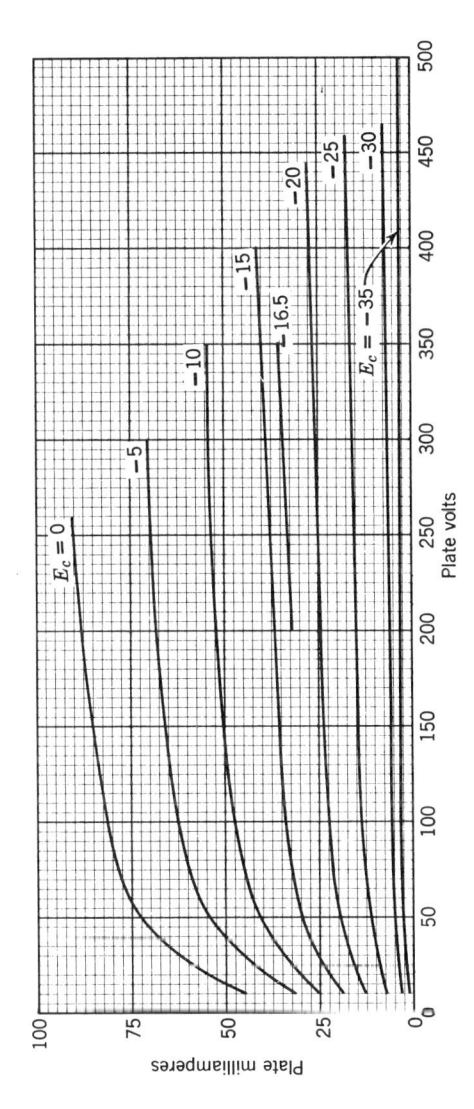

Figure A-8
Average plate characteristics, 6F6 pentode. Class A power amplifier, $E_{bo} = 250$ V, $E_c = -16.5$ V, $I_{bo} = 34$ mA, $r_p = 80{,}000$ Ω, $g_m = 2500$ μmho.

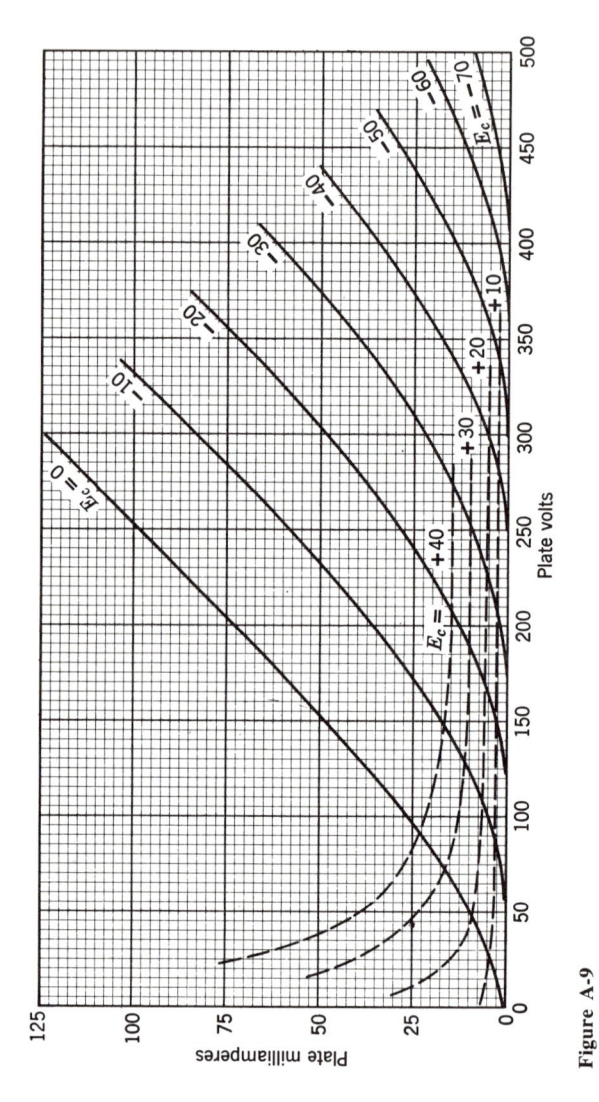

Figure A-9
Average plate characteristics, 6F6 triode (screen connected to plate). Class A power amplifier. $E_{bo} = 250$ V, $E_c = -20$ V, $I_{bo} = 31$ mA, $\mu = 6.8$, $r_p = 2600\ \Omega$, $g_m = 2600\ \mu$mho.

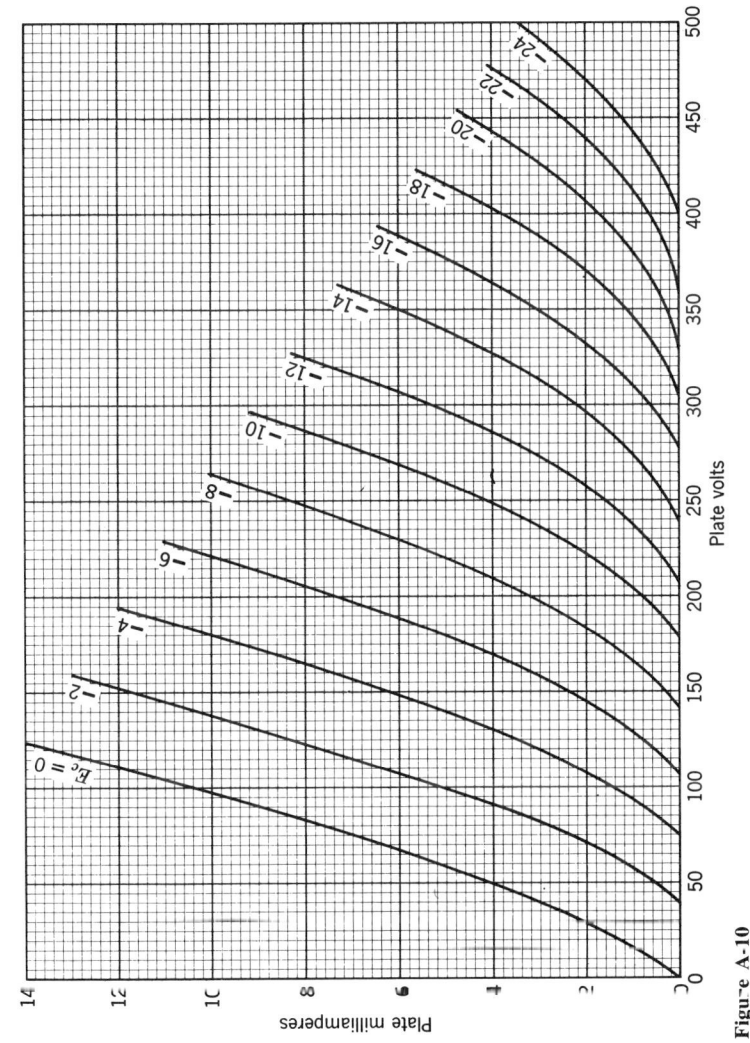

Figure A-10
Average plate characteristics, 6J5 triode. $E_{bo} = 250\,\text{V}$, $I_{bo} = 9\,\text{mA}$, $E_c = -8\,\text{V}$, $r_p = 7700\,\Omega$, $\mu = 20$, $g_m = 2600\,\mu\text{mho}$.

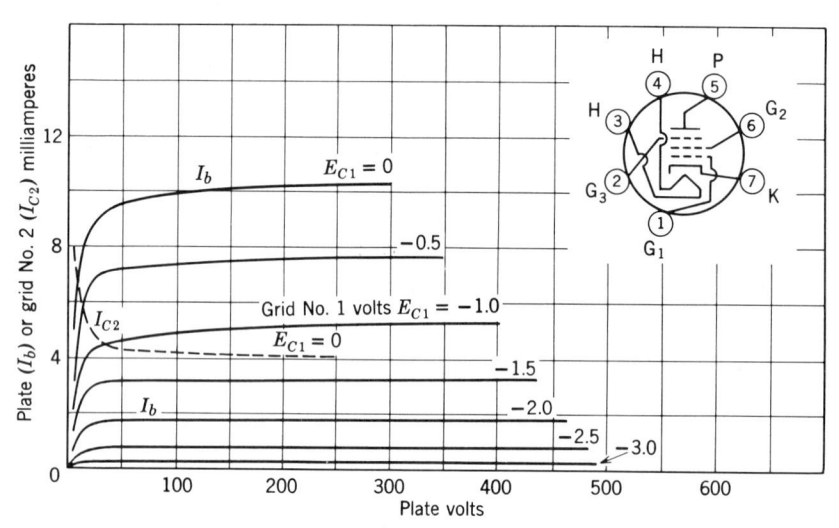

Figure A-11
Average plate characteristics, 6AU6A sharp-cutoff pentode.

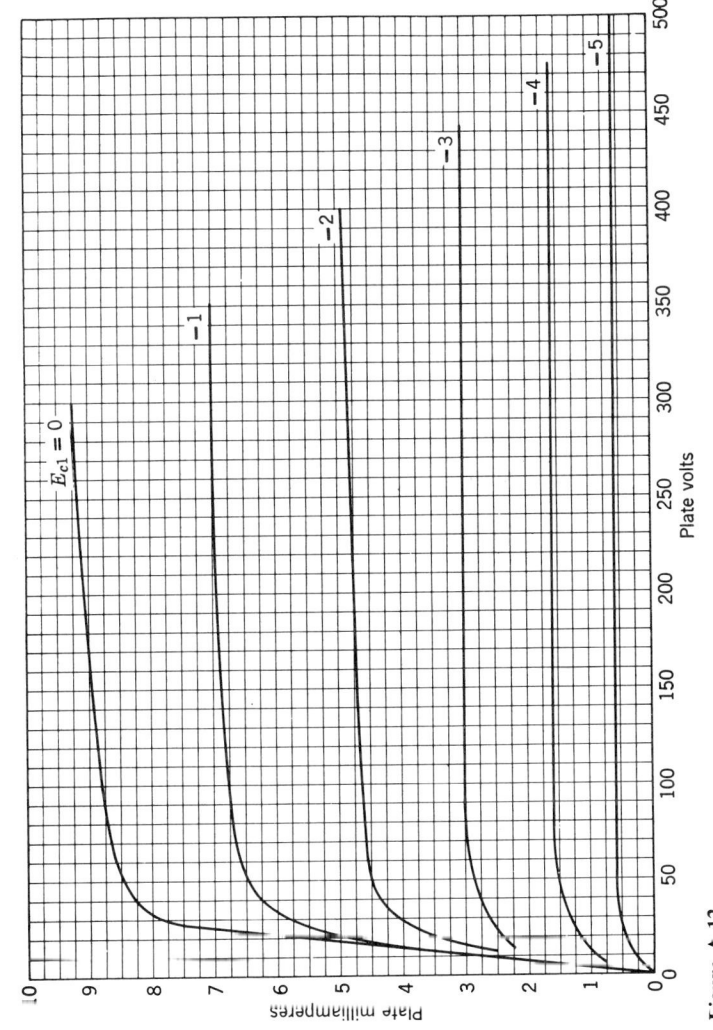

Figure A-12
Average plate characteristics. 6SJ7 pentode. $E_{bo} = 250\,\text{V}$, $E_{c1} = -3\,\text{V}$, $I_{bo} = 3\,\text{mA}$, $r_p = 1\,\text{M}\Omega$ (approx.), $g_m = 1650\,\mu\text{mho}$, $E_{c2} = 100\,\text{V}$, $E_{c3} = 0\,\text{V}$.

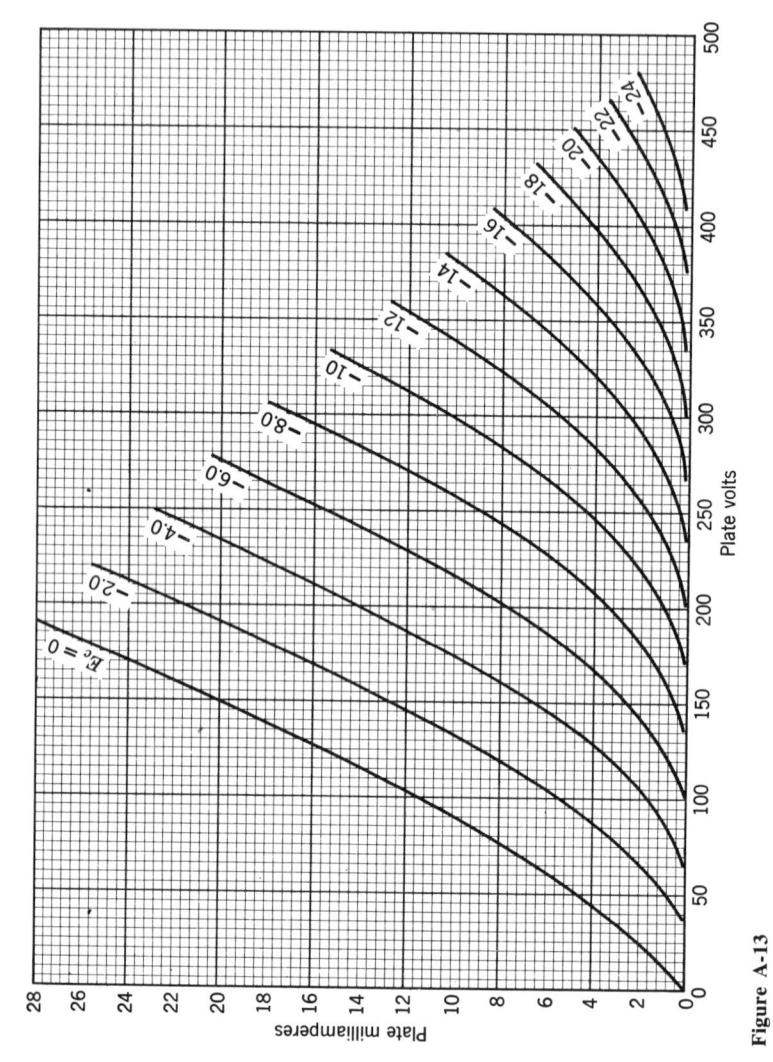

Figure A-13
Average plate characteristics, 6SN7GT. E_f = rated value, $E_{bo} = 250\,\text{V}$, $E_c = -8\,\text{V}$, $I_{bo} = 9\,\text{mA}$, $r_p = 7700\,\Omega$, $g_m = 2600\,\mu\text{mho}$, $\mu = 20$.

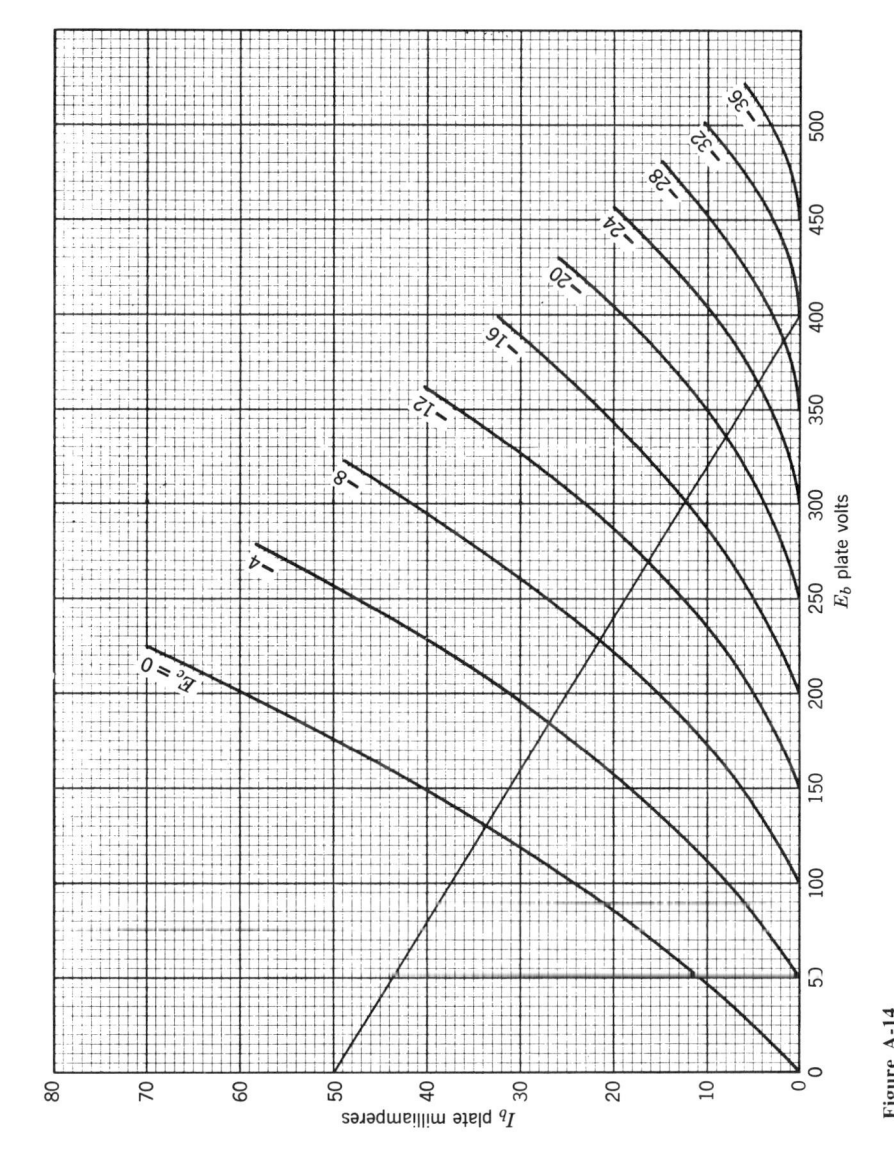

Figure A-14
Average plate characteristics for a 6S4A medium-mu triode.

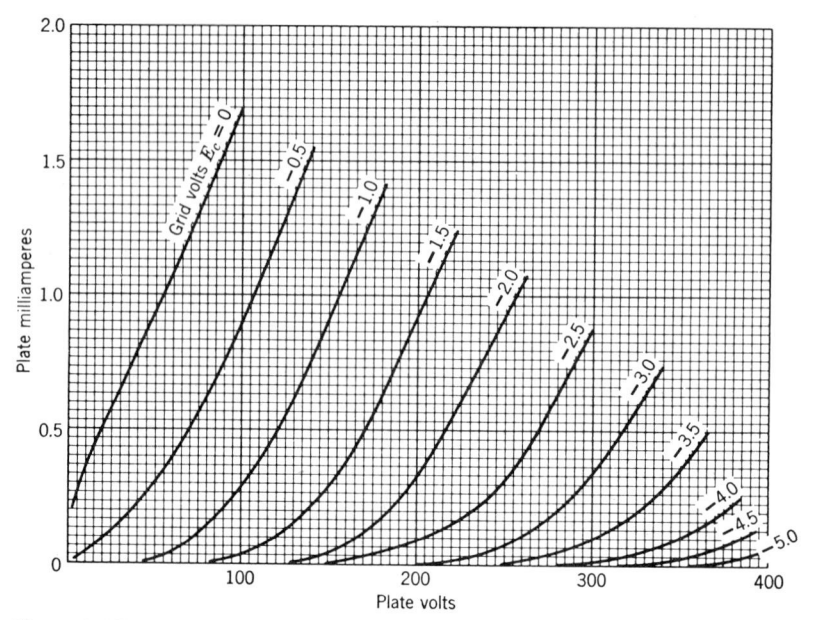

Figure A-15
Average plate characteristics, 6SF5 triode. $E_{bo} = 250$ V, $I_{bo} = 0.9$ mA, $E_c = -2$ V, $r_p = 66,000\ \Omega$, $\mu = 100$, $g_m = 100\ \mu$mho.

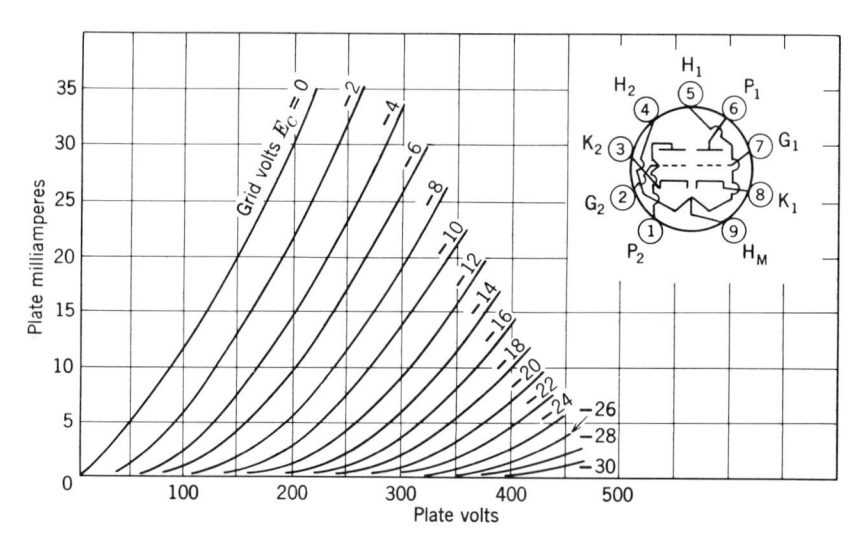

Figure A-16
Average plate characteristics, 12AU7A medium-μ twin triode.

A-12
Sine Wave Template

The REPCO Sine Wave Template (shown in Fig. A-17) is designed to permit rapid sketching of sinusoidal waveforms. In many electronic circuits, the accurate sketching of waveforms helps to show important changes that occur. It is used by many students and practicing engineers for drawing accurate waveforms for examinations and technical reports. The transparent vinyl construction permits easy placement over prior work. The larger profile is based on a 1-in. amplitude and 2 in./cycle, and the smaller profile is half the amplitude on the same angular scale.

The REPCO Sine Wave Template is available at most bookstores serving colleges and technical institutes. It may be obtained at nominal cost from REPCO, Inc., P.O. Box 4002, Lexington, Ky. 40504.

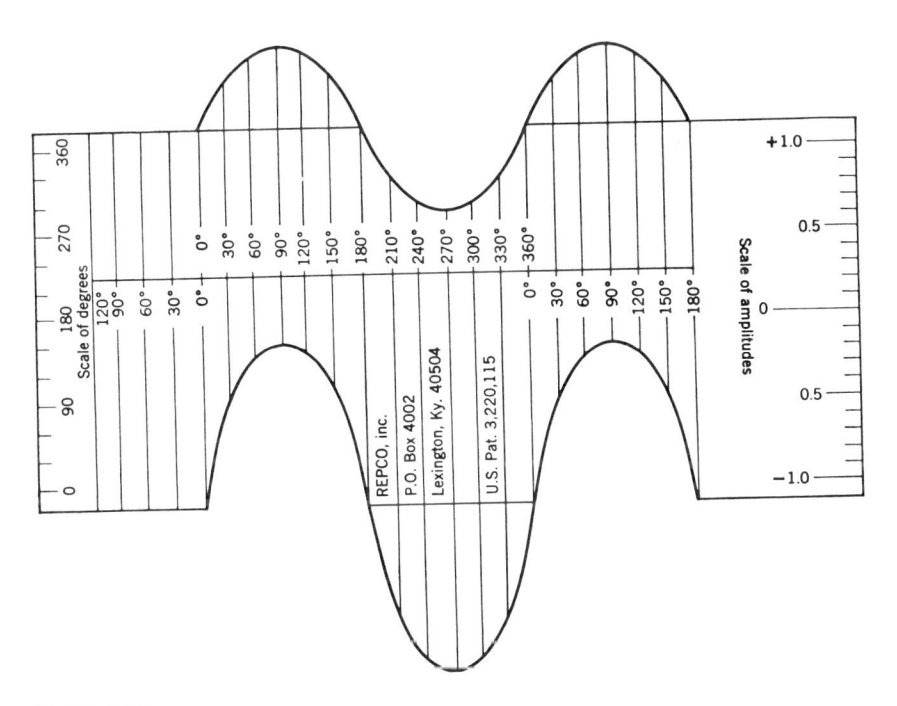

Figure A-17

ANSWERS TO PROBLEMS

Chapter 1

D1-1 Student calculation.
D1-2 Currents also double.
D1-3 Currents reduce to half previous values.
D1-4 60 Hz; 16.7 ms.
D1-5 Exercise.
D1-6 -2 or $j^2 2$; $j2$; $j^2 2$; $j^3 2$.
D1-7 Sketch.
D1-8 (a) $7.33 + j6.5$ (b) $0 - j2 = -j2$.
D1-9 Exercise.
D1-10 (a) $1 \sin(377t + 90°)$ A (b) 60 Hz (c) Sketch.
D1-11 (a) $10,000 \underline{/90°}$ ohms (b) Z increases (c) $100 \underline{/90°}$ V (d) $i_L = 0.01414 \sin 1000t$ A, and $v_L = 141.4 \sin(1000t + 90°)$ V.
D1-12 (a) 10 μF (b) decreases; decreases.
D1-13 $100 + j100 = 141.4 \underline{/45°}$ Ω.
D1-14 Sketch.
D1-15 $15.9 \underline{/-90°}$ ohms.
D1-16 15.9 ohms.
D1-17 0.5 A; 0.125 A.
D1-18 $0.726 \underline{/-84.8°}$ A.
D1-19 Sketch.
D1-20 $5.3 \underline{/-25.8°}$ V drop in direction of I_1.
D1-21 Exercise.
D1-22 Odd-n terms in $(\cos n\omega t + 1)$ equal zero.

517

518 Answers to Problems

D1-23 $-0.0363E_m \cos 6\omega t - 0.021E_m \cos 8\omega t - 0.0134E_m \cos 10\omega t$.
1-1 $1.2 \text{ k}\Omega = 0.0012 \text{ M}\Omega$; $33{,}000 \text{ }\Omega = 0.033 \text{ M}\Omega$; $2{,}200{,}000 \text{ }\Omega = 2200 \text{ k}\Omega$; $0.00015 \text{ }\mu\text{F} = 0.00000000015 \text{ F}$; $20{,}000 \text{ pF} = 0.00000002 \text{ F}$; $120 \text{ }\mu\text{F} = 120{,}000{,}000 \text{ pF}$; 0.000015 H; 0.063 H; $0.148 \text{ mV} = 0.000148 \text{ V}$; $380 \text{ }\mu\text{V} = 0.00038 \text{ V}$; $12{,}000 \text{ mV} = 12{,}000{,}000 \text{ }\mu\text{V}$.
1-2 (a) 1 A (b) 1 A (c) 48 V, 12 V, 60 V.
1-3 (a) 2.5 A, 10 A, 2 A (b) 14.5 A.
1-4 $274 \text{ }\Omega$.
1-5 (a) 10 mA (b) $112.5 \text{ }\Omega$ (c) 3.3 mA left.
1-6 1.2 A from A to B.
1-7 4.22 A; 1.11 A.
1-8 60 Hz; 1/60 s or 16.7 ms; 7.07 A rms.
1-9 (a) 13 A (b) 22.6°.
1-10 (a) $136.6 - j36.6$; $36.6 + j136.6$ (b) $141.4/\!-\!15°$; $141.4/75°$.
1-11 $10.9/\!-\!37.6°$ A; 22.6° lag.
1-12 (a) $20 + j28.66$ (b) sketch (c) sum leads I_1 by 55.1°, lags I_2 by 4.9°, lags I_3 by 34.9°.
1-13 (a) 314 ohms. (b) 314 kilohms (c) 18.8 ohms (d) 50 ohms.
1-14 50 V and 350 V.
1-15 5 kilohm resistor and 1.38 H inductor.
1-16 $1500/0°$ ohms.
1-17 $0.6 + j0.8$.
1-18 $10 - j100$.
1-19 12.2 mA down.
1-20 (a) $20/\!-\!50°$ (b) 12.85 ohms (c) 15.35 ohms (d) R and C.
1-21 (a) $26.2 - j0.75$ (b) $3.83/1.6°$ A (c) 1.6°, current leading.
1-22 $53/45°$ V, + at upper end.
1-23 $5/0°$ A.
1-24 27.5 mA.
1-25 E_{oc} is $8.3/41.7°$ V and Z_{oc} is $25.8/\!-\!81.6°$ ohms.
1-26 $0.239/85.2°$ A; upper end positive.
1-27 (a) 0 and 100 mA, 100 V, 0 V (b) 2500 A/s (c) 40 microseconds, 63.3 mA, 63.3 V, 36.7 V.
1-28 Final 0.1 A in all cases; initial rate of current increase 10,000, 1250, 800 A/s.
1-29 $120 - j50$ ohms; 120-ohm resistor and 7.95 μF capacitor.
1-30 Sketch.
1-31 Sketch.
1-32 (a) 100 kHz (b) 100.
1-33 (a) R 1 V, L 100 V, C 100 V (b) yes.
1-34 (a) 1.98 mH (b) 7200 and 8800 Hz (c) 10.
1-35 (a) 112.5 pF (b) 177.8 kilohms (c) 188.3 (d) 5.62 microamp.

Chapter 2

D2-1 1 eV; 10 eV.
D2-2 1.12×10^{-19} J; 1.76×10^{-19} J.
D2-3 1.7×10^{14} Hz; yes.

Answers to Problems 519

D2-4 2.6×10^{14} Hz; yes.
D2-5 Silicon.
D2-6 1.25 A.
D2-7 77; 212; 302; 392.
D2-8 0.7 V; 0.75 ohm.
D2-9 E_a much greater than 0.7 V; R_{ext} much greater than 0.75 ohms.

2-1 1.12×10^{-19} J.
2-2 1.76×10^{-19} J.
2-3 1 eV; 1.602×10^{-19} J.
2-4 100 eV; 1.602×10^{-17} J.
2-5 1.7×10^{14} Hz.
2-6 Yes.
2-7 Yes.
2-8 -0.117 eV.
2-9 -0.084 eV.
2-10 0.665 eV.
2-11 Germanium.
2-12 Sketch.
2-13 1.5 A; 0.8 mA.
2-14 Silicon.
2-15 1.25 A.
2-16 0.75 V; 0.625 ohm.
2-17 Sketch.
2-18 Graphical plot.
2-19 Graphical plot.
2-20 (a) Sketch (b) sketch (c) 100 V.
2-21 Sketch.
2-22 Sketch.
2-23 Sketches.
2-24 Sketches.

Chapter 3

D3-1 0 V; greater than $e_{s\ max}$; 1314 ohms.
D3-2 None; -10 V; 10 mA.
D3-3 None; $-V$.
D3-4 Graph.
D3-5 (a) 450 ohms (b) 45 V, 100 mA.
D3-6 (a) 68 mA max in both (b) 0.935 W (c) 34 mA (d) 32.5% (e) 170 V.
D3-7 (a) 68 mA max in both (b) 3.74 W (c) 47 mA (d) 65% (e) 340 V.
D3-8 Sketch total 90° angle from 27° to 117° for $\omega R_L C$ about 1.7.
D3-9 Sketch.
D3-10 Indicates double actual rms; $0.707 E_m$.
D3-11 35.5 V; 17.5 V.
D3-12 (a) Current reverses in load (b) double $R_L C$ time constant and reduce load voltage variation.

520 Answers to Problems

D3-13 Graphs.
D3-14 Sketch.
D3-15 Sketch.
3-1 (a) 10 V (b) 0 V (c) $\sqrt{2} \times 117$ V (d) 1555 ohms.
3-2 12.06 V.
3-3 Sketch.
3-4 Load current reverses; double time constant $R_L C$, reducing variation in load voltage; loss of filtering makes half-wave rectifier.
3-5 About 11.5 V on both.
3-6 10.5 mA; 8 V.
3-7 (a) For 1Z6.8, $R = 132$ ohms for 100 mA at 20 V input (b) 1.32 W (c) 100 and 24.2 mA, for 20 V and 10 V inputs (d) 6.8 V if current limited to 100 mA.
3-8 4.5 mA.
3-9 50 mA; 3.75 W or larger.
3-10 (a) 990 ohms. (b) 200 mA, 198 V.
3-11 100 ohms minimum.
3-12 (a) Both 32.7 mA max (b) 0.522 W (c) 16.35 mA (d) 39% (e) 170 V.
3-13 (a) Both 32.7 mA max (b) 2.08 W (c) 23.1 mA (d) 78% (e) 340 V.
3-14 54 Vdc; 60 V rms, 28.4 V rms, 5.66 V rms.
3-15 215 V; 239 V rms.
3-16 (a) 0.0122 (b) 205 V (c) 19.75 W (no bleeder).
3-17 0.033.
3-18 (a) 0.37 (b) 0.49.
3-19 (a) 106 VA both (b) 0.927 A (c) 1630 turns center-tapped; 14 turns.
3-20 (a) 1315 ohms, 1.9 W; 2860 ohms, 3.5 W; 4000 ohms, 2.5 W; 27.8 ohms, 0.324 W; 18.5 ohms, 0.215 W; 92.6 ohms, 1.08 W (b) 97 W (c) 97 W vs. 9.5 W.
3-21 (a) 1080 ohms (b) 100 mA.
3-22 through **3-25** Refer to text.
3-26 Exercise.
3-27 Exercise.
3-28 Reverses load current and voltage.
3-29 Exercise.
3-30 Sketches.
3-31 Sketches.
3-32 Sketch.
3-33 (a) 500 mA max, 159 mA average (b) 132.5 μF (c) 412 mA max, 35 mA average.
3-34 (a) Sketch (b) $RC = 4 \times 10^{-3}$ (c) sketch (d) increase input voltage, increase frequency, or increase RC; practical choice, increase C.
3-35 Sketch.
3-36 C, D, A.

Chapter 4

D4-1, D4-2 Exercises.
D4-3 through **D4-6** Sketches and observations.
D4-7 Sketch.

D4-8 Sketch.
D4-9 (a) No (b) no.
D4-10 Sketch; depends on V_{CC}.
D4-11 Graph.
D4-12 100 μA max.
D4-13 Graph.
D4-14 R_L 143 ohms, R_B 44 kilohms.
D4-15 Cutoff at i_b 100 μA peak.

4-1 No; no.
4-2 Graph.
4-3 Exercise.
4-4 (a) Sketch (b) 5 mA, 10 V; 200 kilohms.
4-5 Equal at Q.
4-6 Exercise.
4-7 (a) 4.8 mA, 4.2 V (b) 15 μA max, saturation.
4-8 Exercise.
4-9 180°.
4-10 Graph.
4-11 (a) 37.5 mW (b) 37.5 mW.
4-12 150 kilohms.
4-13 0.218 V.
4-14 2 megohms, 10 kilohms.
4-15 400 kilohms.
4-16 (a) 800 kilohms, 8 kilohms (b) 30: 0.75 mA; 120: 3 mA (saturation).
4-17 (a) 0.7 to 4.7 mA (b) 40.
4-18 167 ohms.
4-19 67 kilohms.
4-20 Graphical results.
4-21 Graphical results.
4-22 1 kilohm.
4-23 Graph; not quite linear.
4-24 Collector dissipation rating exceeded.
4-25 (a) 40 (b) 40 (c) 1600.
4-26 (a) Sketch (b) sketch (c) 25 μA peak (d) 400 kilohms (e) 12.5 mW.
4-27 340 kilohms; 230 kilohms.
4-28 (a) 167 ohms (b) 1 kilohm.
4-29 60 V.
4-30 (a) Graph (b) graph (c) -19 μA, 1.95 mA, -10.7 V (d) 9.25.
4-31 -70 μA, -1.45 mA, -15 V; 7.04.
4-32 4.3 mA.
4-33 Graphical results.

Chapter 5

D5-1 Exercise.
D5-2 Sketch.
D5-3 Sketch.

522 *Answers to Problems*

D5-4 Sketch.
D5-5 −2 V, 2 mA.
D5-6 i_D from 4.44 mA to 2.22 mA.
D5-7 No; cutoff removes self-bias.

5-1 Sketch.
5-2 Graphical results.
5-3 4 kilohms, 1 kilohm.
5-4 80 μF.

Chapter 6

D6-1 21.2, 9 kilohms, 2400 micromhos; 15.8, 20 kilohms, 1100 micromhos.
D6-2 4.5 mA.
D6-3 1600, 1 megohm, 1600 micromhos; 10.75, 5 kilohms, 2150 micromhos.
D6-4 150 V or larger.
D6-5 32 Hz.
D6-6 150 kilohms, 0.15 μF.
D6-7 Graph.
D6-8 No; cutoff removes self-bias.

6-1 10.3×10^5 m/s.
6-2 (a) 100 (b) 59.3×10^5 m/s.
6-3 (a) 21.2 V (b) 120 V (c) 4.24 mA (d) 26 mA; no.
6-4 Graphical results.
6-5 Graph.
6-6 Graph; amplification factor.
6-7 2400 micromhos.
6-8 10 V max.
6-9 12.5 μF minimum.
6-10 83 ohms.
6-11 4, 880 ohms, 4500 micromhos.
6-12 19.3, 12.5 kilohms, 1550 micromhos.
6-13 Graph.
6-14 36 mA.
6-15 5 mA.
6-16 Increase 44 V.
6-17 10.
6-18 (a) Sketch (b) 2.4 mA, 150 V, −6 V.
6-19 5 mA, 180 V, −6 V; 1 kilohm.
6-20 180 V to 595 V.
6-21 About 2 mA to 6 mA.
6-22 11 V.
6-23 3.3 kilohms.
6-24 133 Hz.
6-25 (a) 16 mA (b) increase 192 V.
6-26 Graphs.

Answers to Problems 523

6-27 Exercise.
6-28 Graphical results.
6-29 16 V increase.
6-30 10 mA.
6-31 2.6 mA, 152 V, −6 V, 0, 0, 0.

Chapter 7

D7-1 h_{ie} 725 ohms, h_{oe} 69.5 micromhos, h_{re} 11 × 10^{-4}, h_{fe} 45.
D7-2 19.5.
D7-3 (a) 320 V, 16 mA (b) −1/20,000 (c) same (d) graph.
D7-4 Graphical results.

7-1 1890, 65.5 dB.
7-2 (a) Sketch (b) sketch (c) 100.
7-3 Sketch.
7-4 h_{ib} 4.35 ohms, h_{ob} 25 micromhos, h_{rb} 10.5 × 10^{-4}, h_{fb} 285.
7-5 h_{ie} 1450 ohms, h_{oe} 8.3 × 10^{-3} mhos, h_{re} 35 × 10^{-4}, h_{fe} 285.
7-6 Graph.
7-7 0.375 mA.
7-8 30 V peak to peak.
7-9 $3/\underline{180°}$ V.
7-10 $-\mu R_{eq}/[2r_p + R_{eq} + R_k(\mu + 1)]$.

Chapter 8

D8-1 Frequency distortion.
D8-2 Exercise.
D8-3 Phase and frequency.
D8-4 Harmonic.
D8-5 (a) Graph (b) 62 (c) −42.
D8-6 Sketch.
D8-7 Exercise.

8-1 (a) 19.1, 180° (b) 16.2, 212°.
8-2 14.3, 139°.
8-3 62 and 18,000 Hz.
8-4 (a) Sketch (b) 25 V, 50.
8-5 (a) 180 V: 57.8 V, 0.67 V (b) 300 V. 63.5, 0.75 V.
8-6 (a) $40.8/\underline{150°}$ (b) 67.8 kHz.
8-7 $19.8/\underline{131.8°}$, 75 Hz.
8-8 (a) Sketch (b) 185 V, 115 V (c) 1.025 mA rms, 0.5 W.
8-9 (a) −444 (b) 117 Hz, 9 kHz (c) 0.225 V.
8-10 (a) −19 (b) 0.29 V (c) 40 kHz (d) 16.8 V.
8-11 29 kHz.
8-12 4950 and 49,800 ohms.
8-13 (a) A_{i1} 41.5, A_{i2} 37.3 (b) R_{i1} 1990, R_{i2} 2075, R_{o1} 33,900, R_{o2} 32,500 (c) $0.9/\underline{26°}$ (d) $1/\underline{0°}$.

8-14 Exercise.
8-15 (a) 14.3 (b) 106 Hz.
8-16 $(1.58 + j1.91) \times 10^{-3}$ mho.
8-17 $\mu/(\mu + 1 + r_p/R_L)$.
8-18 Sketch.
8-19 Yes; harmonic.
8-20 Sketch.
8-21 Exercise.
8-22 Sketch.
8-23 Sketch.
8-24 (a) h_{ie} 1145 ohms, h_{oe} 3.8×10^{-5} mho, h_{re} 5.18×10^{-4}, h_{fe} 38.1. (b) A_{ve} −150, A_{ie} 31.8.
8-25 A_v 0 to −1540; A_i h_{fe} to 0.
8-26 125 μF or larger.
8-27 0.23 μF or larger.
8-28 (a) Sketches (b) Circuit a: 4.1 mA; b: 1.9 mA; c: 1.32 mA (c) a: 10 k; b: 476; c: 31 k.
8-29 51 Hz, 39.2 kHz.

Chapter 9

D9-1 (a) 1.1 W (b) 0.1 W.
D9-2 100 ohms.

9-1 0.28 W.
9-2 15.4 W.
9-3 2.11 W.
9-4 6 W, 30%.
9-5 Exercise.
9-6 (a) Greater than 38.5 W (b) half of 38.5 W; zero.
9-7 Same as input.
9-8 (a) Sketch (b) graph (c) 3 W, 3 W.
9-9 1600 CT to 10.
9-10 40 V, 6.32; 20 V, 25 mA.
9-11 (a) 100 (b) 61.4 mA (c) 10.9.
9-12 Exercise.
9-13 20 V, 0.72, 48 ohms, 1.6 ohms.
9-14 27%.
9-15 171.

Chapter 10

10-1 Sketch.
10-2 Sketch.
10-3 Sketch.
10-4 −167.
10-5 A_i 49, A_v −0.42, A_p 20.58, R_i 120 kilohms, R_o 27 kilohms.

Answers to Problems 525

10-6 A_i −3210, A_v 0.63, A_p 2022, R_i 5.1 megohms, R_o 530 ohms.
10-7 Q_1: 22.7 V, 4.4 µA, 3.9 mA; Q_2: 23.4 V, 0.135 mA, 0.2 mA.
10-8 All 1 megohm.
10-9 Sketch.
10-10 Sketch with $R_f = R_{in}$.
10-11 Exercise.
10-12 0.143 mH, 175 pF, −322.
10-13 20 kHz to 6.66 kHz; 1.42 MHz to 0.82 MHz.

Chapter 11

D11-1 −0.133.
D11-2 −0.04.
D11-3 (a) 0.429 Z_o (b) 2.33 Z_{in}.
D11-4 −24.6; 0.04.
D11-5 20 kilohms.
D11-6 1.6 to 8.0 µF.
D11-7 (a) 0.95 (b) no change necessary.
D11-8 3.95 megohms, 0.395 megohms, 39.5 kilohms.

11-1 Exercise.
11-2 Exercise.
11-3 (a) 200 Hz, 79.6 × 10^{-5}; 2000 Hz, 79.6 × 10^{-6} (b) 796 kilohms (c) 0.001 µF.
11-4 A_{ve} from −100 to −4.88; 0.195.
11-5 2360 ohms change.
11-6 Exercise.

Chapter 12

D12-1 (a) 5 × 10^6 m = 3100 miles (b) 7.5 MHz.
D12-2 Graphs.
D12-3 Sketch.
D12-4 Below 5 kHz.
D12-5 (a) 92.2 MHz (b) 200 kHz (c) 10 (d) 10 (e) graph.

12-1 (a) Fundamental and second harmonic (b) tune to second harmonic (c) 5%.
12-2 Exercise.
12-3 (a) $2\pi \times 100{,}000$, $2\pi \times 101{,}000$; $2\pi \times 99{,}000$ (b) 0.5 (c) each sideband has $m_a^2/4$ times carrier power.
12-4 Sketch.
12-5 3 × 10^6 m; 300 m.
12-6 300 MHz to 60 MHz.
12-7 Less than 15 cm.
12-8 (a) 50% (b) Sketch (c) 0.625 W; 2.25 W.
12-9 18 times.
12-10 (a) 5 kHz (b) depends on value of m_a.
12-11 Sketch.

526 Answers to Problems

12-12 Exercise.
12-13 Exercise.
12-14 (a) 5 V peak for sideband, 10 V peak for carrier (b) 798, 800, 802 kHz.
12-15 (a) 10 kHz (b) 16 kHz (c) 20 kHz.
12-16 (a) FM 10 kHz, AM 2 kHz (b) FM 20 kHz, AM same 2 kHz.
12-17 10 kHz signal amplitude 5 times amplitudes of 2 kHz signal.
12-18 Infinite number; modulating frequency; depends on m_f for amplitudes.
12-19 No.
12-20 Sketch.
12-21 Sketch.

Chapter 13

13-1 90°.
13-2 65°.
13-3 Sketches.
13-4 27 V average, 2.7 A average.

INDEX

AC load line, 257, 273, 329
AC voltmeter, 141
Acceptor, 73, 161
Alpha (α), of transistor, 177
A—M, *see* Amplitude modulation
Amplification, in transistor, 178, 196, 286, 290
   in vacuum triode, 248, 298
Amplification factor, for transistor, 177
   for triode tube, 229, 274
Amplifier, audio, 286
   Classes A, B, C, 282, 336, 343, 347, 417
   common-emitter, 287, 291
   Darlington, 365
   difference, 360
   differential, 357, 487
   operational, 369
   power, 280, 328, 332, 337
   push-pull, 336, 341
   $R-C$ coupled, 286, 297
   transistor, 196, 281
   tuned, 372
Amplitude modulation, 403, 409, 412
AND circuit, 114
Antiresonance, 53
Audio amplifier, 286
Avalanche breakdown, in diode, 89, 93
   in SCR, 448
   in Zener diode, 90

Balanced modulator, 420
Bandwidth, defined, 53
   of amplifier, 309, 384
   in A—M, 413
   in F—M, 428
   reduction factors, 315
   tuned amplifier, 372
Barrier region, 84, 86, 168
Base, of transistor, 159
Bessel functions, in F—M, 429, 497
Beta (β), of transistor, 176
Bias, cathode, 242
   diode, 87, 162
   FET, 207
   forward, 87, 162, 170
   reverse, 87, 91, 163, 168
   self (FET), 212
   transistor, 162, 165, 171, 361
   triode tube, 233, 235
Bias line, 194, 242
Binary code, 436
Bleeder resistor, 139
Breakdown, avalanche, 89, 93
   in SCR, 448
   Zener effect, 90
Bridge rectifier, 129

Capacitor filter, 133, 135, 140
Carrier, majority and minority, 85
   frequency, in modulation, 407
Cathode emission, 102
Channel, of FET, 206, 216
Characteristic curves, diode, 92, 103
   JFET, 208
   MOSFET, 218

528 *Index*

SCR, 452
  transistor, 167, 172, 175
  triode tube, 227
Clamper circuit, 113
Classes of amplifiers, 282
Clipper circuit, 111
Collector, of transistor, 159
  dissipation, 186
Common-base circuit, 169, 267, 291
Common-collector circuit, 169, 267, 291
Common-emitter circuit, 169, 267, 287, 291, 294
Common-mode rejection, 360, 487
Compensation, in semiconductor doping, 75
Complementary symmetry, 349
Complex number, 12, 17
Conduction band, 68
Conductor, 68
Coupling capacitor, 269, 281
Covalent bonding, 71
Crossover distortion, 337
Crystal, lattice, 66
  seed, 79
Cutoff, of transistor, 185
Current gain, of transistor, 177, 280, 288, 290

Darlington amplifier, 365
D'Arsonval meter, as ac voltmeter, 141
dB, defined, 43
DC load line, 182, 212, 242
Decibel (dB), 43
Degenerative feedback, 193, 383
Demodulation, 405
Depletion region, 84, 168, 206
Detection, of modulation, 405
Detector, diode, 143
  envelope, 424
Determinant, 30-35, 289, 390, 490
Deviation ratio, in F–M, 429
Difference amplifier, 360
Differential amplifier, 357, 487
Diode, characteristics, 92, 103
  circuit model, 97
  detector, 143
  grown-junction, 79
  ideal, 97
  logic element, 114
  P–N junction, 84
  ratings, 92
  symbol, 97
  vacuum tube, 101

Zener, 91, 117
Dipole layer, 84
Discriminator, 432
Dissipation, collector, 186
Distortion, crossover, 337
  frequency, 285
  nonlinear, 285
  phase, 285, 384
Donor, 73, 161
Doping in semiconductor, 73
Double tuning, 376
Doubler, voltage, 145
Drain, of FET, 206
Dynamic plate resistance (r_p), 231, 274

Effective (rms) value, 11, 125, 142, 489
Efficiency, power conversion, 125, 335
Emission, thermionic cathode, 102
Emitter, of transistor, 159
Energy bands, 68
Energy levels, 66, 75, 161
Envelope detector, 424
Epitaxial layer, 477
Equivalent circuit, diode, 98
  FET, 275
  Norton, 38, 270
  Thévenin, 37, 270, 386
  transistor, 262, 295
  triode tube, 269, 298

Feedback, current, 381
  degenerative (negative), 193, 383
  effect on impedances, 381, 385, 389
  factor, 382
  inverse (negative), 193, 383
  regenerative (positive), 383
  in transistor amplifier, 389
  voltage, 389
Fermi energy level, 75, 161
Field-effect transistor (FET), 158, 205, 274
Filter, capacitor, 135, 140
  choke-input, 134
  pi-section, 133
  R–C type, 135, 140
  ripple factor, 139
  smoothing, 133
F–M, *see* Frequency, modulation
Forward bias, 87, 164
Fourier series, 40, 126
Frequency, defined, 8
  distortion, 285

half-power, 52, 305, 308, 315, 375, 384
modulation, 404, 427
response, 286, 298ff, 311, 315, 373
translation, 405
Full-wave, phase control of SCR, 459
Full-wave rectifier, 40, 126, 129, 132, 142

Gain, current, 280, 288, 290
in dB, 296
power, 280, 289, 290
voltage, 236, 246, 271, 280, 289, 290
Gain-phase curves, 301, 310
Gate, of FET, 205
of SCR, 446, 450
Grid bias, 233
Grid-leak bias, 235

Half-power frequency, 52, 305, 308, 315, 375, 384
Half-wave rectifier, 40, 122, 132, 135, 143
Heat sink, 350
Hertz, defined, 9
High-frequency response of amplifier, 287, 298, 305, 307ff, 315
Hole, in semiconductor, 72
Holding current, in SCR, 449
h-Parameters, 260

Ideal diode, 97
Impedance, defined, 5, 19
input and output, 280, 289, 293, 362, 386
Impurity, in semiconductor, 73ff
Induced channel, in FET, 216
Input impedance, 280, 289, 293, 362, 386
Insulator, 68
Integrated circuit, 473
Ionization, thermal, 72

JEDEC numbering system, 180
Junction, $P-N$ diode, 79, 84, 161
FET, 205
transistor, 158

Kirchhoff's laws, 5, 28

Laser, 466
Lattice, in crystal, 66
Leakage current, 168, 188
Limiter, diode voltage, 111
in F–M, 432
Load line, ac, 257, 271, 329

dc, 182, 212, 242
Logic circuits, 114, 487
Loop current, 33
Low-frequency response of amplifier, 287, 298, 300ff, 312, 315

Majority carrier, 85
Maximum power transfer, 46, 495
Midband frequency range, 286, 312
Minority carrier, 85
Mixer, frequency, 421
Modulation, amplitide, see Amplitude modulation
factor, in A–M, 410
in F–M, 429
frequency, see Frequency, modulation
pulse code, 405, 436
Modulator, 416, 420
MOSFET, 216, 274
Mutual conductance (g_m), 231, 274

NOR circuit, 487
Norton's theorem, 38, 270
$N-P-N$ transistor, 159, 163
N-type semiconductor, 74

Ohm's law, 4, 28
Operational amplifier, 369
OR circuit, 114
Oscillator, $R-C$ phase-shift, 397, 496
Output impedance, 280, 289, 293, 362, 386
Output power, 273, 334, 414

Parallel resonance, 53
PCM, see Modulation, pulse code
Peak inverse voltage, 91, 93, 126
Peak rectifier, 143
Pentode, 239
Phase angle, 10
Phase control of SCR, 459
Phase distortion, 285, 384
Phase-shift oscillator, 397, 496
Phasor, algebra, 11
polar and rectangular forms, 16
Photomask, 475
Photovoltaic cell, 463
PIV rating, diode, 93
Plate resistance, triode, 231, 274
$P-N$ junction, 79, 84, 161
$P-N-P$ transistor, 159, 163
Polar form of phasor, 16

530 Index

Potential hill, 86, 162
Power, amplifier, 280, 328
  gain, 280, 289
  output, 273, 334, 414
  supply, 121, 132
  transfer theorem, 46, 495
  transformer, 130
P-type semiconductor, 74
Pulse-code modulation, 405, 436
Push-pull amplifier, 336

Q, of tuned load, 51, 56
Q-point, defined, 185, 246
  effect of temperature, 188
  stabilization, 191
  in transistor circuit, 185, 194, 213

Radio waves, 406
Ratings, diode, 92
  FET, 210
  SCR, 452
  transistor, 179
Ratio, common-mode rejection, 360, 487
$R-C$ coupled amplifier, 281, 286, 294, 297
Reactance, capacitive, 23
  inductive, 21
Rectangular form of phasor, 16
Rectifier, bridge, 129
  full-wave, 40, 126, 129, 132
  half-wave, 40, 122, 132, 135, 143
  peak, 143
  semiconductor stack, 146
Regulator, voltage, 117
Resonance, bandwidth, 53
  parallel, 53
  series, 48
Reverse bias, 87, 163
Ripple factor, 139
Ripple, in rectifier output, 133
Root-mean-square value, 11, 125, 142, 489

Saturation, of transistor, 185
SCR, 445
Self-bias, 212, 234
Semiconductor, 68
  controlled rectifier (SCR), 445
  N-type and P-type, 74
  stack rectifier, 146
Series resonance, 48
Shunt-capacitor filter, 135, 140
Sideband frequency, 412, 430

Single sideband (SSB), 420
Small-signal operation of amplifier, 255, 287, 294
Smoothing filter, 133
Solar cell, 463
Stability factor, S, 191
Stack rectifier, 146
Stagger tuning, 377
Substrate, 218, 474
Superposition principle, 35
Symbols, circuit components, 3
  diode, 97
  JFET, 207
  MOSFET, 219
  transistor, 163, 174
  tubes, 103, 224, 228
  Zener diode, 119

Tetrode, 239
Thermal ionization, 72, 189
Thermal runaway, 188
Thermionic emission, 102
Thévenin's theorem, 36, 270, 386
Time constant, 41
Transconductance (g_m), 231
Transformer, power, 130
  turns ratio, 131
Transistor, amplifier, 196, 281
  biasing, 162, 165, 171, 361
  current gain, 280, 288, 290
  equivalent circuit, 262, 295
  field-effect (FET), 158, 205, 274
  junction, 158
  symbols, 163, 174
TRIAC, 462
Trigger circuit, 455
Triode tube, amplifier, 244
  parameters, 231, 274
  symbol, 103, 224, 228
Tuned amplifier, 372
Turns ratio, 131

Universal gain-phase curves, 301, 310

Valence band, 66
Voltage, amplifier, 244, 270, 281, 287, 297
  clamper, 113
  doubler, 145
  feedback, 389
  gain, 248, 271, 280, 289
  limiter, 111

regulator, 117
Voltmeter, 141, 216
Volume unit (vu), 46

Wavelength, 405

Zener diode, 90, 117
Zone refining, 77